PNEUMATICS

PNEUMATICS

Concepts, Design and Applications

Jagadeesha T

Assistant Professor
Department of Mechanical and Production Engineering
National Institute of Technology Calicut
Kozhikode, Kerala

UNIVERSITIES PRESS (INDIA) PRIVATE LIMITED

Registered Office
3-6-747/1/A & 3-6-754/1, Himayatnagar
Hyderabad 500 029 (Telangana), India
E-mail: info@universitiespress.com, www.universitiespress.com

Distributed by
Orient Blackswan Private Limited

Registered Office
3-6-752 Himayatnagar, Hyderabad 500 029 (Telangana), India

Other Offices
Bengaluru / Bhopal / Chennai / Ernakulam / Guwahati / Hyderabad /
Jaipur / Kolkata / Lucknow / Mumbai / New Delhi / Noida / Patna

First published 2015

ISBN 978 81 7371 941 7

Typeset in Times New Roman 10.6pt by
Techastra Solutions Pvt. Ltd., Hyderabad 500 029

Printed in India by
Graphica Printers, Hyderabad 500 013

Published by
Universities Press (India) Private Limited
3-6-747/1/A & 3-6-754/1, Himayatnagar
Hyderabad 500 029 (Telangana), India

This book is dedicated to

all the members of my family, my teachers

and to my students of the past, present and future.

CONTENTS

9. ELECTRO-PNEUMATIC CONTROL

PREFACE

This book is written for students and engineers who wish to understand pneumatics technology and apply it in solving engineering problems. It is primarily designed for the use of graduate and postgraduate students of mechanical, production, automobile and mechatronics engineering as well as those preparing for competitive examinations. Practising engineers, too, will find the book a source of reference for solving many of the problems encountered in the pneumatics industry.

The main objective of writing this book has been to provide a clear understanding of the concepts underlying pneumatics control. I have strived to convey the physical interpretations of various derivations and provide concise, yet comprehensive solution to a variety of complex problems. The parameters kept in mind while developing the content were that the coverage should cater to the syllabi of various Indian universities, and that it be expressed with clarity, aided by a variety of solved and unsolved numerical problems—sourced from competitive examinations as well—and meaningful examples from industry.

Problems are included at the end of each chapter as exercises. While some of these are of the routine kind intended to familiarise the reader with numerical manipulations of magnitudes of various parameters of pneumatics control, there are open-ended problems that will require readers to apply the basic principles to new situations and develop their own designs and circuits.

Organisation of the book

This book presents a comprehensive treatment of the principles of pneumatics. It contains more than enough material for a one-semester course generally offered at the undergraduate or post graduate level, depending upon individual course objectives. The presentation of the subject is along classical lines, with separate discussions for pure pneumatics, electro pneumatics and PLC based pneumatics. Chapter 1 gives a brief introduction to pneumatics and then develops the basic laws of pneumatics. Elementary pneumatic circuit components have been presented along with their advantages and disadvantages. Chapter 2 contains an outline of the aspects of preparation, drying and cooling of compressed air and some practical applications.

Chapter 3 introduces conditioning and distribution of compressed air. Chapter 4 discusses the various pneumatics actuators and air motors. Chapter 5 deals with various types of pneumatics valves, such as directional control valves, flow control valves and pressure control valves. In chapter 6, various single cylinder control methods have been discussed in detail. It also contains design problems and industrial circuits. Chapter 7 discusses the various circuit design method used in pneumatics industries. Cascading and step counter methods are discussed with several examples. Chapter 8 complements Chapter 7 and deals with the KV mapping method to control multi-cylinders. Its systematic procedure is presented in this chapter with the help of a large number of industrial problems. Electro pneumatics systems are dealt with in Chapter 9, and PLC-based pneumatics systems are dealt in Chapter 10. Chapter 11 describes the various circuits and control methods using fluidic control elements. Chapter 12 is concerned with hydropneumatics using both air and oil. Chapter 13 deals with the maintenance of pneumatic power systems.

Multiple-choice questions, review questions and exercises, solved examples, and solutions of problems from various Indian universities examinations are included in all the chapters to make the book useful for teaching and self learning. The salient features of the book may be summed up as

- Presentation of basic theory in simple and readily understandable form
- A balanced approach to mathematical and theoretical concepts. Concise and comprehensive treatment of topics, with strong emphasis on fundamental concepts.
- Large number of solved problems and unsolved problems selected from question papers of various technical institutes and universities of India.
- Inclusion of a number of theoretical questions and unsolved exercises for practice

Jagadeesha T

ACKNOWLEDGEMENTS

During the course of development of this textbook, I benefited from the comments and suggestions received from several of my friends, clients and associates. I am grateful to all of them.

Acknowledgements are due to several of my ex-colleagues and team leaders, both in India and abroad. These include: Ashok Kumar, Head (Operations) and Yogesh Kale, Head (Design), Harith Grammer (TVS-Suzuki Group), Hosur; Louis Kim, Head (Thin Films), Chartered Semiconductor Manufacturing Ltd, Singapore; Guruvaiah, Head (Controls), Shiv Kumar, Head (Fluid Power) and Trivedi, Head (Machine Design), TELCO, Pune; Ricky Tan, Director (Automation) and Md. Johari, Head (Tooling), IBM Singapore Pvt. Ltd; N. Sreekanth, Head (CAD/CAM) and Kenny Kwan, Head (R&D), ASM Technologies, Singapore; Joseph Ong, Head (CVD) and Stanley Teo, Head (PVD), ST Microelectronics, Singapore; James Lee, Director (Operations) and Melvin Leo, Staff Engineer (Applications), APP Systems and Services, Singapore; Young Yee, Staff Engineer (Process), Applied Materials, Korea; Young Yap, Director (Process), Applied Materials, Singapore, Joji Parambath, Joint Director, FTI, Bangalore; and Ganesan, Staff Engineer (Equipment), Infineon, Malaysia. My special thanks to Dr C. K. Subbaraya, Principal, AIT, Chickmagalur, and Dr Joseph Gonsalvis, Principal, SJEC, Mangalore, for their encouragement and help.

I would like to express my deep sense of gratitude to our beloved director, Dr G. R. C. Reddy, a dynamic administrator, who provided an ideal learning atmosphere at NIT Calicut. I am deeply indebted to Dr Allesu Kanjirathinkal, Professor and Head of Mechanical Engineering Department and Dr R. Sridharan, Dean, Faculty Welfare, for their encouragement and guidance. My boundless gratitude is due to Dr T. J. Sarvoththama Jothi, Dr N. Selvaraju and Dr Mahesh Kumar for their help. All my colleagues have been helpful to me, and I express my feelings of gratitude to all of them.

I would like to thank Vasanthi, my wonderful wife, and my son, Ram kumar Jagadeesha for their love, patience and endless sacrifices. My special thanks are due to my sister Prabhavathi T., without whose comforting presence and encouragement this book would have remained merely a good intention.

It takes a team of many people and lots of hard work to create a quality textbook. I am thankful to the members of Universities Press, the publisher, for joining in this endeavour, particularly, Kallol Das who discovered the author in me, and the editorial support of Ramya Sriram, D Savakoor and Sreelatha Menon.

I welcome feedback and suggestions for the improvement of this book from readers. While every attempt has been made to eliminate errors, inadvertent lapses are likely. I would be grateful if you would point them out to me. My email id is jagdishsg@nitc.ac.in

ABOUT THE AUTHOR

Jagadeesha T obtained his BE (Mechanical) degree from NIT, Surathkal, Karnataka; ME (Machine Tools) degree from PSG College of Technology, Coimbatore, Tamil Nadu; Master of Science (Manufacturing and Automation) from National University of Singapore; and Master of Science (AMNS) from Singapore MIT Alliance. He obtained his doctoral degree from Jawaharlal Nehru Technological University. He is a recipient of the prestigious JRD Tata Scholarship and the SMA (Government of Singapore) Scholarship. He has twenty years of work experience that includes working in the industry, teaching, academic research, consultation and managing with excellence projects at reputed organisations. He has worked with TATA Engineering Locomotive Company (India), TVS Suzuki (India), IBM Pvt. Ltd (Singapore), ASM Technology Singapore (Singapore), Applied Materials (Singapore and USA), APP Systems and Services (Singapore), ST Microelectronics (Singapore), Chartered Semiconductor Manufacturing (Singapore), and Sitronics (Singapore). He has developed workbooks on fluid power control, mechanical vibration, machine design, machining science, and IC engines. He is the author of two textbooks: *Fluid Power Control*, published by John Wiley & Sons and *Mechanical Vibration*, published by Tata McGraw Hill. He has also reviewed the following books: *Introduction to Vibration* by Ronald James Anderson, *Mechanical Vibrations and Waves: Analysis, Control and Applications* by Carole Mei, and *The Practice of Engineering Dynamics* by Ronald J. Anderson, all published by Wiley International. He is also a reviewer for the *International Journal of Nanosciences*. He has made significant contribution to the global edition of *Finite Element Analysis: Theory and Application Using ANSYS* by Saeed Moaveni, published by Pearson Education. He is a member of several professional bodies in India and abroad. His innovative research on BPSG CVD process won the best IFIT award at ST Microelectronics, Singapore. He has bagged more than 30 quality suggestion awards at TELCO and the Best Employee award three times at ST Microelectronics, Singapore. Twenty-four of his papers have been published in various journals and conference proceedings.

At present, he is an assistant professor in the Department of Mechanical and Production Engineering at the National Institute of Technology (NIT), Calicut, Kerala, where he teaches solid mechanics, mechanics of machinery, dynamics of machinery, design of machine elements, fluid power control, computational methods, industrial automation and robotics and mechatronics.

1 INTRODUCTION TO PNEUMATICS

Upon completion of this chapter, the reader should be able to

- Explain the meaning of pneumatics
- Describe the desirable properties of the working medium in pneumatic systems
- Explain the advantages and disadvantages of compressed air
- Identify and appreciate the application of pneumatic systems in various industries
- Describe the various gas laws
- List the basic components of a pneumatic system
- Describe the various power transmission systems
- Compare hydraulic, pneumatic and mechanical systems

1.1 Pneumatics and Its Meaning

The word 'pneumatic' and its associate noun 'pneumatics' are derived from the Greek 'pneuma', meaning breath or air. Originally, it was coined to give a name to the science of motion and the properties of air. Compressed air is a vital utility, just like water, gas and electricity, and is used in countless ways in everyday life. Pneumatics is the application of compressed or pressurised air to power machines to control or regulate them. Simply put, pneumatics may be defined as a branch of engineering science dealing with the study of the behaviour and application of compressed air. Pneumatics may also be defined as the branch of fluid power technology that deals with the generation, transmission and control of power using pressurised air. In a pneumatic system, gas behaves like a spring since it is compressible.

Any gas can be used in pneumatic systems but air is the most common, for obvious reasons. Exceptions are most likely in aircraft and space vehicles, where an inert gas such as nitrogen, or one that is generated on board, is preferred. Pure nitrogen may be used if there is a danger of combustion in the work environment like in an air accumulator.

1.2 History of Pneumatics

In its present day form of industrial usage, pneumatics is a relatively young progeny of technology, gaining acceptance in the late 18th century, although it had been used in simpler devices for many centuries before that. For the greater part, these inventions were devised for ritual or martial purposes. Among the various pneumatic devices contained in the encyclopedia edited by Diderot and published in 1774, for example, is a sectional drawing of a pneumatic rifle. Extensive use of pneumatics started only after the industrial revolution. Several instances of modern day pneumatic equipment were introduced within a short period about a century ago, e.g., the pneumatic dispatch system, air brakes, pneumatic riveter, pneumatic rock drill, etc. A pneumatically operated street car appeared, as did a number of pneumatic systems for the railways. While some of these inventions continue to be in use today, although with significant improvements, others have disappeared on account of difficulties, technical or otherwise.

Modern pneumatics, with its versatile scope and possibilities, did not really begin to supplement the established techniques and forms of power and energy in industrial processes until the 1950s. It has since evolved into an extensive and successful branch of engineering. A diversified line of pneumatic equipment of a mature engineering standard is already on the market and can be expected to grow further in size and efficiency in the future. Repeated releases of equipment with newer features and newer application areas demonstrate the continuous growth in the field of pneumatics.

1.3 Pneumatic Systems

Pneumatic control systems typically have the following components:

- Drive elements (e.g., actuators, air motors)
- Final control elements (e.g., directional control valves)
- Signal processing elements (e.g., logic gates)
- Signal input elements (e.g., push button switches, limit switches, sensors)
- Energy elements (e.g., compressors, filters, dryers, regulators, lubricators)

1.4 Choice of Working Medium

Hydraulics and pneumatics are systems that use fluid power to achieve work. In pneumatics, the potential energy stored in compressed air is used as the transmission medium for force, while in hydraulics the working medium is oil. The choice of the working medium depends on the application. Some of the general rules for the selection of a working medium in fluid power systems are listed below:

- When system requirements are high speed, medium pressure (usually 6 to 8 bar) and less than stringent positional accuracy, a pneumatic system is preferred.

- If the system requirements are high pressure and high precision, a fluid system with an oil medium is appropriate.
- When the power requirement is high, like in forging presses, sheet metal presses, etc., an air system will not work and oil hydraulics is the only choice.
- Air is used as the working medium when a quick response of the actuator is required.
- If the temperature variation in the system is large, then air systems may run into condensation problems and oil is the preferred medium.
- If the application requires only medium pressure but high positional accuracy, then a hydro-pneumatic system is preferred.
- Since air is non-explosive, it is preferred in situations where there is a risk of fire and electric hazards. Oil systems being combustible are not recommended in such applications.
- Because air contains oxygen (about 20%) and also since it does not provide adequate lubrication to moving parts and seals, oil is usually introduced into the air stream near the actuator, to provide the required lubrication to prevent wear and oxidation.

Compressed air is expensive to produce and might appear disadvantageous from an energy efficiency perspective. For example, it takes 6 kW of electrical energy to generate 0.76 kW output on an air motor, which translates into an efficiency of only 12%. However, despite this obvious drawback, compressed air is used in many applications because of other advantages listed below.

1. High output-to-weight ratio
2. Ease of storage
3. Easily transferrable from one place to another
4. Air is clean and dry
5. Air is safe and devices can be protected from overload
6. Linear motions are easy to implement with pneumatics
7. Easy to recover heat from air

Air motors are smaller and lighter than electric or hydraulic motors, making them ergonomically superior. This is clearly evident when we consider air-driven tools, e.g., drills, saws, grinders, impact wrenches, sanders or jack hammers, which have a high output to weight ratio that translates to lesser operator fatigue, improved efficiency and hence, better productivity on the shop floor.

The advantages and disadvantages of using compressed air as a working medium in pneumatic systems are given in Table 1.1.

Table 1.1 *Advantages and disadvantages of compressed air*

Advantages of compressed air	Disadvantages of compressed air
Air is available in unlimited quantities.	Compressed air is a relatively expensive means of conveying energy. The higher costs are largely compensated by cheaper elements and simpler and more compact equipment.
Compressed air is easily conveyed in pipelines even over longer distances.	
Compressed air can be stored.	Compressed air requires good conditioning to remove dirt and moisture. Dirt and dust leads to wear on tools and equipment.
After work is completed, compressed air does not need elaborate disposal; it can merely be vented to the atmosphere.	It is difficult to consistently achieve uniform and constant piston speeds with compressed air.
Compressed air is insensitive to temperature fluctuation. This ensures reliable operation even in extreme temperature conditions.	Compressed air is economical only to a certain extent. Under typical conditions, at a pressure of 7 bar and depending on the distance and speed, the limit is about 20 to 50 kN. If the required force exceeds this level, hydraulics is preferred.
Compressed air is clean. This is especially important in food, pharmaceutical, textile, and beverage industries.	Noise levels are high due to the exhaust. Sound suppression solutions can help counter this problem.
Operating elements for compressed air operation are simple and inexpensive.	The oil mist mixed with air to lubricate equipment can escape with the exhaust into the atmosphere.
Compressed air is fast, allowing for high operational speed.	Due to its low conductivity, air cannot dissipate heat as well as hydraulic fluid.
Pneumatics elements can be adjusted to a fine granularity for speed and force.	Unlike mediums in hydraulic systems, air cannot seal the fine gaps between moving parts.
Tools and operating elements are safe; they just stop working when overloaded, without damage to the systems or the user. Straight line movement can be produced directly.	Unlike hydraulic fluid, air is not a good lubricating medium.

1.5 Advantages of Pneumatics over Hydraulics

In many applications, pneumatic systems are preferred over hydraulic systems for the following reasons:

1. Density of air is significantly lower than that of oil; therefore the effect of inertia on pneumatic components is lower than for hydraulic system components. In other

words, the force required to accelerate air is much lower than that required to accelerate an equal amount of oil.

2. Viscosity of air is lower than that of oil. Therefore pneumatic systems have lower friction and power related losses.
3. In pneumatic systems, air can be discharged to the atmosphere after completing the task. Therefore a reservoir for recirculation (like in hydraulic systems) is not required.
4. Pneumatic systems are cheaper compared to hydraulic systems.
5. Pneumatic systems are cleaner and nontoxic, and can be used for food and pharmaceutical industries.
6. Any leakages in pneumatic systems with compressed air are not harmful, with minimal impact on performance.
7. Pneumatic systems are lighter in comparison to hydraulic systems.

1.5.1 Differences between hydraulic and pneumatic systems

- The working medium in pneumatic systems is compressible, whereas the working medium in hydraulic systems is not.
- To do the intended work, pneumatic systems typically use compressed air or nitrogen, whereas hydraulic systems use oil or water.
- Pneumatic systems are designed to be open; the compressed air is discharged to the atmosphere after use. Hydraulic systems are closed, and store and recirculate the liquid (oil or water) after use.

1.6 Characteristics of Pneumatic Systems

The function of pneumatic systems is to compress air, store the energised fluid and use the stored energy to do useful mechanical work. Various characteristics of pneumatic systems are given in Table 1.2.

Table 1.2 *Characteristics of pneumatic systems*

Quantity	Air is available everywhere in unlimited quantity.
Power transmission	Compressed air can be easily transmitted through pipes over long distances.
Storage	Large quantities of compressed air can be stored in a receiver tank.
Acceleration	Compressed air is fast, operating speed is 1–2 m/s.
Energy absorption	Air is compressible and elastic. It is capable of absorbing large amounts of potential energy.
Control	Actuator speed in a pneumatic system can be controlled by the simple use of control valves.

contd...

Table 1.2 *...contd*

Stalling	Because force can be easily controlled in pneumatic systems, mechanical elements driven by compressed air system can be stalled for long periods without any damage.
Over load protection	Compressed air tools and working elements can be subjected to loads even at standstill.
Other advantages	1. A pneumatic system is a simple method of transmitting energy. 2. It is easy to install and maintain. 3. Components are usually compact, lightweight and easily serviceable. 4. Fewer mechanical components make it more efficient and dependable. 5. Air may be exhausted to the atmosphere, minimising pollution. 6. No storage reservoir or return lines are needed, reducing the cost. 7. Air-based equipment is not normally susceptible to sudden failures. 8. Compressed air can be used in hazardous areas where electrical controls cannot be used for safety reasons.
Preparation	Pneumatic systems are vulnerable to dirt and contamination.
Force limitation	Compressed air actuators are economical up to 50 kN.
Uniform speed	Uniform speeds are hard to achieve with compressed air.
Noise	Exhaust air is noisy and will require appropriate noise suppression measures.

1.7 Applications of Pneumatics

Pneumatic systems are used in many applications. New uses for pneumatics are constantly being discovered. In construction, it is an indispensable source of power for tools such as air drills, hammers, wrenches, and even air cushion supported structures, not to mention the many vehicles using air suspension, braking and pneumatic tyres.

In manufacturing, air is used in pneumatic wrenches and riveting machines to power high speed clamping, drilling, grinding and assembly. Plant air is also used to power hoists and cushion support to transport loads through the plant.

Many recent advances in aircushion support are used in the military and in the marine transport industry.

Some of the industrial applications of pneumatics are listed in Table 1.3.

Table 1.3 *Industrial applications of pneumatics*

Material handling	Manufacturing	Other applications
Clamping	Drilling	Aircraft
Shifting	Turning	Cement plants
Positioning	Milling	Chemical plants
Orienting	Sawing	Coal mines

contd...

Table 1.3 *...contd*

Feeding	Finishing	Cotton mills
Ejection	Forming	Dairies
Braking	Quality control	Forge shops
Bonding	Stamping	Machine tools
Locking	Embossing	Door or chute control
Packaging	Filling	Turning and inverting parts
Feeding		
Sorting		
Stacking		

1.8 Properties of Air

1.8.1 Composition

Air is the gaseous form of matter. It has no fixed shape; it conforms to the shape of its container and readily transmits pressure. Gas has no fixed volume; it completely fills up the container. It is highly compressible, while liquids are only slightly so. Gas is also less dense than liquids.

Air is made up of a mixture of gases. By volume it contains about 78% nitrogen, 21% oxygen and about 1% of other gases, including argon and carbon dioxide. Water is the remaining ingredient, and the most important from the perspective of pneumatics. The dilution of oxygen by nitrogen makes air chemically much less active than pure oxygen, but it is still capable of spontaneous combustion, particularly in the presence of oil vapour at high temperatures, as in the case of an air receiver.

Air is colourless, odourless, tasteless, and compressible and has weight. Air has a great affinity for water and unless specifically dried, contains considerable quantities of water vapour, sometimes as much as 1% by weight. Life on earth depends on air for survival and man harness its forces to do useful work. Table 1.4 gives the physical properties of air.

Table 1.4 *Properties of air*

Property	Value
Molecular weight	28.96 kg/kmol
Density of air at 15 and 1 bar	1.21 kg/m^3
Boiling point at 1 bar	−191°C to −194°C
Freezing point at 1 bar	−212°C to −216°C
Gas constant	286.9 J/kg K

1.8.2 Free air and standard air

In pneumatics, the following two conditions of atmospheric air are well accepted:

- **Free air:** Air in atmospheric condition at the point where the compressor is located is defined as free air. Free air will vary with atmospheric conditions such as altitude, pressure and temperature.
- **Standard air:** It is also called normal air. It is defined as the air at sea level conditions (1.01324 bar as per ISO R554, 20°C and relative humidity of 36%). The condition of normal atmosphere is used as the basis for getting average values for compressor delivery volumes, efficiencies and operating characteristics.

1.8.3 Atmospheric pressure, gauge pressure and absolute pressure

Air has mass and exerts pressure on the earth's surface. A barometer consisting of an inverted tube closed at the top will support a column of mercury at exactly 760 mm at sea level when measured at standard conditions. Pressures above one atmosphere ($\sim$1 bar) are positive, whereas a pressure below one atmosphere causes a vacuum to be formed. Both positive pressures and vacuum pressures have useful purposes in pneumatics. Vacuum measurement is usually in given mm of mercury and then converted into a holding force for devices such as suction pads and cylinders of a specified diameter.

Atmospheric pressure Earth is surrounded by air. Since air has weight, it exerts pressure on the earth's surface. The weight of the column of air on one square metre of the earth's surface is known as atmospheric pressure or reference pressure.

The atmospheric pressure varies slightly from day to day. In pneumatic circuit calculations, standard atmospheric pressure (101.325 kPa or 14.7 psi or 760 mm of Hg) is used. The atmospheric pressure is measured using a barometer.

Atmospheric pressure = Density × acceleration due to gravity × height of barometer column (usually mercury)

$$p_{\mathrm{atm}} = \rho g h$$

Using mercury,

$$\rho = 13600\frac{\mathrm{kg}}{\mathrm{m}^2},\ \mathrm{g} = 9.81\frac{\mathrm{m}}{\mathrm{s}^2},\ h = 760 \text{ mm at sea level}$$

$$p_{\mathrm{atm}} = \rho g h = 13600\ \times 9.81 \times 0.760 = 101396\frac{\mathrm{N}}{\mathrm{m}^2} = 1.013 \text{ atm} = 1.013 \text{ bar}$$

Gauge pressure Gauge pressure refers to the pressure measured by a pressure gauge, typically used for measurement in pneumatic applications. Pressure gauges are calibrated to indicate the pressure above atmospheric pressure.

Absolute pressure This refers to the true or total pressure.

Absolute pressure = Atmospheric pressure + Gauge pressure.

Calculations involving formulae associated with gas laws must be made at absolute pressure.

Figure 1.1 shows the difference between gauge and absolute pressure. Let us examine the two pressure levels, p_1 and p_2.

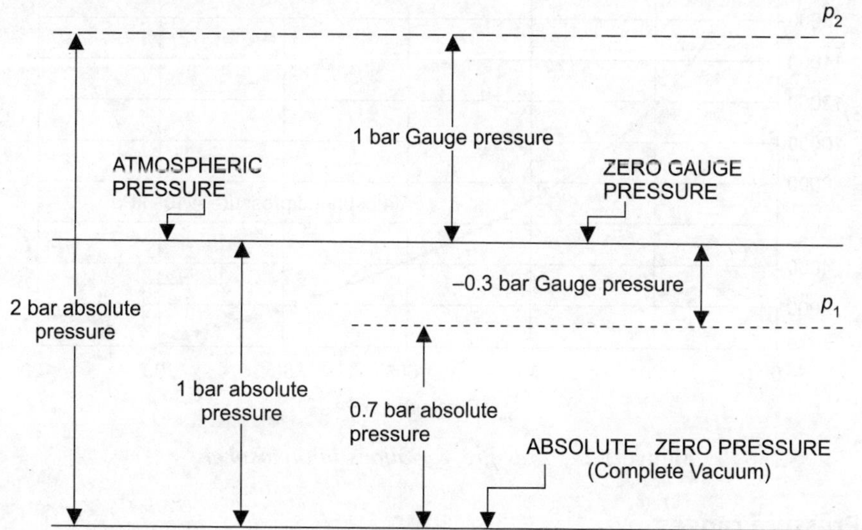

Figure 1.1 *Difference between absolute and gauge pressure*

Relative to a perfect vacuum, they are:

$p_1 = 0.7$ bar (absolute), a pressure less than atmospheric pressure
$p_2 = 2$ bar (absolute), a pressure greater than atmospheric pressure

Relative to the atmosphere, they are:

$p_1 = -0.3$ bar (gauge/suction/vacuum)
$p_2 = 1$ bar (gauge)

It should be noted that vacuum or suction pressures exist in certain sections of fluid power systems (e.g., in the inlet or suction lines of pumps). Therefore, it is important to understand the meaning of pressures below atmospheric pressure. One way to generate a suction pressure is to remove some fluid from a closed vessel originally containing fluid at atmospheric pressure.

1.8.4 Variation of pressure with altitude

According to barometric law, atmospheric pressure decreases exponentially with an increase in altitude. The relation is shown graphically in Fig. 1.2. Variation of pressure with respect to altitude is linear up to 6000 m, and pressure drops about 11 kPa per 1000 m change in altitude.

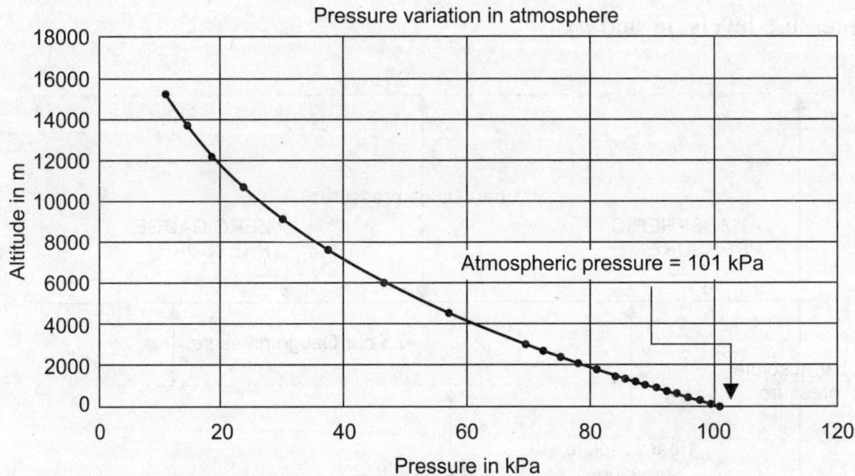

Figure 1.2 *Pressure variations in atmosphere*

1.8.5 Pressure ranges

1. Low pressure range Most industrial pneumatic applications use low pressure air up to 8 bar. The compressors used are;

- One- and two-stage piston compressor
- One-stage screw compressor with oil injection cooling
- Two-stage compressor
- Rotary compressor

2. Medium pressure range Heavy commercial vehicle tyres are filled with compressed air to a pressure of 15 bar. The compressors used are:

- Two-stage piston compressor
- One-stage screw compressor

3. High pressure range Compressors used for starting diesel engines use a pressure of up to 40 bar. Similar ranges of pressures are used for testing pipelines and flushing heavy plastic tanks. The compressors used are:

- Two- and three-stage piston compressor
- Multi-stage screw compressor

4. Very high pressure range One example of the use of high pressure compressed air is for storing breathing air in diving tanks. High pressure compressors are used in power stations, rolling mills and steel works, and for leak testing. Compressors of this type are also used for compressing utility gases such as oxygen.

Three- and four-stage piston compressors are used for this purpose.

The range of pressure is shown schematically in Fig. 1.3.

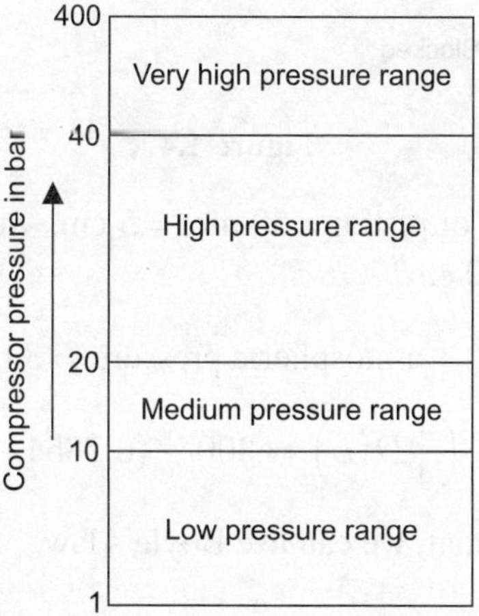

Figure 1.3 *Range of pressure used in pneumatics*

1.9 Gas Laws

Early experiments on the behaviour of air and similar gases were conducted by scientists such as Boyle, Charles and Gay-Lussac. The results of their experiments concluded that the behaviour of gases can be represented by the ideal gas laws.

1.9.1 Boyle's law

English scientist, an Robert Boyle (1627–1691) was among the first to experiment with the pressure–volume relationship of a gas at a constant temperature.

If a given mass of a gas is compressed or expanded at a constant temperature, then the absolute pressure is inversely proportional to the volume.

$$\text{Pressure} \propto \frac{1}{\text{Volume}} \quad \text{(when temperature is held constant)}$$

or $pV = \text{constant}$.

For states 1 and 2 of a substances, $p_1V_1 = p_2V_2$.

Example 1.1 The .50 mm diameter piston of a pneumatic cylinder shown in Fig. 1.4 retracts 130 mm from its present position ($p_1 = 2$ bar (gauge) $V_1 = 300$ cm^3) due to the external load on the rod. If the part at the blind end of the cylinder is blocked, find the new pressure, assuming temperature remains constant.

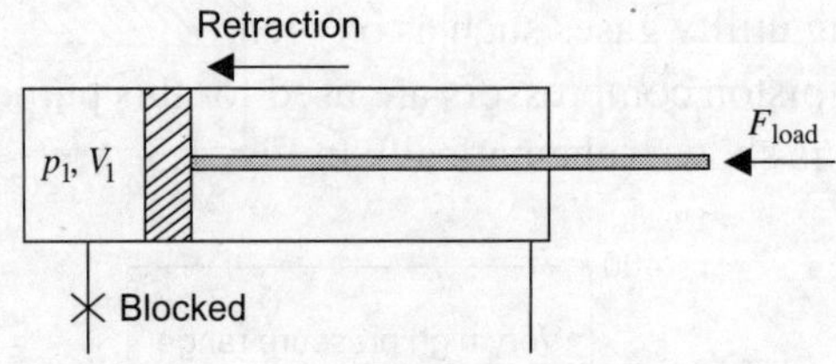

Figure 1.4

Given data: D = diameter of piston = 50 mm = 5 cm; Stroke = L = 130 mm = 13 cm; $p_1 = 2$ bar (gauge); $V_1 = 300$ cm^3

Solution $p_1 = 2$ bar (gauge) + atmospheric pressure = 3 bar (absolute)

$$\text{Final volume} = V_2 = V_1 - \left(\frac{\pi}{4}D^2L\right) = 300 - (0.7854 \times (5^2) \times 13) = 45 \text{ cm}^3$$

Since the temperature is constant, we can use Boyle's law.

$$p_1V_1 = p_2V_2$$

$$3 \times 300 = p_2 \times 45$$

Thus, p_2 = 20 bar (absolute) = 19 bar (gauge).

Example 1.2 Four cubic metres of nitrogen is under a pressure of 7 bars (gauge). The nitrogen is allowed to expand to a volume of 6 cubic m. What is the new gauge pressure?

Given data: $p_1 = 7$ bar (gauge) $= 8$ abs; $V_1 = 4$ m^3; $V_2 = 6$ m^3

Solution Since the temperature is constant, we can use Boyle's law.

$$p_1V_1 = p_2V_2$$

$$8 \times 4 = p_2 \times 6$$

Thus, $p_2 = 5.333$ bar (absolute) $= 4.333$ bar (gauge)

Example 1.3 A piston compresses air at atmospheric pressure to 1/7th the volume, as illustrated in Fig. 1.5. Assuming constant temperature, what is the resultant gauge pressure?

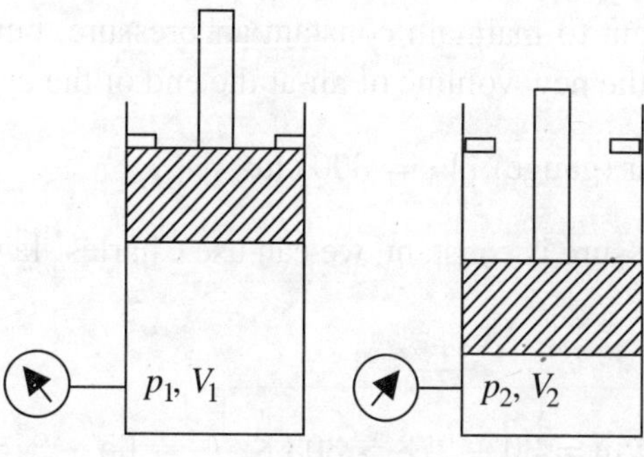

Figure 1.5

Solution

$$\frac{V_2}{V_1} = \frac{1}{7} \quad \text{or} \quad \frac{V_1}{V_7} = 7$$

Since the temperature is constant, we can use Boyle's law.

$$p_1 V_1 = p_2 V_2$$

$$p_2 = p_1 \times \frac{V_1}{V_2}$$

$$p_2 = 1 \times \frac{7}{1} = 7 \text{ bar (absolute)}; \; p_2 = 6 \text{ bar (guage)}$$

1.9.2 Charles' law

Boyle's law assumes conditions of constant temperature. In reality this is rarely the case. Temperature changes continuously and affects the volume of a given mass of gas.

French physicist Jacques Charles (1746–1823) provided much of the foundation for modern kinetic theory of gases. Through experiments, he found that all gases expand and contract proportionally to the change in the absolute temperature, provided the pressure remains constant. The law is stated as follows.

If a given mass of a gas is heated or cooled at a constant pressure, then the volume is directly proportional to the absolute temperature.

Volume $\propto$ Temperature (when pressure = constant)

or $\quad \dfrac{V}{T} = \text{constant.}$

For states 1 and 2 of a substance, $\dfrac{V_1}{T_1} = \dfrac{V_2}{T_2}$.

Example 1.4 The cylinder of Fig. 1.5 has an initial position of ($p_1 = 2$ bar (gauge) $V_1 = 300 \text{ cm}^3$) as controlled by the load on the rod. The air temperature is 30°C. The load on the rod is held constant to maintain constant air pressure, but the air temperature has increased to 65°C. Find the new volume of air at the end of the cylinder.

Given data: $p_1 = 2$ bar (gauge); $V_1 = 300$ cm^2

Solution Since the pressure is constant, we can use Charles' law:

$$\frac{V_1}{T_1} = \frac{V_2}{T_2}$$

T_1 = temperature in Kelvin $= 30 + 273 = 303$ K, $T_2 = 65 + 273 = 338$ K

$$\frac{300}{303} = \frac{V_2}{338}$$

Solving, we get, $V_2 = 145$ cm^3 $= 145 \times 10^{-6}$ m^3.

Example 1.5 A sample of H_2(g) occupies a volume of 70 cm^3 at a pressure of exactly 1 atm when immersed in a mixture of ice and water. When the gas (at the same pressure) is immersed in boiling benzene, its volume expands to 90 cm^3. What is the boiling point of benzene?

Given data: $p_1 = 1$ bar(guage); $V_1 = 70$ cm^3; $V_2 = 90$ cm^3

Solution Since the pressure is constant, using Charles' law:

$$\frac{V_1}{T_1} = \frac{V_2}{T_2}$$

T_1 = temperature in Kelvin $= 0 + 273 = 273$ K,

$$\frac{70}{273} = \frac{90}{T_2}$$

Solving, we get, $T_2 = 351$ K yields the desired result. (The ice-water mixture must be at 273.15 K, the freezing point of water.)

1.9.3 Gay–Lussac's law

Boyle's and Charles' laws lead to a third law called Gay–Lussac's Law. According to Charles' law, doubling the temperature of an ideal gas, with pressure maintained at a constant, results in doubling the volume of the gas. Continuing with this example, according to Boyle's law, maintaining the higher temperature at a constant, we would need to double the pressure to halve the volume (i.e., return the gas to its original volume). Thus, if the volume of gas is to remain the same, doubling the temperature will require doubling the pressure. This law was first stated by the Frenchman Joseph Gay–Lussac (1778 to 1850) as below.

At constant volume, the absolute pressure of an ideal gas will vary directly with the absolute temperature.

Pressure $\propto$ Temperature (when volume = constant)

or $\quad \dfrac{p}{T} = \text{constant.}$

For states 1 and 2 of a substance, $\dfrac{p_1}{T_1} = \dfrac{p_2}{T_2}$.

Example 1.6 A compressed air receiver with a volume of 0.5 m^3 is filled with compressed air at 7 bar (gauge) and at a temperature of 40°C. The temperature then cools to 20°C. What is the final pressure?

Given data: $V_1 = V_2 = 0.5\ \text{m}^3$; $p_1 = 7$ bar (gauge) $= 7 + 1 = 8$ bar (absolute)
$T_1 = 20°\text{C} = 20 + 273 = 293$ K; $T_2 = 40°\text{C} = 40 + 273 = 313$ K

Solution Since the volume is constant, we can apply Gay–Lussac's law.

$$\frac{p_1}{T_1} = \frac{p_2}{T_2}$$

$$\frac{8}{298} = \frac{p_2}{313}$$

Thus, $p_2 = 7.5$ bar (absolute) $= 6.5$ bar (gauge).

1.9.4 General gas equation

For any given mass of gas undergoing changes in pressure, temperature and volume, the general gas equation can be used. Combining Boyle's and Gay–Lussac's laws we get:

$$\frac{p_1 V_1}{T_1} = \frac{p_2 V_2}{T_2}.$$

Example 1.7 As seen in Fig. 1.6, a gas at 80 bar (gauge) and 50°C is contained in a cylinder of volume 1290 cm^3. A piston compresses the volume to 1000 cm^3 while the gas is heated to 120°C. What is the final pressure of the cylinder?

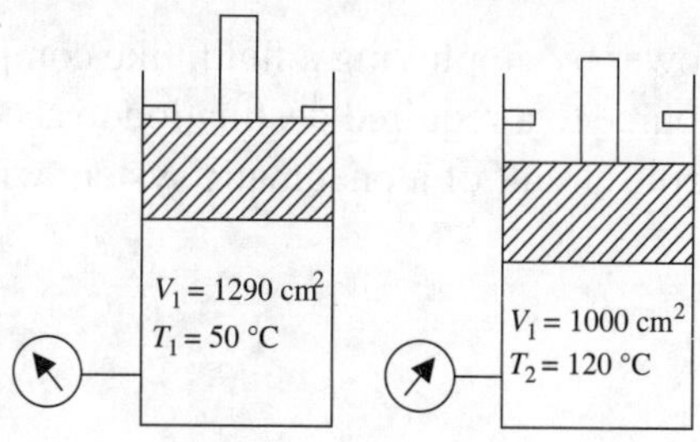

Figure 1.6

Given data: $p_1 = 80$ bar (gauge); $V_1 = 1290\ \text{cm}^3$; $V_2 = 1000\ \text{cm}^3$; $T_1 = 50°\text{C}$; $T_2 = 120°\text{C}$.

Solution Since the volume is constant,

$$p_1 = 80 \text{ bar (gauge)} = 81 \text{ bar (absolute)}$$

$$T_1 = 50°\text{C} = 50 + 273 = 323\ \text{K};\quad T_2 = 120°\text{C} = 120 + 273 = 393\ \text{K};$$

$$V_1 = 1290\ \text{cm}^3,\ V_2 = 1000\ \text{cm}^3.$$

Using the general gas law

$$\frac{p_1 V_1}{T_1} = \frac{p_2 V_2}{T_2},$$

$$\frac{81 \times 1290}{323} = \frac{p_2 \times 1000}{393}.$$

Solving, we get: $p_2 = 127.13$ bar (absolute) $\cong 126.13$ bar (gauge).

Example 1.8 Two cubic metres of a gas at 5 bar and 300 K are compressed to a volume of 1 cubic metre and then heated to a temperature of 420°K. What is the new gauge pressure?

Given data: $p_1 = 5$ bar (gauge); $V_1 = 2\ \text{m}^3$; $V_2 = 1\ \text{m}^3$; $T_1 = 300$ K; $T_2 = 420$ K

Solution Using the general gas law

$$\frac{p_1 V_1}{T_1} = \frac{p_2 V_2}{T_2},$$

$$\frac{6 \times 2}{300} = \frac{p_2 \times 1}{420}.$$

Solving, we get $p_2 = 16.8$ bar (absolute) $= 15.8$ bar (gauge).

1.10 Basic Components of Pneumatic Systems

Pneumatic systems carry power by employing a fluid, like compressed gas (generally air), to transmit energy from a source to a required destination to accomplish useful work.

Figure 1.7 shows the simple circuit of a pneumatic system with the basic components.

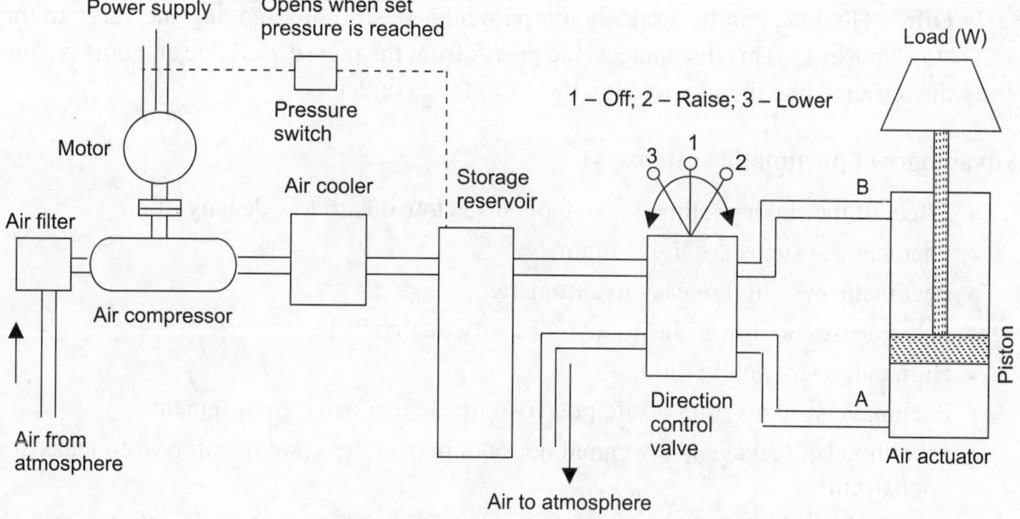

Figure 1.7 *Components of a pneumatic system*

The functions of the components in a pneumatic system are listed below.

1. A pneumatic actuator converts fluid power into mechanical power to do useful work.
2. A compressor is used to compress air drawn from the atmosphere.
3. A storage reservoir is used to store a given volume of compressed air.
4. Valves are used to control the direction, flow rate and pressure of the compressed air.
5. An external power supply (e.g., motor) is used to drive the compressor.
6. A piping system carries the pressurised air from one location to another.

Air is drawn from the atmosphere through an air filter and raised to the required pressure by an air compressor. As the pressure rises, the temperature also rises, and hence an air cooler is provided to cool the air along with some preliminary treatment to remove moisture content. Then the treated pressurised air is stored to maintain the pressure. The storage reservoir is fitted with a pressure switch to start and stop an electric motor to maintain the pressure levels.

The movement of the cylinder is controlled by a pneumatic valve. One side of the pneumatic valve is connected to the compressed air and to the silencers for the exhaust system, and the other side of the valve is connected to ports A and B of the cylinder.

The positions of the valve are as follows:

1. **Raise:** In this position, port A is connected to the compressed air supply and port B is connected to the exhaust line, resulting in lifting of the load.
2. **Lower:** To lower the load, the valve is positioned at the "Lower" position wherein the compressed air line is connected to port B, and port A is connected to exhaust line.

3. **Off:** The load can be stopped at a particular position by moving the valve to the "Off" position. This disconnects the port A from the pressurised line and port B from the retrieval line thereby locking the air in the cylinder.

Advantages of pneumatic systems

- Effect of inertia in pneumatic components is low due to low density of air.
- Pneumatic systems are lightweight.
- Pneumatic systems are easy to automate.
- Power losses are lower due to low viscosity of air.
- High output-to-weight ratio.
- Pneumatic systems offer a safe power source in explosive environments.
- Likelihood of leakage is lower and does not impact the systems. Moreover, leaks are not harmful.

Disadvantages of pneumatic systems

- Suitable only for low pressure and low force applications.
- Compressed air actuators are economical only up to 50 kN.
- Generation of compressed air is expensive in comparison to electricity.
- Air exhaust systems are noisy and require silencing components.
- Rigidity of the system is poor.
- High weight to pressure ratio.
- Low precision; hard to achieve uniform speed due to compressibility of air.
- Pneumatic systems are vulnerable to dirt and contamination.

1.11 Comparison of Hydraulic and Pneumatic Systems

Usually hydraulic and pneumatic systems do not compete. They are so dissimilar that any problems in deciding between the two systems are usually readily resolved after careful consideration of all the factors. But in less than ideal circumstances, it may be a harder decision; some of the power resources may not be available and compressed air has to be used to do something for which oil might be the better choice, and vice versa. Availability is certainly one of the important factors of selection and often outweighs other factors. In numerous instances, for example, air is preferred due to certain unalterable conditions. For example, hot spots where there is an open furnace or other potential ignition hazards, or in operations where motion is required at extremely high speeds. It is often more efficient to use a combined circuit in which oil is used in one section, and air in another, within the same machine or process. The comparison between hydraulics and pneumatics is given in Table 1.5.

Table 1.5 *Comparison between hydraulic and pneumatic systems*

Sl. No	Hydraulic system	Pneumatic system
1	Employs pressurised liquid as the fluid	Employs compressed gas (usually air) as the fluid
2	Oil hydraulic systems operate at pressures up to 700 bar.	Pneumatics systems usually operate at 5 to 10 bar.
3	Generally designed for closed systems	Usually designed as open systems
4	Leaks slow down the system.	Leaks do not affect the system much.
5	Valve operations are difficult.	Valve operations are easy.
6	Heavier in weight	Lighter in weight
7	Pumps are used to provide pressurised liquids.	Compressors are used to provide compressed gas.
8	System is vulnerable to fire hazards.	System is secure from fire hazards.
9	Automatic lubrication is provided.	Special arrangements for lubrication are needed.

1.12 Comparison of Different Power Systems

The three basic modes of transmitting power are electrical, mechanical, and fluid systems. Usually a combination of the three systems is used for optimum efficiency. The strengths and weaknesses of each system determine its best use. For example, fluid systems can transmit power farther than mechanical systems but electrical systems are the best choice when distance is the factor.

1.12.1 Electrical power transmission

Electrical energy in the form of current is transmitted and distributed through wires and cables. The energy medium is controlled using switches, pushbuttons, relays, contactors, timers, sensors, pressure switches, etc. The final actuator in this system is an electric motor which converts electrical energy into mechanical energy in the form of rotary motion to perform useful work. Electrical systems are suitable for power transmission over long distances. The limitations of electric power include magnetic saturation, which may limit the torque capacity of the motor, and material limitation which may limit the speed and heat dissipation. Switching contacts may arc or corrode. Electric arcs are a fire hazard in explosive environments. The response time of the electro mechanical solenoid is too slow for today's controllers. The switching time of control elements is usually greater than 10 milliseconds. Electrical systems need to provide for heat dissipation like most systems which generate heat by pressure or friction.

1.12.2 Mechanical power transmission

Mechanical power is transmitted by employing a variety of kinematic mechanisms such as belts, chains, pulleys, sprockets, gear trains, bar linkages and cams. They are suitable for the transmission of motion and force over relatively short distances. The disadvantages of mechanical power transmission include lubrication problems, limited speed and torque control capabilities, uneven force distribution, and a relatively large space requirement.

1.12.3 Hydraulic power transmission

In hydraulic transmission of energy, a pump is used to raise the pressure of oil, and the energy stored in the oil is transmitted through pipes and hoses to perform useful work. Hydraulic systems are suitable for power transmission over intermediate distances; they can be employed over greater distances than mechanical systems but cannot cover as much distance as electrical systems.

Hydraulic systems are mechanically stiff, and can be designed for fast operation and to move very heavy loads. They can easily generate linear motion using liner actuators (also called cylinders). Speed control is casy and precise motion of the actuator is possible.

The disadvantages of hydraulic systems include fluid leakage, containment and fire hazards with flammable hydraulic fluids.

1.12.4 Pneumatic power transmission

All gases are readily compressible and it is this property which differentiates them from liquids as a power transmission medium. In pneumatic transmission of energy, a compressor is used as the power source to raise the pressure of the air to the required level quite slowly. Pneumatic systems are suitable for power transmission over intermediate distances. They use simple equipment, have small transmission lines, and do not present a fire hazard.

The disadvantages of pneumatic system include a high fluid compressibility and a small power-to-size ratio of the components. Pneumatic systems are unsuitable for uniform motion. The operating pressure of pneumatic systems is around 6 to 8 bar, making them capable of generating only medium forces. The switching time of control elements is usually greater than 5 milliseconds, and the speed of the control signal is 10 to 50 m/s. Table 1.6 gives a comparison of all the three systems.

Summary

1. Pneumatics is the branch of engineering science which studies the behaviour and application of compressed air. Pneumatics provides the industry with a valuable and economic medium of power for production and control.
2. Pneumatics has many uses in the industry. Air at 6–8 bar is universally available at construction sites and plants. Pneumatic systems use pressurised air as their medium to transmit and control power.

3. Pneumatic systems are used when the system requirement is high speed, medium pressure (usually 6 to 8 bar) and lower positional accuracy. They are unsuitable for industries like forging and sheet metal presses where the power requirements are high, making oil hydraulics the only choice.
4. Weight density of air is low compared to oil; therefore the effect of inertia on pneumatic components is less of an issue than in hydraulic system components. Air has much lower viscosity than oil, therefore pneumatic systems have lower frictional and power losses.
5. Air is a mixture of gases containing about 78% nitrogen, 21% oxygen and 1% other gases such as argon and carbon dioxide. Standard air is air at sea level having a pressure of 1.013 bar, a relative humidity of 36% and a temperature of 20°C.
6. Absolute pressure is the sum of gauge pressure and atmospheric pressure. Atmospheric pressure decreases with an increase in altitude.
7. Boyle's law states that the volume of a gas varies inversely with the pressure exerted on it, provided the temperature remains constant.
8. Charles' law states that the volume of a gas increases proportionally to the temperature, provided the pressure remains constant.
9. Gay–Lussac's Law states that at constant volume, the absolute pressure of an ideal gas will vary directly with the absolute temperature.
10. The basic components required for any pneumatic system are air compressors, air tanks, prime movers to drive the compressor, control valves, pneumatic actuators and piping.

Table 1.6 *Comparison of electrical/mechanical, pneumatic and hydraulic transmission systems*

Property	Electrical/Mechanical	Pneumatic	Hydraulic
Energy source	IC engines, electrical energy is used to drive motors	Electrical energy is used to drive compressor and other equipment	IC engines, electric motors, air turbines are used to drive hydraulic pumps.
Working medium	There is no medium, energy is transferred through levers, gears, shafts	Compressed air/gas in pipes and hoses	Pressurised liquid in pipes and hoses
Energy storage	Batteries	Reservoir, air tank	Accumulators
Regulations	Variable frequency drives	Pneumatic valves	Hydraulic valves

contd...

Table 1.6 *...contd*

Transmitters	Transmitted through mechanical components like levers, gears, cams and screws	Transmitted through pneumatic cylinders, rotary drives and rotary actuators	Transmitted through hydraulic cylinders, and hydraulic rotary actuators
Distribution system efficiency	Good	Limited (up to 1000 m)	Good (up to 100 m)
Operating speed	Low	Limited (up to 1.5 m/s)	Limited (up to 0.5 m/s)
Positioning accuracy	Precision to within a few micron can be achieved	Precision to within a few mm (usually 0.1 mm) can be achieved	Precision to within a few micron can be achieved
Stability	Good stability is possible using mechanical elements	Low stability due to high compressibility of air	Good stability is possible due to low compressibility
Forces	Mechanical elements break down if overloaded. Poor overloading capacity	Protected against overload with system pressure of 6 to 8 bar, forces up to 50 kN can be generated	Protected against overload with high system pressure of 600 bar, very large forces can be generated.
Cost of energy	Lowest	Highest	Medium
Linear actuators	Short stroke length using mechanical/electrical transmission elements	Pneumatic actuators (cylinders) can be used to produce medium force.	Hydraulic actuators (cylinders) can be used to produce heavy force
Rotary actuators	AC, DC, servo and steeper motors can be used	Pneumatic rotary actuators can be used	Hydraulic motors and vane motors can be used
Controllable force	Possible with solenoid and DC motors, needs cooling which adds complexity	Medium force can be controlled easily	High force can be controlled
Work-place hazards	Dangerous because of electric shock	High noise levels	Dangerous, unsightly and risk of fire hazard in case of leakage

EXERCISES

Fill in the blanks

1. When the power requirement is ——, like in forging and sheet metal presses, oil hydraulic systems are preferred.
2. Pneumatic systems have —— output-to-weight ratio, whereas electrical systems have —— output-to-weight ratio.
3. Positioning accuracy of a pneumatic actuator is —— compared to hydraulic systems.
4. Compressed air actuators are economical up to —— force only.
5. Low inertia effect of pneumatic components is due to —— density of air.

State whether true or false

1. Pneumatic systems are used for high-pressure and low-speed applications.
2. Pneumatic systems are very useful in hazardous environments.
3. Compressed air is expensive.
4. According to barometric law, atmospheric pressure increases exponentially with an increase in altitude.
5. If a given mass of gas is compressed or expanded at a constant temperature, then the absolute pressure is directly proportional to the volume.

Review questions

1. Define pneumatics.
2. List the broad rules used in the selection of a working medium.
3. List the advantages and disadvantages of compressed air.
4. Why is air used as the fluid medium in pneumatic systems?
5. List seven reasons for considering pneumatic systems instead of hydraulic systems.
6. Name five characteristics of pneumatic systems.
7. List some applications of pneumatics.
8. Differentiate between free air and standard air.
9. Differentiate between absolute pressure and atmosphere pressure.
10. Discuss the various pressure ranges used in pneumatic applications.
11. State the following gas laws:
    (a) Boyle's law (b) Charles' law (c) Gay–Lussac's law (d) General gas law
12. Name the six basic components of a pneumatic system.
13. With a simple sketch, explain the functions and workings of the basic components of a pneumatic system.

14. List five advantages and disadvantages of pneumatics.
15. Differentiate between hydraulics and pneumatics.
16. Compare and contrast hydraulic, pneumatic and electrical power transmission systems.

Problems

1. If an empty 0.076 m^3 water tank, where the pressure gauge initially reads 1.38 bar, is half filled with water as shown in Fig. 1.8, what will be the new reading on the pressure gauge attached to the tank?

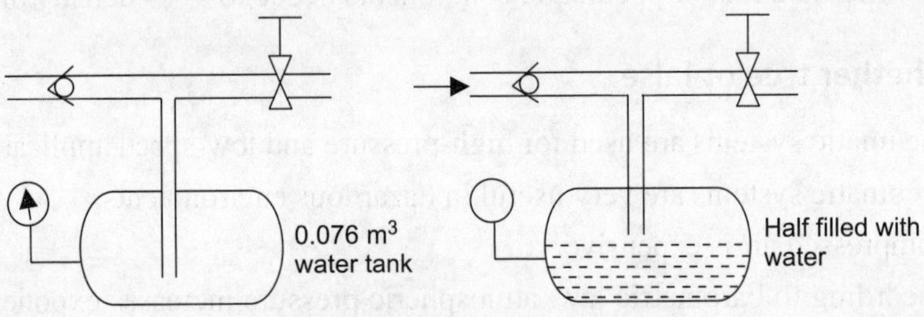

Figure 1.8

2. If the accumulator in Fig. 1.9, using a deadweight ballast against an initial volume of 0.025 m^3 of a gas, is heated from 300°K to 366°K, what volume will the heated gas occupy?

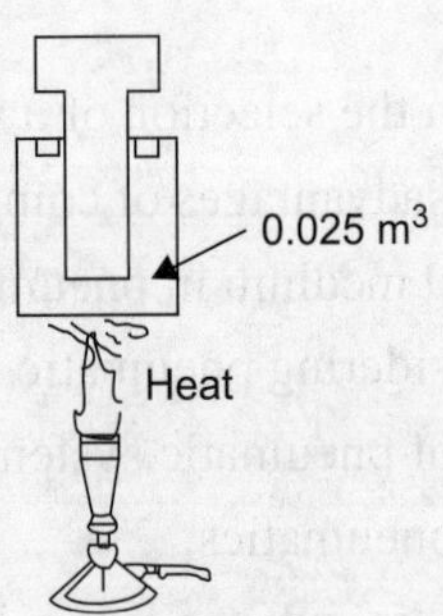

Figure 1.9

3. The fixed volume vessel shown in Fig. 1.10 has a pressure gauge reading of 138 bar. It is heated from 300°K to 395°K. What is the new pressure gauge reading?

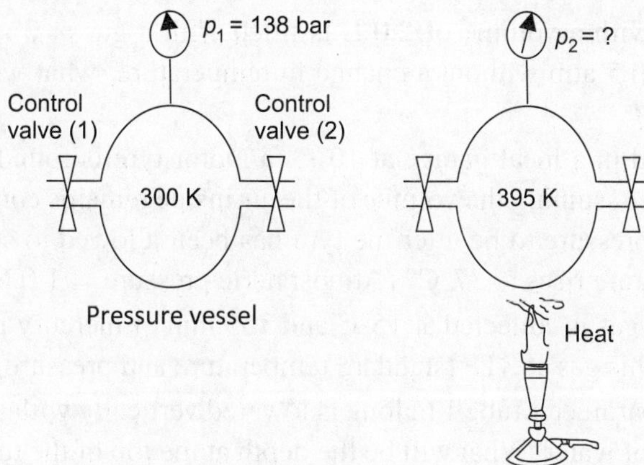

Figure 1.10

4. Gas in a 0.025 m^3 cylinder at 138 bar is reduced in volume to 0.016 m^3 while being heated from 297°K to 395°K. What is the final gauge pressure in the cylinder?

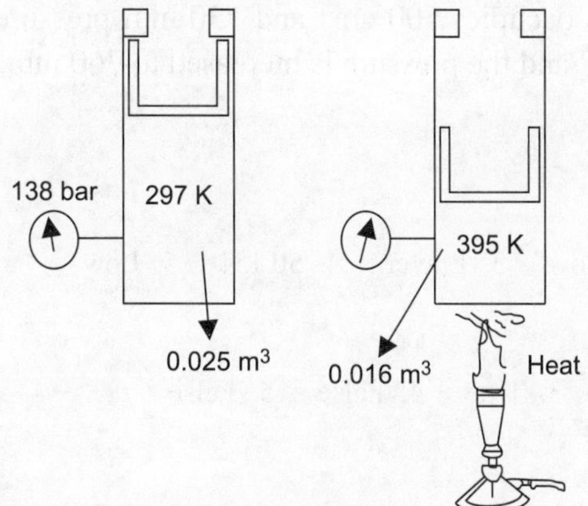

Figure 1.11

5. A fixed quantity of gas at a constant pressure occupies a volume of 8.50 L and has a temperature of 29°C.
   (a) What volume will the gas occupy if the temperature is increased to 125°C?
   (b) At what temperature will the volume be 5.00 L?
6. What is the number of moles of an ideal gas with $V = 2.50$ L, $T = 37$°C and $p = 725$ torr? Assume gas constant $R = 0.08206$ L-atm/mol-K
7. What is the density of SF4 vapour at 650 torr and 100°C?
8. A sample of helium gas at 25°C is compressed from 200 cm^3 to 0.240 cm^3. Its pressure is now 3.00 cm Hg (mercury). What was the original pressure of the helium?

9. A balloon with a volume of 2.0 L is filled with a gas at 3 atm. If the pressure is reduced to 0.5 atm without a change in temperature, what would be the volume of the balloon?
10. When tested in a local garage at 10°C, a motor tyre is found to have a pressure of 1.2 N/cm^2. Assuming the volume of the air inside remains constant, what would you expect the pressure to be after the tyre has been allowed to stand in the sun so that the temperature rises to 37°C? (Atmospheric pressure = 1.0 N/cm^2)
11. 125 cm^3 of gas is collected at 15°C and 755 mm of mercury pressure. Calculate the volume of this gas at STP (standard temperature and pressure).
12. An empty barometer tube 1 m long is lowered vertically with into mouth downwards into a tank of water. What will be the depth at the top of the tube when the water has risen 20 cm inside the tube? (Atmospheric pressure may be assumed equal to 10.4 m head of water.)
13. Find the volume of 0.250 moles of gas at 200 kpa and 300 K temperature. Assume $R = 8.314$ J/K mol.
14. Find the temperature of a 250 ml cylinder containing 0.50 moles of gas at 153 kPa.
15. At 25°C, a gas occupies 500 cm^3 and 730 mm pressure. If the temperature is increased to 50°C and the pressure is increased to 760 mm, what is the new volume?

Answers

Fill in the blanks

1. High 2. High/low 3. Lower 4. 50 kN 5. Low

True or false

1. False 2. True 3. True 4. False 5. False

Problems

1. $p_2 = 3.76$ bar (guage)

2. $V_2 = 0.0305\ \text{m}^3 = 30500\ \text{cm}^3$

3. $p_2 \cong 183$ bar (absolute) $= 182$ bar (gauge)

4. $p_2 = 288.85$ bar (absolute) $= 287.85$ bar (gauge)

5. (a) $V_2 = 11.2\ \text{L} = 11200\ \text{cm}^3$, (b) $T_2 = 177.65$ K

6. Number of moles $(n) = 0.0938$ moles

7. Density $(\rho) = 3.02 \frac{\text{g}}{\text{L}}$

8. $p = 3.60 \times 10^{-2}$ mm Hg $= 4.74 \times 10^{-5}$ atm

9. Volume $(V) = 5.33$L

10. $p_2 \cong 2.41 \frac{\text{N}}{\text{cm}^2\ \text{(absolute)}} = 1.41 \frac{\text{N}}{\text{cm}^2\ \text{(gauge)}}$

11. $V_2 = 118\ \text{cm}^2$

12. Top of tube 1.8 m below surface.

13. Volume $(V) = 3.12$ L

14. Temperature $(T) = 9.2$ K

15. $V_2 = 520.5\ \text{cm}^3$

2 PREPARATION OF COMPRESSED AIR

Upon completion of this chapter, the reader should be able to

- Explain the various stages of air preparation
- Describe the working of various compressors
- List the advantages and disadvantages of various compressors
- Carry out thermodynamic analysis of compressors
- Compare various types of compressors
- Understand the selection criteria for compressor
- Explain the importance of secondary treatment of compressed air
- List the various functions of an aftercooler
- Describe the functions of air and water coolers
- Explain the theory of air dryers
- Describe various types of air dryers used in pneumatic industry
- List the various functions of air receivers
- Recommend the size of air receivers and compressors

2.1 Air Preparation

Pneumatic control systems operate on a supply of compressed air, which must be made available in sufficient quantity and at a pressure that suits the capacity of the system. Impurities in the compressed air such as scale, rust and dust as well as the liquid constituents in the air which deposit as condensate can cause a great deal of damage in pneumatic systems. These impurities adversely affect the functioning and performance of the pneumatic components. In order to eliminate these effects, compressed air has to be prepared with great care.

There are four distinct stages in the preparation of air.

Stage 1: This stage comprises the air intake system whose role is to ensure that the intake air is free of moisture or pollution. The quality of air is judged by its temperature, humidity and cleanliness.
Stage 2: This stage involves compressors (some with drives and controls), inter-cooling, compressor cooling, waste heat recovery and air inlet filtration
Stage 3: In this stage, the outlet temperature of the compressor is reduced, solid contaminants large than 100 microns are removed, and the air is dried to reduce to its humidity. This stage includes conditioning equipment consisting of air receivers, aftercoolers, separators, traps (also frequency called drain traps or drains), filters and air dryers.
Stage 4: In this stage, moisture and fine dirt particles are removed, pressure is regulated and a fine oil mist is introduced to lubricate the machine. It involves air distribution subsystems, including main trunk lines, valving, additional filters and traps (drains), air hoses and possible supplement air conditioning equipment.

2.2 Classification of Air Compressors

A compressor is a machine that compresses air or any other type of gas from a low inlet pressure (usually atmospheric pressure) to a higher desired pressure level.

The classification of compressors is shown in Fig. 2.1.

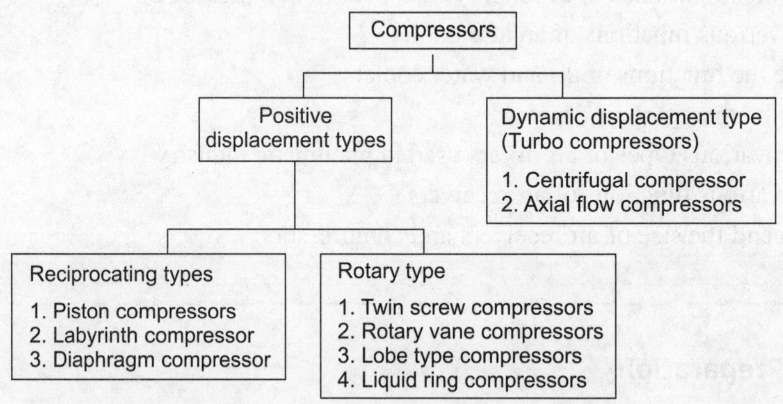

Figure 2.1 *Classification of compressors*

2.2.1 Reciprocating compressors

Reciprocating compressors are the most widely used type of compressors for industrial plant air systems.

2.2.1.1 Piston compressors

Piston type compressors are the oldest and most commonly used compressor in the pneumatic industry because of their flexibility, high pressure capability, ability to rapidly dissipate heat of compression and on account of thier being oil-free. They are built for either stationary or portable services.

A. Single cylinder compressor

The construction and working of a piston type reciprocating compressor is very similar to that of an internal combustion engine.

Construction The piston type compressor consists of a cylinder, cylinder head, a piston with piston rings, inlet and outlet spring-loaded valves, connecting rod, crankshaft and bearings.

Operation Compression is accomplished by the reciprocating movement of a piston within a cylinder. This motion alternately fills the cylinder and then compresses the air. A connecting rod transforms the rotary motion of the crankshaft into the reciprocating motion of piston in the cylinder. The schematic diagram of a single cylinder compressor is shown in Fig. 2.2.

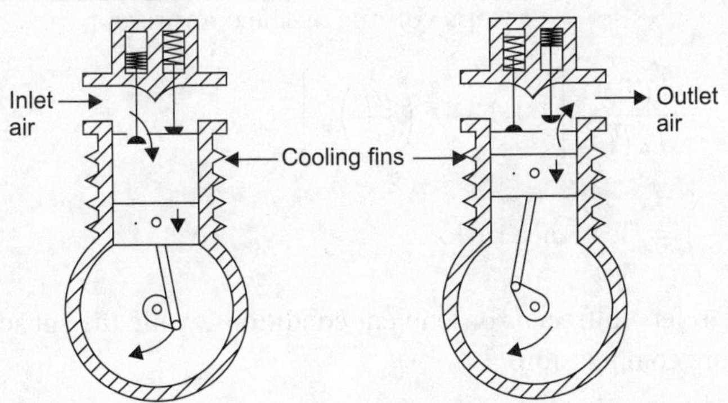

Figure 2.2 *Single cylinder compressors*

Inlet stroke The suction or inlet stroke begins with the piston at the top dead centre (a position providing a minimum clearance volume). During the downward stroke, the piston motion reduces the pressure inside the cylinder below the atmospheric pressure. The inlet valve then opens against the pressures of its spring and allows air to flow into the cylinder. The air is drawn into the cylinder until the piston reaches a maximum volume position (bottom dead centre). The discharge valve remains closed during this stroke.

Outlet stroke During the compression stroke, the piston moves in the opposite direction (bottom dead centre to top dead centre), decreasing the volume of the air. As the piston starts moving upwards, the inlet valve is closed and pressure starts to increase continuously until the pressure inside the cylinder is above the pressure of the delivery side which is connected to the receiver. Then the outlet valve opens and air is delivered to the receiver during the remaining upward motion of the piston.

Isothermal efficiency In a reciprocating compressor, the work is minimum when compression follows the isothermal process. The ratio of isothermal work done to the actual work done is called isothermal efficiency.

$$\eta_{\text{isothermal}} = \frac{\text{Isothermal work}}{\text{Actual work}}$$

Volumetric efficiency The amount of atmospheric air that can be sucked by the compressor at the suction or inlet of the compressor at one atmospheric pressure, 20°C, 100% dry air with the compressor motor running at 100% of its rated value is called **free air delivery (FAD)**. FAD measures the capacity of a compressor in terms of the air flow it can handle and is an important parameter to consider while buying a compressor. It is used to compare different compressors. It is important to note that the induced mass per cycle must equal the delivered mass per cycle as per the law of conservation of mass, even though the induced and delivered volumes are different.

$$\eta_{\text{volumetric}} = \frac{\text{Actual volume of air taken referred to free air conditions}}{\text{Swept volume of the compressor}}$$

$$= \frac{p_1 (T_a)}{p_a T_1}\left[1 + k - k\left(\frac{p_2}{p_1}\right)^{\frac{1}{n}}\right]$$

where $k = \left(\frac{V_c}{V_s}\right) = \text{clearance ratio.}$

The subscript "a" refers to free air or ambient conditions, while the subscript "1" refers to the condition before compression.

Example 2.1 A compressor delivers 4 m^3 of free air per minute at a pressure of 7 bar gauge. Assuming that the compression follows the law $p\,V^{1.3}$ = constant, determine the theoretical work done.

Given data: n, the polytropic index for air = 1.3;

$V_1 = 4$ m^3; $p_1 = 1$ bar (absolute); $p_2 = 7$ bar (gauge)= 8 bar (absolute)

Solution

$$p_1 V_1^{1.3} = p_2 V_2^{1.3}$$

$$V_2 = V_1\left(\frac{p_1}{p_2}\right)^{1.3} = 4\left(\frac{1}{8}\right)^{1.3} = 0.807 \text{ m}^3\text{/min.}$$

$$\text{Work done during compression} = \frac{n}{n-1}(p_2V_2 - p_1V_1) = \frac{1.3}{0.3}(8 \times 0.807 - 1 \times 3) \times 10^5.$$

Solving, we get:
Work done = 14.98×10^5 N.m/minute = 0.2496×10^5 N.m/s = 0.25×10^5 watts = 25 kW.

B. Multi-stage piston compressor

In a single-stage compressor, the entire compression of air takes place in a single stroke of the piston. In a multi-stage compressor, compression takes place in stages. For maximum compressor efficiency, it is desirable to cool air after each stage using an inter-stage cooler. In a two-stage compressor, the initial compression takes place in the low pressure cylinder. Air from this stage (low pressure cylinder) is passed through the inter cooler to reduce temperature. Then the cooled air is compressed in high pressure cylinder.

Working Figure 2.3 shows a two-stage (inline type) reciprocating air compressor. When the prime mover is connected to the crank shaft, the crank rotates and the piston in the first stage reciprocates. It sucks the air through the suction filter and inlet valve. The air, compressed to a certain degree passes from the left cylinder to right cylinder through the intermediate cooler. The compression ratio in the first stage is determined by the degree of cooling required.

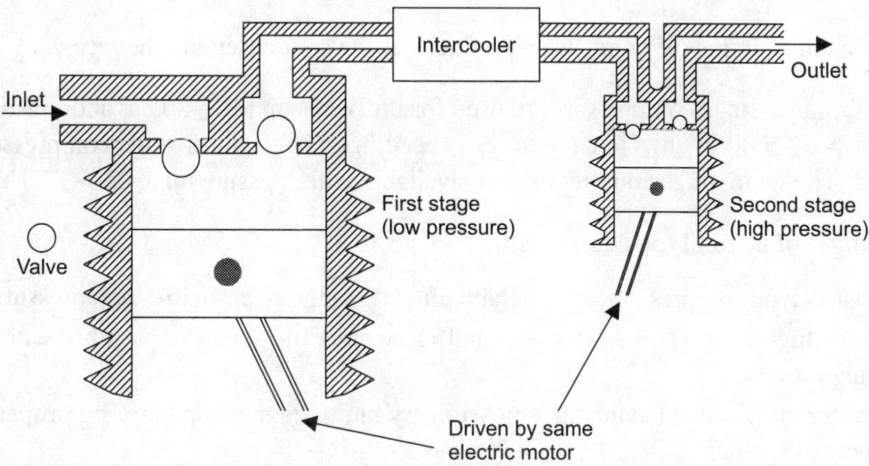

Figure 2.3 *Multi stage piston air compressor with intercooler*

Figure 2.4 shows the various parts of a three-stage (V type) reciprocating air compressor with a receiver (air tank). A pressure switch is connected to an electric motor. When the desired pressure in the air tank is reached, it stops the motor and hence the compressor. The safety valve opens when the pressure in the air tank exceeds the set safe pressure.

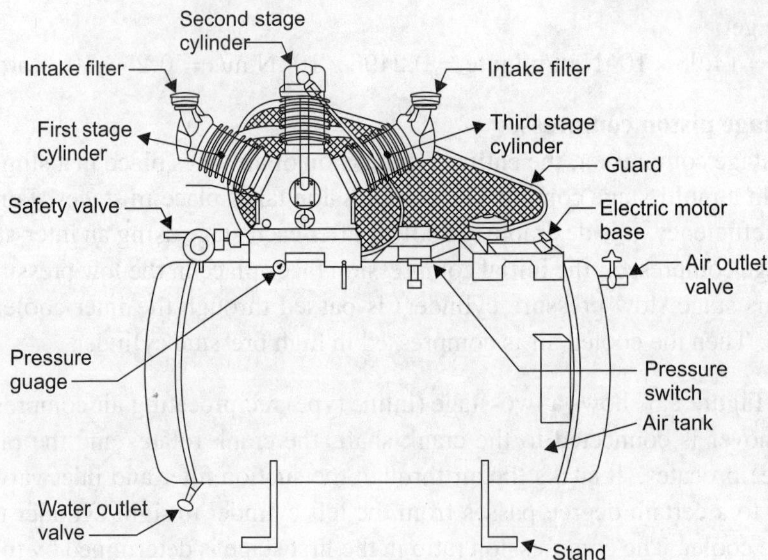

Figure 2.4 *Various parts of a three-stage compressor*

The drain valve drains the condensate produced at the condenser and the receiver.

Range Single-stage compressors are used for pressures up to 4–30 bar and low delivery volumes (< 10000 m^3/h). For pressures exceeding 30 bar, multi-stage compressors are required. The multi-stage compressors are available with pressure up to 250–350 bar.

Advantages of piston type compressors

1. Piston type compressors are available in wide range of capacities and pressures.
2. Very high air pressure (250 bar) and air volume flow rate is possible with multi-staging.
3. Better mechanical balancing is possible by multi-stage compressor by proper cylinder arrangement.
4. High overall efficiency compared to other compressors.

Disadvantages of piston type compressors

1. Reciprocating piston compressors generate inertia forces that shake the machine. Therefore, a rigid frame fixed to solid foundation is often required.
2. Reciprocating piston machines deliver a pulsating flow of air. Properly sized pulsation damping chambers or receiver tanks are required.
3. They are suited for small volumes of air at high pressures.

2.2.1.2 Labyrinth compressors

Labyrinth piston compressors do not use piston rings. Unlike the oil-free compressors of traditional design with dry running piston, no friction occurs in the cylinder. The same applies, as a rule, to the piston rod stuffing box. Instead of piston rings, a labyrinth piston is provided with a large number of grooves that generate a labyrinth sealing action with regard to the cylinder wall.

2.2.1.3 Diaphragm compressors

In piston compressors, there is a likelihood of small amounts of lubricating oil from the piston walls contaminating the lubricating air. This very small contamination may prove very harmful in food, pharmaceutical and chemical industries. For applications in such industries, diaphragm compressors may be used as a power source.

Operation The operation of diaphragm compressor can be explained using Fig. 2.5. There are two strokes, namely, suction stroke and delivery stroke.

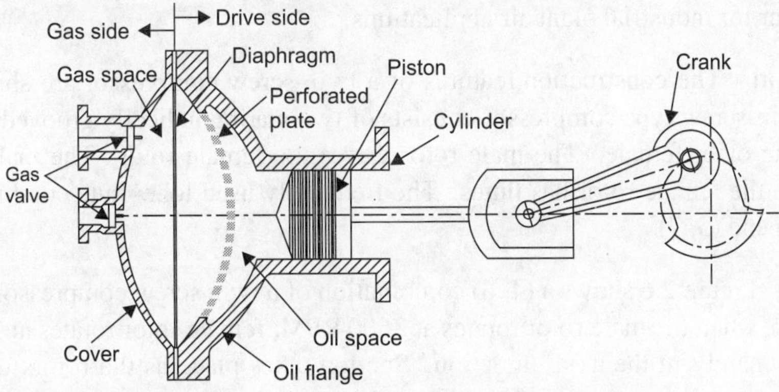

Figure 2.5 *Operation of diaphragm compressors*

Suction stroke On the drive side, when the piston moves from left to right, the movement of piston and hydraulic oil causes the diaphragm also to move towards right till it touches the perforated plate. This movement of diaphragm creates a vacuum on the gas side and the inlet valve opens and suction of air takes place on the gas side of the compressor.

Delivery During the compression stroke, the piston in the drive side presses the hydraulic oil into the diaphragm through the perforated plate, and the diaphragm pushes the air through the delivery valve on the gas side.

Advantages of a diaphragm compressor

1. Oil-free clean air can be obtained from a diaphragm compressor.
2. Hermitically sealed, resulting in exceptionally low leakage rates
3. High efficiency of the order of 80 to 85%
4. High stage pressure ration, allowing for very high final pressures

Disadvantages of a diaphragm compressor

1. Excessive back pressure can rupture the diaphragm.
2. Smooth pulse-free discharge is not possible.
3. Low suction capacity (1000 m^3/hour)
4. Lateral forces caused by the crank drive reduce the lifetime of sealing components.

2.2.2 Rotary compressors

The term 'rotary' describes a class of compressors that operate on positive displacement principle and employs rotary motion to transfer energy. Popular rotary compressors are twin screw compressors, rotary vane compressors, lobe type compressors and liquid ring compressors.

2.2.2.1 Twin screw compressors

The twin screw compressor, introduced commercially in the 1960s, has become increasingly popular for industrial plant air applications.

Construction The construction features of a twin-screw compressor are shown in Fig. 2.6. The twin-screw type compressor consists of two mating helically grooved rotors, one male and the other female. The male rotor drives the female rotor. The male rotor has lobes, while the female rotor has flutes. The frequently used lobe–flute combinations are (4, 6), (5, 6) and (5, 7).

Operation Figure 2.6 shows a (4, 6) combination of a twin screw compressor. For (4, 6) combination, when the male rotor rotates at 3600 RPM, female rotor rotates at 2400 RPM. The flow is mainly in the axial direction. Suction takes place as the rotors unmesh and delivery takes place as they mesh. Meshing and unmeshing takes place depending upon the location of the suction port, delivery port and geometry of the compressor.

Range Single stage screw compressors are generally designed to operate for capacities up to 100 cfm and pressures less than 10 bar. Higher pressures can be attained by multi-stage compression and higher capacity by parallel operation.

Advantages of screw type compressors

1. Simple design and low-to-medium initial cost
2. There is no surface-to-surface contact, so there is no cooling needed.
3. Low noise levels
4. Small loss of efficiency
5. Fewer moving parts rotating at a high constant speed
6. Supply steady delivery of compressed air without pressure fluctuation.
7. Normally they rotate at high speeds and can handle fairly large amount of air

8. Compared to reciprocating compressors, they are more balanced and give less vibration.
9. Easy to install and low maintenance cost

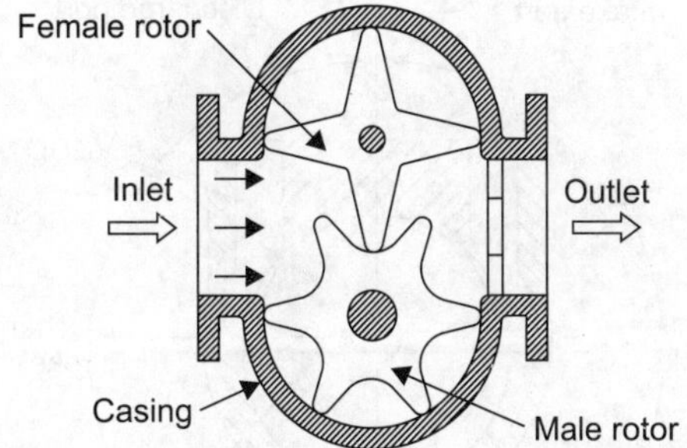

Figure 2.6 *Operation of twin screw compressors*

Disadvantages of screw type compressors

1. Shorter life expectancy compared to other designs
2. They are difficult to operate in dirty environments
3. Single stage designs have lower efficiency
4. Higher rotational speeds.
5. For two stages, oil-free designs have higher initial cost.

2.2.2.2 Rotary vane compressors

Rotary vane compressors are used for low pressure and moderately high volume applications. They are commonly used for instruments and in laboratory works. Rotary vane units generally have lower pressure ratings than piston units because of the difficult sealing problems and greater sensitivity to thermal effects.

Construction The rotor is mounted eccentrically, i.e., not in the center of the casing. It consists of a number of slots with sliding vanes as shown in the Fig. 2.7. During the running of the compressor, the sliding vanes, which are normally made of non-metallic materials, are held against the cylinder due to centrifugal forces.

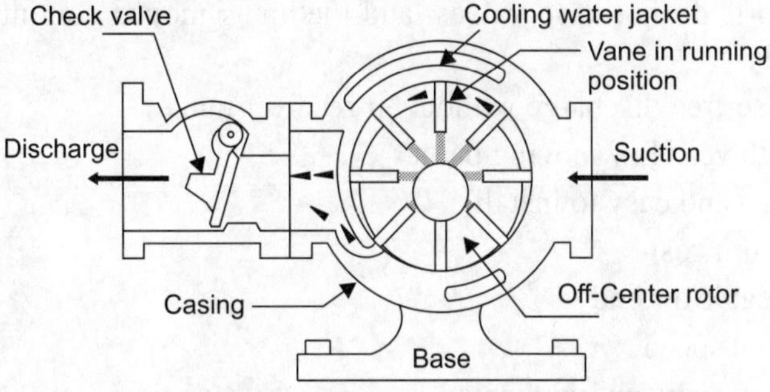

Figure 2.7 *Construction of a rotary vane compressor*

Operation The operation of the vane compressor is explained with the help of Fig. 2.8. For the sake of clarity, only four vanes are shown.

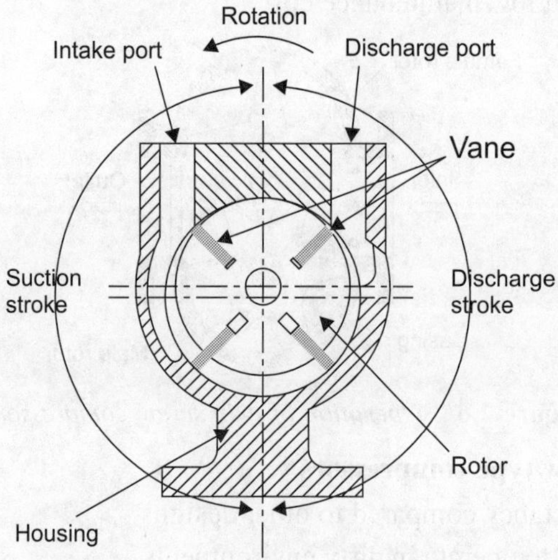

Figure 2.8 *Operation of a vane compressor*

Suction As the rotor rotates, the vanes move outward and are held against the cylinder body by centrifugal and pressure loading forces. This creates a series of air compartment of unequal volume (because of the rotor's eccentricity). The compartments formed between the adjacent vanes gradually become larger during the suction part of the cycle, and air is drawn into the compartment from the inlet port.

Delivery During the discharge portion of the cycle, the compartment volume gradually becomes smaller, compressing the air. When the rotating compartment reaches the discharge port, the compressed air escapes to the delivery system.

Advantages of rotary vane compressors

1. The suction and exhaust flows are relatively free of pulsation because the inlet and discharge ports do not have valves, and the air is moved continuously rather than intermittently.
2. Smooth, pulse-free discharge without air receiver tanks
3. Compact with very few moving parts
4. Simple design and easy to install
5. Low to medium cost
6. Long serviceable air end
7. Low rotational speed
8. Can operate in dirty environments

Disadvantages of rotary vane compressors

1. Single stage designs have lower efficiency
2. Oil injected designs have oil carry-over
3. They cannot be used for high pressure
4. Oil-free designs are unavailable

2.2.2.3 *Lobe type compressors*

The lobe type air compressor is very simple with no complicated moving parts. Single or twin lobes are attached to the drive shaft driven by the prime mover. The lobes are displaced by 90 degrees. Thus, if one of the lobes is in horizontal position, the other at that particular instant will be in vertical position. The air gets trapped in between these lobes and as they rotate, gets compressed and delivered to the delivery line. Lobe type compressors have higher efficiency at moderate compression ratios and are most efficient for compression ratios of 1.1 to 2. They find use in applications that require relatively constant flow rate at varying discharge pressures. They are generally available for capacities of 150 LPM to 150000 LPM for pressures up to 1 bar in single stage construction. They are extensively used in applications such as pneumatic conveying, aeration applications, cement plants and water treatment plants.

2.2.2.4 *Liquid ring compressors*

Liquid ring compressors belong to the positive displacement compressor family. They use liquid that is centrifuged in a specifically shaped casing. The liquid can be water, sulphuric acid, crude oil, gasoline or any other liquid that does not react with the gas that has to be compressed. This specific feature reduces the wearing off of the metallic parts theoretically to zero, and therefore maintain their efficiency for a long time.

2.2.3 Dynamic displacement compressors (turbo compressors)

These compressors raise the pressure of air or gas by imparting velocity or energy and converting it to pressure energy. First, rapidly rotating impellers (similar to fans) accelerate the air. Then the fast flowing air passes through a diffuser section that converts its velocity head into pressure by directing it into a volute. These compressors can be axial or centrifugal type.

2.3 Comparison of Different Compressors

Flow rate, efficiency and pressure rise within the compressor are the three most-used parameters employed in defining the performance of a compressor and in its selection. A typical comparison is given in the Table 2.1.

Table 2.1 *Comparison of different compressors*

Item	Reciprocating	Rotary vane	Rotary screw	Centrifugal
Efficiency at full load	High	Medium-high	High	High
Efficiency at part load	High due to staging	Poor: below 60% of full load	Poor: below 60% of full load	Poor: below 60% of full load
Efficiency at no load (power as % of full load)	High (10%–25%)	Medium (30%–40%)	High-Poor (255%–60%)	High-medium (20%–30%)
Noise level	Noisy	Quiet	Quiet	Quiet
Size	Large	Compact	Compact	Compact
Oil carry over	Moderate	Low-medium	Low	Low
Vibration	High	Less	Less	Less
Maintenance	Many wearing parts	Few wearing parts	Very few wearing parts	Sensitive to dust in the air
Capacity	Low-high	Low-medium	Low-high	Medium-high
Pressure	Medium-very high	Low-medium	Medium-high	Medium-high

2.4 Control of Compressors

There is a growing variety of control systems available for compressed air installations. These most often concern electric driver controls and compressor controls. The compressor control systems are listed below.

1. Continuous capacity regulation
2. Load/unload regulation
3. Start/stop and load/unload controls
4. Pressure relief valve control
5. Bypass regulation
6. Throttling the intake
7. Pressure relief with throttled intake

2.5 Selection and Specification of Compressors

The important criteria to be considered while selecting compressor for a pneumatic system are listed below:

1. Pressure The operating pressure of the systems has to be determined first. If long pipelines are required, then pressure losses along the length of the pipe must be taken into account. For example, if a particular operation requires 5.5 bar, then single stage compressor that operates between 6.5 to 8.5 will work, assuming that the compressor produces enough air volume.

2. Capacity The volume of air should be large enough to cater to the operation of all devices at a time. If the equipments are used intermittently, then depending upon the nature of device/operations, the capacity must be estimated. The capacity of a compressor is usually expressed in CFM (cubic feet per minute).

3. Compressor configuration and cylinder geometry

- The type of the compressor can be horizontal or vertical.
- Arrangement of cylinders can be vertical, V type, W type or L type.

4. Duty cycle The term duty cycle refers to the percentage of time an air compressor can run before needing time to cool down. The duty cycle of an air compressor is often expressed as a percentage of a certain amount of time, usually 10 minute periods. That is, the duty cycle measures the proportion of time an air compressor can run against the amount of time needed for it to cool.

For example, if an air compressor has a 30% duty cycle, it can be run for 3 minutes, after which it must rest for 7 minutes. However, there are air compressors with 100% duty cycle that can be run continuously without a cool down time.

5. Power source It is always recommended to operate air compressors on 230 V–460 V, three-phase power when it is available. If the electrical power outlet is rated at 115 V and 20 A, then this limits the size of air compressor motor to about 2 HP.

2.6 Importance of Secondary Treatment

Compressed air is an essential power source that is widely used throughout production industry. This safe, powerful and reliable utility is an important part of the production process.

The three desirable qualities of compressed air are listed below.

- It must be cleaned.
- It must be cooled.
- It must be dried.

2.7 Aftercooler

The compressed air discharged from an air compressor is hot and contains large quantities of water in vapour form. Aftercoolers are heat exchangers for cooling the discharge from an air compressor. They use either air or water and are an effective means of removing moisture from the compressed air.

2.7.1 Functions of compressed air aftercoolers

- Cools air discharged from air compressors via the heat exchanger
- Reduces risk of fire (hot compressed air pipes can be a source of ignition)
- Reduces compressed air moisture level
- Increases system capacity
- Protects downstream equipment from excessive heat

2.7.2 Types of aftercoolers

There are two basic types of air aftercoolers:

1. Air-cooled
2. Water-cooled

Compressor manufacturers may include aftercoolers within the compressor package. In general, these compressors are referred to as integral aftercoolers. A stand-alone or free standing aftercooler is a separate unit installed downstream of the compressor.

2.7.2.1 Air-cooled aftercooler

Air-cooled aftercoolers provide economical cooling by using ambient air to cool the hot compressed air from air compressors. They cool the hot compressed air leaving the compressor at a temperature of approximately 100°C to 150°C to the desired inlet temperature of an air dryer, which is approximately 35°C to 50°C. As the compressed air cools, about 75% of the water vapour present condenses into liquid water, which should be immediately removed from the system with a separator. Air-cooled aftercoolers can be sized to cool the hot compressed air to within −15°C to −5°C of the ambient air temperature. They are available in capacities ranging from 500 LPM to100000 LPM.

2.7.2.2 Water-cooled pipeline aftercooler

The most common style for compressed air service is a shell and tube heat exchanger. The pipeline aftercooler consists of a shell with a bundle of tubes fitted inside. Typically, the compressed air flows through the tubes in one direction as water flows on the shell side in the opposite direction. The heat from the compressed air is transferred to the water. As the compressed air cools, water vapour forms. The moisture is removed by the moisture separator and drain valve.

The tube bundles can be fixed or removable. Fixed tube bundles cost less but are more difficult to maintain than bundles that can be removed for cleaning or servicing.

The disadvantages of a water-cooled aftercooler include high water usage and difficulty in heat recovery. The advantages of a water-cooled aftercooler include better heat transfer and no requirement of electric supply.

2.8 Drying of Compressed Air

Drying of compressed air lowers the dew point of the compressed air by removing the moisture from it. For simple applications, to remove excess humidity, we need a simple aftercooler, an air receiver, and a filter with condensate traps. However, to get high quality compressed air, additional means of dehydration must be provided using dryers.

2.8.1 Types of air dryers

Generally four basic types of air dryers are used in industries.

1. Absorption type dryer
2. Adsorption type dryer
3. Refrigeration dryer
4. Membrane dryer

2.8.1.1 Absorption type dryer

Absorption drying is a purely chemical process. The moisture in the compressed air forms a compound with a drying agent like phosphorous pentaoxide in the tank. This causes the drying agent to break down. It is then discharged in the form of a fluid at the base of the tank. A schematic diagram of absorption dryer is shown in Fig. 2.9.

Oil vapour and oil particles are also separated in the absorption dryer. Large quantities of oil affects the efficiency of the dryer. Therefore, it is advisable to include a fine filter in front of the dryer.

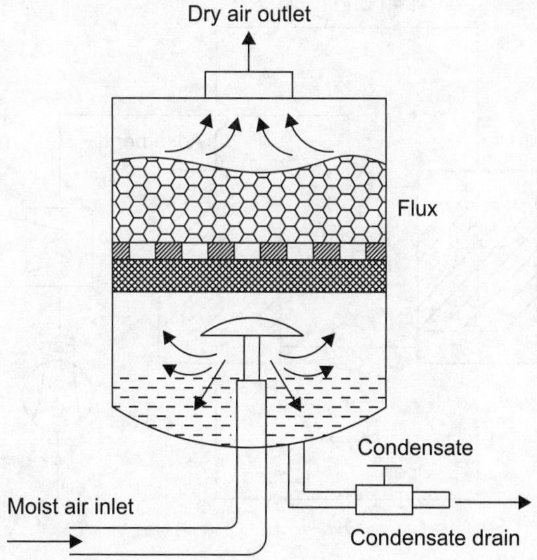

Figure 2.9 *Absorption dryer*

Advantages of absorption dryers

1. Simple to install
2. Low mechanical wear because there are no moving parts
3. No external energy requirement

Disadvantages of absorption dryers

1. Maintenance cost is high
2. Low efficiency
3. Consumable cost is high

2.8.1.2 *Adsorption type dryer*

Adsorption is a physical process of moisture removal from the porous surface of certain granular materials. Gaseous molecules are attracted to certain solid surfaces by van der Waals forces and this causes adsorption. The degree of attraction or adsorption depends on the properties of the gaseous molecules and the desiccant. Most commonly used desiccants are activated alumina, molecular sieves and silica gel.

Figure 2.10 shows the various parts of an adsorption dryer. Wet incoming compressed air, after passing through a pre-filter, is directed to the adsorption chamber containing the desiccant. Water vapour in the compressed air is absorbed by the desiccant. Thereafter, dry air is allowed to pass to the application through the after-filter.

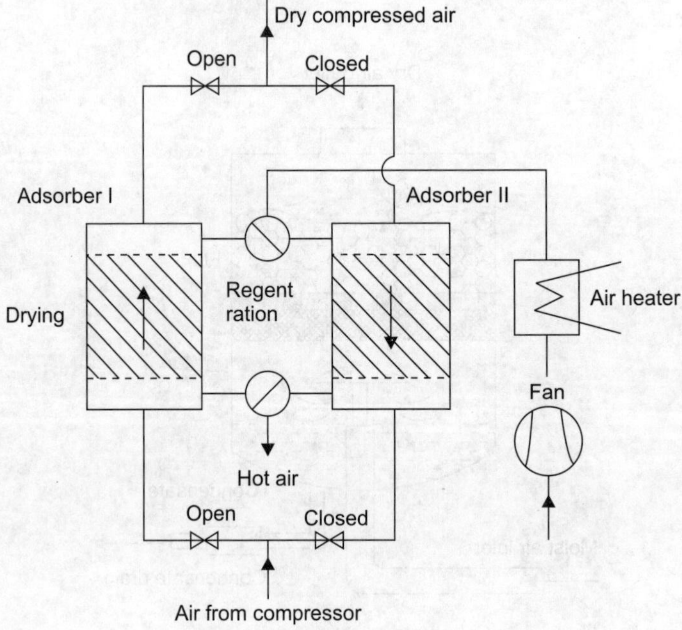

Figure 2.10 *Adsorption type dryer*

Adsorption dryers usually have two desiccant-filled chambers with interconnecting piping and switching valves. The valves permit removal of the collected moisture from one chamber, while the other chamber is used to purify the compressed air. The twin tower design facilitates simultaneous compressed air drying and saturated desiccant regeneration for non-stop production. The contaminated desiccant bed can be regenerated by either elevating its temperature or by decreasing its pressure and purging.

The capacity of the desiccant bed is limited owing to abrasion and contamination of the adsorption medium by oil and other substances. Under normal conditions, it is required to replace the drying agent once in every 2 to 3 years.

2.8.1.3 *Refrigerated dryer*

The layout of a typical refrigerated air dryer is shown in Fig. 2.11. It is composed of a heat exchanger (Stage 1) and a refrigerating unit (Stage 2) to reduce the temperature of the compressed air. The incoming warm and humid air is first passed through the air-to-air heat exchanger, and then through the refrigerating unit to reduce the temperature of the compressed air to as low as +2°C. This drying method is based on the principle that if the compressed air is cooled to a temperature below the dew point, condensation takes place and water is precipitated. Almost all the water and oil particles get condensed and collected in the water traps provided at appropriate points. The cooled compressed air is then filtered to remove from it the suspended solid particles and most of the oil mist. The pressure dew point of 2°C is possible with this type, which is sufficient enough for the smooth operation of the most of the industrial and process applications.

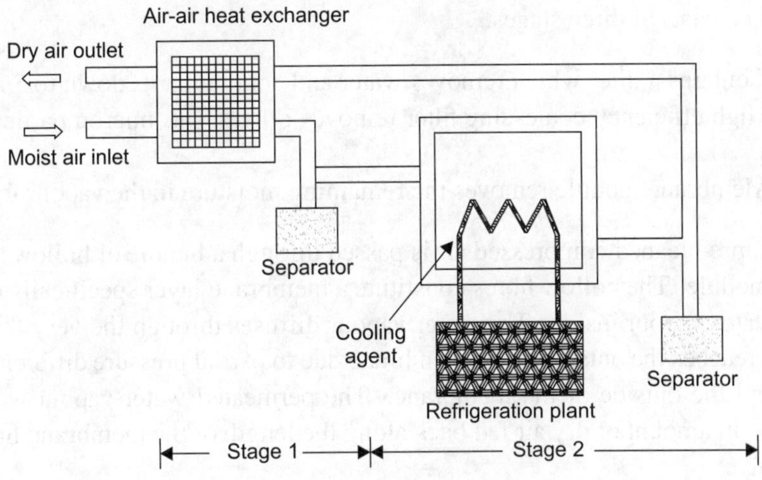

Figure 2.11 *Refrigerated dryer*

In most cases, the type of dryer required is determined by the pressure dew point needed in the systems. Table 2.2 gives the summary of advantages and disadvantages of these dryers.

Table 2.2 *Comparison of three types of dryers*

Type	Advantages	Disadvantages
Absorption	Pressure dew point +16°C	Inlet temperature must not exceed 30°C.
	Low capital cost	Drying agents are consumables and therefore must be regularly replenished. Highly corrosive chemicals are used. They are not environment friendly.
Refrigeration	+3°C pressure dew point	Output dew point will vary with the approach temperature at the inlet and cleanliness of the heat exchanger
	Input temperature can be as high as 16°C	
Adsorption	Achievable pressure dew point of −40°C	High capital cost
		High operating cost
		Use of micro-filters to prevent the residue from chemicals adds to the cost.

2.8.1.4 Membrane dryer

Membrane dryers are yet another type of dryers that remove moisture from compressed air. The process consists of three stages.

Stage 1: Contains a filter which removes water and contaminants down to 5 microns
Stage 2: High efficiency coalescing filter removes oil and sub-micron particles down to 0.01 micron
Stage 3: Membrane module removes the remaining moisture in the vapour form

In this dryer, pre-cleaned compressed air is passed through a bundle of hollow fibres in the membrane module. The hollow fibres constitute a membrane layer specifically designed to attract the water vapour inside. The water vapour diffuses through the very thin selective layer until it reaches the outside of the membrane due to partial pressure difference between the inside and the outside of the membrane. The permeated water vapour is then swept away by a small amount of dry air fed back along the length of the membrane fibre through a purging valve.

Advantages of membrane dryers

1. Typically maintains a pressure dew point of 0°C
2. Simple and compact
3. Runs almost noiselessly

4. No need for regeneration because membranes never gets saturated
5. Do not require electric supply
6. Low operating cost

2.8.2 Theory of drying

Air taken from the atmosphere contains moisture, which in normal circumstances is precipitated as condensate within the system through which it passes, namely the compressor aftercooler, air receiver, collecting points in the piping and the filter of the service unit ahead of the air consumers.

Compressed air will always contain as much water as it is capable of absorbing at the lowest temperature assumed by the air on its passage from the compressor to the consumption point. Typical values of water vapour saturation capacity of air at selected atmospheric temperatures are given in Table 2.3.

Table 2.3 *Water vapour saturation capacity of air at selected atmospheric temperatures*

Temperature °C	−10	0	5	10	15	20	30	50	70	90
Water vapour, g/m^3	2.1	4.9	7	9.5	13	17	30	83	198	424

For example, at 20°C, one cubic metre of compressed air will still contain 17 grams of moisture. Existing in the form of water vapour, this moisture is not separated in the filter of the service unit; rather, it is entrained by the air into all the downstream control and operating components. Due to the centrifugal action taking place in the filter and increased flow velocity resulting therefrom, the air stream is cooled by a certain amount relative to the ambient temperature, which causes a small portion of the condensate to be separated in the filter.

Since the control and operating elements of the pneumatic systems will normally be at ambient temperature, no further condensate will be precipitated at these points. Therefore, the remaining moisture passes out with the external air released to the atmosphere. As observed above, the low residual moisture content of the air will hardly constitute a potential hazard to the standard pneumatic control and operating components.

In the case of specialised applications of compressed air, however, as in spray painting, intricate low-pressure control systems, chemical and pharmaceutical industries, food industry, pneumatic instruments and pneumatic conveying, the situation is different. Everywhere that the compressed air comes in direct contact with the process medium in such instances, simple conditioning of the air by the means prescribed usually will not be sufficient. It is then necessary to provide additional means of dehydration and filtering of the compressed air using air dryers.

2.8.2.1 Humidity and relative humidity

Water vapour is constantly evaporating from lakes, rivers and seas and is absorbed by the atmosphere and carried across vast distances by winds, finally being deposited in the form of rain, mist, etc. Atmospheric air, therefore, is nature's way of transmitting large quantity of water vapour all over the earth.

Relative humidity(RH) It is the ratio of amount of water present in a given quantity of air to the maximum possible amount which it can contain under the same conditions of pressure and temperature. This ratio is usually expressed as percentage. The amount of moisture condensing out of compressed air is a function of the relative humidity of the intake air and temperature.

$$\text{Relative humidity of air} = \frac{\text{Absolute humidity}}{\text{Humidity at saturation}} \times 100$$

In other words, relative humidity is the amount of water vapour present in a given volume of air, whereas the humidity at saturation is the total amount of water vapour that the same volume of air can absorb at the given temperature. Table 2.4 gives the mass of water in kg per 100 m^3 of free saturated air. Table 2.5 gives the amount of condensate in g/m^3 of air at various temperatures and relative humidity.

Table 2.4 *Mass of water in kg per 100 m^3 of free saturated air*

Temperature °C	Gauge pressure (bar)				
	0	2	4	6	8
0	0.48	0.17	0.10	0.07	0.05
20	1.73	0.576	0.346	0.247	0.192
40	5.10	1.7	1.02	0.728	0.567
60	12.95	4.32	2.59	1.85	1.44
80	29.04	9.68	5.81	4.15	3.23
100	56.00	19.33	11.76	8.40	6.53
120	0.00	36.73	22.04	15.74	12.24
140	0.00	0	37.74	29.96	20.97

Table 2.5 *Amount of condensate in g/m^3 of air at various temperatures and RH*

Temp °C	Percentage relative humidity (RH)									
	10	20	30	40	50	60	70	80	90	100
−12	0.179	0.354	0.533	0.709	0.888	1.066	1.2503	1.421	1.600	1.780
0	0.483	0.965	1.451	1.933	2.426	2.906	3.387	3.867	4.348	4.830
10	0.934	1.865	2.790	3.730	4.670	5.606	6.520	7.460	8.400	9.337
15	1.313	2.617	3.816	5.063	6.386	7.795	9.029	10.158	11.631	17.957

contd...

Table 2.5 *...contd*

21	1.826	3.661	5.469	7.300	9.131	11.946	12.793	14.600	16.431	18.698
27	2.494	4.980	7.516	10.069	12.521	15.012	17.582	20.024	22.757	25.634
32	3.307	6.614	10.061	13.548	16.835	20.276	23.486	27.0044	30.451	33.721
35	3.936	7.872	11.808	15.744	19.681	23.617	27.554	31.489	35.426	39.248
37.8	4.531	9.039	13.571	18.102	22.611	27.141	31.673	36.489	39.900	45.248
43.3	6.018	12.037	18.056	24.075	30.094	36.112	42.131	48.150	54.160	60.371
49	7.895	15.790	23.686	31.580	39.476	47.370	55.267	63.166	78.910	78.928
55	10.161	20.322	30.583	40.650	50.805	61.200	71.127	81.3	91.504	101.656

The condensate, as the precipitated water in the compressed air line is called, causes damage if it is not removed properly. It causes corrosion in pipes and tubes, corrosion in control and working elements, and corrosion in machine parts. If the condensate gets into pneumatic equipment, proper functioning may be prevented. Solid particles such as dust, rust, and scale can also have an adverse effect on the functioning of the various items of pneumatic equipment.

Oil residues from the compressor can produce, together with the compressed air, a mixture of oil mist and air (gas mixture) that can cause explosions at higher temperatures (above 353 K).

Example 2.2 A compressor delivers 400 m^3 of free air per hour at a pressure of 6 bar gauge and a temperature of 40°C. Atmospheric air at compressor intake has a relative humidity of 80% and a temperature of 20°C. Determine the amount of water that can be extracted from the compressor plant per hour.

Solution Refer to Table 2.4. At 20°C and zero bar gauge pressure, 100 m^3 of free saturated air contains 1.73 kg of water. From the definition of RH,

$$\text{Relative humidity} = \frac{\text{Amount of water actually present in air}}{\text{Amount of water present in saturated air}} \times 100$$

$$80 = \frac{\text{Amount of water actually present in air}}{1.73} \times 100$$

Amount of water actually present in air $= 1.384$ kg.

Since 400 m^3 is delivered, water content of air entering the compressor $= 1.384 \times 4 = 5.536$ kg.

From Table 2.4, corresponding to 40°C and 6 bar compressor output pressure, the amount of water per 100 m^3 of free saturated air is given by 0.728 kg.

Since 400 m^3 is delivered, water content of air leaving the compressor is

$$0.728 \times 4 = 2.912 \text{ kg}.$$

Therefore, the amount of water extracted from the compressor plant per hour is

$$5.536 - 2.912 = 2.62 \text{ kg}.$$

2.8.2.2 Dew point temperature

The temperature at which air is fully saturated with moisture (that is, 100% humidity) is the dew point. Simply put, dew point is the temperature where condensation begins. Cooling below dew point will cause condensation of water vapour. Lower the dew point, the less moisture the air is able to absorb or hold. For example, 1 m^3 of air has 17 grams of water at 20°C, and at −10°C it is 2.1 grams. The capacity of holding water in air is a function of volume and temperature; it does not depend on pressure. But it is still necessary to consider the working pressure of systems when comparing different facilities for dehydrating air. This brings in the term 'pressure dew point'.

Pressure dew point The temperature representing the dew point at the respective operating pressures of the dryer is known as the pressure dew point. Air is a compressible gas, and the dew point temperature changes with pressure. More precisely, it is not the pressure but the volume that matters. When a gas is pressurised, the volume is reduced and the air has less capacity to hold moisture.

Atmospheric dew point As discussed above, compression and expansion of air affects its dew point. Generally speaking, compression increases dew point, and expansion (i.e., de-compression) lowers dew point. For example, consider compressed air leaving a dryer at 15 bar with a pressure dew point of −40°C @ 15 bar. If the pressure is eventually reduced to 7.5 bar, the pressure dew point will fall to −45°C @ 7.5 bar. For this reason, term 'pressure dew point (PDP)' is commonly used. This term usually refers to the dew point of the compressed air at full line pressure. Conversely, the term 'atmospheric dew point' refers to what the dew point would be if fully depressurised to atmospheric conditions. Figure 2.12 shows the relationship between atmospheric dew point and pressure dew point for various pressures ranging from 0 MPa to 1.5 MPa.

Example 2.3 Find the atmospheric dew point where the pressure is 0.7 MPa and pressure dew point is 5°C.

Solution Refer to Fig. 2.12. Corresponding to 0.7 MPa and pressure dew point of 5°C, we get atmospheric dew point as −20°C.

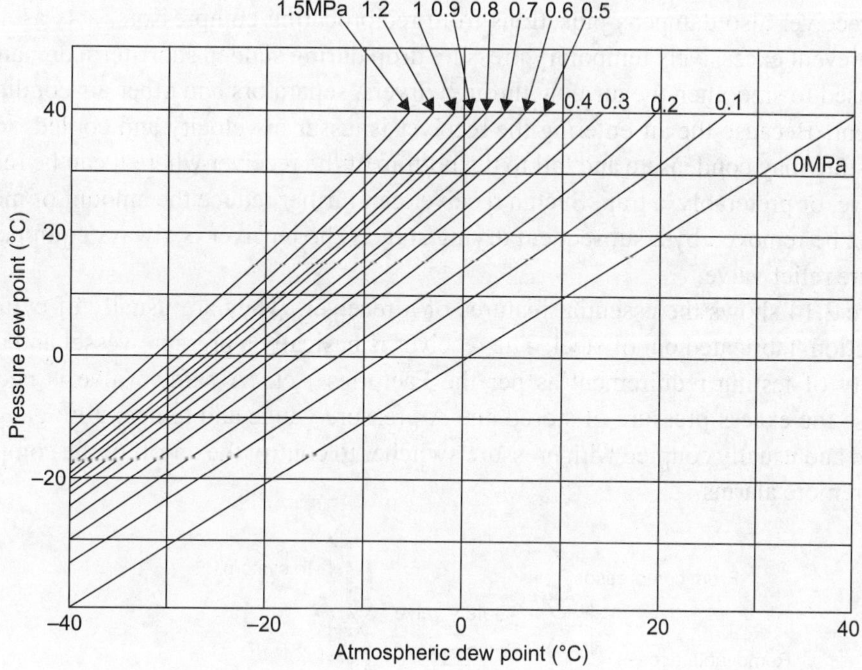

Figure 2.12 *Atmospheric dew point versus pressure dew point*

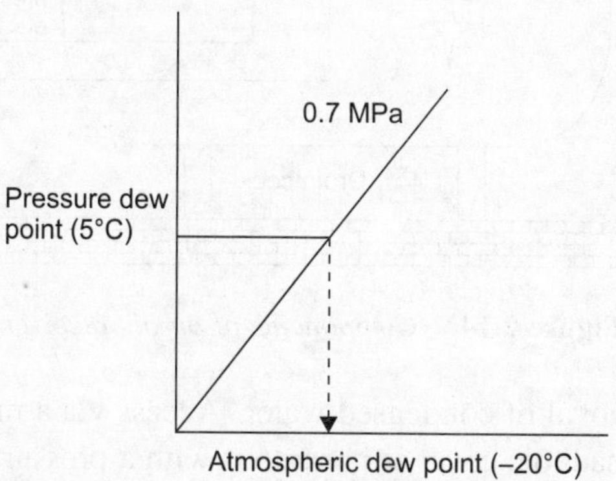

Figure 2.13 *Pressure dew point versus atmospheric dew point*

2.9 Storage of Compressed Air

2.9.1 Air receivers

Receivers perform several functions in compressed air systems. Firstly, they provide larger system capacity, which increases the cycle time of the compressor control systems. This makes the elimination of unstable and over correcting control cycles less difficult.

The receiver also dampens pulsations from reciprocating compressors, acts as a reservoir to prevent excessively temporary pressure drop during sudden short-term demand, and can be used to smoothen the air flow through dryers, separators and other air-conditioning equipment. Because the air entering the receiver is lesser in velocity and cooled, some of the moisture may condensate and fall to the bottom of the receiver where it can be removed by a valve, or preferably, a trap. Such a receiver can further reduce the amount of moisture that must be removed by a subsequent drying stage. The receiver is always equipped with a pressure relief valve.

Figure 2.14 shows the essential features of a receiver. They are usually of cylindrical construction, fabricated out of steel. The receiver is basically a pressure vessel and attracts the safety of testing requirement as per the Factories Act. A safety valve is necessary to release the excess pressure of stored air. A pressure gauge and temperature gauges are provided and usually coupled with pressure switches to control the on-off of the compressor and for remote alarms.

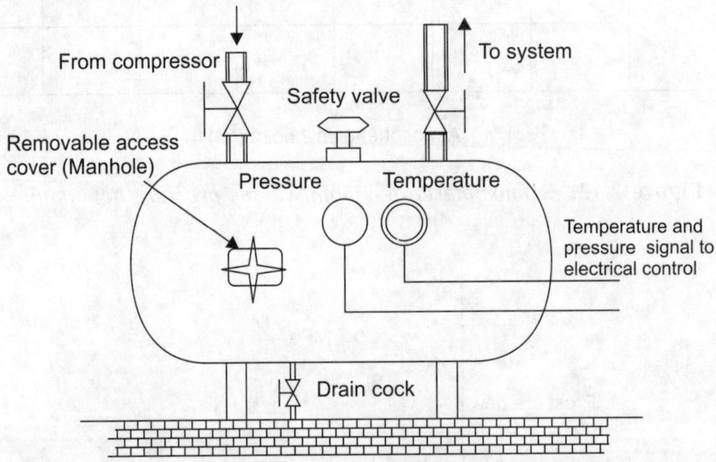

Figure 2.14 *Components of an air receiver*

A drain cock allows removal of condensed water. Access via a manhole allows cleaning. Obviously, removal of manhole cover is hazardous with a pressurised receiver, and safety routines must be defined and followed to prevent accidents.

The installation of an accumulator within a pneumatic system will depend upon the specific needs of the air consumers and will only be necessary when large volumes of air are consumed in short periods, that is, when there is an intermittent peak load situation.

Figure 2.15 shows the diagram of a compressed air installation with air receiver and accumulator.

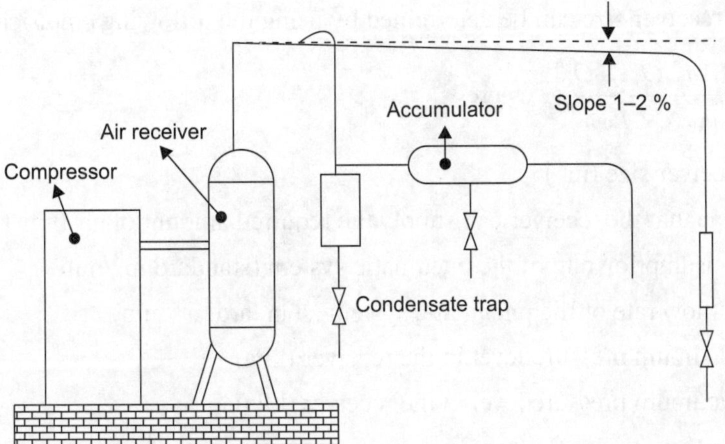

Figure 2.15 *Compressed air distribution system with a branch line containing an accumulator*

2.9.2 Sizing of air receivers

The receiver size, in cubic feet, should normally be at least 12 seconds of the compressor capacity in small units to not less than 8 seconds of the compressor capacity for larger units. Empirical rule suggests that the receiver volume should one-tenth the FAD per minute to one-sixth the FAD per minute for small compressors. Sometimes, small compressors are mounted on receivers and the receivers serve as a basic mounting frame for the assembly of the compressors and their accessories. Under these and some other circumstances, the only pressure relief valve is frequently located on the receiver. When this mounting procedure is employed, care must be taken to ensure that no other valve comes between the compressor and the receiver, and that support members or frames are not welded to the receiver tank seam welds.

2.9.2.1 Empirical relations to size the compressor

The sizing of air receivers requires taking into account parameters such as system pressure and flow rate requirements, compressor output capability, and the type of duty of operations. Basically, a receiver is an air reservoir. The receiver must be capable of handing transient demand.

Case 1: No air is supplied to the receiver during the time interval in which air is being drawn off.

In such cases, we can find the capacity of the receiver as

$$V_r = \frac{101\,t\;[Q_r]}{[p_{\max} - p_{\min}]}.$$

Case 2: Air is supplied to the receiver during the time interval in which air is being drawn off.

Then, the air receiver size can be determined by using the following empirical equation:

$$V_r = \frac{101\,t\ [Q_r - Q_c]}{[p_{\max} - p_{\min}]},\ \text{where}$$

V_r = receiver size (m^3)

t = time that the receiver can supply the required amount of air (min)

Q_r = consumption rate of the pneumatic system (standard m^3/min)

Q_c = outflow rate of the pneumatic system (standard m^3/min)

$p_{\max}$ = maximum pressure level in the receiver (kPa)

$p_{\min}$ = maximum pressure level in the receiver (kPa).

Case 3: Compressor with offloading/loading regulation
The following relation applies when dimensioning the receiver's volume. Note that this relation only applies for compressors with off loading/loading regulation.

$$V_r = \frac{0.25\ T_r\ Q_{\text{FAD}}}{f_{\max}\,[p_{\max} - p_{\min}]\,T_c},\ \text{where}$$

V_r = receiver size (litre)

T_r = temperature of the compressed air in receiver (K)

Q_{FAD} = compressor FAD (litre/second)

T_c = maximum ambient temperature, compressor inlet temperature (K)

$p_{\max}$ = maximum pressure level in the receiver (bar)

$p_{\min}$ = maximum pressure level in the receiver (bar)

$f_{\max}$ = maximum loading frequency, usually 1 cycle every 30 seconds

2.9.2.2 *Analysis of power requirement for driving the compressor*

Another important design consideration is to determine the power required to drive an air compressor to meet system pressure and flow rate requirements. The following empirical relationship can be used to find the theoretical power requirement.

$$\text{Theoretical power} = \frac{p_1 Q}{17.1}\left(\left(\frac{p_2}{p_1}\right)^{0.286} - 1\right)$$

p_1 = Inlet pressure of the compressor (kPa)

p_2 = Outlet pressure of the compressor (kPa)

Q = flow rate (m^3/min)

To find the actual power, the theoretical power is divided by the overall compressor efficiency.

$$\text{Actual power} = \frac{\text{Theoretical power}}{\text{Overall efficiency}}$$

$$\text{Actual power} = \frac{\dfrac{p_1 Q}{17.1}\left(\left(\dfrac{p_2}{p_1}\right)^{0.286} - 1\right)}{\text{Overall efficiency}}$$

Example 2.4 Air is used at a rate of 1 m^3/min from a receiver at 40°C and 1000 kPa (gauge). If the atmosphere pressure is 101 kPa (abs) and the atmospheric temperature is 20°C, how many m^3/min of free air (standard 1 m^3/min) must the compressor provide?

Solution

$$p_2 = 1000 \text{ kPa (gauge)} = 1101 \text{ KPa (absolute)};\ p_1 = 101 \text{ KPa (absolute)}$$

$$T_2 = 40°\text{C} = 40 + 273 = 313 \text{ K},\ \ T_1 = 20°\text{C} = 20 + 273 = 293 \text{ K}$$

$$V_2 = 1\frac{\text{m}^3}{\text{min}}$$

Using general gas law $\dfrac{p_1 V_1}{T_1} = \dfrac{p_2 V_2}{T_2}$,

$$\frac{101 \times V_1}{293} = \frac{1101 \times 1}{313}.$$

On solving, we get $V_1 = 10.20$ standard $\dfrac{\text{m}^3}{\text{min}}$.

Example 2.5 Calculate (a) the required size of the receiver that must supply air to pneumatic system consuming 0.850 m^3/min for 10 minutes between 828 kPa and 690 kPa before the compressor resumes operation. (b) What size is required if the compressor is running and delivering at 0.170m^3/min?

Solution The air receiver size can be determined using the following equation.

$$V_r = \frac{101\, t\ [Q_r - Q_c]}{[p_{\max} - p_{\min}]}$$

(a) The following are given:

V_r = receiver size (m^3)

t = time that receiver can supply required amount of air, (min) = 10 min

Q_r = consumption rate of pneumatic system (standard $\frac{m^3}{min}$) $= 0.850\frac{m^3}{min}$

Q_c = outflow rate of compressor (standard $\frac{m^3}{min}$) $= 0\frac{m^3}{min}$

p_{max} = maximum pressure level in receiver (kPa) = 828 kPa

p_{min} = maximum pressure level in receiver (kPa) = 690 kPa

$$V_r = \frac{101 \times 10\ [0.850 - 0]}{[828 - 690]}$$

On solving, we get $V_r = 6.22\ m^3$.

(b) The required size of the compressor when the compressor is running and delivering air at 0.170 m^3/min:

$$V_r = \frac{101 \times 10\ [0.850 - 0.170]}{[828 - 690]}$$

Solving we get $V_r = 4.977\ m^3 \cong 5\ m^3$.

Example 2.6 Determine the output pressure of a compressor operating with the following data:

1. The actual power required to drive the compressor is 20 kW.
2. The overall efficiency of compressor is 75%.
3. The compressor delivers 4 standard m^3/min.
4. The compressor inlet pressure is 100 kPa abs.

Solution We know that

Theoretical power = Actual power × Overall efficiency $= 20 \times 0.75 = 15$ kW.

$$\text{Theoretical power} = \frac{p_1 Q}{17.1}\left(\left(\frac{p_2}{p_1}\right)^{0.286} - 1\right).$$

$$15 = \frac{100 \times 4}{17.1}\left(\left(\frac{p_2}{100}\right)^{0.286} - 1\right).$$

On solving, we get $p_2 = 566$ kPa (absolute).

Example 2.7 Determine the actual power required to drive a compressor that delivers air at 5.668 m^3/min at 828 kPa(gauge). The overall efficiency of the compressor is 72%.

Solution We know that $p_2 = 828 + 101 = 929$ kPa.

$$\text{Actual power} = \frac{\frac{p_1 Q}{17.1}\left(\left(\frac{p_2}{p_1}\right)^{0.286} - 1\right)}{\text{Overall efficiency}}$$

$$= \frac{101\times 5.668}{17.1}\left(\left(\frac{828+101}{101}\right)^{0.286}-1\right)\div 0.72$$

Solving, we get actual power $= 41.21$ kW.

Summary

1. The operational reliability and service life of a pneumatic system depends, to a large extent, on the preparation of the compressed air. The stages involved in the preparation of compressed air are the air intake treatment, the compressor stage, the conditioning stage and the distribution stage.
2. Compressors are used to compress air of 6–8 bar. There are two types of compressors: positive displacement and dynamic displacement compressors. Piston type compressors are the most commonly used compressors.
3. A displacement compressor is described by enclosing a volume of gas or air and then increasing the pressure by reducing the area of the enclosed volume.
4. Diaphragm compressors are mechanically or hydraulically actuated compressors. Mechanical diaphragm compressors are used with a small flow and low pressure or as vacuum pumps. Hydraulic diaphragm compressors are used for high pressure.
5. The screw compressor is not equipped with valves and has no mechanical forces that cause unbalance. It can work at high shaft speed and combines a large flow rate with smaller external dimensions.
6. Dynamic compressors are available in axial (also called centrifugal) and radial (turbo) designs. A dynamic compressor works with a constant pressure.
7. Various control and regulations systems are available to control compressors. These include continuous capacity regulation and unload/load control.
8. The important criteria to be considered while selecting compressors for a pneumatic system are pressure, duty cycle, the number of cylinders, cylinder arrangement, capacity and power source.
9. Atmospheric air contains water vapour, more at high temperatures and less at lower temperatures. When air is compressed, the water concentration increases.
10. The term 'pressure dew point' (PDP) is used to describe the water content in the compressed air. It is the temperature at which water vapour transforms into water at the current working pressure. Low PDP values indicate small amounts of water vapour in the compressed air. It is important to remember that atmospheric dew point can not be compared with PDP when comparing different dryers.
11. An aftercooler is a heat exchanger which cools hot compressed air to precipitate water that would otherwise condense in the pipe system. It is water cooled or air cooled, generally equipped with a water separator with automatic drainage, and it should be placed next to the compressor.

12. In refrigerant drying, the large amount of water condensed during cooling can be separated. Over-compression is perhaps the easiest method to dry compressed air. Air is first compressed to a higher pressure than the intended working pressure, which means the concentration of water vapour increases. Thereafter the air is cooled, whereby water is separated. Finally the air is allowed to expand to the working pressure, whereby a lower PDP is attained. However, this method is only suitable for very small air flow rates due to the high energy consumption.
13. Absorption drying is a chemical process where water vapour is bound to the absorption material. The absorption material can either be a solid or a liquid. Sodium chloride and sulphuric acid are frequently used, which means the possibility of corrosion must be taken into consideration. This method is unusual and has a high consumption of absorption material. The dew point is only lowered to a limited degree.
14. Adsorption is a physical process of moisture removal on the porous surface of certain granular materials. Gaseous molecules are attracted to certain solid surfaces by van der Waals forces, and this causes the adsorption.
15. The function of an air receiver is to store the high-pressure compressed air from the compressor. The air receiver reduces pressure fluctuations that arise from changes in load and from compressor switching. The air coming from the compressor is hot. A large surface area of the receiver dissipates this heat into the surrounding atmosphere. Any moisture left in the air from the compressor will condense at the bottom of the receiver. Therefore, it is advisable to take air to the system from the top of the receiver.

EXERCISES

Fill in the blanks

1. ——— types of reciprocating compressors are most commonly used.
2. Compared to positive displacement type compressors, dynamic compressors are much —— in size and produce much ——— vibration.
3. If the air compressor has a duty of ——, and the air compressor running for 10 minutes, then it should run for a combined maximum of 6 minutes ON and 4 minutes OFF.
4. It is not possible in practice to achieve isothermal process as the compressor must run at very ———. In practice, compressors run at high speeds results in ——— process.
5. In ——— control regulation, pressure relieved air is cooled and returned to the compressor intake.
6. In a single acting reciprocating compressor, the suction, compression and delivery of air takes place in ———— of the piston.
7. Intercooling in multi-stage compressor is done to ——— the work of compression.

8. The ratio of work done per cycle to the stroke volume of the compressor is known as ———
9. The volume of air delivered by the compressor is called —————.
10. In a centrifugal compressor, an increase in speed at a given pressure ratio causes increase in flow and ——— in efficiency.
11. About —— of the moisture can be removed using an aftercooler.
12. Adsorption is a ——— process of moisture removal on the porous surface of certain granular materials.
13. Absorption drying is a purely —— process. The moisture in the compressed air forms a compound with drying agent like phosphous pentoxide in the tank.
14. ——— is to lower the dew point of the compressed air by removing the moisture from it.
15. Temperature representing the dew point at the respective operating pressure of the dryer is known as ——— dew point.

State whether true or false

1. A dynamic compressor works at a constant pressure.
2. Positive displacement compressors are selected for larger volume of gas and higher pressure ratios. Dynamic compressors are selected for lower volume of gas fluid and smaller pressure ratios.
3. Compressors will have higher efficiency if the compression follows an isothermal process.
4. Running at the same operating speed, the two-lobed blowers can deliver larger air flow and volume than the conventional three-lobe type.
5. Continuous capacity regulation is the most common type of control and regulation used in compressors.
6. The total heat rejected in a reciprocating air compressor is equal to the sum of the heat rejected during polytropic compression per kg of air and heat rejected in the intercooler per kg of air.
7. In a four-stage compressor, if the pressure at the first and third stage is 1 bar and 16 bar, then the delivery pressure at the fourth stage will be 64 bar.
8. The volumetric efficiency for reciprocating air compressors is about 60 to 70%.
9. The minimum work required for a two-stage reciprocating air compressor is double the work required for each stage.
10. The clearance volume in the compressor is kept minimum because it affects the volumetric efficiency.
11. In adsorption process, gaseous molecules are attracted to certain solid surfaces by van der Waals forces.

12. High pressure dew point values indicate small amounts of water vapour in the compressed air.
13. It is important to remember that atmospheric dew point can be compared with PDP when comparing different dryers.
14. Installation of accumulator within a pneumatic system will depend on specific air consumers and will only be necessary for intermittent peak load.
15. Membrane dryers do not require electric supply.

Review questions

1. Explain the different stages of preparation of compressed air.
2. Explain the effect of type of compression in reciprocating air compressors.
3. Define isothermal efficiency and derive an expression for it.
4. Explain the effect of clearance volume in a single-cylinder single-stage compressor.
5. Define volumetric efficiency of an air compressor and derive an expression for it.
6. Explain the effect of increasing delivery pressure on the volume of air delivered.
7. Derive an expression for the intermediate pressure which gives minimum power in a two stage compressor with perfect inter cooling
8. What are the advantages of multi-stage compressors?
9. Differentiate between positive and dynamic displacement compressors.
10. Name three types of positive displacement compressors.
11. Explain the working principle of a screw compressor. What are its advantages and disadvantages?
12. What is a diaphragm compressor? List its types and advantages.
13. Explain on/off regulation of compressors with a neat sketch.
14. Explain pressure relief and bypass regulation of compressor control with neat sketches.
15. List all the factors to be considered in the selection of a compressor for a given application.
16. State the adverse effects of moisture content in compressed air.
17. What is the function of an aftercooler?
18. What are the different methods of drying compressed air?
19. Explain the working of an absorption dryer with a neat sketch.
20. Explain the working of an adsorption dryer with a neat sketch.
21. Explain the working of a regenerative dryer with a neat sketch.
22. Why are two drying chambers used in adsorption dryer?
23. Explain the functional and constructional features of a refrigerated dryer.
24. List the advantages and disadvantages of absorption dryers.
25. List the advantages and disadvantages of membrane dryers.

26. List the advantages and disadvantages of refrigeration dryers.
27. List five functions of air receivers.
28. Why must the receiver tank be drained periodically?
29. Name the three fittings used on all air receivers.
30. What are the important maintenance activities to be carried out for air receivers?
31. What are the thumb rules we follow in selecting air receivers?

Problems

1. The bore and stroke of a single acting air compressor are both 10 cm, clearance volume 80 cc, index of compression and expansion 1.23, suction pressure 1 bar, delivery pressure 7 bar and speed 4000 RPM. If the temperature at the beginning of the compression is 22℃. Find the amount of free air dealt with by the compressor per minute at a pressure of 1.05 bar and temperature of 16℃.
2. A single acting stage air compressor is required to deal with 17 m^3/min of air measured at 1 bar and 15°C. The pressure and temperature at the end of suction is 0.98 bar and 30°C. The delivery pressure is 6.3 bar. The RPM of the compressor is 500. Assuming a clearance volume of 5% of the stroke volume, laws of compression and expansion as $pV^{1.32} =$ constant, calculate the necessary (i) stroke volume, (ii) temperature of the air delivered and (iii) power of the compressor.
3. A single-stage, single-cylinder air compressor has a cylinder diameter 15 cm, and a stroke of 22.5 cm. Air is drawn into the cylinder at a pressure of 1.05 bar and a temperature of 15 . It is then compressed adiabatically to 6.3 bar. Find the theoretical power required to drive the compressor if its speed is 100 RPM. Find also the weight of the air compressed per minute.
4. A single-stage reciprocating compressor takes 1 m^3 air per minute at 1.013 bar and 15°C and delivers it at 7 bar. Assuming that the law of compression is $pV^{1.3} =$ constant and clearance is negligible, calculate the indicated power.
5. A two-stage air compressor compresses 1 kg/min of air from 1 bar to 42.18 bar. Initial temperature is 15°C. At intermediate pressures, the intercooling is perfect. The compression takes place according to law $pV^{1.35} =$ constant. Neglecting the effect of clearance, determine the minimum power required to run the compressor. Also find the mass of cooling water required in the intercooler if the temperature rise of water is limited to 5°C.
6. A compressor delivers 500 m^3 of free air per hour at a pressure of 8 bar gauge and a temperature of 60°C. Atmospheric air at compressor intake has a relative humidity of 70% and a temperature of 20°C. Determine the amount of water that can be extracted from the compressor plant per hour.
7. Find the atmospheric dew point where the pressure is 0.7 MPa and pressure due point is 10°C.

8. Find the amount of condensate per shift in a plant if air is compressed to 8 bar (gauge) receiver pressure. Assume the initial condition of air at sea level at 21°C and 70% relative humidity (RH). Assume 35°C to be the temperature of compressed air receiver.
9. Find the amount of condensate in one hour if a 22 kW compressor operates under the following conditions: air at 60% relative humidity and 30°C ambient temperature pressurised to 6 kg/cm^2 (6 bar) and then cooled to 25°C. The compressor output is 3 Nm^3/min at 6kg/cm^2 (6 bar).
10. Air is used at a rate of 2 m^3/min from a receiver at 40°C and 1000 kPa (gauge). If the atmosphere pressure is 101 kPa (abs) and the atmospheric temperature is 20°C, how many m^3/min of free air (standard m^3/min) must the compressor provide?

Answers

Fill in the blanks

1. Piston	2. Smaller/less	3. 60/40	4. Slowly/polytropic
5. Bypass	6. Two strokes	7. Minimise	8. Mean effective pressure
9. Compressor capacity	10. Decrease	11. 70–85%	12. Physical
13. Chemical	14. Aftercooler	15. Pressure	

True or false

1. True	2. False	3. True	4. False	5. False	6. True	7. True	8. False
9. True	10. True	11. True	12. False	13. False	14. True	15. True	

Problems

1. $V_2 = 520.5$ cm^3
2. (i) V_s = swept volume = 20.11 m^3/ min = 0.0402 m^3/cycle; (ii) $T_2 = 475.71$ K (iii) power = 70 kW
3. (i) Actual power = 65.7 kW; (ii) Isothermal power = 51.92 kW (iii) Weight of air per minute = $0.00505 \times 100 = 0.505$ kg/min
4. Power = 4.22 kW 5. (i) Power = 3.3 kW; (ii) $m_{\text{water}} = 0.1432$ kg/s
6. 1.145 kg 7. −13°C 8. 81 kg 9. 17.1 grams /Nm^3
10. $V_1 = 20.40$ standard $\dfrac{\text{m}^3}{\text{min}}$

3 CONDITIONING AND DISTRIBUTION OF COMPRESSED AIR

Upon completion of this chapter, the reader should be able to

- Explain the importance of fluid conditioners
- List the major functions of filters and their selection criteria
- List applications of the seven qualities of air as per ISO
- Describe the functions of various types of air regulators
- List the functions of an FRL (Filter–Regulator–Lubricator) unit
- Explain the various compressed air network systems
- List factors for the selection of components of a pneumatic network
- Determine size of pipe diameter and pressure drop for a given pneumatic network

3.1 Fluid Conditioners

The atmospheric air taken into the compressor contains contaminants like dirt, smoke and water vapour. Contaminated air can lead to excessive wear on, and failure of pneumatic components. Therefore, fluid conditioners are used to provide clean, dry and contamination-free compressed air.

The operating instructions issued for pneumatic components almost always contain a note recommending the installation of an air filter, a pressure regulator and a lubricator upstream of the components. This is to ensure that only suitably conditioned air reaches the consumer.

The purpose of fluid conditioners is to make the compressed air an acceptable fluid medium for the pneumatic system as well as the operating personnel. The following fluid conditioners are used in pneumatic systems:

1. Air filters 2. Air regulators 3. Air lubricator

3.1.1 Air filters

The purpose of an air filter is to clean the compressed air of all impurities and any condensate it contains.

Function of air filters

- To remove foreign matter and enable clean, dry air to flow unrestricted to the regulator and onto the lubricator
- To condense and remove water from the air
- To arrest fine particles and all solid contaminants in the air

Filters are available in a wide range; from a fine mesh wire cloth (to strain heavy foreign particles) to filters made of synthetic material (to remove very small particles).

Inline filter elements can remove contaminants in the 5–50 microns range.

Sources of contamination

Contaminants in a compressed air system are usually attributed to the following:

- The quality of air drawn into the compressor
- The type and operation of the air compressor
- The compressed air storage devices and distribution systems

Types of contamination in a compressed air system

- Atmospheric dirt
- Water vapour, condensed water and water aerosols
- Rust and pipe scale
- Micro-organisms
- Liquid oil and oil aerosols
- Oil vapour

Factor affecting selection of filters

The following factors should be taken into account while selecting filters.

- Size of particles to be filtered from the system
- Capacity of the filter
- Accessibility and maintainability
- Life of the filter
- Ability to drain condensate

Construction The construction of a typical cartridge type filter, along with graphical symbols, is shown in Fig. 3.1. It consists of a filter cartridge, a deflector, a bowl and a water drain valve. The filter bowl is usually made of plastic and is transparent. For pressures higher than 10 bar, the bowl may be made of brass.

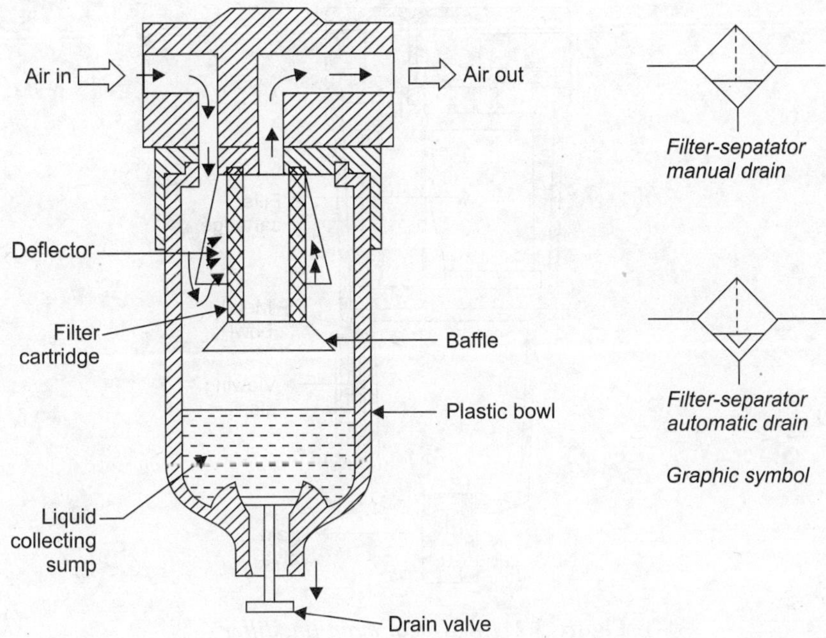

Figure 3.1 *Construction of an air filter*

Operation Air enters the inlet port of the air filter through angled louvers, which cause the air to spin as it enters the bowl. The centrifugal action of the rotating air causes the larger particles of dirt and water to be thrown against the inner wall of the filter bowl. The contaminants then flow down to the bottom of the filter bowl.

A baffle prevents turbulent air from splashing water on to the filter element. The air which has been pre-cleaned in this manner then passes through the filter element, where the fine dirt particles are filtered out. The size of the dirt particles which can be filtered out depends on the mesh size of the filter element (usually 5–50 microns). The compressed air then exits through the outlet port.

The pressure difference between the inlet and outlet indicates the degree to which the filter element is clogged. Commercially available filters have additional features like automatic drain facility, coalescing type filter element, service life indicator, etc.

Main line filter This is a high capacity filter that removes dirt and oil and water vapour and is situated after the air receiver unit. This filter has a very small pressure drop and can remove oil from the compressor that would otherwise emulsify with water condensation. There is no deflector, as it cannot function without a minimum pressure drop. A drain is provided to remove the accumulated water.

The filter is generally a quick-change cartridge type. Figure 3.2 shows the schematic diagram of a main line filter.

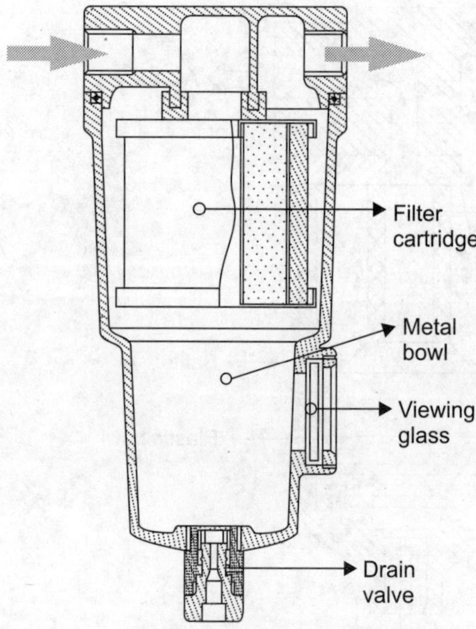

Figure 3.2 *A typical main line filter*

3.1.2 Air regulator

Function The function of an air pressure regulator is to maintain the working pressure at a constant level despite line pressure and air flow fluctuations. When the pressure is too low, it results in poor efficiency, and when the pressure is too high it leads to wasted energy and degradation of equipment performance.

In a pneumatic system, pressure fluctuations occur due to variations in the supply or load pressures, and it is therefore essential to regulate the pressure to a constant value that matches the load requirements.

Regulation criteria

In pneumatic systems, the pressure is generally regulated in two places:

- At the receiver tank
- In the load circuits

Pressure regulation at the receiver tank is required as a safety measure for the system. In load circuits, pressure regulation is provided for the downstream components such as valves and actuators.

Types of pressure regulators

There are two types of pressure regulators:

- Diaphragm type regulator
- Piston type regulator

The diaphragm type regulator is commonly used in industrial pneumatic systems. There are two types of diaphragm type regulators:

- Non-relieving or non-venting
- Relieving or venting

The relieving or venting type is commonly used and is explained below.

3.1.2.1 Relieving or venting type pressure regulator

A relieving type pressure regulator is shown in Fig. 3.3. The outlet pressure is sensed by a diaphragm pre-loaded with an adjustable pressure setting spring. The compressed air flowing through a controlled cross section of the valve seat acts on the other side of the diaphragm. The diaphragm has a large surface area exposed to the secondary (outlet) pressure and is quite sensitive to its fluctuations. The movement of the diaphragm regulates the pressure.

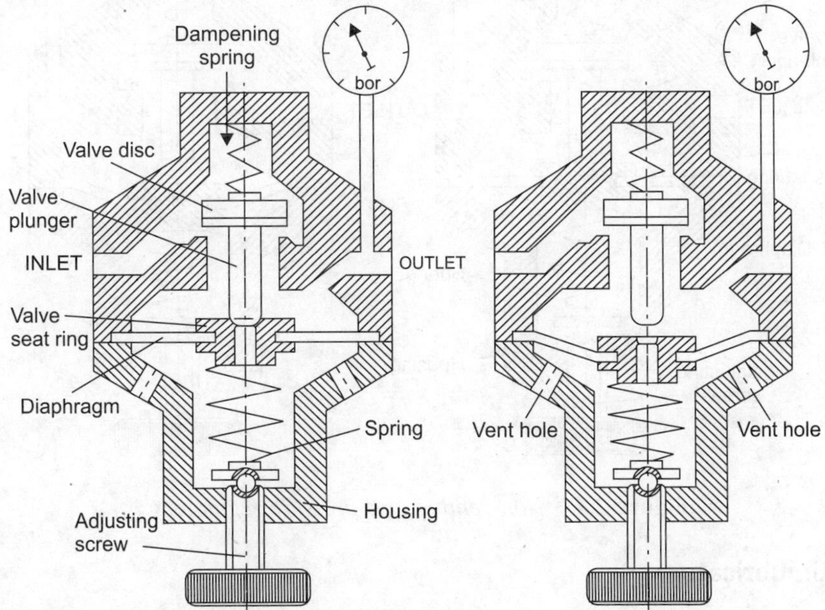

Figure 3.3 *Relieving/venting type pressure regulator*

Low outlet pressure When a large quantity of compressed air is consumed on the secondary or load side, the load pressure reduces, resulting in a lower force on the diaphragm. The opposing higher spring force pushes the diaphragm, thereby increasing the valve disc movement to permit more air to flow to the secondary side, which increases the pressure again.

High outlet pressure When a smaller amount of compressed air is consumed on the secondary or load side, the load pressure increases resulting in a higher force on the diaphragm. The opposing higher spring force pulls down the diaphragm, reducing the valve disc movement and permitting air to flow out of the vent, and thus decreases the pressure again.

3.1.2.2 Non-relieving or non-venting pressure regulator

In this configuration, the diaphragm does not have a vent for trapped air. In the event of high back pressure, compressed air cannot escape to the atmosphere because there is no vent hole, and therefore it is not commonly used in pneumatic control. Figure 3.4 shows the non-relieving, non-venting type pressure regulator.

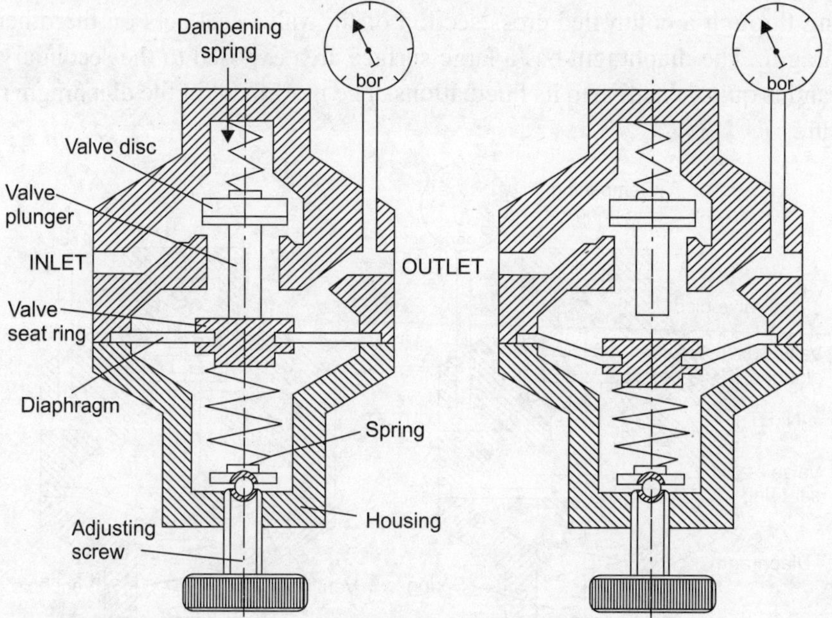

Figure 3.4 *Non-venting type pressure regulators*

3.1.3 Air lubricator

Function The function of an air lubricator is to add a controlled amount of oil to the air to ensure proper lubrication of the internal moving parts of a pneumatic system. Lubricants are used to:

- Reduce wear on moving parts
- Reduce frictional losses
- Protect equipment from corrosion

The lubricator adds oil in the form of a fine mist. Excessive lubrication is undesirable and may result in:

- Component malfunction
- Components seizing and sticking after a prolonged downtime
- Environmental pollution

Operation The operation of the air lubricator is similar to the principle of a carburettor. The schematic diagram is shown in Fig. 3.5. As air enters the lubricator, its velocity is increased by a venture ring. The pressure at the venture ring is lower than the atmospheric pressure, while the oil is at atmospheric pressure. Due to this pressure difference between the upper and lower chambers, oil is drawn up the riser tube. Oil droplets mix with the incoming air and form a fine mist. The needle valve is used to adjust the pressure differential across the oil jet and control the oil flow rate. The air/oil mixture is swirled as it leaves the central cylinder so that large oil particles are forced back to the bowl and only the mist goes to the outlet.

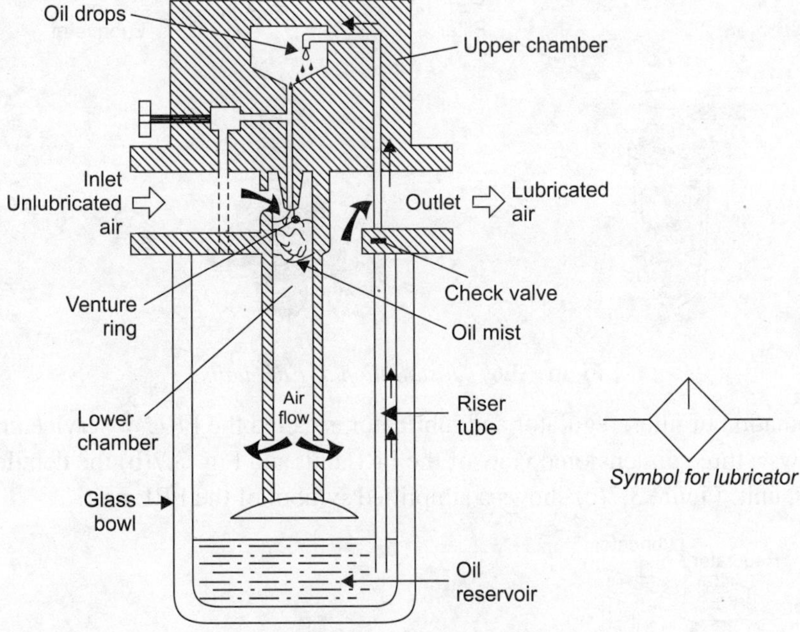

Figure 3.5 *Air lubricator*

The lubricator operates only when there is sufficient flow of air. If too little air is drawn in, the flow velocity at the nozzle is not sufficient to produce the vacuum necessary to draw the oil out of the vessel.

3.1.4 Filter–regulator–lubricator (FRL) unit

In most pneumatic systems, the compressed air is first filtered and then regulated to a specific pressure before passing through the lubricator for oil treatment. So a filter, regulator and lubricator are placed in that order along the inlet line of each air circuit. They may be installed as separate units, but more often, they are in the form of a combined unit. Figure 3.6 shows the schematic arrangement of a filter, regulator and lubricator unit.

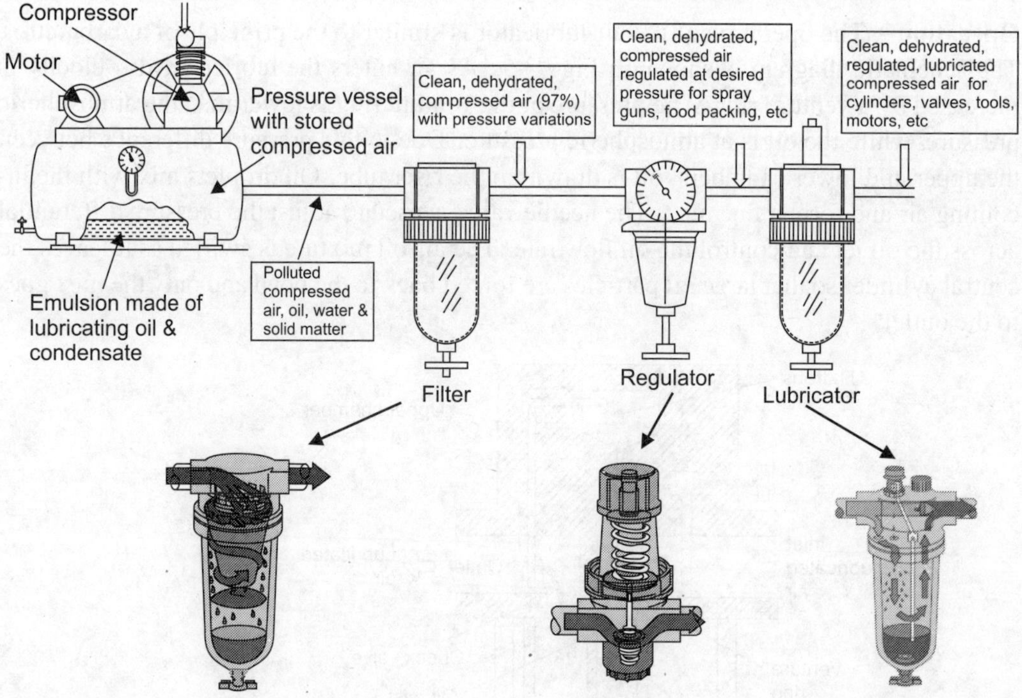

Figure 3.6 *Installation of FRL units*

The combination of filter, regulator and lubricator is called the FRL or service unit. Figure 3.7(a) shows a three dimensional view of the FRL unit and Fig. 3.7(b) the detailed symbol of the FRL unit. Figure 3.7(c) shows a simplified symbol of the FRL unit.

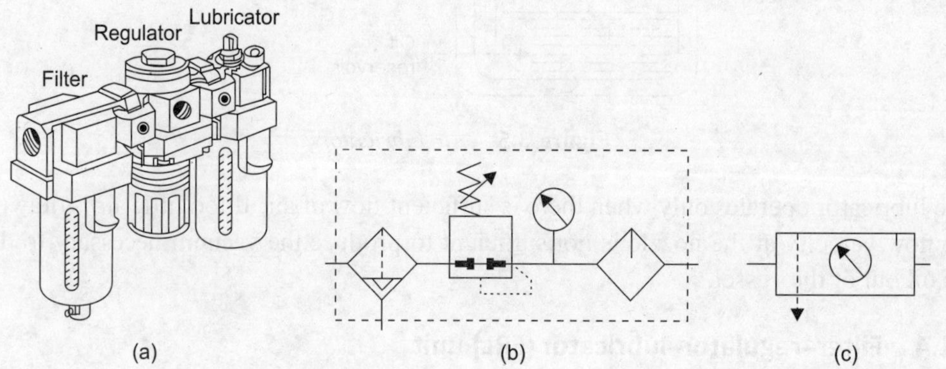

Figure 3.7 *An FRL unit: (a) 3-D view, (b) detailed symbol, (c) simplified symbol of FRL*

3.2 Air Distribution System

The main objective of an air distribution system is to provide a leak-proof distribution channel for compressed air while keeping the pressure drop within permissible limits. The air distribution system consists of conductors and fittings which interconnect various components of a pneumatic system. Figure 3.8 shows a typical air distribution system. It consists of a compressor, water-based aftercooler, air receiver, dryer and a ring main system. The air main takes the shape of a ring. Air from the main header is drawn by sub-headers, which may have their own accumulators.

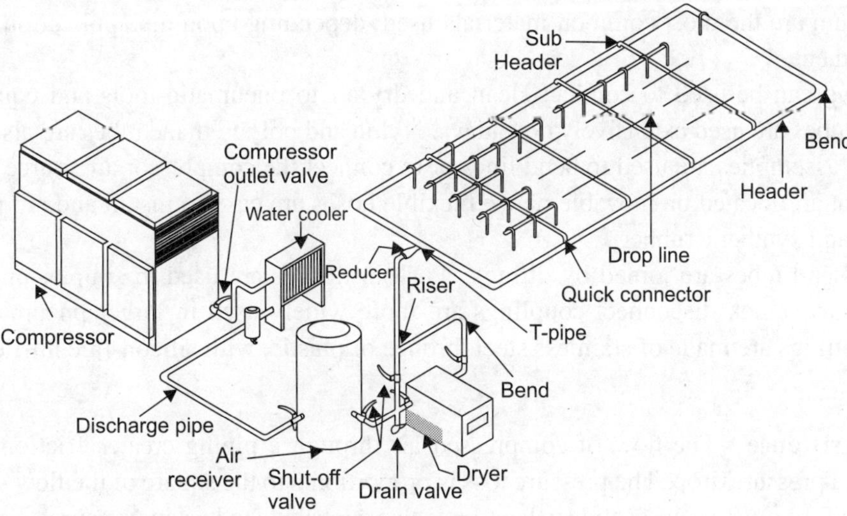

Figure 3.8 *Typical air distribution system (ring type)*

The air distribution should take into account the following parameters:

1. Choice of fluid conductor
2. Flow resistance
3. Correct sizing of pipes
4. Correct sizing of fittings
5. Pipe layout

Choice of fluid conductor Fluid conductors used in pneumatic applications carry air from the reservoir to the actuator and then to the atmosphere. Fluid conductors are generally divided into three classes:

- Rigid pipe
- Semi rigid tubing
- Flexible hose

The choice of the conductor depends on:

1. Permissible pressure drop
2. Leak proof operation
3. Resistance to corrosion
4. Cost of pipes per metre
5. Flexibility for future expansion
6. Installation and maintenance costs

The main air distribution system is made of rigid pipelines. Copper, steel, iron and aluminium are the most common materials used, depending upon the application and the environment.

Tubing can be used to conduct clean and dry air to pneumatic tools and equipment. Plastic tubes are used extensively nowadays. Nylon and polyurethane tubes are also used.

Hose assemblies are used to bend lines or to connect the compressor air source to actuators that are located on movable parts. Flexible hoses are easy to install and are made of natural and synthetic rubber.

Pipes and tubes are joined by fittings. Push-in fittings are used in simpler pneumatic circuits and quick disconnect couplings are more widely used in larger pneumatic systems. Fittings are made of stainless steel, bronze or plastic, with silicon-free nitrile rubber O-rings.

Flow resistance The flow of compressed air through a piping creates friction, which leads to a pressure drop. The pressure loss is proportional to the square of the flow velocity. Elbows, T, two-way valves, slide valves, etc., also contribute to loss in pressure. A pressure drop cannot be avoided, but can be considerably reduced by routing pipes properly and assembling the fittings correctly.

Correct sizing of pipes The pipe size should be selected properly to keep the pressure drop to a minimum over the entire air distribution system. Selection is governed by the following factors:

- Delivery volume
- Length of piping
- Number of restrictions in the system
- Operating pressure in the system
- Permissible pressure drop
- Number of fittings
- Flow velocity of air

Lines carrying compressed air may have internal diameters ranging from a few millimetres up to 250 mm. Pipes used to convey air may be constructed of rubber, plastic or metal.

The air mains are the piping system into which compressed air is routed from the receiver. It is a permanently installed system of inter-connected pipes carrying compressed air to various consumers. The important criteria to be considered are flow velocities and pressure drops in the piping and tight joints throughout the main.

In order to start and maintain the flow of air in a pipeline, a certain pressure drop to overcome the frictional resistance of the pipe walls and fittings is required. The size of the pressure drop depends on the flow rate, pipe diameter, pipe length, pipe geometry and the surface roughness. Pressure loss is considered an operating cost and a proper piping design balances this cost against the cost of piping. Empirical formulas or tabular or graphical data can be used to determine this balance for a straight pipe of any given length. Pressure drop through the valves and fittings are measured in terms of equivalent pipe length, i.e., the corresponding length of straight pipe of similar diameter that results in an identical pressure drop. Determining the relationship between the flow rate and the pressure drop leads to complex formulas which are tedious to manipulate.

The salient factor in determining the internal diameter of the piping is the overall air consumption plus a margin to cover additional pneumatic devices that will almost inevitably be added later. Empirical data on the maximum flow velocities and allowable pressure drops in piping under optimum economic conditions are available.

Air consumption rate should be determined in advance by the planner. Flow velocity and pressure drops are closely related, and the surface of the inside walls of the piping and the number of installed valves and fittings also affect the pressure drop.

Recommended velocity of air in the mains is 6 to 10 m/s, and every effort should be made to keep the velocity less than 10 m/sec. Pipe elbows, valves, reducers, and hose fittings increase the flow velocity to beyond the permissible levels at many points. A temporary increase in the flow velocity also occurs on actuating devices using air at a high rate.

Pressure drop should preferably not exceed 0.1 bar. Another rule of thumb is that the pressure drop should not exceed 5% of the working pressure in the system (e.g., pressure drop should be less than 0.3 bar if the working pressure is 6 bar).

Pressure drop through the bends and fittings can only be determined accurately by empirical tests since it is specific to the internal geometry involved. The determination is time consuming and rather complicated. It is generally more convenient and simpler to calculate the pressure drop (Δp) through fittings in terms of equivalent length of a straight pipe, i.e., the length of a straight pipe of the same bore size as the fitting and which will have the same Δp as the fitting. Flow restrictions are formed in air mains by valves, bends and tees. To calculate the inside pipe diameter, these restrictions must be converted into the equivalent pipe length in the main. Table 3.1 gives the equivalent pipe lengths for the various types of restrictions.

Table 3.1 *Equivalent length for various fitting*

Valve or fitting	Figure	Equivalent pipe length in metres										
		Pipe internal diameter in millimetre										
		25	40	50	80	100	125	150	200	250	300	400
Gate valve fully open		0.3	0.5	0.6	1	1.3	1.6	1.9	2.3	3.2	3.9	5.2
Gate valve half open		5	8	10	16	20	25	30	40	50	60	80
Diaphragm valve fully open		1.5	2.5	3.0	4.5	6	8	10	–	–	–	–
Angle valve fully open		4	6	7	12	15	18	22	30	36	–	–
Globe valve fully open		7.5	12	15	24	30	38	45	60	–	–	–
Swing check valve fully open		2	3.2	4	6.4	8	10	12	16	20	24	32
Bend $R = 2d$		0.3	0.5	0.6	1.0	1.2	1.5	1.8	2.4	3.0	3.6	4.8
Bend $R = d$		0.4	0.6	0.8	1.3	1.6	2.0	2.4	3.2	4.0	4.8	6.4
Mitre bend		1.5	2.4	3.0	4.8	6.0	7.5	9	12	15	18	24
Run of tee		0.5	0.8	1.0	1.6	2.0	2.5	3	4	5	6	8
Side outlet tee		1.5	2.4	3.0	4.8	6.0	7.5	9.0	12	15	18	24
Reducer		0.5	0.7	1.0	2.0	2.5	3.1	3.6	4.8	6.0	7.2	9.6

A variety of empirical formulas have been used by engineers to calculate the pressure drop, a popular one being:

$$\Delta p = \frac{1.6 \times 10^3 \times Q^{1.85} \times L}{d_p^5 \times p_{in}}$$

where

Δp = pressure drop in Pa
L = length of piping in metres/equivalent length in metres
d_p = diameter of pipe in metres
p_{in} = inlet pressure of air at entrance of pipe in Pa
Q = flow rate in $\frac{\text{m}^3}{\text{s}}$.

It can be seen from the formula that pipe length is relatively insignificant as a design control parameter compared with bore size, since the pressure drop is inversely proportional to d_p^5. This means that a small change in bore size can cause a large change in the pressure drop (Δp) whereas even doubling the pipe length only results in doubling the pressure drop (Δp).

Another empirical formula, known as the Harris formula, is also used:

$$\Delta p = \frac{f \times Q^2 \times L}{d_p^5 \times p_{\text{in}}},$$

where f = an empirical coefficient.

Compressor manufacturers have already done the ground work for calculation of air piping by issuing nomograms to help find unknown quantities by simple means. Using this nomogram (Fig. 3.9), the known variables provide the necessary pipe diameter by simple alignment. Starting on the right hand side of the nomogram, at the intersection of flow rate (air consumptions rate) and working pressure, a further intersection is obtained from the given overall pipe length and finally from the permissible pressure drop.

Correct sizing of joints and fittings Air mains are usually made with steel pipes with welded joints since welds are more durable than screwed joints over long term. A disadvantage of welding is the formation of scales during the process, leading to rust over time. The advantages of welding pipes are tight sealing of joints and lower costs. Pipes made of commercial grade DIN 2448 and DIN 2458 are used.

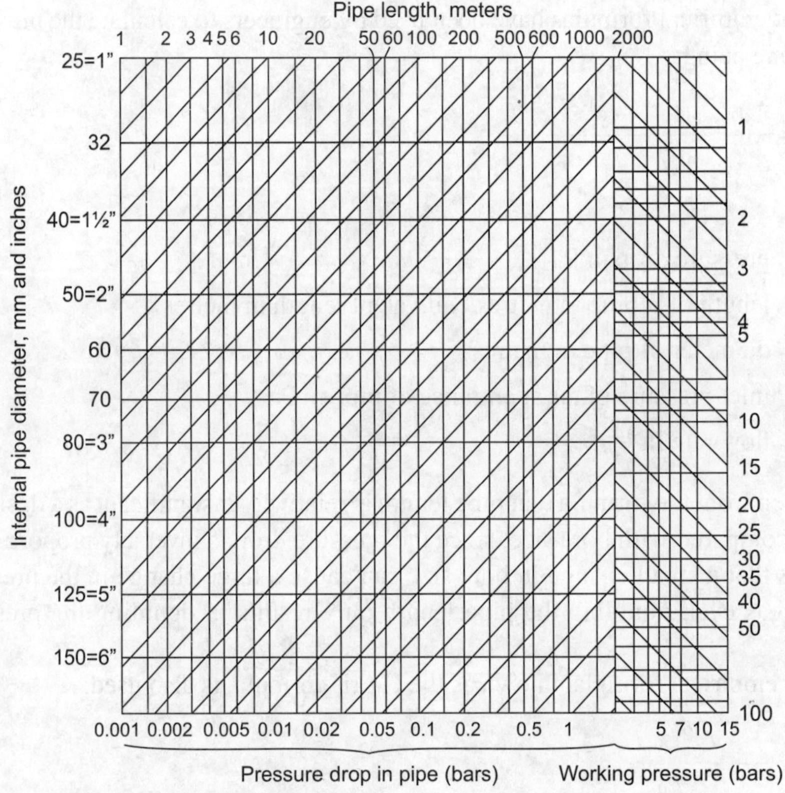

Figure 3.9 *Nomogram of pressure drop versus pipe diameter*

An alternative means of making a pipe connection is to use a screw union or a bite sleeve or compression sleeve (Fig. 3.10). These are fitted on thin-walled, seamless, drawn steel pipes. They allow for easy disconnecting or rearranging of sections of the overall piping layout. Attention should be paid to ensure an accurate fit and a tight sealing of the screw unions.

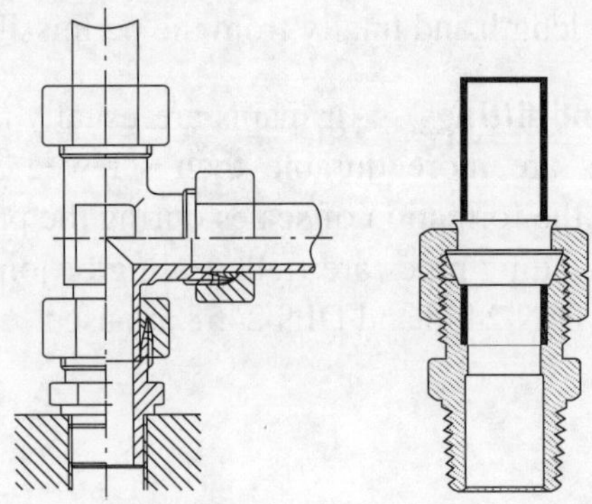

Figure 3.10 *Screw unions for pipe*

Plastic piping is used in many applications. New types of unions are now available, made wholly of plastic, similar to the pipes (Fig. 3.11). A collar is formed on the pipe at a short distance from the end, with a simple cold-upsetting tool giving it a tight seal when squeezed between the stud and nut while assembling the union.

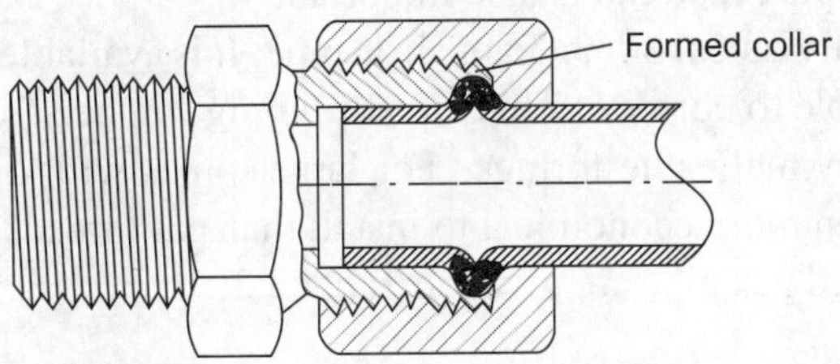

Figure 3.11 *Screw unions for plastic piping*

Apart from compression sleeves, a number of other fittings are available. Plastic hoses are used for flexible air lines. Table 3.2 shows the common fittings used in pneumatic industry.

Table 3.2 *Common fittings used in the pneumatic industry*

Elbow male adaptor	Equal cross connector	Straight connector	Straight female fitting	Straight male adaptor
Straight male fitting	Swivel elbow	Swivel Tee adaptor	Bulk head connector	Rigid manifold
Equal parallel V connector	Swivel elbow adaptor	Straight adaptor	Male to female union	Swivel angle fitting
Swivel V adaptor	Side T adaptor	Swivel V female adaptor	Swivel elbow adaptor	Swivel Tee adaptor

Pipe material Air mains make up a large portion of the initial cost of a compressed air installation. Reducing the pipe diameter lowers the investment cost but results in a higher air pressure drop which increases operational expenses, more than offsetting the initial savings. So, selecting the optimum diameter involves balancing these factors. In addition, selection of the proper pipe material is also important.

The air main is usually a steel or malleable iron pipe. It is available in black or galvanised form, which is less liable to corrode. This type of piping can be screwed to accommodate the range of proprietary malleable fittings. For larger pipes with diameters over 80 mm, welded flanges are often more economical to install than cut threads. The specifications of the carbon steel standard gas pipe are given in Table 3.3.

Table 3.3 *Specifications of the carbon steel standard gas pipe*

Nominal diameter		Outside diameter	Thickness	Mass
Mm	Inch	(mm)	(mm)	(kg/m)
6	1/8	10.5	2	0.419
8	1/4	13.8	2.3	0.652
10	3/8	17.3	2.3	0.851
15	1/2	21.7	2.8	1.310
20	3/4	27.2	2.8	1.680
25	1	34	3.2	2.430
32	1½	42.7	3.5	3.380
40	1½	48.6	3.5	3.890
50	2	60.3	3.65	5.100
65	2½	76.1	3.65	6.510
75	3	88.9	4.05	8.470
100	4	114.3	4.5	12.100

Stainless steel pipes are used in long straight main lines when larges diameters are required. Copper tubes are used where corrosion and heat resistance, as well as high rigidity are required. Copper tubing up to a diameter of 40 mm can be used, but is it more expensive than 28 mm diameter tubing. Compression fittings used with annealed quality tubing are easy to install.

Rubber hose or reinforced plastic is suitable for air actuated hand tools as it offers freedom of movement for the operator. The dimensions of pneumatic rubber hose are given in Table 3.4.

Table 3.4 *The specifications of the rubber air hose (cloth wrapped)*

Nominal width (inches)	Outside diameter (mm)	Inside diameter (mm)	Inner sectional area (mm^2)
1/8	9.2	3.2	8.04
1/4	10.3	6.3	31.2
3/8	18.5	9.5	70.9
1/2	21.7	12.7	127
5/8	24.10	15.9	199
3/4	29	19.0	284
1	35.4	25.4	507
1¼	45.8	31.8	794
1½	52.1	38.1	1140
1¾	60.5	44.5	1560
2	66.8	50.8	2030
2¼	81.1	57.1	2560
2½	90.5	63.5	3170

Rubber hose is used in tools and applications where the tube is exposed to mechanical wear. Plastic tubing is commonly used to connect pneumatic components, and is easy to cut and fit for quick connections as long as the operational conditions are within its temperature tolerance range. If tighter bends or constant movement demand greater flexibility, a softer grade of nylon or polyurethane is an option, but it is limited by lower safe working pressures.

Pipe layout Apart from correct sizing of piping and the quality of pipe material, accurate layout is the decisive factor in determining cost effective operation of the compressed air system. Lines for the distribution of air power must be properly installed and must be sized to provide adequate supply of air for each work station. The two principle factors affecting the pressure drop in air systems are leaks and friction.

Generally air distribution is arranged in two types of layouts:

1. Manifold layout
2. Ring layout

Manifold layout This layout is simple and less expensive. It is also called a leader system and is effective for adequate distribution of air to several tools simultaneously. This layout has the advantage of distributing compressed air over the shortest possible run, reducing the pressure drop and economising on piping cost. Its chief disadvantage is its lack of flexibility. A manifold type pneumatic network is shown in Fig. 3.12.

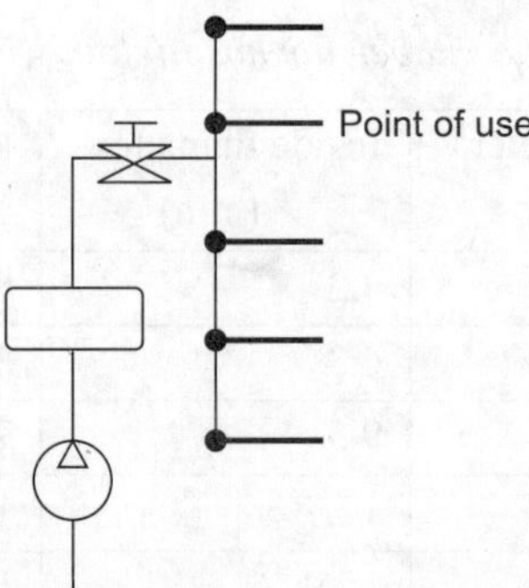

Figure 3.12 *Manifold type pneumatic network*

Ring layout As actuating devices consume air, pressure is decreased downstream. Using the ring main layout helps counter this effect. Demand is met from two directions and this ensures uniform pressure conditions in the air network. The inside diameter of the pipe in a ring main may be about a third smaller than that of a branch line. A ring type pneumatic network is shown in Fig. 3.13.

It is useful to have the layout of a distribution system in the form of a ring main. A ring main takes the form of closed circuit with the advantage that the velocity of the air in the main is lower and hence the pressure drop is lower. The ring main itself may have a significant volume allowing it to act as a reservoir, which helps in maintaining pressure when there is a sudden draw-off. Also, if the system is extended by the addition of a branch, this branch can draw its supply from both ends of the main, reducing velocity and minimising the pressure drop. A ring main offers maximum flexibility with functional advantages and is generally the preferred choice, but care has to be taken to keep the longer length leak-free.

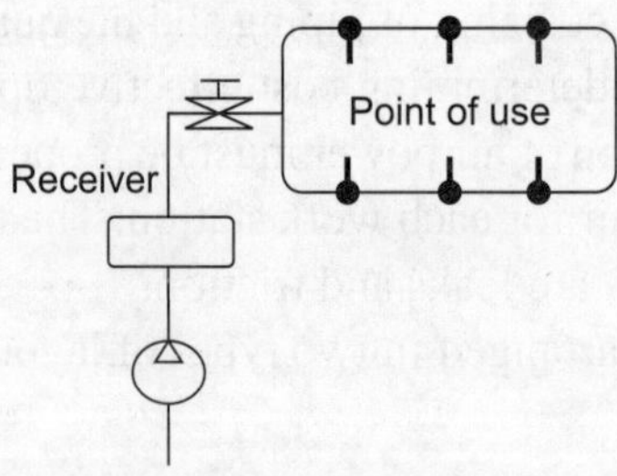

Figure 3.13 *Manifold type pneumatic network*

With an interconnected network systems shown in Fig. 3.14, parts of the ring can be separated using shut-off valves for maintenance, repair and extension of network without disrupting the rest of the system.

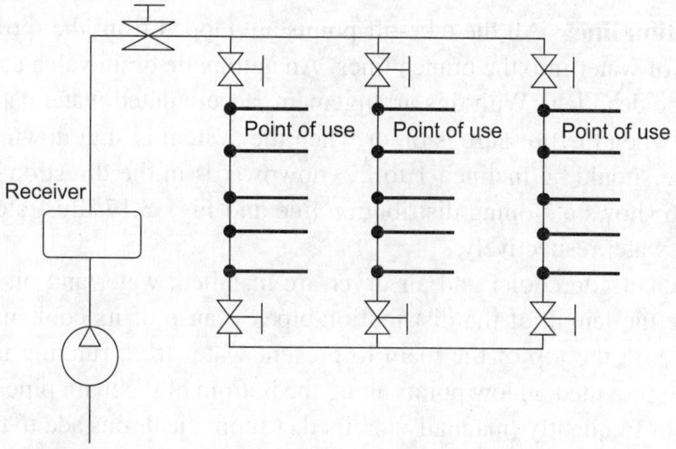

Figure 3.14 *Interconnected ring type pneumatic network*

Permanently installed air mains piping should be, wherever possible, accessible from all sides. Installation in brickwork or in tight pipe ducts is therefore ruled out. Access is necessary in order to keep things simple and allow for the inspection of the piping main for leaks. Horizontal runs of air pipe should slope downwards 1–2% in the direction of flow. Vertical main lines should extend beyond consumer take-offs so that condensate precipitated in the main can collect at the lowest point of the branch line and drain off instead of travelling to the next station.

A branch main must always lead away from the top of the main line, sloping up before curving downwards, and the inner radius of the bend should be at least twice the pipe diameter.

There are three ways to remove water from the air in the distribution system:

1. Bent drop leg
2. Sloping distribution line
3. Automatic drain in line

Bent drop leg It is better to run a pipe or hose line with a drop leg that bends the tube or pipe upwards rather than directly to the point of use unit. This way, water gets separated by the action of gravity. The collected water can be periodically drained using a manual or automatic drain. A schematic diagram is shown in Fig. 3.15.

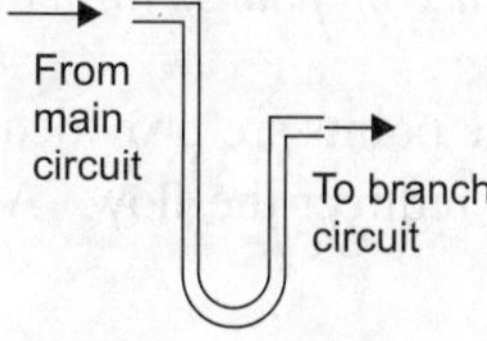

Figure 3.15 *Bent drop*

Sloping distribution line All the take-off points are tapped from the top of the pipe to prevent the entry of water into the branch lines. An automatic drain valve can be provided for terminating the dead leg. With this arrangement, accumulated water can be automatically drained off when the pressure is on or when the system is shut down. For effective drainage, the pipe should be inclined 1 to 2% downwards in the direction of the area of flow. Figure 3.16 shows a sloping distribution line and Fig. 3.17 illustrates the take-off points for air and water respectively.

Unless an efficient aftercooler and air dryer are installed, water and oil condense and accumulate along the length of the distribution pipe because of its cool surface. Branch lines are provided on the top of the main to prevent water from running into them, and drainage tubes are provided on low points along the bottom of the main pipe. These should be automatically or frequently (manually) drained. Automatic drains add to the initial cost, but save on manpower costs later and prevent contamination problems that might be caused by manual error or neglect.

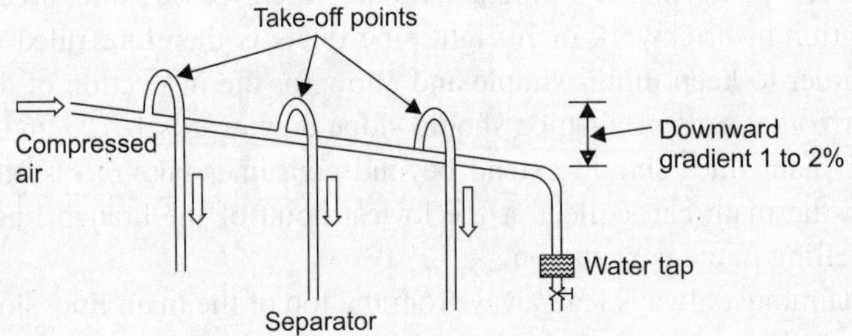

Figure 3.16 *Sloping distribution line*

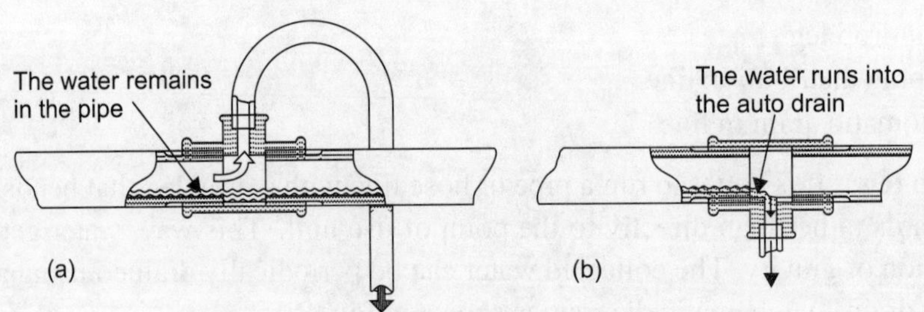

Figure 3.17 *Take off points: (a) for air, (b) for water*

Automatic drain Here, drainage points are provided using equal tees. Water separation is achieved by arranging them to redirect the flow. A schematic illustration is shown in Fig. 3.18.

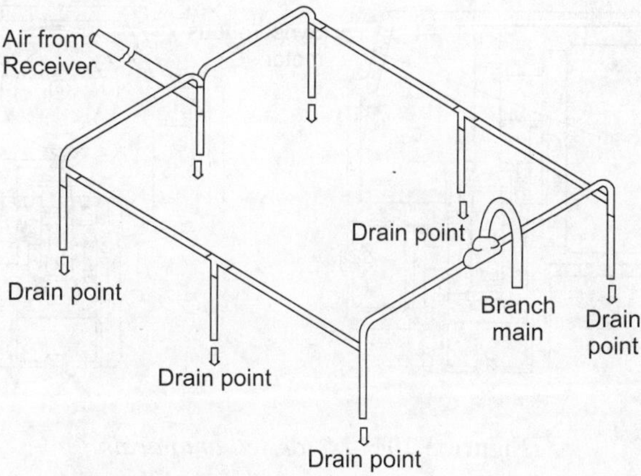

Figure 3.18 *Automatic drains in the ring*

Figures 3.19 and 3.20 illustrate two types of automatic drains. Figure 3.19 shows a float type drain where the float is guided by a tube and connected to the outside with a filter, a relief valve and a spring loaded piston with a stem for manual operation. When enough condensate accumulates at the bottom of the housing, it lifts the float from its seat, transmitting the pressure in the housing to the piston which then opens the drain valve seat to expel the water. The float then lowers to shut off air supply to the piston. The relief valve limits the piston pressure when the float shuts the nozzle. The preset value ensures that the piston resets consistently, allowing the trapped air to vent through the relief valve.

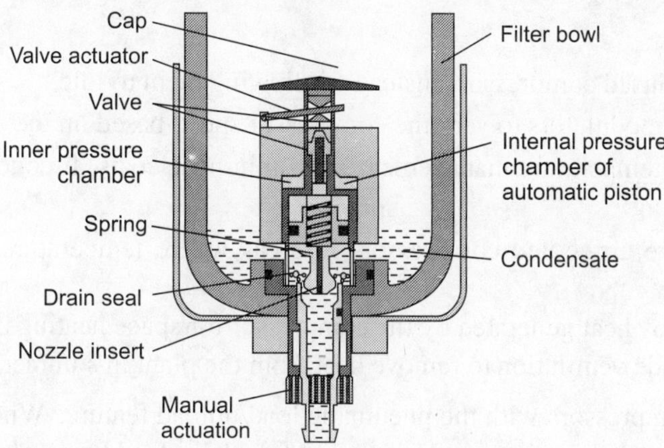

Figure 3.19 *Float type auto drain*

Figure 3.20 shows an electrically motorised drain with a rotating cam wheel that trips a lever-operated poppet valve to periodically expel the accumulated condensate. It is vibration resistant and works in any orientation, making it suitable for mobile compressors and vehicular pneumatic systems.

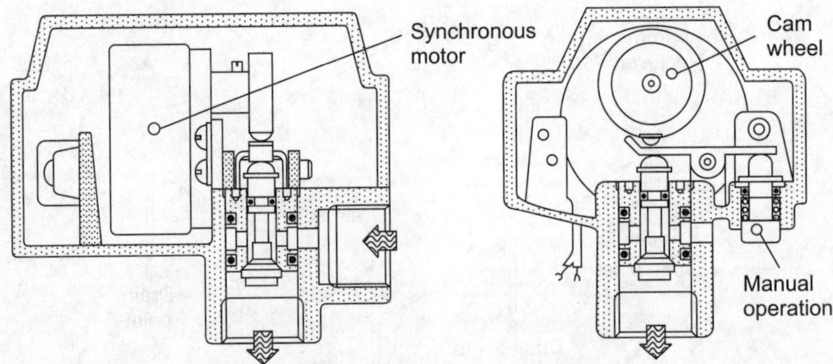

Figure 3.20 *Motorised auto drain*

3.3 Guidelines for Component Selection

The nature of demand and consumption helps in selecting the type and number of compressors. If demand is consistent over time, a single large air compressor is the logical choice in terms of cost and efficiency. If, however, demand is significantly lower at nights and during weekends, a configuration using several compressors is more appropriate. The compressors together meet the high daytime demand, and a smaller subset is active to meet the lower night/weekend demand with the rest on standby to save energy. Other energy saving measures are detailed below.

Compressor

1. Shut off unused compressors instead of allowing them to idle.
2. Use speed modulators to vary the compressor speed based on the demand.
3. Intelligent temperature management helps minimise costs. If conditions allow it, do the following:
   - Locate air-cooled compressors in areas where temperatures will not exceed 40°C.
   - Deploy heat generated by the compressor for space heating in winter.
   - Provide ventilation to remove heat from the plant in summer.
4. Use air compressors with the pneumatic load/unload feature. When fully unloaded, they consume approximately 15 percent of the base load horse power.
5. When economically feasible, use multi-stage compression with inter-cooling.
6. A properly designed and robust foundation for the compressor minimises vibrations for safe and stable operation.

Aftercooler with separator The degree of drying required downstream determines the selection of an aftercooler. The discharge air temperature of the aftercooler influences the size of the dryer, and both capital and operational costs can be reduced by accounting for these factors.

A well designed ventilation system allows hot exhaust from air-cooled aftercoolers to be used for space heating during winter and for heat removal and temperature control during summer.

Filters and dryers

1. The aim should be to improve the air quality only to the required degree at a specific point of use, even if it means using different filters and dryers at different points because requirements vary at the branch lines.
2. The size and type of a dryer affects the operating cost. An accurate determination of the dew point at the point of use helps to select the best dryer configuration that can achieve a dew point at least 7°C below the lowest ambient temperature. If there is a wide temperature variation by season, the dryer should provide dew point adjustment.
3. Ambient temperature affects drying efficiency and wherever possible, dryers should be located where temperatures do not rise beyond 40°C.
4. The dryer and aftercooler combination should be able to keep the air temperature at the intake as low as possible, and the pressure as high as possible. Accurate monitoring and measuring of ambient temperature and pressure is important because even small variations can significantly affect the operating costs.

Air leakage

1. Air leakage rates should be no higher than 10% of the installed system's flow rate. Pipe joints, hose and equipment connections should be monitored and regularly maintained by trained personnel to keep leaks to a minimum.
2. To be safe and leak free, a compressed air system should be built with high quality materials and workmanship and streamlined to keep piping and joints to a minimum.

3.4 Guidelines for Compressor Sizing

1. Collect the maximum air consumption of individual tools from the air consumption charts. Refer to Table 3.5.
2. Calculate the standard pressure. Standard pressure is normally 2 to 5 bars higher than the highest working pressure.
3. Calculate the air consumption at standard pressure by applying gas laws.
4. Calculate the average air consumption taking into account the use factor.
5. Convert the average consumption in terms of FAD.
6. From the relevant air compressor specification charts, select the correct compressor size. Refer to Table 3.6.

Table 3.5 *Air consumption for typical applications*

Pneumatic tools	Specifications	Air consumption (LPM)
Spray gun	*Nozzle* 0.7 mm 1.3 mm 1.6 mm	 40.70 50.90 110.14
Air screw driver	*Thread size* 3 mm PT 5 mm PT 6 mm PT	 140 250 600
Impact wrench	*Bolt size* 6mm 12 mm 25 mm	 150.3 700 1200
Chipping hammer	*Bore size* 24 mm 28 mm 30 mm	 380 400 500
Riveting Hammer	*Rivet size* 16mm 26 mm 41 mm	 600 800 1070
Break Hammer	*Hammer weight* 20 kg 30 kg 40 kg	 1200 1400 1700
Rock drill	*Hammer diameter* 48 mm 57 mm 62 mm	 800 1400 1700

Pneumatic tools	Specifications	Air consumption (LPM)
Rotary drill	*Capacity* 6.3 mm	 300
Air grinder	*Wheel diameter* 25mm 100 mm 200 mm	 300 1000 1400
Rotary saw	*Saw diameter* 203 mm 254 mm 305	 1300 1400 1800
Tapper air bust	*Tap size* *Nozzle* 3.8 mm 1.5 mm	 350 140
Air cylinder	*Diameter* 70 mm 150 mm 305 mm	 4.6 22 88
Air hoist	*Capacity* 1/4T 1 T 2 T	 1000 1540 1960
Air winch	*Capacity* 100 kg 250 kg 500 kg	 580 1100 2100

Table 3.6 *Example of a two-stage compressor specification*

Model	Motor (kW)+	Bore × stroke (mm)	No. of cylinders	Working pressure (Bar)	Compressor speed (RPM)	Displacement LPM	FAD LPM	Tank capacity (litre)
M1	1.5	L:82 × 70	L:1	16	550	203	154	129
		H:42 × 70	H:1					
M2	2.2	L:82 × 70	L:1	16	960	355	270	129
		H:82 × 70	H:1					
M3	3.7	L:95 × 85	L:1	16	900	542	406	195
		H:50.8 × 85	H:1					
M4	5.5	L:82 × 85	L:2	16	900	808	606	195
		H:65 × 85	H:1					
M5	7.5	L:95 × 85	L:2	16	870	1048	786	283
		H:77 × 85	H:1					
M6	11	L:165 × 100	L:1	12	710	710	1064	283
		H:95 × 100	H:1					
M7	22	L:210 × 127	L:1	12	800	800	2183	283
		H:125 × 127	H:1					

Example 3.1 Refer to the relevant air consumption and compressor specifications charts (Tables 3.5 and 3.6) and select a suitable compressor for a machine tool workshop having the following air tools (see Table 3.7).

Table 3.7 *Details of air tools used in workshop*

Item no	Air tools	Quantity	Working pressure (bar) (absolute)	Utilisation factor
1	Cylinder (70 mm)	10	6	0.5
2	Cylinder (150 mm)	5	6	0.5
3	Spray gun (0.7mm)	3	7	0.3
4	Rotary drill (6.3 mm)	1	7	0.2

Solution First calculate the total demand as shown in Table 3.8.

Table 3.8 *Calculation of air demand*

Air tools	Quantity	Working pressure (bar) (absolute)	Utilization factor	Air demand LPM	Air demand at STD @9 bar	Total air demand
	A		**C**		**B**	**A × B × C**
Cylinder (70 mm)	10	6	0.5	4.6	3.1	$10 \times 3.1 \times 0.5 = 15.5$
Cylinder (150 mm)	5	6	0.5	22	14.6	$5 \times 14.6 \times 0.5 = 36.5$
Spray gun (0.7mm)	3	7	0.3	40.7	31.6	$3 \times 31.6 \times 0.3 = 28.4$
Rotary drill (6.3 mm)	1	7	0.2	300	233.3	$1 \times 233.3 \times 0.2 = 46.6$
Total demand = 127 LPM @9 bar						

Step 1: Find standard pressure (STD) $= 7 + 2 = 9$ bar

Step 2: Find the air demand at STD.

$$\frac{p_1 Q_1}{T_1} = \frac{p_2 Q_2}{T_2}$$

$$6 \times 4.6 = 9 \times Q_2$$

$$Q_2 = 3.1 \text{ LPM}$$

To find Q_3 at FAD (at FAD, $P_2 = 1$bar):

$$\frac{p_2 Q_2}{T_1} = \frac{p_3 Q_3}{T_2}$$

$$p_2 Q_2 = p_3 Q_3$$

$$9 \times 127 = 1 \times Q_3$$

Therefore, $Q_3 = 1143$ LPM @ 1 bar.

The compressor can now be selected from the compressor specification chart. Specification details are given in Table 3.9.

Table 3.9 *Compressor specification*

Model number	M7
Power	22 kW
Working pressure	12 bar
Compressor speed	800 RPM
Tank capacity	283 litre

Summary

1. Compressed air that is not suitably conditioned may severely impair or totally obstruct the working of pneumatic components.
2. The purpose of air filters is to clean the compressed air of all impurities and the condensate it contains.
3. Pressure regulators are reducing valves assigned the duty of maintaining a largely constant operating pressure, regardless of fluctuations in the air main pressure and air consumption rate. The inlet pressure must always be higher than the outlet pressure.
4. Different types of pressure regulators may be used, vented and non-vented. If the outlet pressure is reduced by adjusting the set screw on a non-vented regulator, air must first be consumed downstream before the previously higher pressure can come down. On the vented type of regulator, however, excess pressure is released to the atmosphere through vent ports so that the newly adjusted outlet pressure is obtained right away without any need for downstream consumption of air.
5. A lubricator ensures adequate supply of lubricant to the pneumatic equipment. The oil injected into the air stream should be dispersed as a sufficiently fine vapour to make sure that it will not be precipitated at the lubricating points or reduce the flow areas in extensive pneumatic systems.
6. Apart from correct sizing of the piping and the quality of the pipe material, correct pipe layout is the decisive factor in determining the economical operation of the compressed air system.
7. A ring main ensures largely constant pressure conditions. Pressure fluctuations in the pneumatic network make it necessary to ensure that the pipes are mounted securely, avoiding leaks at screwed and brazed connections.

8. In order to allow maintenance, repair or extension of the network without interfering with the overall air supply, it is advisable to sub-divide the network into individual sections by means of shut-off valves. Branches with T-pieces and a manifold with plug-in couplings make it possible to supply additional consumption devices as the need arises, rapidly and easily.
9. In order to protect the consuming devices from condensate from the main line, branch lines must be laid with an upward incline.
10. Despite the best water separation in pressure generating systems, pressure drops and external cooling may produce condensate in the pipe system. In order to discharge this condensate, the pipes should be 1–2% inclined. The condensate can then be discharged from the system via water separators at the lowest point.

EXERCISES

Fill in the blanks

1. Air filter, pressure regulator and lubricator are built as a packaged combination known as FRL unit or ——— unit.
2. The pressure difference between the inlet and outlet will indicate the degree to which the air filter element is ———
3. In pneumatic systems, pressure fluctuations occur due to variation in ——— pressure or load pressure.
4. The main objective of an air distribution system is to provide a distribution channel for compressed air without any leaks and to keep the ——— within permissible limits.
5. Equivalent pipe length is the length of straight pipe of the same ——— size, giving the same pressure drop.

State whether true or false

1. Usually, inline filter elements can remove contaminants in the 0.005–0.0005 micron range.
2. A manifold type pneumatic network has the advantage of being in the form of a closed circuit so that the velocity of the air in the main will be reduced and the pressure drop will be lower.
3. Two quality levels of air are required in pneumatic production and distribution systems.
4. In case of the non-venting type pressure regulator, compressed air cannot escape to the atmosphere in the event of high back pressure acting on the diaphragm as there is no exit path provided in the diaphragm for the trapped air.
5. A single, large air compressor is more efficient and less expensive than the several smaller units if the demand is fairly constant.

Review questions

1. State the importance of fluid conditioning.
2. Describe the function of an air filter.
3. With help of a neat sketch, explain the working of an air filter.
4. State clearly nine qualities of filtered air and its applications.
5. Describe the function of an air pressure regulator.
6. With the help of a neat sketch, explain the working of an air regulator.
7. Explain the difference between venting and non-venting types of pressure regulators.
8. Describe the function of an air lubricator.
9. With the help of a neat sketch, explain the working of an air filter.
10. List the commonly used oils in an air lubricator.
11. State the advantage of a main ring pneumatic network.
12. List five important considerations in pipe layout in a pneumatic network.
13. With the help of a neat sketch, explain the ring and manifold types of pneumatic networks.
14. List the guidelines for selection of pneumatic components and compressors.
15. Discuss three ways to remove water from the air in an air distribution system.

Problems

1. Refer to the relevant air consumption and compressor specification charts and select a suitable compressor for a machine tool workshop having the air tools listed in Table 3.10.

Table 3.10

Item no	Air tools	Quantity	Working pressure (bar)(absolute)	Utilization factor
1	Cylinder (70 mm)	5	6	0.3
2	Spray gun (1.3 mm)	4	5	0.3
3	Air screw driver (5 mmPT)	2	6	0.2

2. Refer to the relevant air consumption and compressor specification charts and select a suitable compressor for a machine tool workshop having the air tools given in Table 3.11.

Table 3.11

Item no	Air tools	Quantity	Working pressure (bar)(absolute)	Utilization factor
1	Rotary drill	1	8	0.25
2	Spray gun(1.3mm)	2	7	0.10
3	Air cylinder (70 mm)	10	6	0.30

3. A 24 metre pipe with an inner diameter of 80 mm leads a compressed air flow of 150 l/s. The pipe is routed with 8 elbows that all have a bending radius equal to that of the pipe inner diameter. What will be the pressure drop across the pipe if the initial pressure is 8 bar (abs)?
4. In a machine shop, there are four compressors installed for compressed air requirements. All the four are in operation at any point of time. The rated capacities of the compressors are 2.84 m^3/minute, whereas the actual output of each are given below in Table 3.12.

Table 3.12

Compressor 1	2.17 m^3/min
Compressor 2	2.23 m^3/min
Compressor 3	2.75 m^3/min
Compressor 4	2.50 m^3/min

The power consumption of these compressors is 29 kW, 27 kw, 22 kW and 25 kW. The compressors are continually operated at 7 bar pressure. It was observed that the leakage losses constitute around 45% of the actual output of the compressor. Noting that the compressors have on and off control, estimate the following:

(a) Power loss due to leakage.
(b) The power saving if 50% of leakages are arrested.

5. A reciprocating two-stage air compressor of 13.93 m^3/hour capacity is operating in a pneumatic shop. The FAD test was conducted by pump-up method. The parameters measured during the test are given in Table 3.13.

Table 3.13

Receiver diameter	1.78 m
Receiver height	0.8 m
Volume of interconnecting pipelines	0.09 m^3
Time taken to fill up the receiver	1.48 minutes

contd...

Table 3.13 *...contd*

Initial pressure	Atmospheric
Receiver pressure	7.75 bar
Ambient temperature	30°C
Air temperature in the receiver	32°C

(a) Calculate the actual FAD delivered by the compressor and express it in terms of percentage of design FAD.

(b) Also calculate the isothermal efficiency and specific power consumption if the actual power consumption by the compressor is 89.7 kW. Assume motor efficiency of 90%.

Answers

Fill in the blanks

1. service 2. clogged 3. supply 4. pressure drop 5. diameter

True or false

1. False 2. False 3. False 4. True 5. True

Problems

1.

Model Number	M6
Power	11 kW
Working pressure	12 bar
Compressor speed	710 RPM
Tank capacity	283 litre

2.

Model Number	M6
Power	10 kW
Working pressure	16 bar
Compressor speed	870 RPM
Tank capacity	283 litre

3. 0.0054 bar 4. (a) 46.4 kW (b) 23.2 kW

5. (a) 13.93 m^3/hour, 79% (b) 48.06%, 8.2 kW m^3/hour

4 PNEUMATIC ACTUATORS AND AIR MOTORS

Upon completion of this chapter, the reader should be able to

- Explain the meaning of a pneumatic actuator
- Classify the various types of pneumatic actuators
- Explain the working of various pneumatic actuators
- Understand the importance of cushioning
- Study the seals used in the pneumatic actuators
- Understand the working of rod-less cylinders
- List the advantages and disadvantages of various cylinder mountings
- Study the effect of buckling in cylinders
- Explain the working of limited angle rotary actuators
- Explain the working and application of air motors

4.1 Pneumatic Actuators

Pneumatic actuators are devices used to convert the pressure energy of compressed air into mechanical energy to perform useful work. In other words, actuators are used to perform the task of exerting the required force at the end of the stroke, or used to create displacement by the movement of the piston. Pressurised air from the compressor is stored in the reservoir from where it is supplied to the pneumatic actuator to do work.

Their chief limitation is that the elastic nature of compressed air makes the actuators unsuitable for powering movement where a constant force or motion is required against a fluctuating load, or where extreme accuracy of the supply is vital. The air cylinder is also inherently limited in thrust output by the relatively low supply pressure, so high output forces can only be produced by large sized cylinders.

4.2 Types of Pneumatic Actuators

Pneumatic cylinders can be used to produce linear, rotary and oscillatory motion. There are three types of pneumatic actuators:

- Linear actuators or pneumatic cylinders
- Rotary actuator or air motors
- Limited angle actuators

4.2.1 Types of pneumatic cylinders (Linear actuators)

Pneumatic cylinders are devices that convert air pressure into linear mechanical force and motion. They are used for single purpose applications such as clamping, stamping, transferring, branching, allocating, ejecting, metering, tilting, bending, turning, etc.

Pneumatic cylinders are classified according to the parameters shown below:

1. **Cylinder application**
   (i) Light duty air cylinders
   (ii) Medium duty air cylinders
   (iii) Heavy duty air cylinders
2. **Cylinder action**
   (i) Single acting cylinder
   (ii) Double acting cylinder
      - Single rod double acting cylinder
      - Double rod double acting cylinder
3. **Cylinder movement**
   (i) Rotating air cylinder
   (ii) Non-rotating air cylinder
4. **Cylinder design**
   (i) Telescopic cylinder
   (ii) Tandem cylinder
   (iii) Rod-less cylinder
      - Cable cylinder
      - Sealing band cylinder with slotted barrel
      - Cylinder with magnetically coupled slide
   (iv) Impact cylinder
   (v) Duplex cylinders
   (vi) Cylinders with sensors

4.2.1.1 Classification based on cylinder application

Air cylinders can be classified according to their intended use as light, medium or heavy duty cylinders. This classification indicates the strength of the cylinder, and thus the typical choice of material and the form of construction. A comparison is given in Table 4.1.

Table 4.1 *Materials of construction for light, medium and heavy duty cylinders*

Components	Type of cylinder		
	Light duty	**Medium duty**	**Heavy duty**
Cylinder tubes	Hard drawn seamless aluminium or brass tubes, plastics	Hard drawn seamless brass tubes, aluminium, brass, iron or steel castings	Hard drawn seamless tubing, brass, bronze, iron or steel casting
End covers	Aluminium alloy castings, fabricated aluminium, brass, bronze	Aluminium brass, bronze, iron or steel castings, fabricated brass, bronze,	High tensile castings
Pistons	Aluminium alloy castings	Aluminium alloy castings, castings, brass, cast iron	Aluminium alloy brass, cast iron
Piston rods	EN 8 or similar steel ground and polished, or chrome plated	EN 8 steel, ground and polished, or chrome plated, ground and polished stainless steel	Ground and polished stainless steel
Mounting brackets	Aluminium alloy casting	Aluminium, brass, iron castings	High tensile castings or fabricated brackets

4.2.1.2 Classification based on cylinder action

Based on the cylinder action, cylinders are classified as single acting and double acting. Single acting cylinders have a single air inlet line. Double acting cylinders have two air inlet lines. The advantages of double acting cylinders over single acting cylinders are:

1. In single acting cylinders, compressed air is fed only on one side.
2. In single acting cylinders, the stroke length is limited by the compressed length of the spring. In principle, the stroke length is unlimited in a double acting cylinder.
3. When the piston moves forward in a single acting cylinder, air has to overcome the pressure of the spring and hence some power is lost before the actual stroke of the piston starts. This is not a problem in double acting cylinders.

(A) Single acting cylinders

A single acting cylinder has one working port. Forward motion of the piston is obtained by supplying compressed air to the working port. Return motion of the piston is obtained by the spring placed on the rod side of the cylinder. A schematic diagram of a single acting cylinder is shown in Fig. 4.1.

Single acting cylinders are used in applications such as clamping, feeding, sorting, locking, ejecting, braking etc., where force is required to be exerted only in one direction.

Single acting cylinders are usually available in short stroke lengths (maximum length up to 80 mm) due to the natural length of the spring. They exert force only in one direction and require only about half the air volume consumed by a double acting cylinder for one operating cycle.

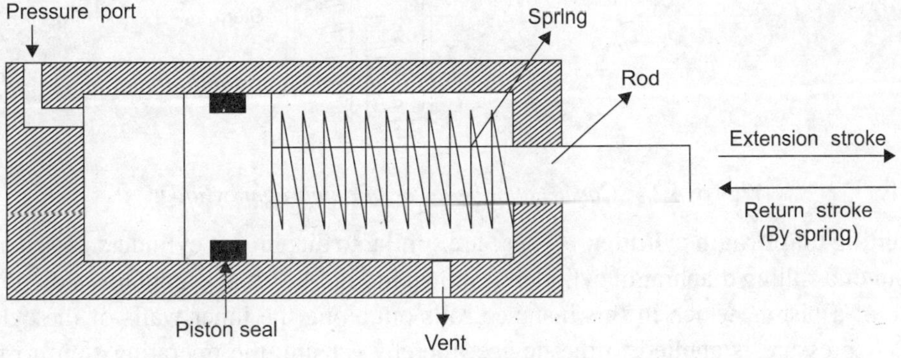

Figure 4.1 *Construction features of a single acting cylinder*

There are varying designs of single acting cylinders including:

(i) Diaphragm cylinder
(ii) Rolling diaphragm cylinder
(iii) Gravity return single acting cylinder
(iv) Spring return single acting cylinder

(i) Diaphragm cylinder This is the simplest form of a single acting cylinder where the piston is replaced by a diaphragm of hard rubber, plastic or metal clamped between the two halves of a metal casing expanded to form a wide, flat enclosure. A schematic diagram of a diaphragm cylinder is shown in Fig. 4.2. The operating stem which takes the place of the piston rod in a diaphragm cylinder can also be designed as a surface element that acts directly as a clamping surface. Only short operating strokes, up to a maximum of 50 mm, can be executed by a diaphragm cylinder. This makes the diaphragm type of the cylinder particularly adaptable to clamping operations. The return stroke is accomplished by a spring built into the assembly or by the tension of diaphragm itself in the case of a very short stroke. Diaphragm cylinders are used for short stroke applications like clamping, riveting, lifting, and embossing.

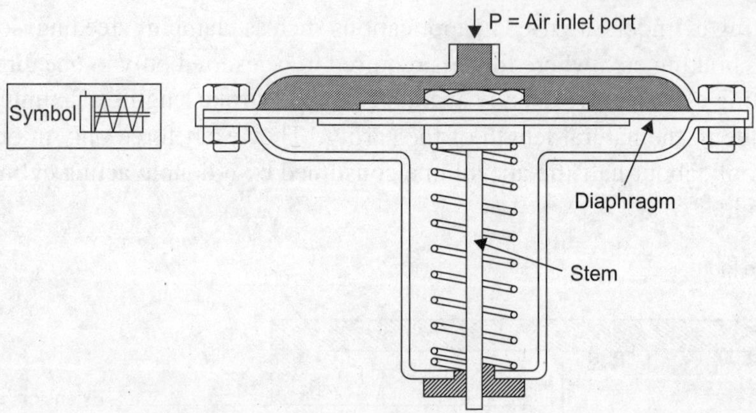

Figure 4.2 *Construction features of diaphragm cylinder*

(ii) Rolling diaphragm cylinder These are similar to diaphragm cylinders. A schematic diagram of a rolling diaphragm cylinder is shown in Fig. 4.3. It also contains a diaphragm instead of a piston, which in this instance rolls out along the inner walls of the cylinder when air pressure is applied to the device, thereby causing the operating stem to move outwards. Compared with the standard diaphragm type, a rolling diaphragm cylinder is capable of executing appreciably longer operating strokes (averaging from 50 mm to 800 mm). A separate guide for the stem is not normally provided in these designs, since the component being actuated by the cylinder usually cannot break out of set limits of motion. Any off-centre displacement is countered by the rolling diaphragm with no loss of power. Materials used for rolling diaphragms in present-day designs ensure durability under normal operating conditions. This is important because even a small crack or cuts in the diaphragm can lead to early failures once the flexible material undergoes high stress as it unrolls at each stroke. If the actuator needs to be dismantled for any reason, it must be inspected carefully for any burrs or sharp edges inside. Metal cuttings also constitute a hazard if they enter the cylinder housing.

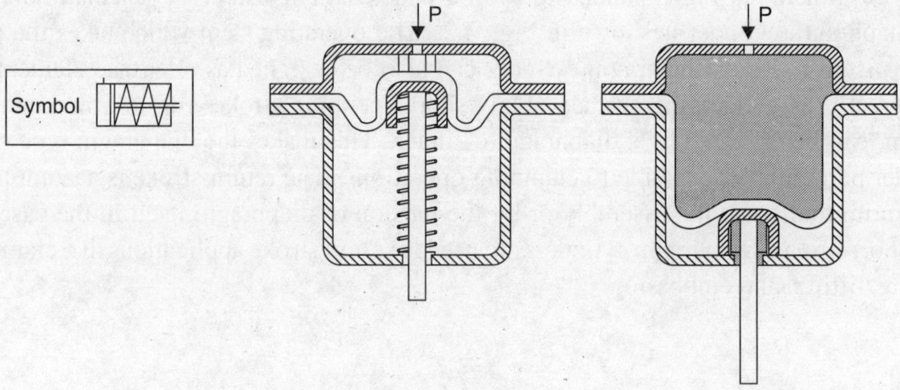

Figure 4.3 *Construction features of a rolling diaphragm cylinder*

(iii) Gravity return single acting cylinder Figure 4.4 shows a gravity return single acting cylinder. In push type (a), the cylinder extends to lift a weight against the force of gravity by applying oil pressure at the blank end. The air is passed through the blank end port or the pressure port. The rod end port or vent port is open to the atmosphere so that air can flow freely in and out of the rod end of the cylinder. To retract the cylinder, the pressure is simply removed from the piston by connecting the pressure port to the tank. This allows the weight of the load to push the fluid out of the cylinder and back into the tank.

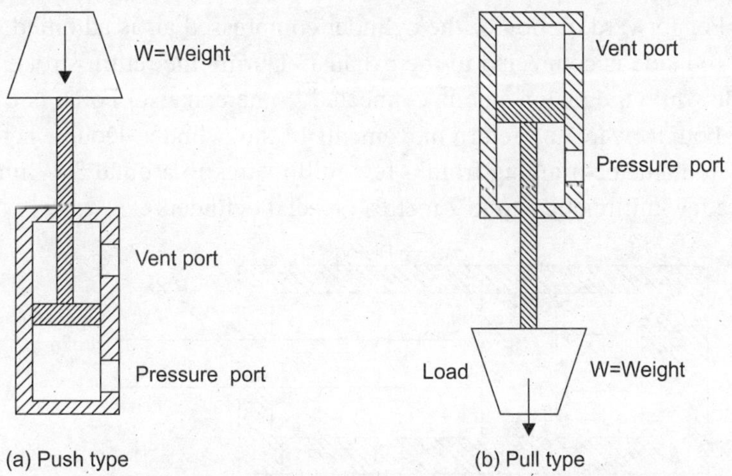

Figure 4.4 *Gravity return single acting cylinder*

In the pull type (b), the cylinder lifts the weight by retracting. The blank end port is the pressure port and the blind end port is now the vent port. The cylinder automatically extends whenever the pressure port is connected to the tank.

(iv) Spring return single acting cylinder A spring return single acting cylinder is shown in Fig. 4.5. Part (a) is push type, where the pressure is sent through pressure port situated at the blank end of the cylinder. When the pressure is released, the spring automatically returns the cylinder to the fully retracted position. The vent port is open to the atmosphere so that air can flow freely in and out of the rod end of the cylinder.

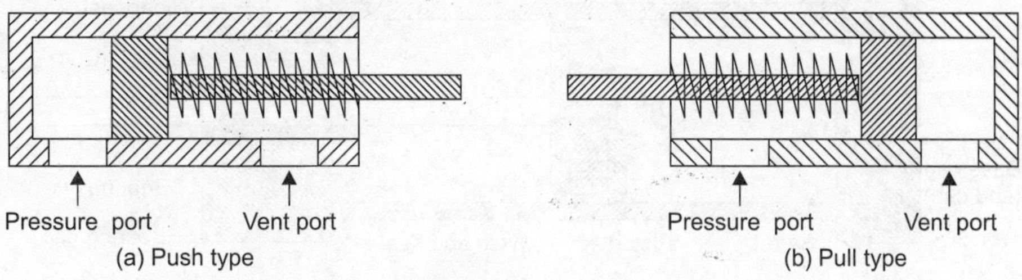

Figure 4.5 *Spring return (push and pull type) single acting cylinder*

Part (b) shows the pull type spring return single acting cylinder. In this design, the cylinder retracts when the pressure port is connected to the pump flow, and extends when the pressure port is connected to the tank. Here the pressure port is situated at rod end of the cylinder.

(B) Double acting cylinders

A schematic diagram of a double acting cylinder is shown in Fig. 4.6. Double acting cylinders are equipped with two working ports, one on the piston side and the other on the rod side. For forward motion of the cylinder compressed air is admitted on the piston side, and the rod side is connected to the exhaust. During the return stroke, air admitted in the rod side while the piston side is connected to the exhaust. Force is exerted by the piston during both forward and return movements of the cylinder. Double acting cylinders are available in diameters ranging from a few millimetres to around 300 mm, and stroke lengths from a few millimetres up to 2 metres (special cylinders).

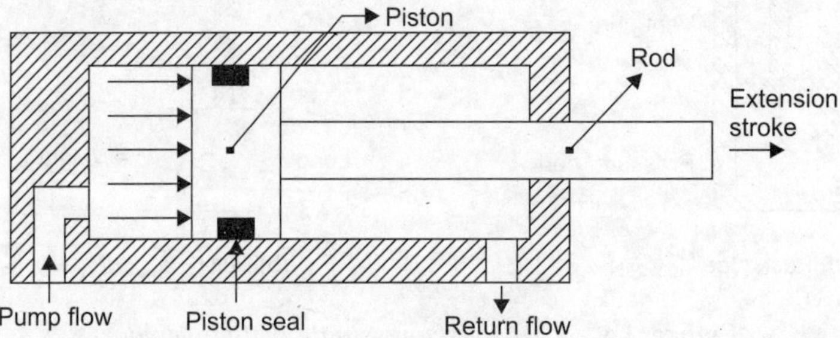

Figure 4.6 *Double acting cylinders*

Construction of a double acting cylinder The construction of a double acting cylinder is shown in Fig. 4.7 and is similar to that of a single cylinder except that there is no return spring. In a double acting cylinder, air pressure can be applied to either side (supply or exhaust) of the piston, thereby providing a pneumatic force in both directions. Double acting cylinders are most commonly used in applications where a longer stroke length is required.

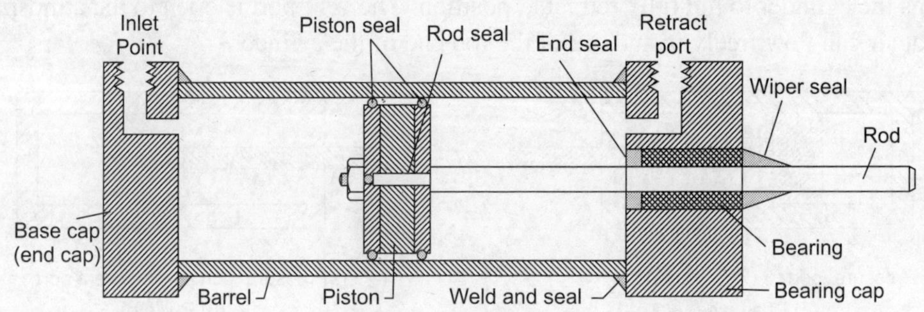

Figure 4.7 *Construction features of double acting cylinder*

The seven parts of a double acting cylinder are:

- Base cap with a port connection
- Bearing cap with a port connection
- Cylinder barrel
- Piston
- Piston rod
- Scraper rings
- Seals

The base and bearing caps are made of cast aluminium or malleable cast iron. The two caps can be fastened to the cylinder barrel by tie rods, threads or flanges.

The cylinder barrel is usually made of a seamless drawn steel tube to increase the life of the sealing components, and the bearing surfaces of the cylinder are precision machined. For special applications, the cylinder barrel can be made of aluminium, brass or steel tubes with a hard chromed bearing surface. These special designs are used where operation is infrequent or where there are corrosive influences.

The piston rod is made preferably from heat treated steel. A certain percentage of chrome in the steel protects against rusting. Generally the threads are rolled to reduce fractures. Piston seals are provided between the piston and the barrel to avoid leakage. A sealing ring is fitted in the bearing cap to seal the piston rod. The bearing bush guides the piston rod and may be made of sintered bronze or plastic coated metal. In front of this bearing bush is a scraper or wiper ring. It prevents dust and dirt particles from entering the cylinder space. Bellows are therefore not usually required.

The materials for the double cup packing sealing are:

- Perbunan (−20°C to +80°C)
- Viton (−20°C to +100°C)
- Teflon (−80°C to +200°C)
- O rings (for static sealing)

There are two types of double acting cylinders:

(i) Double acting cylinder with piston rod on one side
(ii) Double acting cylinder with piston rod on both sides

(i) Double acting cylinder with piston rod on one side Figure 4.8 shows the operation of a double acting cylinder with a piston rod on one side. To extend the cylinder, pump flow is sent to the blank end port as in Fig. 4.8(a) and the fluid from the rod end port returns to the reservoir. To retract the cylinder, the pump flow is sent to the rod end port and the fluid from the blank end port returns to the tank as seen in Fig. 4.8(b).

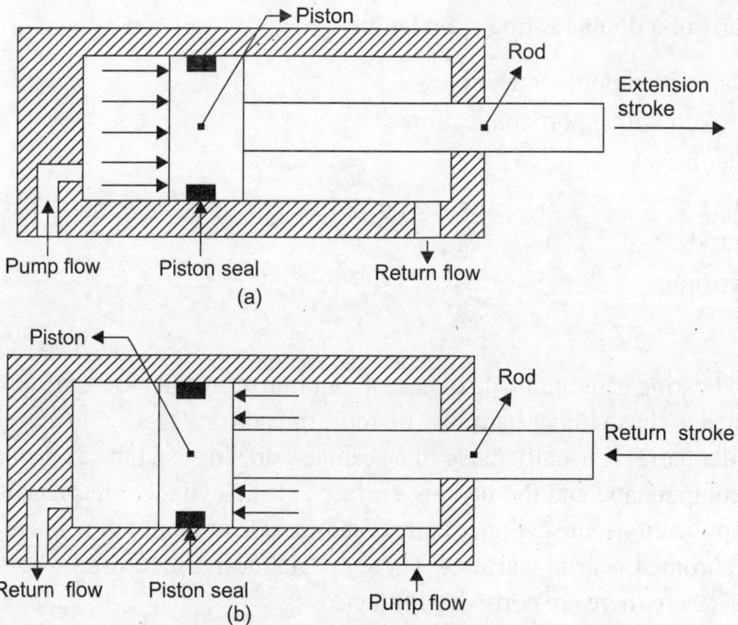

Figure 4.8 *Double acting cylinder with piston rod on one side*

(ii) Double acting cylinder with piston rod on both sides A double acting cylinder with a piston rod on both sides (Fig. 4.9) has a rod extending from both ends. Work can be done by both ends of the cylinder, thereby making it more productive. Double rod cylinders can withstand higher side loads because they have an extra bearing on each rod to withstand the load. They are used when there is a bending load and accurate alignment and maximum strength is required. A further advantage is that the rod is precisely located and may be used to guide the machine member coupled to it, dispensing with external guides or bearings in many cases. Most standard production models are available either in single rod or double rod configurations. A disadvantage of the double rod configuration is that there is a reduction in the maximum thrust due to the blanking effect of the rod cross-section on the piston area.

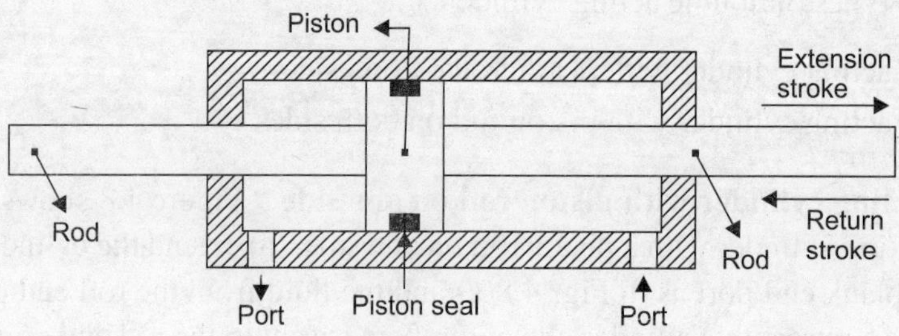

Figure 4.9 *Double acting cylinder with piston rod on both sides*

4.2.1.3 *Classification based on cylinder action*

The rotating type of cylinder is used in applications where the cylinder body is connected to a rotating member, and the air connection to the cylinder is in a stationary housing. They are not widely used.

Non-rotating cylinders are widely used in industries. The cylinder body is connected to the air connection which is mounted on stationary housing and the piston rod moves inside the cylinder.

4.2.1.4 *Classification based on cylinder design*

In the industry, a distinction is made between specially designed regular cylinders and special duty cylinders designed for a specific purpose. Special design cylinders are basically natural variations of single or double acting cylinders, with changes during production to swap indifferent shapes or materials for certain parts. Special duty cylinders on the other hand, are designed from the start for non-standard conditions of service or application. The following section deals with some of the commonly used special design and special duty cylinders.

(A) Telescopic cylinder

A telescopic cylinder (shown in Fig. 4.10 (a) and (b)) is used when a long stroke length and a short retracted length are required. It extends in stages, each stage consisting of a sleeve that fits inside the previous stage. Figure 4.10 (c) shows the construction of a typical double acting telescopic cylinder with two pistons (two stages).

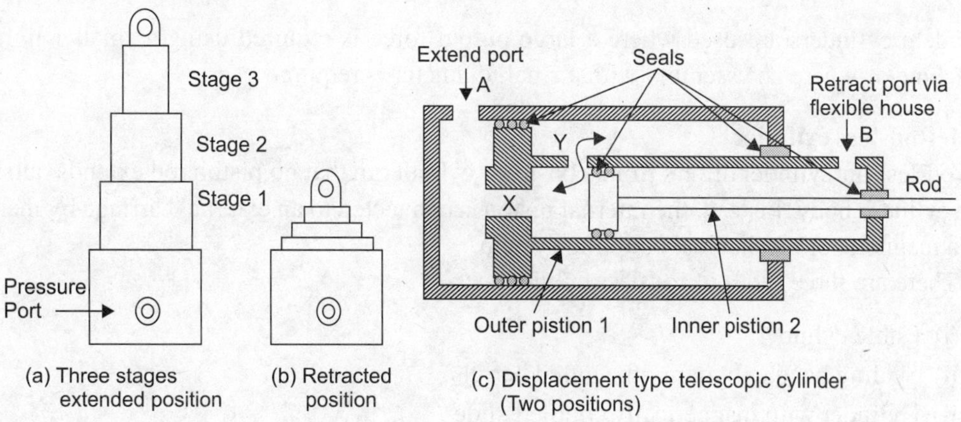

Figure 4.10 *Telescopic cylinder*

Extension stroke When pressure is applied at port A, air flows through ports A and X and pressure is applied on both sides of piston 1. But a difference in the areas causes piston 1 to move to the right. Once piston 1 fully extends, the inner piston 2 extends.

Retraction stroke To retract, air is applied to port B. Air pressure acts on the annulus of the inner piston 2 and moves it to the left. This starts to close port A whereupon air from port B goes to the annular side of piston 1 and pushes piston 1 to the left.

(B) Tandem cylinder

A schematic diagram of a tandem cylinder is shown in Fig. 4.11. Tandem cylinders are two separate double acting air cylinders arranged in line as one cylinder body, so that the power generated by them works in combination approximately doubling the piston output. A tandem cylinder is used in applications where a large amount of force is required from a small diameter cylinder. Basically, a tandem cylinder is simply two or more separate cylinders stacked end to end in a unit, with all the pistons mounted on a common piston rod. Pressure is applied to both pistons, resulting in an increased force because of the larger working area. The drawback is that these cylinders must be longer than a standard cylinder to achieve equal speed, because the flow must go to both pistons.

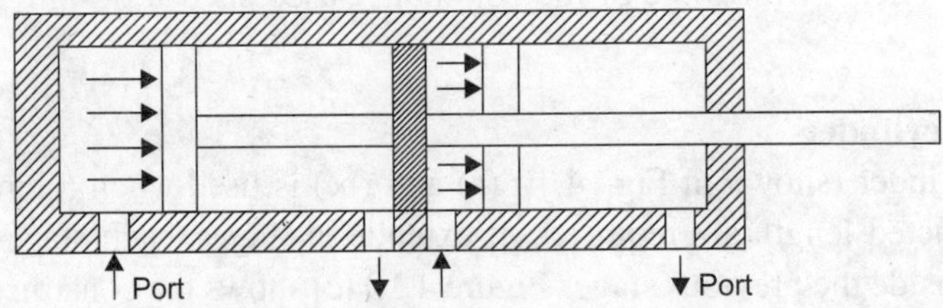

Figure 4.11 *Tandem cylinder*

Tandem cylinders are used where a large output force is required using a smaller lighter machine or where an assembly with a small diameter is required.

(C) Rod-less cylinder

A rod-less air cylinder differs from a basic air cylinder in that no piston rod extends outside the cylinder body. Instead, the internal piston is connected to an external carriage by means of a magnetic or mechanical coupling system.

There are three types of rod-less cylinders:

(i) Cable cylinder
(ii) Sealing band cylinder with slotted barrel
(iii) Cylinder with magnetically coupled slide

(i) Cable cylinder It is used for very long strokes of up to 2000 mm. It consists of a nylon jacketed cable which enters the cylinder barrel, is attached to one end of the internal cylinder and exits through the gland seal to re-enter the other end of the internal cylinder via another gland seal, as shown in Fig. 4.12. When compressed air enters the cylinder, the piston moves from end to end. The cables attached to either side of the piston and extending out to the ends of the cylinder, move as well. Depending on the direction of the piston, both the carriage and any on-board tooling move towards one or the other end of the air cylinder.

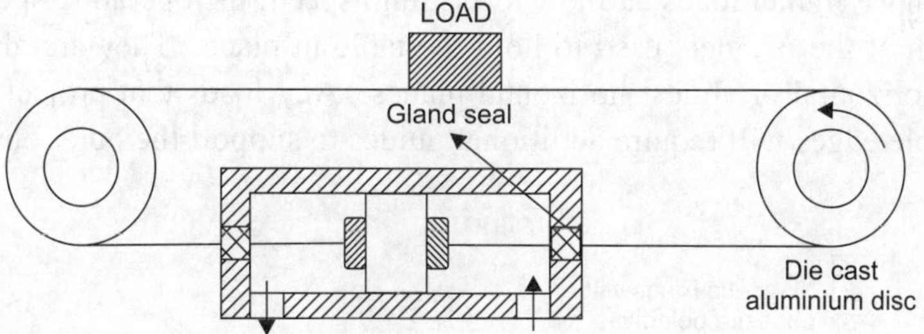

Figure 4.12 *Cable cylinder*

(ii) Sealing band cylinder with slotted barrel A schematic diagram of a sealing band cylinder with a slotted barrel is shown in Fig. 4.13. Common components of the band cylinder are:

- End caps
- Cylinder barrel
- Cylinder piston
- Carriage
- Mechanism connecting carriage to piston
- Sealing bands/strips

As the carriage moves in either direction, the two band sealing strips alternately open in front of, and then close behind the moving carriage. Air is trapped inside the cylinder barrel by the piston seals which seal the inner band tight against the barrel.

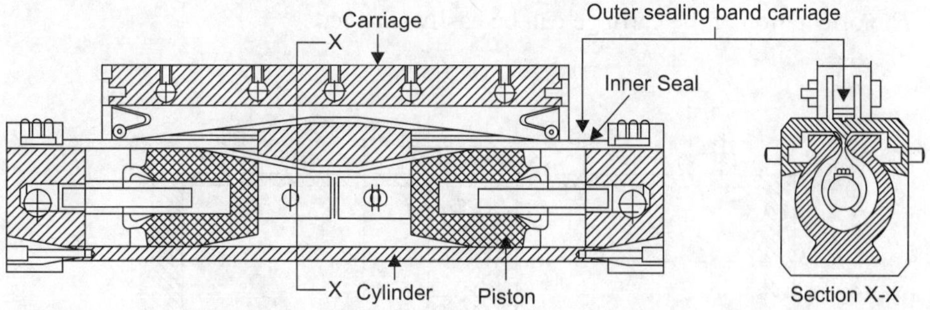

Figure 4.13 *Sealing band cylinder with slotted barrel*

The carriage on top of the band cylinder has a wiper assembly at both ends to brush off debris from its path, and to secure the top seal against the outside slot of the cylinder barrel to prevent compressed air leaks.

(iii) Cylinder with magnetically coupled slide Construction features of a magnetically coupled cylinder are shown in Fig. 4.14(a). Magnetically coupled rod-less cylinders have the ability to move lighter loads at a low to medium speed; the operating speed is limited by the strength of the magnets used to hold the table in place. They are also limited to operating in horizontal or almost horizontal planes. Any loads that project significantly beyond the table edges will require additional guides to support the outer envelope of the cylinder.

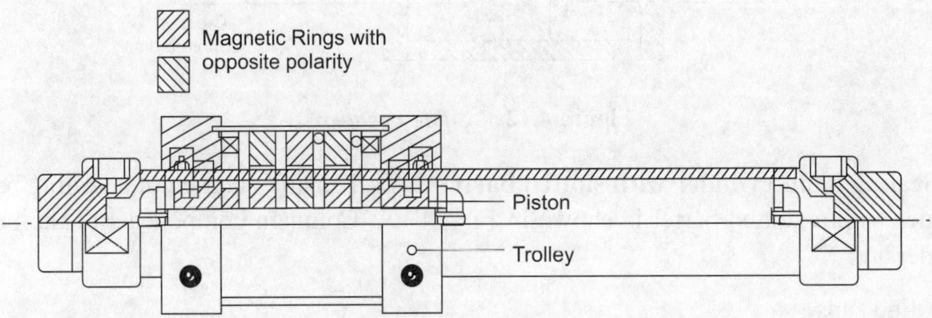

Figure 4.14 *(a) Construction features of cylinder with magnetically coupled slide*

Figure 4.14(b) shows the schematic diagram of a magnetic piston cylinder. The piston has a powerful magnet that bonds the piston inside the cylinder to another powerful magnet contained in the outer carriage. The magnetic cylinders are available up to about 40 mm diameter and with stroke lengths ranging from 50 mm to 4000 mm. They can operate at a speed of about 3000 mm/sec. The major advantages of this type of cylinder are:

(i) There is no leakage
(ii) There is no direct contact between moving elements, reducing wear and tear
(iii) The orientation of the carriage can be easily changed

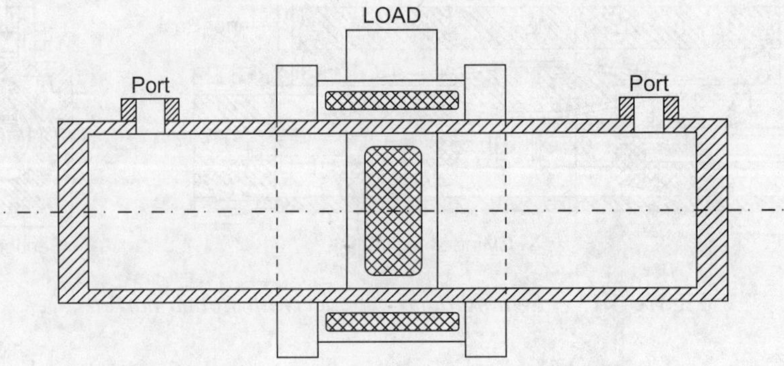

Figure 4.14 *(b) Schematic of cylinder with magnetically coupled slide*

(D) Impact cylinders

Impact cylinders are used for high energy applications. The schematic diagram of an impact cylinder is shown in Fig. 4.15. It comprises a normal pneumatic cylinder allied to a reservoir. Without using larger valves or a higher supply pressure, it provides a 15-fold increase in operating speed and a 200-fold increase in the energy rating. Impact cylinders find extensive applications in high energy rate forming; a 200 mm bore impact cylinder may accomplish what might have previously required a 50 tonne press.

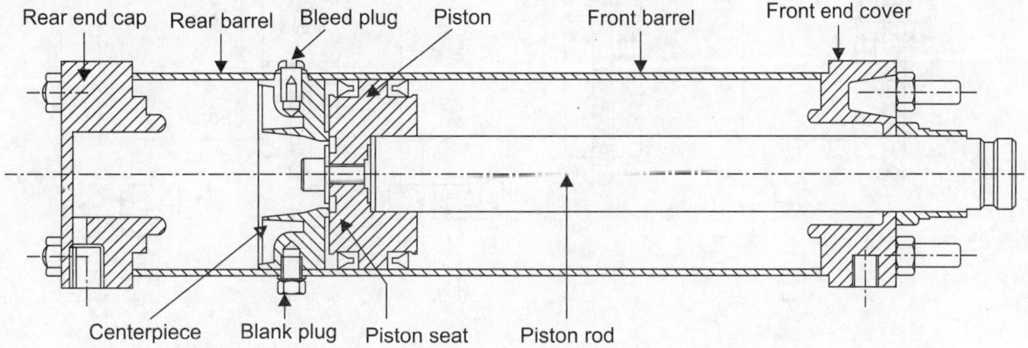

Figure 4.15 *Schematic diagram of an impact cylinder*

If normal cylinders are used in forming operations, the thrust force of the compressed air is limited, whereas impact cylinders provide high kinetic energy. In accordance with the formula for kinetic energy (KE), the obvious means of obtaining high impact energy is by increasing the velocity.

$$KE = \frac{1}{2}mv^2,$$

where m = mass in kg and v = speed in m/s.

If the forming stroke is large, the velocity diminishes rapidly and hence also the impact energy. This type of cylinder is therefore not suitable for large forming strokes; rather it is well suited for stamping, punching, cutting, riveting, beading, embossing etc.

The sequence of operation is shown in Fig. 4.16. A reservoir is added to the top of a conventional cylinder which opens into the main cylinder via a poppet valve (centre piece) controlled by the piston movement. Ratio of the piston area to the poppet valve area is of the order of 9:1, so pressure in the reservoir is approximately $1/9^{\text{th}}$ the pressure in the exhaust end of the cylinder.

To charge the cylinder, the reservoir is connected to the line and the pressure builds up to the line pressure. The movement of the piston is opposed by the exhaust pressure until it falls to $1/9^{\text{th}}$ of the line pressure. Forces acting on the top and bottom of the piston are then in balance and the piston starts to move away from the poppet seat (centre piece).

The poppet valve opens immediately and the reservoir pressure is applied over the entire piston area, resulting in very rapid acceleration of the piston up to the maximum speed. The energy released by the swiftly expanding reservoir air is converted into kinetic energy by the piston assembly and is used to deliver an impact blow.

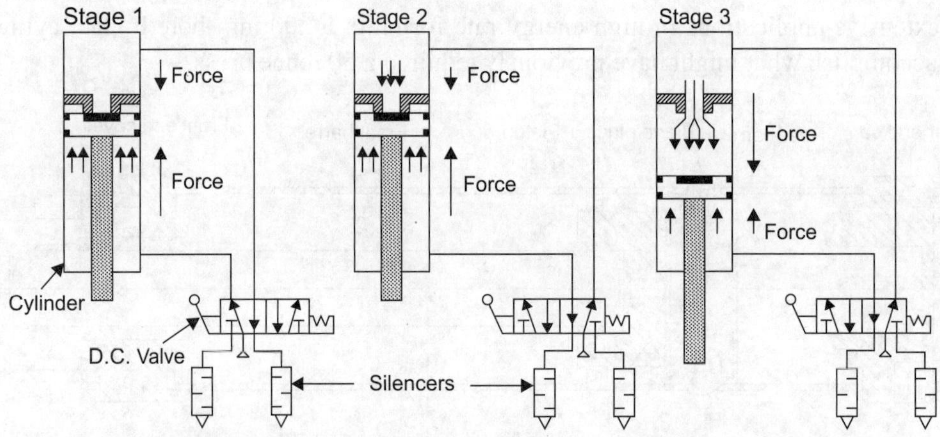

Figure 4.16 *Sequence of operation of an impact cylinder*

(E) Duplex cylinder (also known as opposite thrust or multi-position cylinder)

A special duty tandem arrangement is also used in a duplex cylinder, although in this case the (two) pistons are mounted on separate connecting rods, one operating the other. Each piston is thus capable of independent motion via its own separate air supply control. A duplex cylinder may be used for towing motion duties–e.g., clamping a workpiece in position by the motion of one piston, followed by a shearing, punching, pressing or similar operation powered by the other piston.

Opposite thrust or multi-position cylinders are similarly a combination assembly of at least two double acting cylinders with the pistons and rods in opposed arrangement for a four position cylinder as shown in Fig. 4.17 (b). The characteristic of opposed thrust cylinders is the availability of more than two fixed operating positions. Theoretically, it is possible for a multiple number of pistons and rods to be combined into the one cylinder assembly so as to obtain an actuator with six or eight positions. Three position cylinders as illustrated in Fig. 4.17(a) can be obtained as standard units. Designs with up to 12 operating positions are possible.

Application:

- Filling shelves from a conveyor
- Lever actuation
- Sorting (accept/reject/rework)

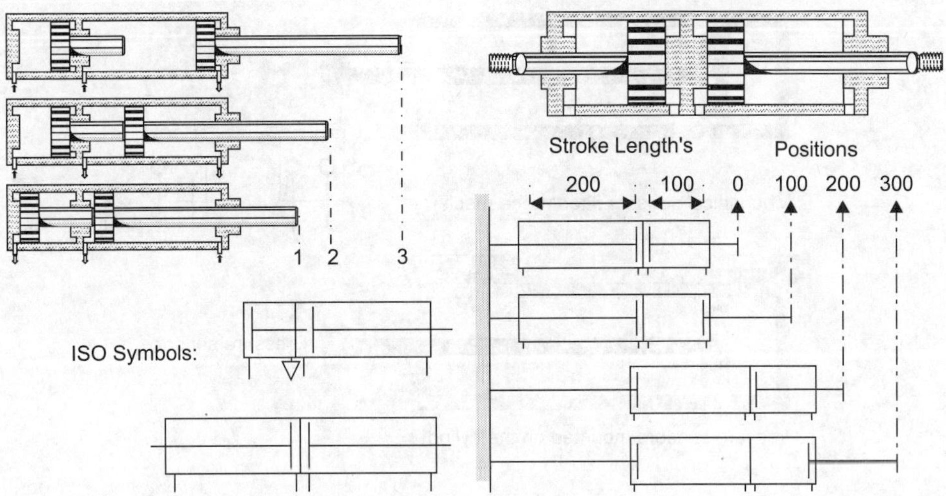

Figure 4.17 *Duplex cylinder (a) three position (b) four position*

(F) Position sensor for cylinders

Automation applications often require a signal to be generated when the cylinder piston reaches a particular position along its stroke, so that the control system can initiate the next phase of the operation. This position may be at either end of the stroke or at some intermediate point between the ends. A schematic diagram of cylinders with position sensors is shown in Fig. 4.18. There are several ways in which this can be done:

- The piston rod trips a micro-switch or pneumatic valve.
- Pressure threshold sensors respond to a drop in exhaust pressure when the piston stops moving.
- Magnetic sensors mounted directly on the cylinder barrel sense the magnetic field created by a permanent magnet incorporated into the piston, and trigger the reed switch.
- Hall effect sensors are triggered by a magnetic piston.
- Pneumatic reed valves are triggered by a magnetic piston.
- Miscellaneous mechanisms such as photoelectric, inductive and capacitive detectors can be used.

Magnetic trip switches are the traditional method for position sensing and are still widely used and reliable. Pressure threshold sensors respond less to actual position than to piston velocity, so they actually detect the end of the piston movement and are thus unreliable with slow moving pistons. Both these are being gradually superseded by the two types of magnetic sensors.

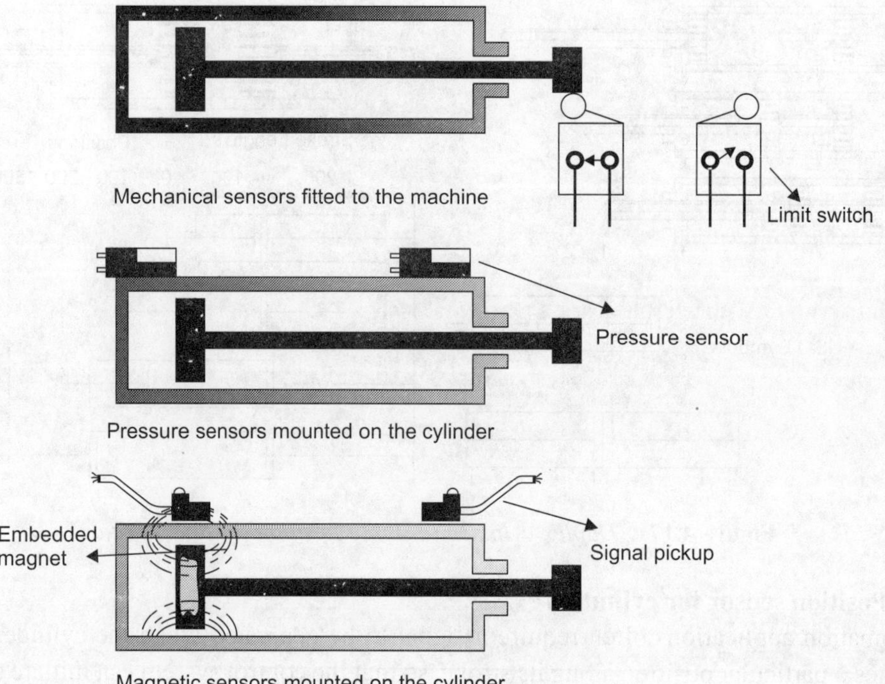

Figure 4.18 *Cylinder with sensors (a) Mechanical (b) Pressure (c) Magnetic*

The other types of sensors in use are electronic sensors with magnetic detection. They are directly mounted on a magnetic cylinder tube. A permanent magnet embedded in the piston creates a magnetic field. When the piston moves, the magnet actuates the electronic system of the sensor and provides the desired signal. These electronic position sensors work at 10–24 volts. The maximum current is around 150 mA. The leakage current is around 10 mA at 24 volts and the internal voltage drop is less than 0.5 volts for 100 mA. They are capable of working between −10°C to +60°C.

4.3 Standard Metric Cylinders

British Standards BS: 5785 1980 provides tables (Table 4.2) of preferred sizes for the cylinder bore and rod diameter of metric cylinders. Most cylinder manufacturers base their standard range of metric cylinders on these recommendations. Higher sizes (>100 mm) are used for special applications.

Table 4.2 *Standard cylinder sizes*

Piston diameter(mm)		40	50	63	80	100	125	140	160	180	200	220	250	280	320
Piston rod diameter(mm)	small	20	28	36	45	56	70	90	100	110	125	140	160	180	200
	large	28	36	45	56	70	90	100	110	125	140	160	180	200	220

A number of combinations have a piston rod to piston diameter ratio in the region of 0.7, which gives an annulus area of approximately one-half the full bore area. This area ratio is of use in regenerative circuits to achieve similar rates of speed and thrust on both the extension and retraction strokes.

4.3.1 Graphic symbols for cylinders

There are several symbol systems and conventions in use around the world for pneumatic cylinders. ISO 1219–1 is recognised by most standard bodies and commonly used. Table 4.3 lists the symbols for the most common cylinders in the industry.

Table 4.3 *Symbols for pneumatic cylinders*

Sl No	Symbol	Explanation
1.	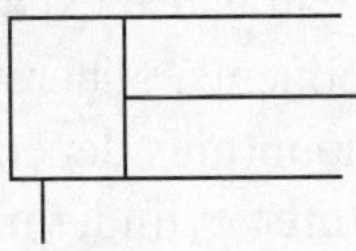	**Single acting cylinder with unspecified return**: Air pushes the piston in one direction, the piston return is unspecified. External dock or lever may be used for pushing.
2.	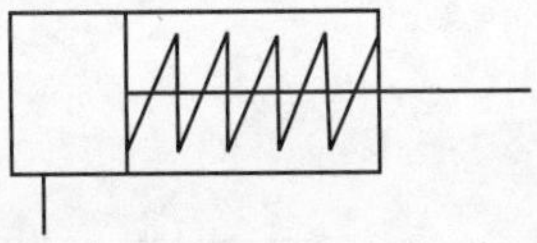	**Single acting cylinder with spring return**: Air pushes the piston in one direction and piston returns by spring on the rod side.
3.	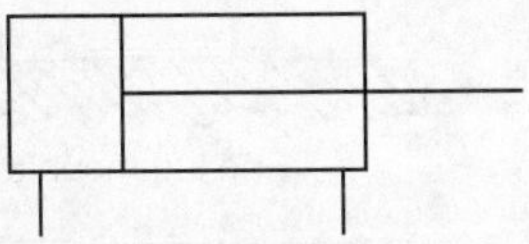	**Double acting cylinder with single piston rod**: Force exerted by compressed air moves the piston in both directions.
4.	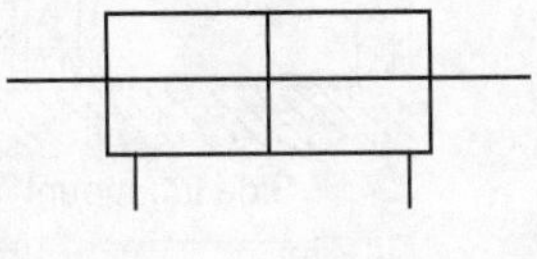	**Double acting cylinder with double piston rod**: It has piston rods extending from both ends of the cylinder and produces equal force and speed on both sides of the cylinder.
5.	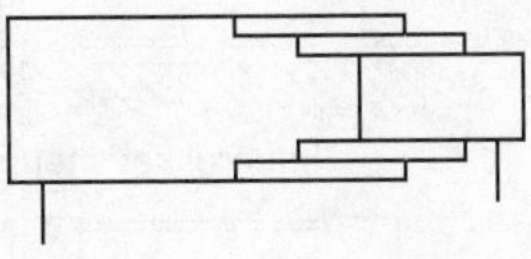	**Telescopic cylinder – double acting**: This is used where space is a constraint. Used in long stroke applications like pneumatic cranes, dump trucks, fork lift trucks, dipper wagon.
7.	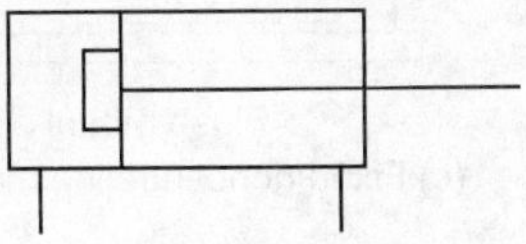	**Double acting cylinder**: It has a fixed cushion on one side, and cushioning is used in the end position to prevent sudden impact which may damage parts.

contd...

Table 4.3 *...contd*

8.		**Double acting cylinder–variable cushion on one side**: It has a fixed cushion on one side. Cushioning is variable in one direction by adjusting the orifice opening.
9.		**Double acting cylinder–variable cushion on both sides**: It has a fixed cushion on one side. Cushioning is variable in both directions.

4.4 Cylinder Mountings

The way in which the cylinder is mounted affects the service life, maintenance frequency and the success of the entire installation. Poor mounting design can cause excessive side loads and stresses which bring about early failure of vital components. There are three main categories of cylinder mountings. The selection of these mountings depends on the application and the machine configuration. Figure 4.19 shows all three cylinder mountings.

1. Fixed centreline mountings
2. Pivoted centreline mountings
3. Fixed non-centreline mountings

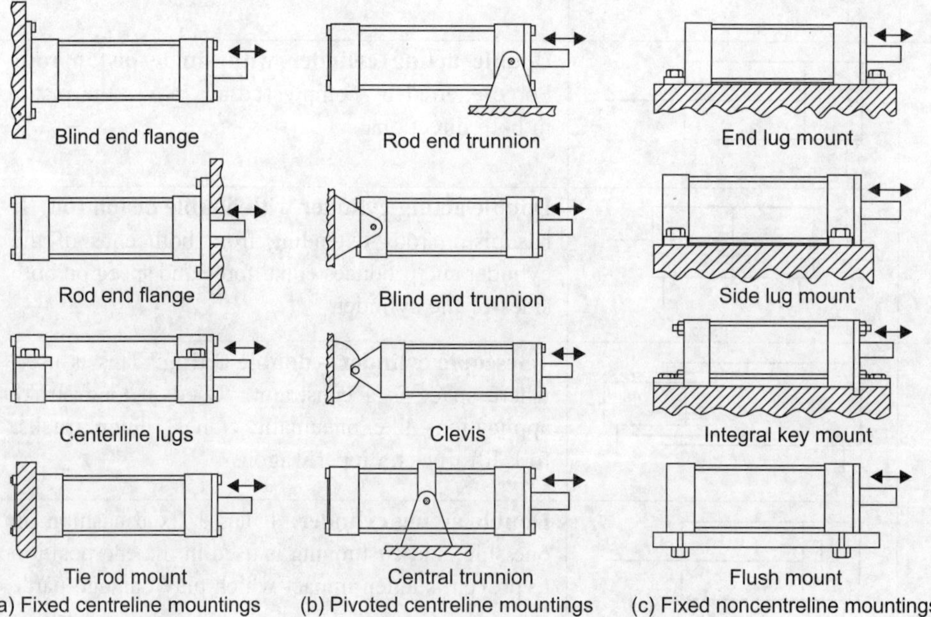

Figure 4.19 *Types of mounting: (a) fixed centreline, (b) pivoted centreline, (c) fixed non-centreline*

4.5 Cylinder Force, Velocity and Power

The output force (F) and the piston velocity (V) of double acting cylinders are not the same for extension and retraction strokes.

During the extension stroke shown in Fig. 4.20(a), the fluid pressure acts on the entire circular piston area A_p. During the retraction stroke, the fluid enters the rod end and its pressure acts on the smaller annular area between rod and cylinder bore ($A_p - A_r$) represented by the shaded area in Fig. 4.20 (b), where A_r is the area of the piston rod. This difference in the cross-sectional area changes the velocity of the piston. Since A_p is greater than ($A_p - A_r$), the retraction velocity (V_{ret}) is greater than the extension velocity (V_e) for the same flow rate.

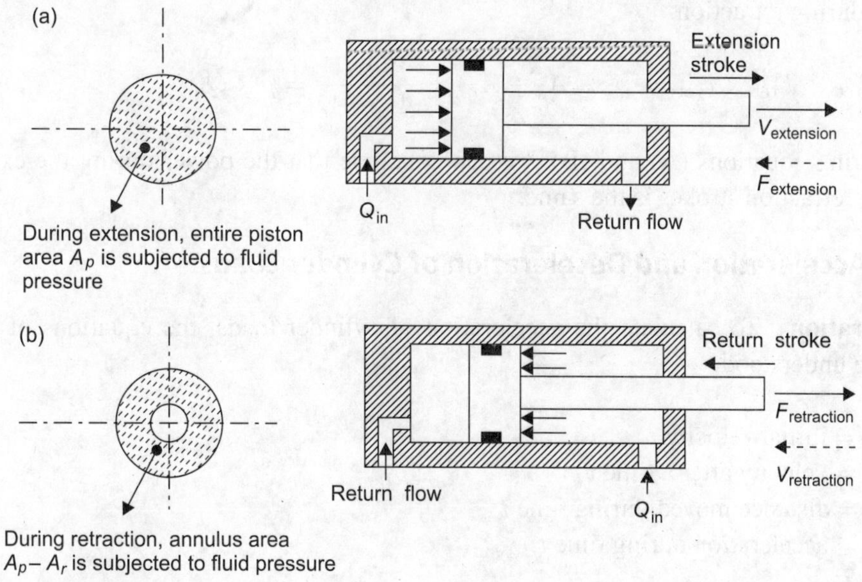

Figure 4.20 *Effective areas during extension and retraction strokes*

Force and velocity during the extension stroke

$$v_{ext} = \frac{Q_{in}}{A_p} \tag{1}$$

$$F_{ext} = p \times A_p \tag{2}$$

Force and velocity during the retraction stroke

$$v_{ret} = \frac{Q_{in}}{A_p - A_r} \tag{3}$$

$$F_{ret} = p \times (A_p - A_r) \tag{4}$$

Power developed by a pneumatic cylinder (during both extension and retraction)

$$\text{Power} = \text{Force} \times \text{velocity} = F \times v \tag{5}$$

In metric units, the kW power developed for either the extension or the retraction stroke is

$$\text{Power (kW)} = v_p \text{ m/s} \times F\text{ (kN)} = Q_{\text{in}} \text{ m}^3\text{/s} \times p(\text{kPa}) \tag{6}$$

Power during extension:

$$P_{\text{ext}} = F_{\text{ext}} \times v_{\text{ext}} = p \times A_p X \frac{Q_{\text{in}}}{A_p} = p \times Q_{\text{in}} \tag{7}$$

Power during retraction:

$$P_{\text{ret}} = F_{\text{ret}} \times v_{\text{ret}} = p \times (A_p - A_r) \times \frac{Q_{\text{in}}}{A_p - A_r} = p \times Q_{\text{in}} \tag{8}$$

Comparing equations (A) and (B), we can conclude that the power during the extension and the retraction strokes is the same.

4.6 Acceleration and Deceleration of Cylinder Loads

Acceleration To calculate the acceleration of cylinder loads, the equations of motion must be understood.
Let:

u = initial velocity
v = velocity after a time t
s = distance moved during time t
a = acceleration during time t.

The standard equations of motion are:

$$v = u + at \tag{9}$$

$$v^2 = u^2 + 2as \tag{10}$$

$$s = ut + \frac{1}{2}at^2 \tag{11}$$

and $s = 1/2(u + v)t$.

The force F to accelerate a weight W horizontally with an acceleration a is given by

$$\text{Force} = \text{mass} \times \text{acceleration}$$

$$F = (W/g)a$$

where g is the acceleration due to gravity (9.81 m/s^2). The force P required to overcome friction is given by $P = \mu W$, where μ is the coefficient of friction.

Dynamic cylinder thrust In dynamic applications the load inertia, seal friction and load friction must be accounted for in calculating the dynamic thrust.

As a first approximation, the dynamic thrust can be taken as 0.9 times the static thrust. (**Note:** This is only an approximation and can vary considerably, depending on the load conditions and the associated circuitry.)

Cylinder seal friction varies with the seal and cylinder design. The data on the pressure required to overcome seal friction is not readily available from most cylinder manufacturers. Therefore the seal friction breakout pressure can be assumed to be 5 bar for calculation purposes. Seal friction goes down when the piston starts to move. The pressure required to overcome seal friction decreases as the cylinder bore size increases, and varies according to the seal design.

Example 4.1 A pneumatic cylinder is required to move a 1000 N load 150 mm in 0.5s. What is the output power?

Solution Velocity:

$$V = d/t = 0.15/0.5 = 0.3 \text{ m/s}$$

Power:

$$P = F \times V = 10^3 \times 0.3 = 300 \text{ W} = 0.3 \text{ kW}$$

Example 4.2 A pneumatic cylinder is required to extend at a minimum speed of 0.75 m/s in a system with a flow rate of 50 LPM. What size of cylinder is required?

Solution

$$Q = 50 \text{ LPM} = \frac{50}{1000}\text{m}^3/\text{min} = \frac{50}{1000 \times 60} = 8.33{\times}10^{-4}\text{m}^3/\text{s}$$

$$Q = A_p \times v$$

$$8.33{\times}10^{-4} = \frac{\pi}{4} d_p^2 \times 0.75$$

Solving, we get:

$$d_p = 0.376 \text{ m} = 37.6 \text{ mm.}$$

Example 4.3 A pneumatic cylinder with 8 cm diameter has a 4 cm diameter rod. If the cylinder receives a flow of 100 LPM at 6 bar, find (a) the extension and retraction speeds and (b) the extension and retraction load carrying capacities.

Solution

$Q_{\text{in}} = 100\text{ LPM} = 100/(1000 \times 60) = 1/600 \ (\text{m}^3/\text{s})$

$d_p =$ diameter of cylinder $= 8$ cm $= 8 \times 10^{-2}$ m

$d_r =$ diameter of piston rod $= 4$ cm $= 4 \times 10^{-2}$ m

$p = 6$ bar $= 6 \times 10^5$ N/m^2 or Pa

$$\text{(a)}\ v_{\text{ext}} = \frac{Q_{\text{in}}}{A_p} = \frac{\frac{1}{600}}{\frac{\pi d_p^2}{4}} = 0.3315 \text{ m/s}$$

$$v_{\text{ret}} = \frac{Q_{\text{in}}}{(A_p - A_r)} = \frac{1/600}{\pi\left(d_p^2 - d_r^2\right)/4} = 0.442 \text{ m/s}$$

$$\text{(b)}\ F_{\text{ext}} = p \times a_p = 6\times10^5 \frac{\pi\left(8\times10^{-2}\right)^2}{4} = 2513.3 \text{ N} = 2.51 \text{ kN}$$

$$F_{\text{ret}} = p \times (A_p - A_r) = \frac{5\times10^5\times\pi\left[\left(8\times10^{-2}\right)^2 - \left(4\times10^{-2}\right)^2\right]}{4}$$

$$= 1885 \text{ N} = 1.89 \text{ kN}$$

4.7 Linear Motion Using Pneumatic Cylinders

A cylinder must produce a force equal to the load that the cylinder is required to overcome. It may be placed with its axis vertical, horizontal or inclined depending on the load to be actuated.

4.7.1 Vertical cylinder

In vertical cylinders, the load to be actuated is in the vertical direction. The cylinder load F is equal to the weight W of the object, acting in the vertical direction. The schematic diagram of a vertical cylinder is shown in Fig. 4.21.

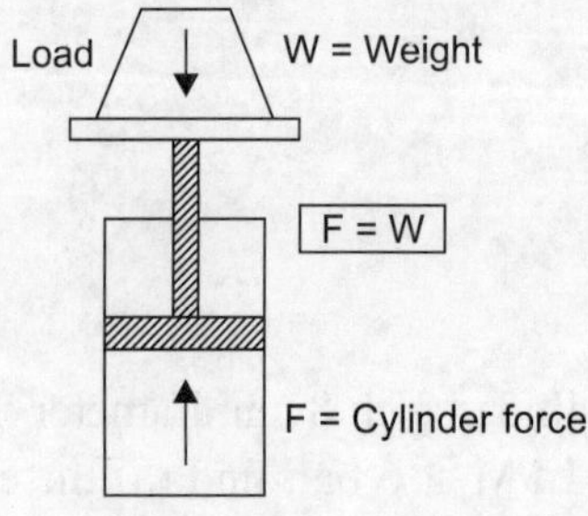

Figure 4.21 *Vertical cylinder with load*

4.7.2 Horizontal cylinder

In a horizontal cylinder, the load is theoretically zero since no component of the object's weight is acting along the axis of the cylinder. However when the object slides across the horizontal surface, the cylinder must overcome the frictional force created between the object and the horizontal surface. A schematic diagram of a horizontal cylinder is shown in Fig. 4.22.

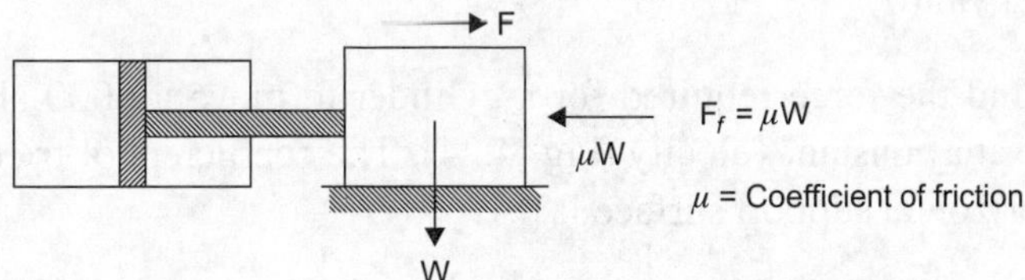

Figure 4.22 *Horizontal cylinder with load*

4.7.3 Inclined cylinder

In an inclined cylinder, the load is the sum of the component of the object's weight acting along the axis of the cylinder and any frictional force if the object is sliding along an inclined surface. A schematic diagram of an inclined cylinder is shown in Fig. 4.23.

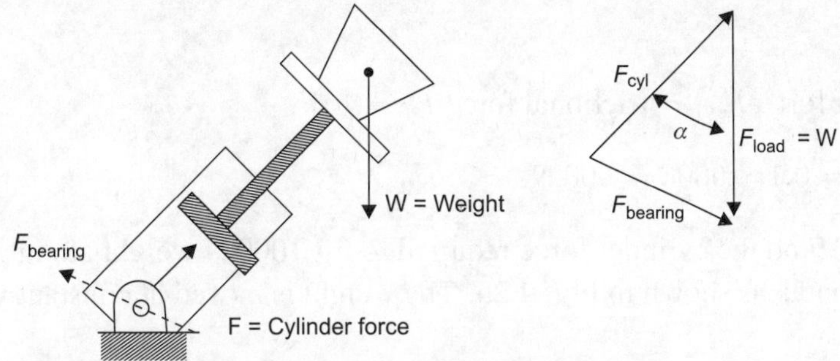

Figure 4.23 *Inclined cylinder with load*

The load that an inclined cylinder must overcome is less than the weight of the object to be moved, if the object is not sliding on an inclined surface. The cylinder loads calculated in the figure above are based on moving an object at a constant velocity. But when the object has to be accelerated from zero velocity to a steady state velocity, then an additional force called inertia force must be added to the weight component along with any frictional force involved.

Let:

$F_{\text{load}} = W$ = weight or load acting vertically downwards

F_{cyl} = load acting on cylinder

F_{bear} = force on the bearings

α = angle between load W and the axis of the cylinder.

Then:

$$F_{cyl} = F_{load} \cos(\alpha).$$

$$F_{bear} = F_{load} \sin(\alpha).$$

Example 4.5 Find the force required for a cylinder to move a 6000 N weight along a horizontal surface at a constant velocity (Fig. 4.24). The coefficient of friction between the weight and the horizontal support surface is 0.1.

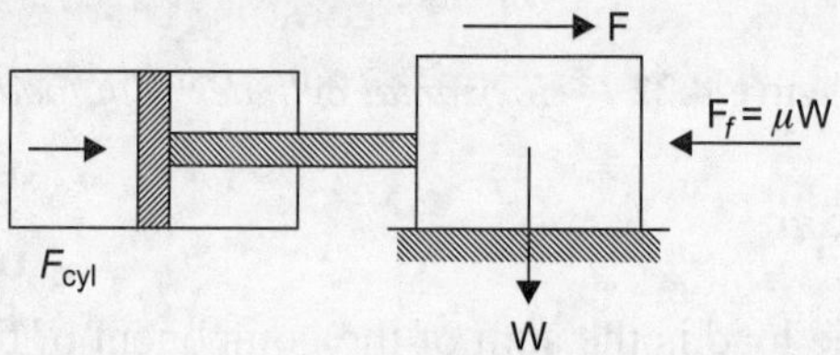

Figure 4.24

Solution

The cylinder force F_{cyl} = frictional force $F_f = \mu W$.

Hence $F_{cyl} = 0.1 \times 6000 = 600$ N.

Example 4.6 Find the cylinder force required to lift 1000 N weight along an incline 30° from the horizontal, as shown in Fig. 4.25. The weight is moved at constant velocity.

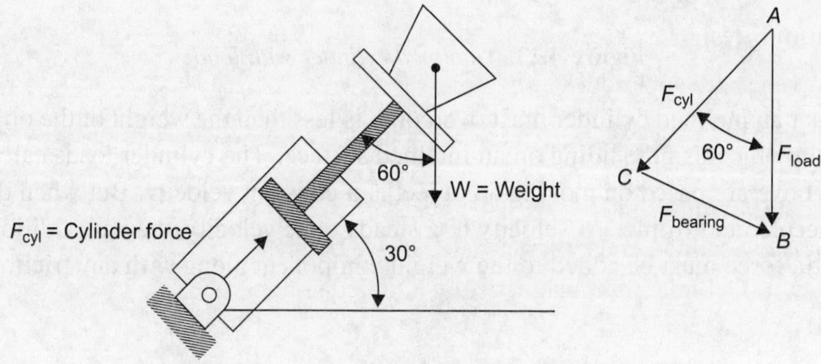

Figure 4.25

Solution Let:

$F_{load} = W =$ weight or load acting vertically downwards

$F_{cyl} =$ load acting on cylinder

$F_{bear} =$ force on the bearings

$\alpha =$ angle between load W and the axis of the cylinder.

Then, $F_{cyl} = F_{load}\cos\alpha$

$F_{bear} = F_{load}\sin\alpha.$

Hence $F_{cyl} = 1000 \times \cos 60 = 500$ N .

Example 4.7 A 1000 N weight is to be lifted upward in a vertical direction for the system shown in Fig. 4.26. Find the cylinder force required to:

(a) Move the weight at a constant velocity of 0.75 m/s
(b) Accelerate the weight from zero to 0.75 m/s in 0.5 s

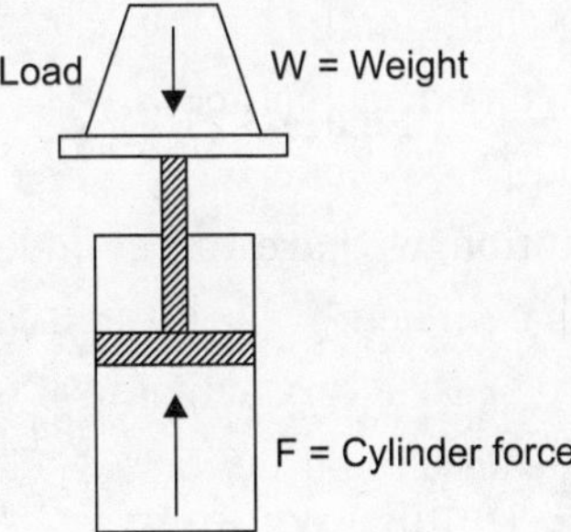

Figure 4.26

(a) For a constant velocity, the cylinder force required:

$F_{cyl} = W = 1000$ N.

(b) Acceleration:

$$a = \frac{0.75 - 0}{0.5} = 1.5 \text{ m/s}^2$$

Force required to accelerate the weight is:

$$F_a = m \times a = \frac{1000 \times 3.5}{9.81} = 152.9 \text{ N}.$$

Cylinder force F_{cyl} required equals the sum of the weight and acceleration force:

$F_{cyl} = 1000 + 152.9 = 1152.9$ N.

Example 4.8 A 1000 N weight is to be lowered by a vertical cylinder as shown in the Fig. 4.25. The cylinder has a 75 mm diameter piston and a 50 mm diameter rod. The weight is to decelerate from 100 m/min to a stop in 0.5 seconds. Determine the required pressure at the rod end of the cylinder during the deceleration motion.

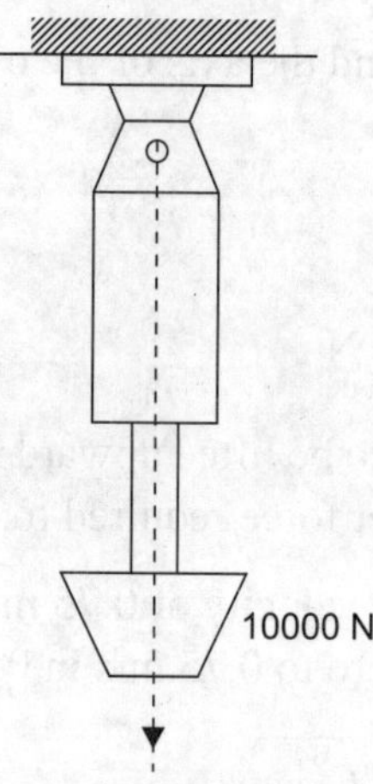

Figure 4.27

According to Newton's law of motion, we have:

$$\sum F = ma$$

where $a = \dfrac{100\ (\text{m/min}) \times \dfrac{1}{60}\ (\text{min/s})}{0.5\ \text{s}} = \dfrac{1.67(\text{m/s})}{0.5\ \text{s}} = 3.34\ \text{m/s}^2.$

Summing the forces on the 1000 N weight:
Pneumatic force load = mass × acceleration

$$p\,(A_p - A_r) - 1000\ \text{N} = \frac{1000\ \text{N}}{9.81\ (\text{m/s}^2)} \times 3.34\ \text{m/s}^2$$

$$p \times \frac{\pi}{4}\left(0.075^2 - 0.050^2\right)\text{m}^2 - 1000\ \text{N} = 340.8\ \text{N}$$

Solving, we get:

$$p = 546.3 \times 10^3\,\frac{\text{N}}{\text{m}^2} = 546.3 \times 10^3\text{Pa} = 546.3\ \text{kPa}.$$

Example 4.9 A 2700 N weight is being pushed up an inclined surface at a constant speed by a cylinder as shown in Fig. 4.28. The coefficient of friction between the weight and inclined surface equals 0.15. For a pressure of 689.4 kPa:

(a) Determine the required cylinder piston diameter considering the friction force.
(b) Determine the required cylinder piston diameter, if the weight is to be accelerated from zero to a velocity of 1524 mm/s.

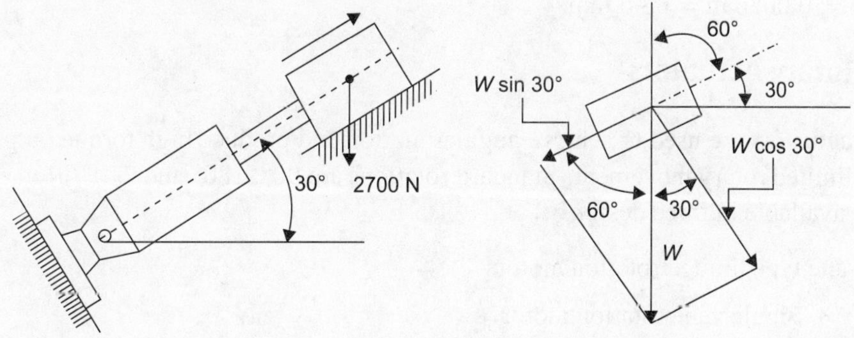

Figure 4.28

Part (a) The component of the weight W acting along the axis of the cylinder is $W\sin 30°$. The component of weight W acting along normal to the incline surface is $W\cos 30°$. The frictional force equals the coefficient of friction times the force normal to the sliding surface. Therefore the frictional force F_f acting along the axis of the cylinder is:

$$F_f = \mu \times W\cos 30 = 0.15 \times 2700 \times \cos 30 = 350.7 \text{ N}.$$

The cylinder force (F_{cylinder}) $= F_f + W\sin 30 = 350.7 + 2700 \times \sin 30 = 1700.7\,\text{N}$

$$F_{\text{cylinder}} = pA_p$$

$$F_{\text{cylinder}} = 689400 \times \frac{\pi}{4}D^2 = 1700.7\,\text{N}$$

Solving, we get $D = 0.05605\text{m} = 56$ mm.

Part (b) According to Newton's law of motion, we have:

$$\sum F = ma$$

where $a = \dfrac{1524\ (\text{mm/s})}{0.5\ \text{s}} = 3048\ \text{mm/s}^2 = 3.048\ \text{m/s}^2.$

Using the values determined in part (a) and summing the forces on the 2700 N weight:

$$pA_p - 1700.7 = \frac{2700 \times 3.048}{9.81}$$

$$689400 \times \frac{\pi}{4}D^2 - 17007 = \frac{2700 \times 3.048}{9.81}$$

$$541453.5\,D^2 - 1700.7 = 838.9.$$

Solving, we get:

$$D = 0.0685 \text{ m} = 68.5 \text{ mm}.$$

4.8 Rotary Actuators

Rotary actuators are used to achieve angular motion and produce high torque output while having limited rotary movement. Standard rotations are 90°, 180° and 270°. Rotary actuators are available in three designs:

(i) Vane type limited rotation motors
- Single vane rotation motor
- Double vane rotation motor

(ii) Rack and pinion type rotary actuator

(iii) Helix spine rotary actuator

4.8.1 Vane type actuators

Vane type actuators are essentially mechanically connected linear actuators that convert a linear force to an output torque and rotational movement. These devices are capable of providing an output motion of more than one revolution but not continuous rotation. There are two types of vane type actuators. They are single vane limited rotary actuator and double vane limited rotary actuator.

(i) Single vane limited rotation actuators The single vane actuator consists of a cylindrical housing, through which passes a central shaft with the vane rigidly attached to it. The housing has shoe or a stopper fixed to the internal diameter of the housing as shown in the Fig. 4.29(a), which divides the interior space into two chambers. Pressurised air enters through port A and rotates the vane in the clockwise direction, and air in the other chamber moves out of port B. Similarly, when air pressure is applied to port B, the vane rotates in an anti-clockwise direction and air in the other chamber moves out of port A. Design geometry normally limits the rotary movement of a single vane actuator to a maximum of about 280 degrees.

(ii) Double vane limited rotation actuators The design can be modified to have two vanes fixed to the output shaft 180 degrees apart. The two fixed stoppers in the casing provide two separate operating halves, each with two chambers as shown in Fig. 4.29(b). This gives twice the maximum torque output of a single vane device, for the same supply pressure. The maximum angle of rotation is limited to about 100 degrees because of the second stopper.

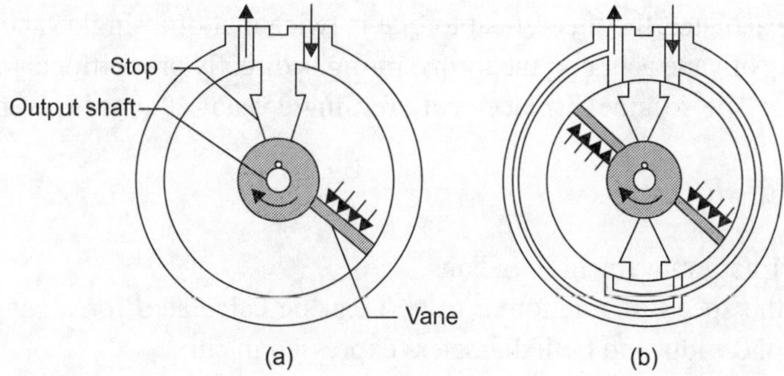

Figure 4.29 *Vane actuators (a) single vane (b) double vane*

(iii) Analysis of the torque capacity of a single vane limited rotation actuator

R_R = Outer radius of the output shaft (m)

R_V = Outer radius of the vane (m)

L = Width of the vane (m)

p = Hydraulic pressure (Pa)

F = Hydraulic force acting on the vane (N)

A = Surface area of vane in contact with oil (m^2)

T = Torque capacity (Nm)

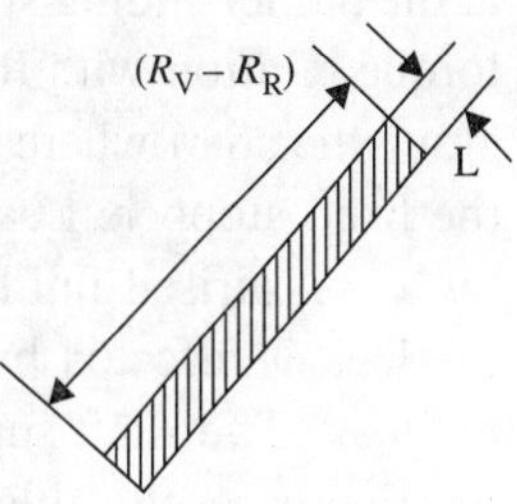

The force on the vane equals pressure times the vane surface area:

$$F = pA = p(R_V - R_R)L$$

The torque equals vane force times the mean radius of the vane:

$$T = p(R_V - R_R)L\frac{R_V + R_R}{2}$$

Rearranging, we have :

$$T = p(R_V^2 - R_R^2)L. \tag{1}$$

A second equation for torque can be developed based on the following relationship for volumetric displacement V_D:

$$V_D = \pi(R_V^2 - R_R^2)L. \tag{2}$$

Combining equations 1 and 2:

$$T = \frac{p \times V_D}{2\pi}.$$

For a two vane actuator, the theoretical torque is twice that of a single vane. For the given design and style of vane actuator, the torque rating is directly proportional to the maximum supply pressure. The volume displacement of a single vane actuator is given by:

$$Q = \phi \left(R_{\mathrm{V}}^2 - R_{\mathrm{R}}^2\right) \times \frac{L}{8}$$

where ϕ is the rotary movement in radians.

From this, the speed of rotation ω (rad/s) can be calculated for a supplied flow rate Q(mL/s), with the width and both diameters expressed in cm:

$$\omega = \frac{8Q}{L\left(R_{\mathrm{V}}^2 - \mathrm{R}_{\mathrm{R}}^2\right)}.$$

Although the total volume in a two vane actuator is the same for the same body size, its speed is halved for the same flow rate because two chambers must be filled. If necessary, speed can be controlled by throttling, or by fitting orifice plugs to control inflow or outflow. The most critical feature of the vane actuator is the sealing required along the length of the vane, between the end faces and the internal bore of the casing. Since this is a moving seal, there is usually a compromise between good sealing with low leakage and the resulting higher friction. The sealing surface with the bore is also at the largest radius, and operating at the higher sliding speed. Seals are usually pressure activated, and may have a non-linear torque relation with the pressure. In addition to the vane seal, further seals are required along the shaft where it emerges from the body of the actuator, and between the shaft and the fixed stopper. Leakage may be a problem where the actuator is locked with the load by a pressurised fluid column, as in the case of a linear cylinder. The effectiveness of the lock is affected by internal leakage, unless extra fluid is supplied to compensate for the loss. The shaft support may be plain bearing, roller bearing or the needle roller type. Friction is responsible for the difference between the starting torque and dynamic torque. This starting torque can be estimated as 80% of the dynamic torque. The overall efficiency of a vane actuator is between 70% and 90%.

4.8.2 Rack and pinion rotary actuator

A schematic diagram of a rack and pinion type rotary actuator is shown in Fig. 4.30. These are special duty rotary actuators with a high torque and a small footprint. They have double pistons that transmit turning motion to the output shaft. The toothed piston rods act on the output shaft in a rack-and-pinion arrangement to produce an even turning moment throughout the rotation. The drive shaft is supported by self-lubricating bushings. The output shaft has keyways with a stub at the other end for positioning and manual operation. The rotational movement of a rotary actuator is limited by stop lugs to protect against the effects of excessive load inertia. Compressed air is fed into the piston chambers via a connection plate and galleries drilled into the central part and the end covers. The cylinder block is made of natural anodised aluminum while the end covers are made of black anodized aluminum. Rotary actuators are available in five sizes with ranges of 20 Nm to 200 Nm of turning motion. All sizes have a standard turning angle of either 90° or 180°.

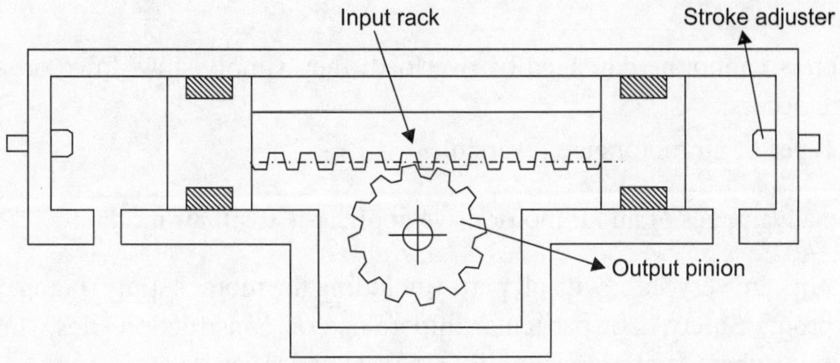

Figure 4.30 *Rack and pinion type rotary actuator (single rack)*

4.8.3 Helix spine rotary actuator

Figure 4.31 shows a simplified cross-sectional view of a spiral-shaft rotary actuator

It consists of a hollow piston with internal spiral grooves. Only rotational movement of the spiral shaft is possible. If fluid is supplied to the CW port, the direction of rotation of shaft is clockwise. If fluid is supplied to the CCW port, the direction of the rotation of shaft is counter clockwise.

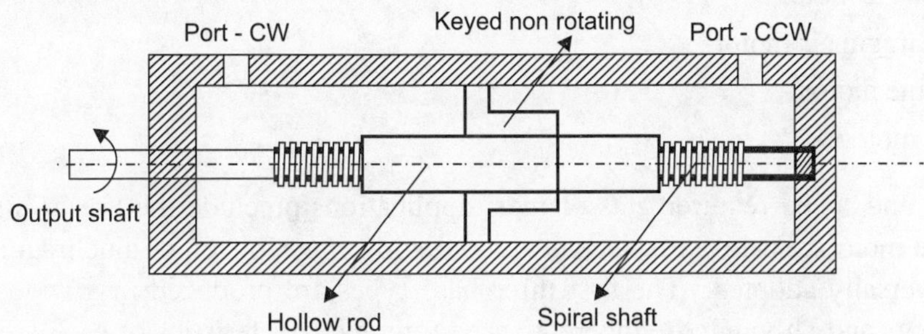

Figure 4.31 *Helix spine rotary actuator*

4.9 Air Motors

Air motors are high power, high weight, compact and dependable units requiring little maintenance over long periods of use.

Advantages of air motors

- Completely safe in conditions where electric motors might create a risk of fire or explosion.
- Can be readily be adjusted for speed by varying the pressure of the air supply via a regulator.
- Air motors cannot be damaged by overload; they simply slow down or stop before damage occurs.
- Many types of air motors are reversible

The principle advantages of an air motor drive applicable to all air motors:

1. Air motors are very safe, with all parts (including the motor supply) being completely flame proof. Safety is of particular importance on construction sites, ship building yards, etc., where temporary supplies are used under severe conditions.
2. Motors can be overloaded and stalled without damage.
3. Simple design and construction, easily maintained and overhauled.
4. Operates satisfactorily in adverse conditions, such as heat, moisture, dirt, etc.
5. Emergency supplies can be easily stored.

Types of air motors

- Vane air motor
- Piston air motor
- Diaphragm air motor
- Turbine motor
- Gear motor

The torque and speed requirements of most applications preclude the use of high speed turbine type motors; positive displacement motors like vane, piston or diaphragm type are almost universally adopted. The first three base types are produced in a wide range of power outputs and although the inherent speed/torque characteristics of each are particularly suited to certain applications, in practice there is a considerable overlap due to the extensive use of gearing in the vane and piston types. The general speed/torque characteristics of the first three types is shown in Table 4.4.

Table 4.4 *Speed and torque characteristics of different air motors*

Motor type	Torque	Speed
Vane	Low	High
Piston	Medium	Medium
Diaphragm	High	Low

Vane type motors are normally preferred for applications requiring low or medium power, particularly as they can be made simple and compact in design. They are used extensively in portable tools, multiple tool assemblies for mass production lines, conveyor belts, hoists, pumps, tractors, etc.

Piston type motors are less subject to changes in speed with varying loads than vane motors, and are usually preferred for heavier loads or where powerful low speed torques are required. They may be designed as single or multi-cylinder layouts, and can be readily produced in both reversible and non-reversible designs. The piston type motor is less compact and somewhat heavier than the vane type for a similar duty. It may be used in similar applications as the vane type, but is also suitable for additional duties where the characteristically low torque of the vane motor is a distinct disadvantage.

Diaphragm motors are quite distinct and are basically oscillators rather than motors, deriving a rotary output via a ratchet engaging with a toothed wheel. They are particularly suitable for applications demanding extreme accuracy of control at high and low torque, such as valve operation.

Due to the inherent flexibility of the motors in terms of their motor/gear ratio combinations, one type can often work just as satisfactorily as another. Therefore, both technical and economic factors have to be taken into consideration to select the most suitable type of motor for a particular application.

4.9.1 Vane motor

The vane type motor is produced in a variety of power outputs, varying from fractional to approximately 15 kW. The fractional kilowatt units are used extensively in portable tools such as drills, grinders, chippers etc., and also in multiple tool mass production assemblies like nut and stud runners. The large kilowatt motors are frequently used to power pumps, valves, hoists, winches, soot blowers etc.

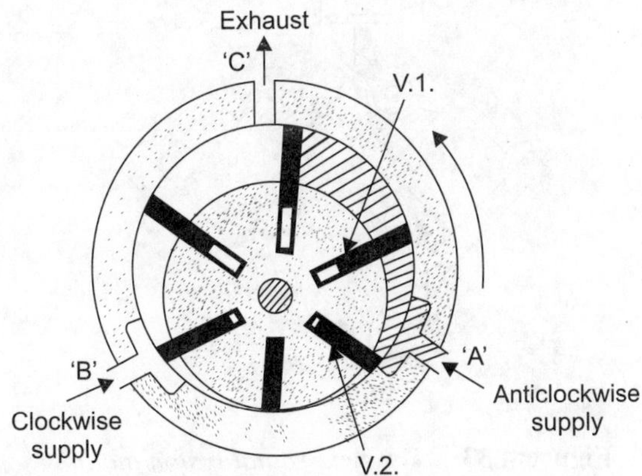

Figure 4.32 *Vane motor*

Figure 4.32 illustrates the operational principle of a reversible internal vane motor. A rotor shaft carrying vanes is mounted eccentrically in the power chamber with the vanes arranged to slide radially to seal against the power chamber wall. The vanes are usually mechanically or pneumatically sprung against the casting to provide sealing force for starting; this force is increased by centrifugal action when the motor speed rises. Air is supplied to the rotor chamber via either port A or port B, depending on the direction required.

4.9.2 Piston motor

Many applications require a motor that is capable of large power outputs combined with good torque characteristics at low speeds. The piston type motor possesses these characteristics and is manufactured in a wide range, from fractional to approximately 20 kW. The smaller types of motors sometimes use gearing to increase the torque. The gearing is usually an integral part of the motor, and the complete assembly forms a very compact unit.

The two most common cylinders are:

(i) Vee type radial piston motor
(ii) Radial piston motor

Vee type radial piston motor: The vee type units have four cylinders arranged in two banks at 90° to each other. Drop forged two-throw crank shafts are normally used, with two connecting rods on each crank pin. Rotary or oscillating distribution valves are located between the cylinders and operated either by gearing or by eccentric cams. The valve timing can be controlled to provide equal power in both directions, or alternatively, may be biased to provide increased power in one direction with a corresponding decrease in the opposite direction. Figure 4.33 shows a typical arrangement of a vee motor.

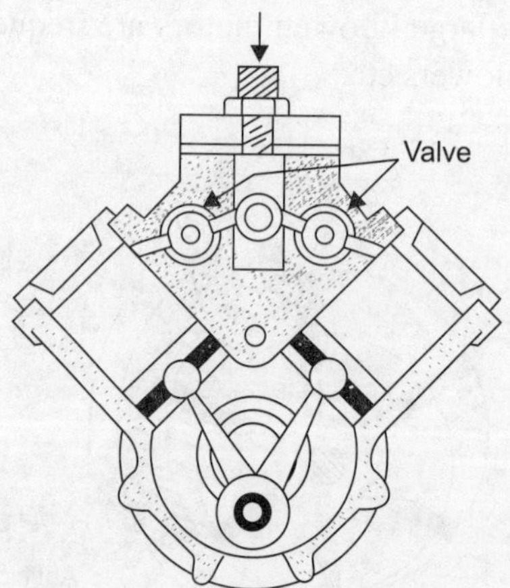

Figure 4.33 *Vee type radial piston motor*

Radial piston motor: The radial design is frequently adopted on the large kilowatt motors and three, four, five, or six cylinder models are produced. A schematic diagram of a piston type radial motor is shown in Fig. 4.34. All connecting rods operate on a common crank pin as found in a standard radial motor practice. The power and exhaust strokes are controlled by rotary distributing valves and a two-stage exhaust is frequently adopted. The inlet valve opens at top dead centre and remains open for approximately 160 degrees. The inlet valve then closes and the large primary exhaust port opens quickly to release the high pressure air into the atmosphere. At approximately 200 degrees, the primary exhaust valve closes and a secondary exhaust port opens to evacuate the remaining air. A reversal of direction is obtained by interchanging the inlet and secondary exhaust directions.

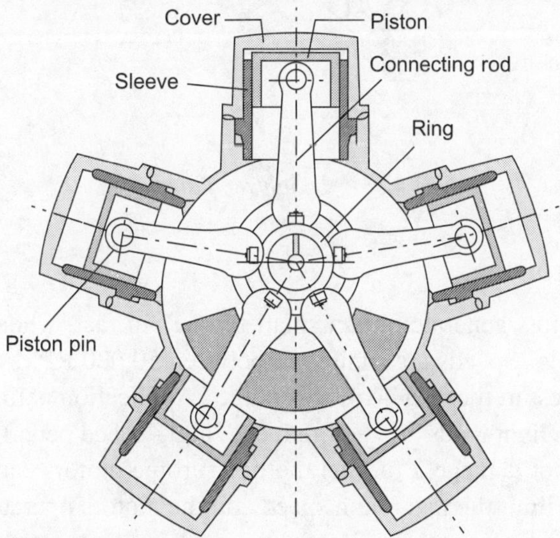

Figure 4.34 *Radial type radial piston motor*

Piston type motors are suitable for control applications due to their robust and variable speed and torque characteristics.

4.9.3 Diaphragm air motor

The schematic diagram of a diaphragm air motor is shown in Fig. 4.35. Diaphragm air motors are used when high torque at very low speeds is required. This motor operates directly on the valve shaft via a ratchet and foot wheel mechanism. The motor contains a diaphragm to produce the necessary thrust, and a snap-over valve mechanism operated by the motor shaft alternately admits air to the power chamber and then vents it. This produces continuous oscillations of the motor shaft and hence rotations of the valve spindle.

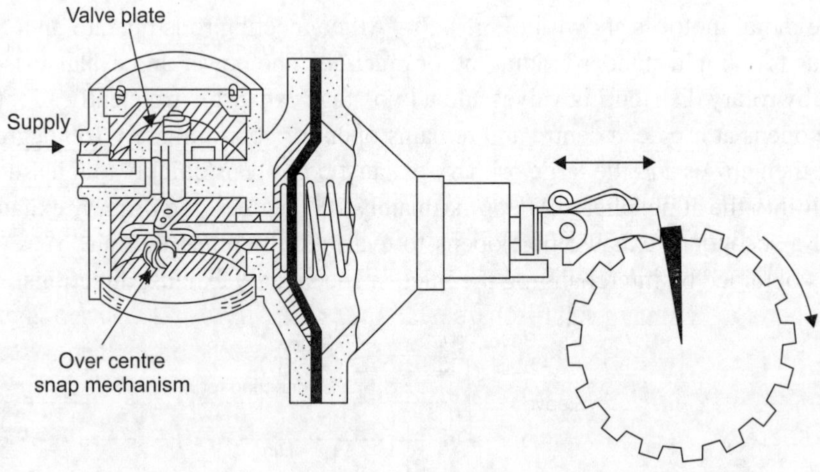

Figure 4.35 *Diaphragm air motor*

4.9.4 Turbine motor

These high speed motors generate only a small torque. In fact, when stalled, the torque is practically negligible. Speeds are limited to between 50,000 RPM to 80,000 RPM and the torque is the highest in the middle of this range. Application is limited to where high speeds combined with light loads are required, e.g., high- speed pencil grinders and dental drills. Stalling does not damage a turbine motor. Turbine motors can be fitted with mechanical governors to limit the maximum speed and the motors operate without sliding or rubbing (except for the bearings). Bearing friction is therefore the only limiting factor for speed. Only the bearings require lubrication, otherwise turbines can process dry air with no lubrication.

4.9.5 Gear motor

These motors are categorised as dynamic motors. The mating gears are of three types: straight, helical or double helical. These are pure displacement motors, i.e., they don't use the expansion properties of compressed air and have low volumetric efficiency. Gear motors are used in heavy industrial applications requiring power capacities up to or higher than 30 kW.

Advantages of gear motors

1. Very high speeds are possible.
2. The motor is oil free.
3. There is no wear.
4. They are suitable for continuous operation.

Summary

1. Pneumatic power is converted into straight line, reciprocating and rotary motion by pneumatic cylinders and pneumatic motors.
2. In single acting cylinders, compressed air is applied only to one side and they can produce work in only one direction. Air is used for movement only in one direction; either a built-in spring or an external force moves the piston in the other direction. These cylinders are used in clamping, ejecting, pressing, lifting, feeding, etc.
3. In diaphragm cylinders, a built-in diaphragm (made of rubber, plastic or even metal) replaces the function of the piston. The piston rod is attached to the centre of the diaphragm. There is no sliding seal, and the only friction is due to the stretching of the material. The construction and working of a rolling diaphragm cylinder is similar to a diaphragm cylinder.
4. The force exerted by compressed air moves the piston in a double acting cylinder in two directions. A force is applied on both advance and return movements. Double acting cylinders are most suitable in applications where the piston is required to perform a work function on both the advance and return movements. In principle, the stroke length of the cylinder is unlimited, although buckling and bending of the extended piston must be allowed for.
5. If large masses are moved by a cylinder, cushioning is used in the end position to prevent a hard impact and damage. Before the end point is reached, a cushioning piston interrupts the direct flow of air to the outside. Instead, a very small and often adjustable escape aperture is opened.
6. A cylinder with a double sided piston rod has better guidance because of two bearing positions. A small lateral load can also be applied. Special members can also be attached to the free side of the piston rod. The force is the same in both directions.
7. A tandem cylinder design involves two double acting cylinders which have been joined to form a single unit. In this arrangement, because of the simultaneous loading of both pistons, the force on the piston rod is almost doubled and is used for applications where a large force is required, but the cylinder diameter is limited by space.
8. A multi-position cylinder consists of two or more double acting cylinders connected together. Depending on the pressure applied, the individual cylinders travel out.
9. Impact cylinders are used in forming operations, where the thrust force of compressed air provided by normal cylinders is limited. An impact cylinder produces high kinetic energy.
10. Rotary operating pneumatic elements transform pneumatic energy to a mechanical rotary movement. Such rotary working elements are called pneumatic motors.

EXERCISES

Fill in the blanks

1. Single acting cylinders can produce work in ———— direction.
2. In ———— type of cylinder, the force on the piston rod is almost doubled.
3. Telescopic cylinders are normally used for ———— stroke lengths.
4. For heavy-duty cylinders , mounting brackets are usually ———.
5. A cylinder with —— mounting generally requires strong machine frames to be able to resist bending movements.
6. The single acting cylinder converts compressed air energy into —— in the form of force and linear movement in one direction only.
7. Pneumatic —— element converts the energy in compressed air into force and motion. The pneumatic drive elements can move in a linear, reciprocating or rotating motion.
8. Typically, the piston can have diameters of as much as — mm.
9. ———— acting cylinders are used in assembling and packing automated lines to move, lift, feed, eject, press or push objects or to clamp parts. They are suitable for oil-free operation.
10. Compressed air is applied only on the bottom side of the piston, which is why the cylinder can move loads or perform mechanical work only in a forward direction and the effective force is reduced by the ———.
11. The working speed of a piston is in the range of 50 mm/s to —— mm/s.
12. A ——— type of cylinder is equipped with a compressible chamber with flexible sides that can be expanded to draw air in and compressed to force the air out.
13. An alternative construction is a single acting cylinder, spring —— with a spring in the piston area causing the piston to extend. Such pistons are used in the automotive industry and are mounted in the air brakes in vehicles and trains.
14. Another special form of single acting cylinder is the —— return spring where the piston's return stroke is caused by external forces or by its own weight.
15. Short-stroke cylinders are used when short strokes are required together with high ——.

State whether true or false

1. A telescopic cylinder is used when a long stroke length and a short retracted length are required.
2. Double acting cylinder with double rod can be used for doing work at both ends of the cylinder.
3. Telescopic cylinders are used for short stroke length and a long retracted length.

4. In single acting cylinders, stroke length is unlimited.
5. In pneumatic cylinders, dynamic thrust can be taken as 0.1 times the static thrust.

Review questions

1. What is the function of a pneumatic actuator?
2. How are pneumatic actuators classified?
3. How do hydraulic actuators differ from pneumatic actuators?
4. What is the function of a pneumatic cylinder?
5. Explain the working of a double acting double rod cylinder with a neat sketch.
6. Explain the working of tandem and telescopic cylinders with a neat sketch.
7. Mention some applications of pneumatic cylinders.
8. Differentiate between single acting and double acting cylinders.
9. What is meant by a limited rotation air motor?
10. Mention some applications of air motors.
11. What are the main advantages of a magnetic cylinder?
12. How are air motors classified?
13. What are the advantages and disadvantages of air motors?
14. Explain the working of piston air motors.
15. Compare and contrast air motors, hydraulic motors and electric motors.

Problems

1. A pneumatic cylinder has a bore of 200 mm and a piston rod diameter of 140 mm. For an extend speed of 5m/min, calculate:
   (a) The supply flow rate
   (b) The flow rate from the annulus side on the extend
   (c) The retract speed using Q_E
   (d) The flow rate from the full bore end on retract.

2. A compressor supplies air at 0.002 m^3/s to a 50 mm diameter double acting cylinder with a rod diameter of 20 mm. If the load is 1000 N both in extending and retracting, find:
   (a) Piston velocity during the extension stroke and retraction stroke
   (b) Pressure during the extension stroke and retraction stroke
   (c) Power during the extension stroke and retraction stroke.

3. For the crane system given in Fig. 4.36, find the cylinder force required to lift a load of 1000 N.

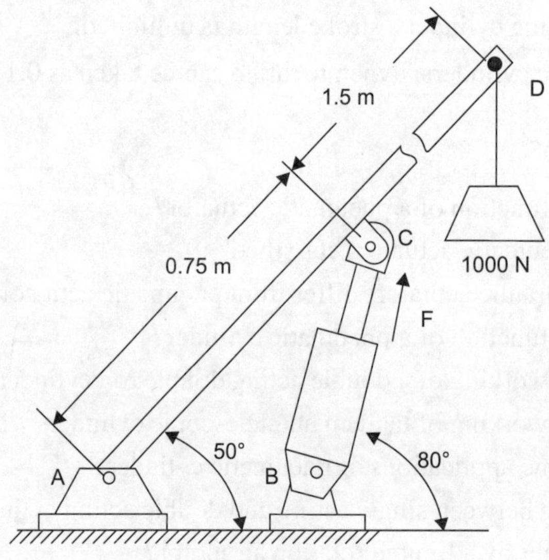

Figure 4.36

4. For the bent lever system shown in Fig. 4.37, the cylinder force is 400 N. How much load can the system drive?

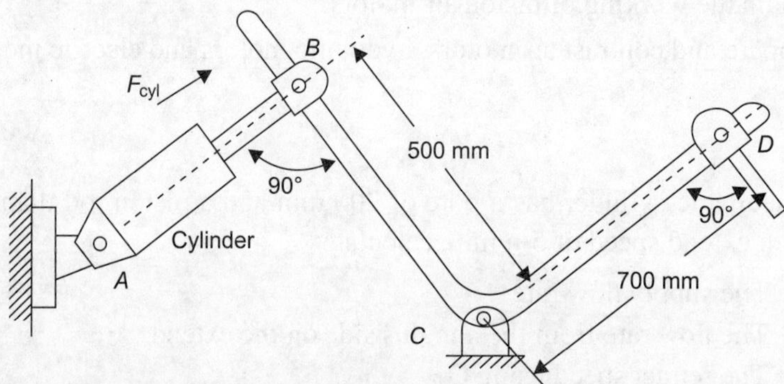

Figure 4.37

5. A two-stage telescopic cylinder is used to tilt the body of a car. When the car is fully loaded, the cylinder has to exert a force equivalent to 4000 N at all points in its stroke. The outside diameters of the tubes forming the two stages are 75 mm and 100 mm. If the compressor powering the cylinder delivers 12 LPM, calculate the extend speed and the pressure required for each stage of the cylinder when tilting the fully loaded car.

Answers

Fill in the blanks

1. only one	2. tandem	3. long	4. fabricated	5. non-center line
6. mechanical energy	7. drive	8. 100 mm	9. single	10. return spring
11. 500	12. diaphragm	13. extend	14. without	15. forces

True or false

1. True 2. True 3. False 4. False 5. False

Problems

1. (a) 157 LPM (b) 80 LPM (c) 9.8 m/min (d) 309 LPM
2. (a) 1m/s, 1.2 m/s (b) 5.1 bar, 6.06 bar (c) 1.02, 1.21 kW
3. 3.75 bar 4. 7.43 bar 5. 500 N

5 PNEUMATIC CONTROL VALVES

Upon completion of this chapter, the reader should be able to

- Define the function of a valve
- Classify valves
- Identify directional control valves (DCVs) as per ISO designation
- Explain the different types of directional control valves
- Explain the various method of valve actuation
- Describe the function of various non-return valves
- Understand the working of quick exhaust valves
- Differentiate between pressure control valves and sequence valves

5.1 Valves

Valves are defined as devices to control or regulate the commencement, termination, direction, pressure and speed of the flow of fluid, delivered by a compressor or vacuum pump, or stored in a reservoir.

Valves of different types perform three main functions in pneumatic installations.

- They control the supply of air to power units, e.g., cylinders
- They provide signals which govern the sequence of operations
- They act as interlock and safety devices

For the most part, the type of valve used is of little importance in a pneumatic control. What is important is the function that can be initiated by the valve, its mode of actuation, and the line connection size which determines the flow size of the valve. Valves used in pneumatics mainly have a control function i.e., they act on some process, operation or quantity to be stopped. A control function requires control energy, the intent being to achieve the greatest possible effect with the least effort. The form of control energy is dictated by the valve's mode of actuation and may be manual, mechanical, electrical, hydraulic or pneumatic.

Valves available for pneumatic control can be classified into four principal groups according to their function:

1. Direction control valves
2. Non-return valves
3. Flow control valves
4. Pressure control valves

5.2 Direction Control Valves

Like hydraulic systems, pneumatic systems also require control valves to direct and regulate the flow of fluid from the compressor to various devices like air actuators and air motors. In order to control the movement of air actuators, compressed air has to be regulated, controlled and reversed with a predetermined sequence. The pressure and the flow rate of the compressed air are controlled to obtain the desired level of force and speed of air actuators.

The function of directional control valves (DCV) is to control the direction of flow in a pneumatic circuit. DCVs are used to start, stop and regulate the direction of air flow and to help in the distribution of air along the line.

5.2.1 Types of direction control valves

Directional valves control the way the air passes, and are used principally for controlling commencement, termination and direction of air flow. The classification scheme of pneumatic valves based on different factors is given below:

1. Construction

   (i) Poppet or seat valves

   - Ball seat valve
   - Disc seat valve
   - Diaphragm Valves

   (ii) Sliding spool valves

   - Longitudinal slide valve
   - Suspended spool valves
   - Rotary spool valves

2. Number of ports

   (i) Two-way valves
   (ii) Three-way valves
   (iii) Four-way valves

3. Methods of actuation
   (i) Mechanical
   (ii) Electrical
   (iii) Pneumatic

4. Size of the port
The valve's port size is a factor and port sizes are designated as M5, G1/8, and G1/4 etc., where 'M' refers to Metric thread and 'G' refers to British standard pipe (BSP) thread.

5. Mounting styles
   (i) Sub base
   (ii) Manifold
   (iii) In-line
   (iv) Valve island

5.2.1.1 ISO designation of direction control valves

Valves are represented by symbols because the actual construction is quite complex. A symbol specifies the function of the valve, method of actuation and the number of ports and ways. Pneumatic symbols have been standardised in ISO 1219–1:2006 *(Fluid power systems and components – Graphic symbols and circuit diagram).* Another standard ISO 1219–2:1995 establishes the rules for drawing diagrams of fluid power systems using symbols from ISO 1219–1. Port designations are described in ISO 5599.

Port markings As per ISO 5599, ports are designated using a number system. Earlier, a letter system was used to designate a port. Table 5.1 shows port markings.

Table 5.1 *Port markings of direction control valves*

Port	Old (Letter) System	ISO (Number) System	Remarks
Pressure port	P	1	Supply port
Working port	A	2	3/2 DCV
Working ports	A, B	4, 2	4/2 or 5/2 DCV
Exhaust port	R	3	3/2 DCV
Exhaust ports	R, S	5, 3	5/2 DCV
Pilot ports	Z or Y	12	Pilot line (flow 1-2)
Pilot ports	Z	14	Pilot line (flow 1-4)
Pilot ports	Z or Y	10	Pilot line (no flow)
Internal pilot ports	Pz, Py	81, 91	Auxiliary pilot line

Ports and position DCVs are described by the number of port connections or ways that they control (two-way, three-way, four-way valves). Table 5.2 shows commonly used DCVs with old and new designations.

Table 5.2 *Port designation of DCV*

Port and position	
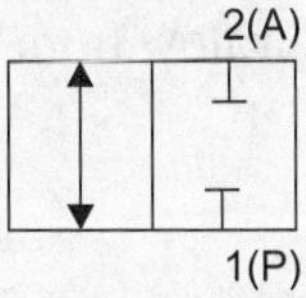 2(A) 1(P)	2/2 Directional control valve 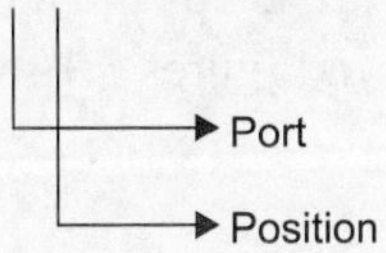 (Port, Position)
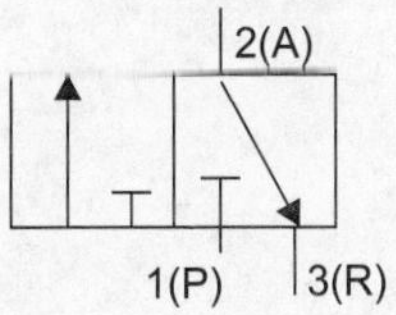 2(A) 1(P) 3(R)	3/2 Directional control valve (normally closed)
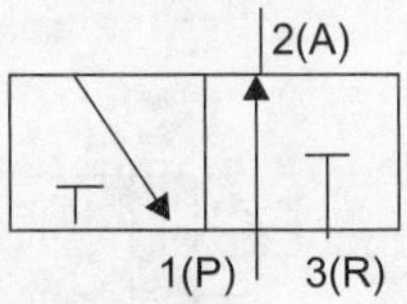 2(A) 1(P) 3(R)	3/2 Directional control valve (normally open)
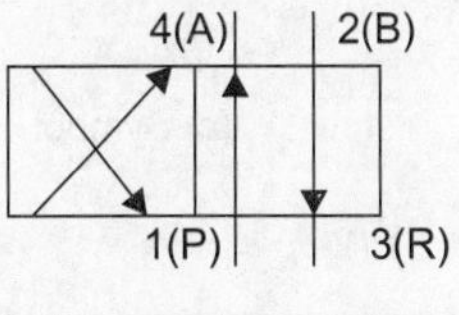 4(A) 2(B) 1(P) 3(R)	4/2 Directional control valve
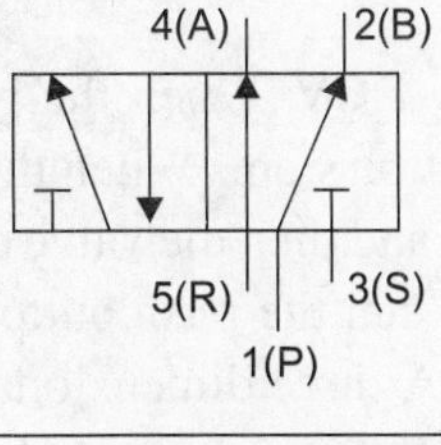 4(A) 2(B) 5(R) 3(S) 1(P)	5/2 Directional control valve
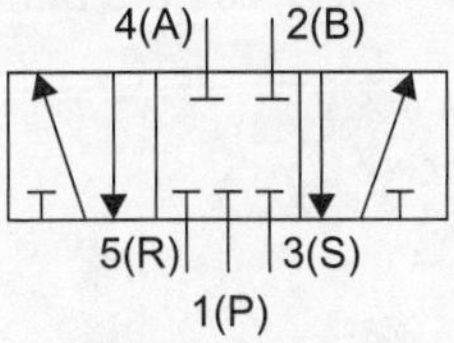 4(A) 2(B) 5(R) 3(S) 1(P)	5/3 Directional control valve

5.2.1.2 *Poppet direction control valves*

There are two different types of poppet valves, namely ball seat valve and disc seat valve.

(A) Ball seat valve

A poppet valve uses discs, cones or balls to control flow. Figure 5.1 shows the construction of a simple 2/2 normally closed valve. When the push button is pressed, the ball lifts off from its seat and allows air to flow from port P to port A. When the push button is released, spring force and air pressure keeps the ball back and closes air flow from port P to port A. Valve position are shown in Figures 5.1 (a), (b), (c).

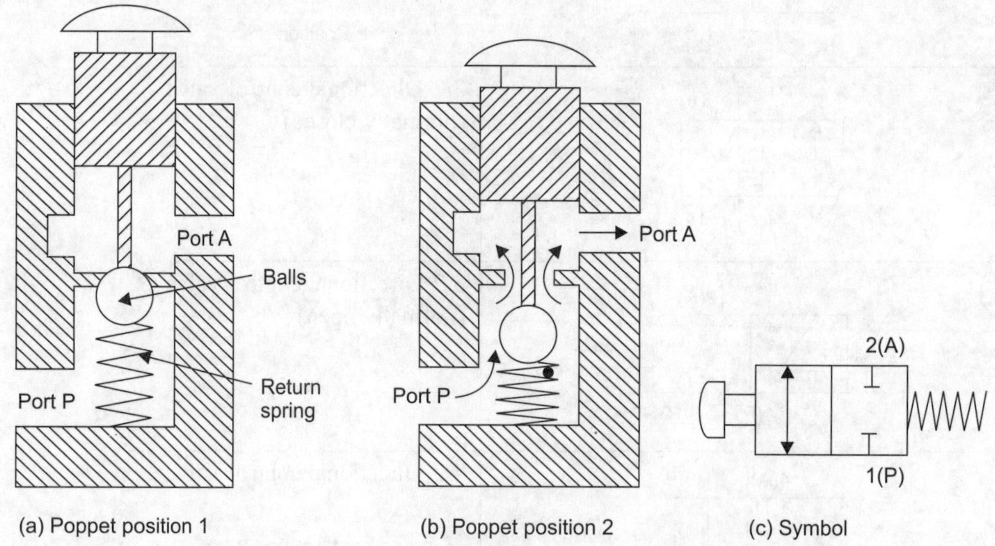

Figure 5.1 *2/2 Ball seat poppet valve*

(B) Disc seat poppet valve

Figure 5.2 shows the construction of a disc type 3/2 way DCV. When the push button is released, ports 1 and 3 are connected via a hollow pushbutton stem. When the push button is pressed, port 3 is first blocked by the moving valve stem and then the valve disc is pushed down to open the valve, thus connecting port 1 and 3. When the push button is released, the spring and air pressure from port 1 closes the valve. A three dimensional view of the disc seat poppet valve is shown in Fig. 5.3 and a comparison between the ball seat and disc seat valve is given in Table 5.3.

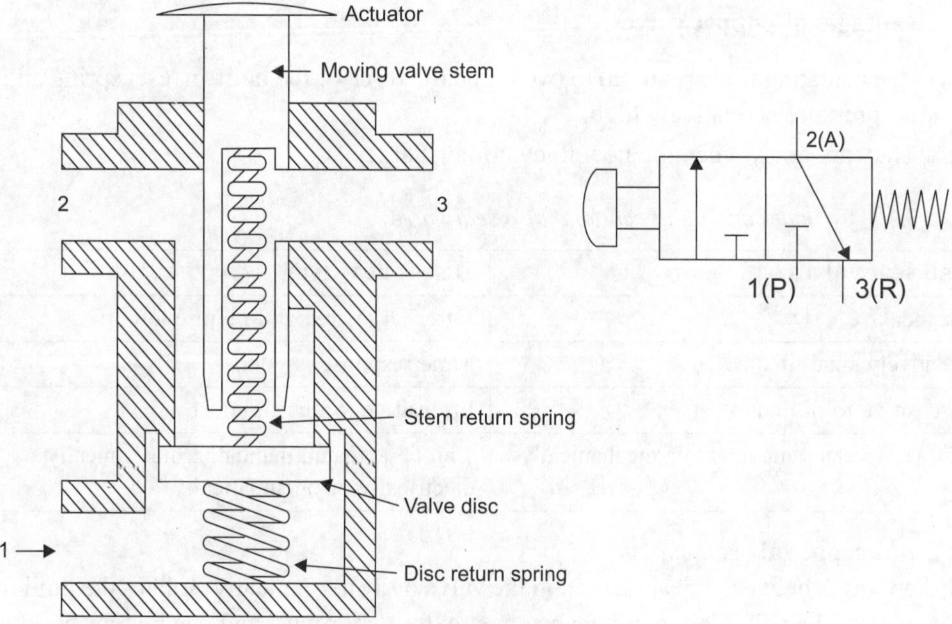

Figure 5.2 *Disc seat poppet valve*

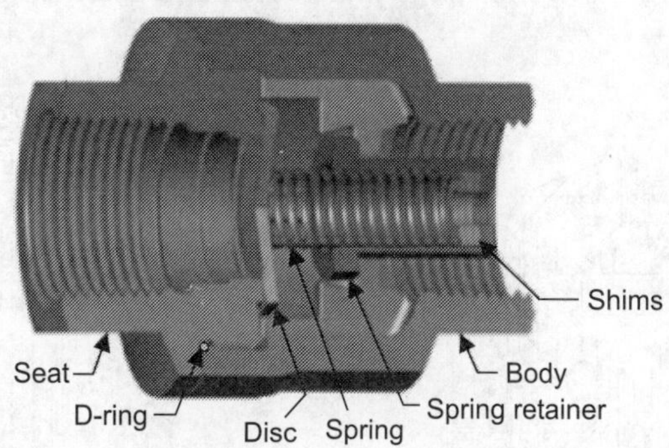

Figure 5.3 *Three dimensional view of a disc type poppet valve*

Advantages of poppet valves

(i) The response of a poppet valve is very quick, requiring only a short stroke to trigger the valve to open fully.

(ii) The valve seats are usually simple elastic seals so wear is minimised.

(iii) Valves are robust and not vulnerable to dust and dirt; the seats are self-cleaning.

(iv) Valves are inexpensive.

(v) Maintenance is simple and inexpensive.

(vi) They have longer service lives as a result of short strokes and fewer parts prone to wear.

Disadvantages of poppet valves

(i) The actuating force required to overcome the force of the built-in reset spring and the air pressure is relatively high.
(ii) They are noisy when the flow fluctuation is large.

Table 5.3 *Comparison of ball seat and disc seat valves*

Ball seat valves	Disc seat valves
Inexpensive	Offer large area and require low lift
Relatively small in size	Time response is good
Insensitive to dirt and dust	Insensitive to dirt and dust
Can be operated manually or mechanically	Can be actuated manually, mechanically, electrically or pneumatically

(C) Diaphragm valves

The diaphragm between the actuator and the valve body hermetically isolates the fluid from the actuator. The valves are maintenance free, extremely robust and can be retrofitted with a comprehensive range of accessories like electrical position feedback, stroke limitation or manual override. Figure 5.4 shows the unactuated and actuated positions of diaphragm valves.

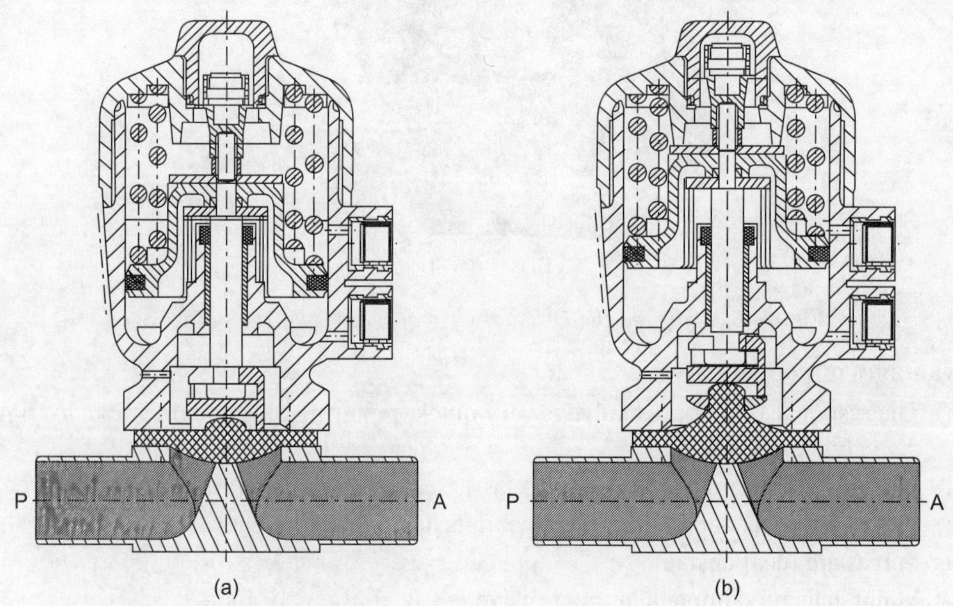

Figure 5.4 *Diaphragm valve: (a) unactuated position, (b) actuated position*

Closed position When de-energised, the valve is closed by spring action.

Open position If the actuator is pressurised by the control pressure, it simultaneously lifts the control piston and the valve spindle to open the valve.

5.2.1.3 Spool direction control valves

(A) Hand operated 3/2 DCV

The cross-sectional views of a 3/2 DCV (normally closed) based on a spool design is shown below in Fig. 5.5. When the valve is not actuated, port 2 and 3 are connected and port 1 is blocked. When the valve is actuated, port 2 and 1 are connected and port 3 is blocked.

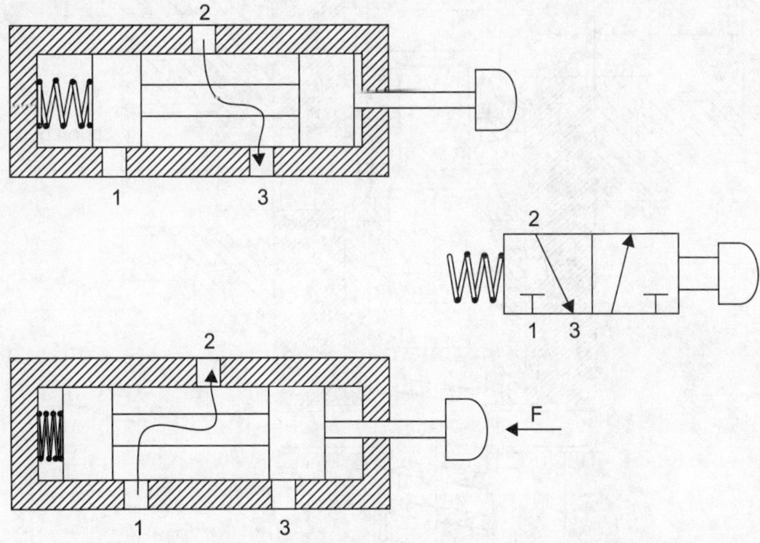

Figure 5.5 *3/2 Directional control valve (normally closed)*

Figure 5.5 shows a schematic diagram of a 3/2 spring operated valve. There are three ports: common port, normally open (NO) port and normally closed (NC) port. When the valve is not actuated, there is flow from the NO port to the common port. When the valve is actuated there is flow from the NC port to the common port. Figure 5.6 shows the detailed construction of a 3/2 valve.

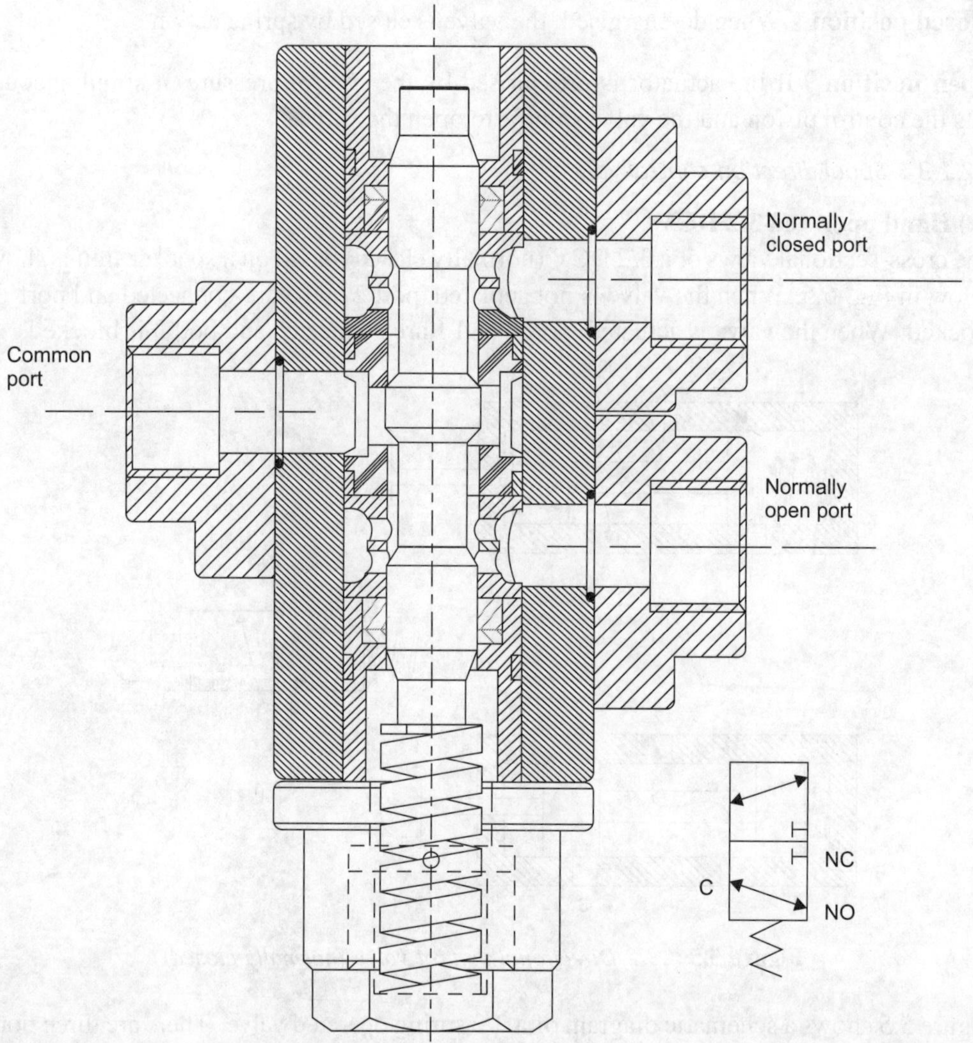

Figure 5.6 *Construction features of a 3/2 spring operated valve*

(B) Pneumatically actuated 3/2 DCV

The cross–sectional views of a pneumatically actuated NC 3/2 DCV in normal and actuated positions are shown in the Fig. 5.7.

In the normal position, the working port (2) is closed to pressure port (1) and open to the exhaust port (3). When compressed air is applied through the pilot port (12), the spool is moved against the spring. In the actuated position, the working port (2) is open to the pressure port (1) and closed to exhaust port (3). Thus, the application of compressed air to port 12 causes the pressure port (1) to be connected to the working port (2).

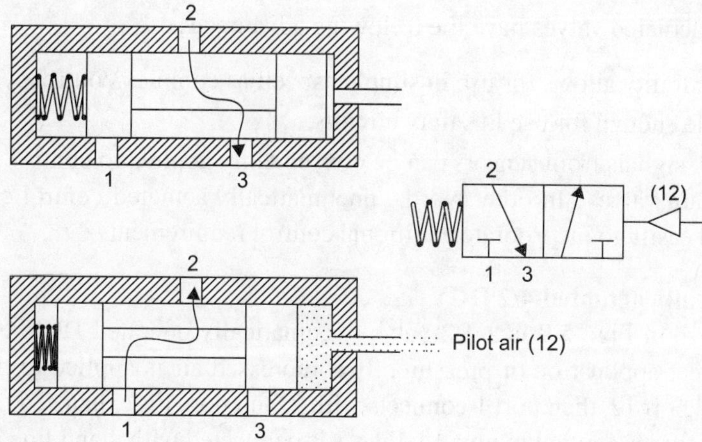

Figure 5.7 *3/2 Directional control valve (pneumatically operated)*

Figure 5.8 shows the detailed construction features of a 3/2 pilot operated valve. There are three ports: common port, normally open port and normally closed. When the valve is not actuated, there is flow from the NO port to the common port. When the valve is actuated there is flow from the NC to the common port.

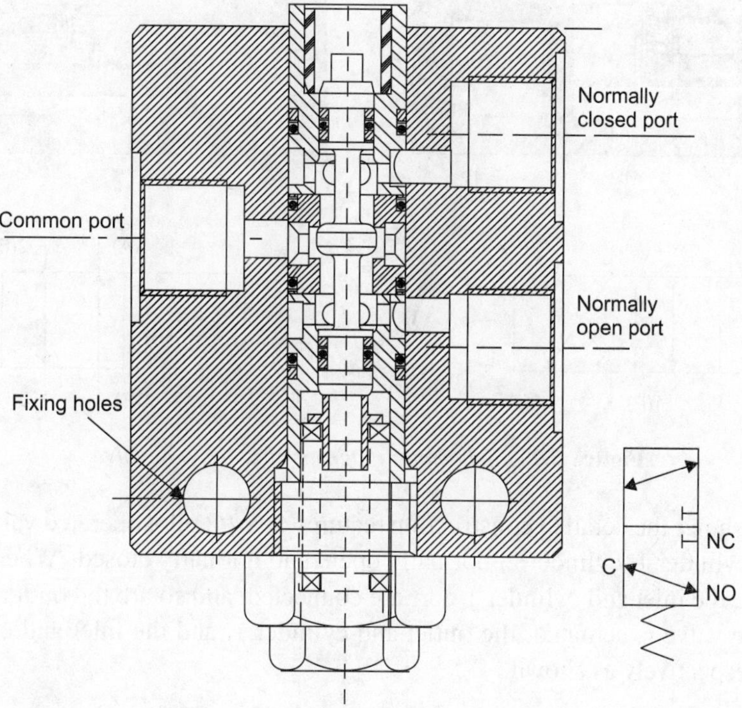

Figure 5.8 *Construction features of a 3/2 pilot operated valve*

Pneumatically actuated valves have the following advantages.

(i) Great flexibility allows for use in simple as well as complex control systems.
(ii) Adaptable enough for use in safety circuits.
(iii) Feedback signals from sensors can be conveniently applied to control the pilot ports of the main valves. In other words, pneumatically actuated control circuits can be modified easily to incorporate additional control requirement.

(C) Pneumatically actuated 4/2 DCV

The valve shown in Fig. 5.9 is a 4/2 valve pneumatically operated DCV. Switch-over is effected by direct application of pressure. If compressed air is applied to the pilot spool through control port 12, then port 1 connects with 2, and port 4 is exhausted through port 3. If the pilot pressure is applied to port 14, then 1 is connected with 4 and line 2 is exhausted through port 3. On disconnecting compressed air from the control line, the pilot spool remains in its current position until the spool receives a signal from the other control side.

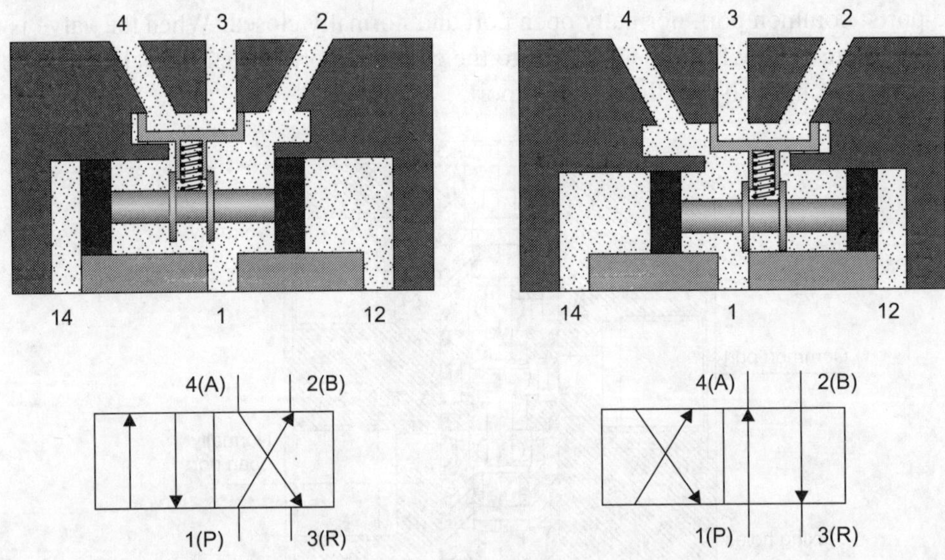

Figure 5.9 *Schematic diagram of a 4/2 way valve*

Figure 5.10 shows the detailed construction features of a 4/2 pilot operated valve. There are four ports: cylinder 1, cylinder 2, normally open, and normally closed. When the valve is not actuated, the inlet and cylinder 1 port are connected, and so are the outlet and cylinder 2. When the valve is actuated, the outlet and cylinder 1, and the inlet and cylinder 2 are connected respectively as shown.

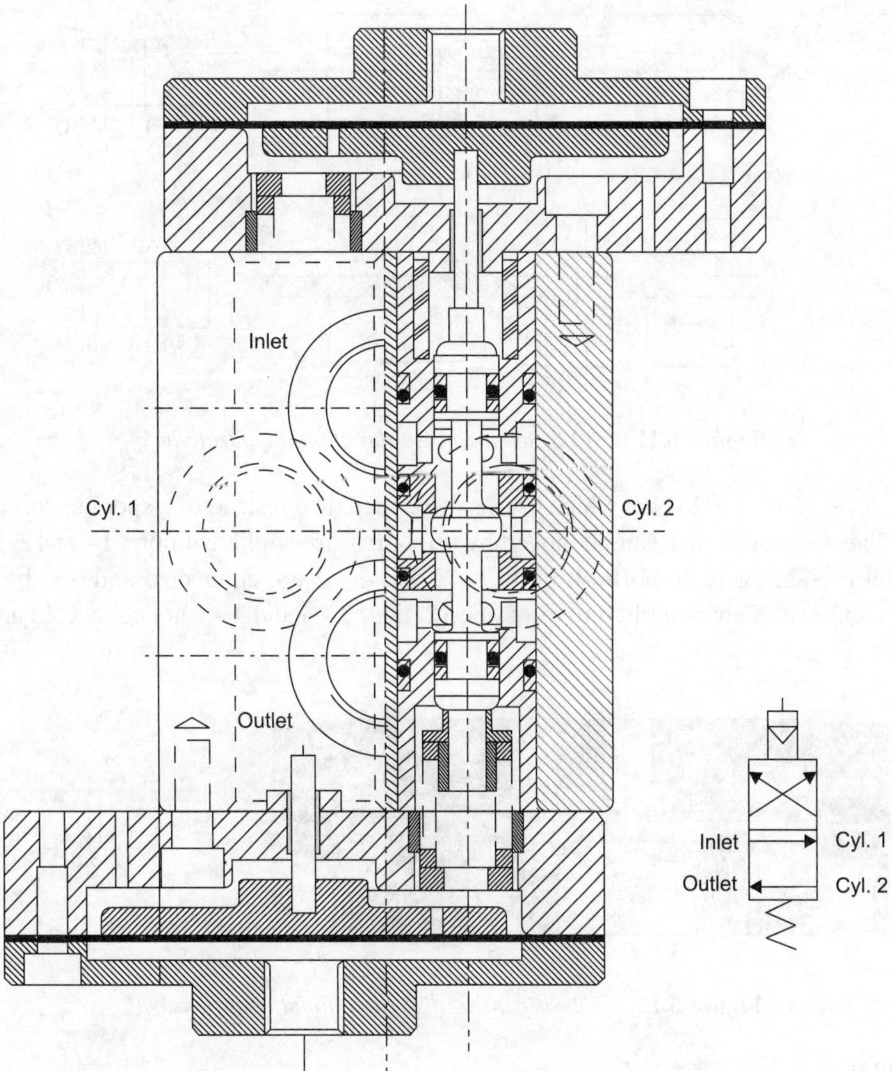

Figure 5.10 *Schematic diagram of 4/2 way valve*

(D) Suspended disc direction control valves

This valve is quite similar to a 4/2 spool valve. A schematic diagram is shown in Fig. 5.11. In this design, a disc is used instead of a spool to control the opening and closing of ports. The suspended disc can be moved by pilot pressure or a solenoid or by mechanical means. The main disc connects port 1 to either port 4 or to port 2. The secondary seat discs seal the non-functional exhaust port 3. These valves are generally provided with a manual override to allow for manual moving of the cylinder.

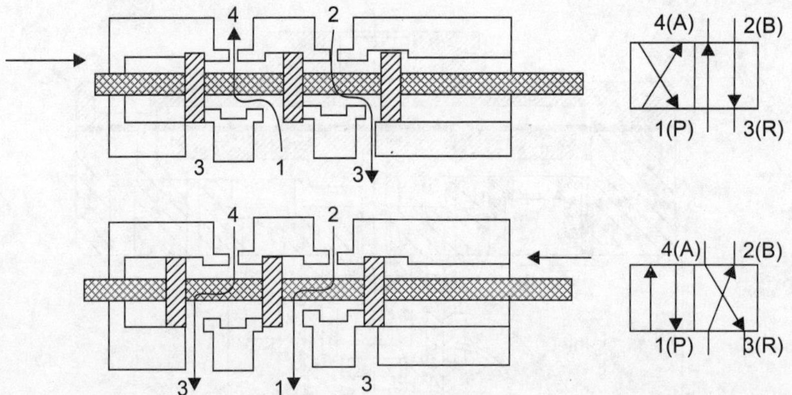

Figure 5.11 *4/2 Suspended disc directional control valve*

Figure 5.12 shows a 5/2 valve which uses a suspended disc instead of a spool to control the ports. The suspended disc can be moved by pilot pressure applied at ports 14 or 12. When the pilot pressure acts on port 14, ports 1 - 2 and 4 - 5 are connected and 3 is blocked. When compressed air is applied to pilot line 12, then 2 - 3 and 4 - 1 are connected and 5 is blocked.

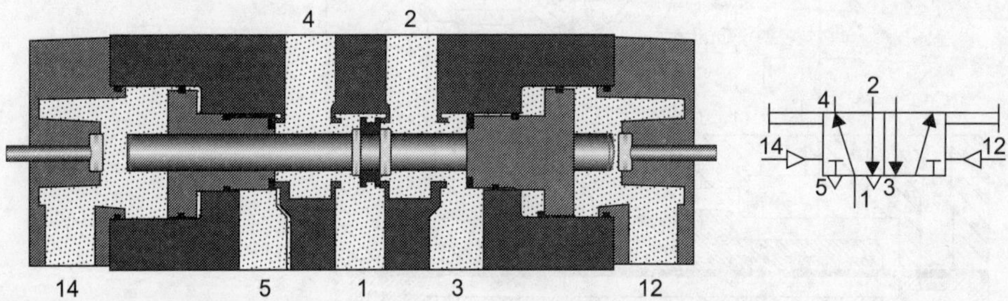

Figure 5.12 *5/2 Suspended disc directional control valve*

Advantages

(i) Actuation movements are shorter.

(ii) Quicker operation because of small switching movements.

(iii) There is no signal conflict. If signals are applied to both ports simultaneously, then first signal will dominate and perform the required operation.

Disadvantages

(i) Construction of valves can be complex.

(ii) Valves can be expensive.

(E) Rotary valves

Figure 5.13 illustrates a rotary spool directional control valve. It has a round core mounted within a fixed sleeve, with one or more passages through it. The sleeve has ports connected to lines in the fluid system, and as the core rotates within the sleeve, the passages either align with or block these ports.

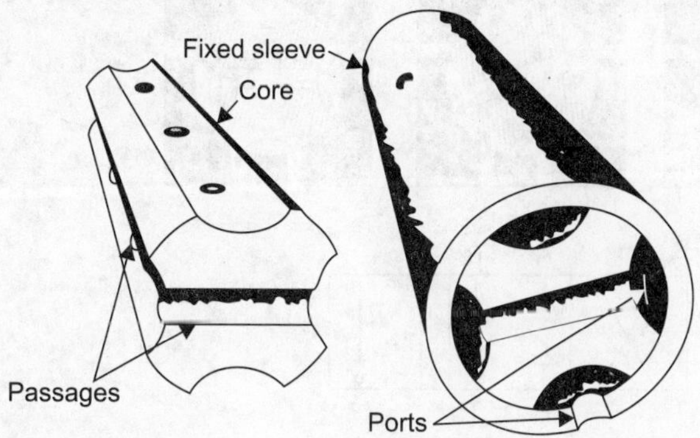

Figure 5.13 *Parts of a rotary spool directional control valve*

Figure 5.14 shows the construction of a rotary spool directional control valve. The ports are connected by rotating the handle whereby the core is aligned with different holes for the required valve configuration. This type of valve can be directly mounted on a panel using a bolt.

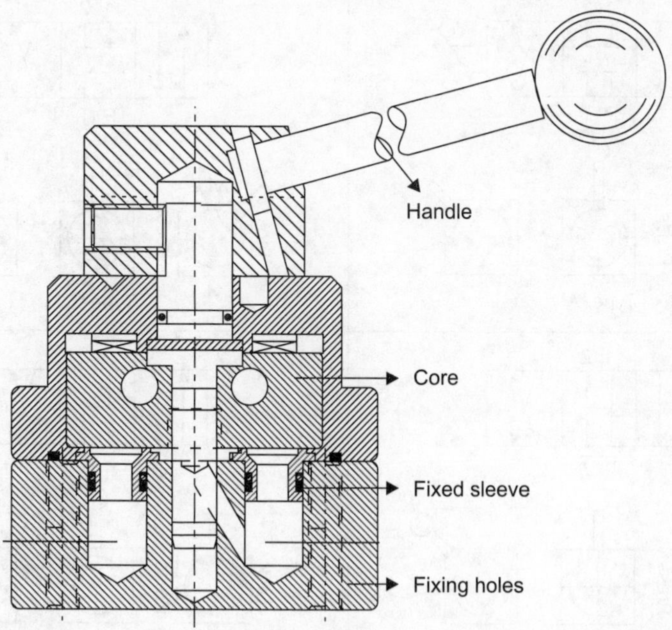

Figure 5.14 *Construction of a rotary spool directional control valve*

Figure 5.15 shows three different positions of the core when the handle is rotated. The left most envelope of the DCV connects ports P to B and A to T. The middle envelope of the DCV blocks all the ports. The right most envelope of the DCV connects ports P to A and T to B.

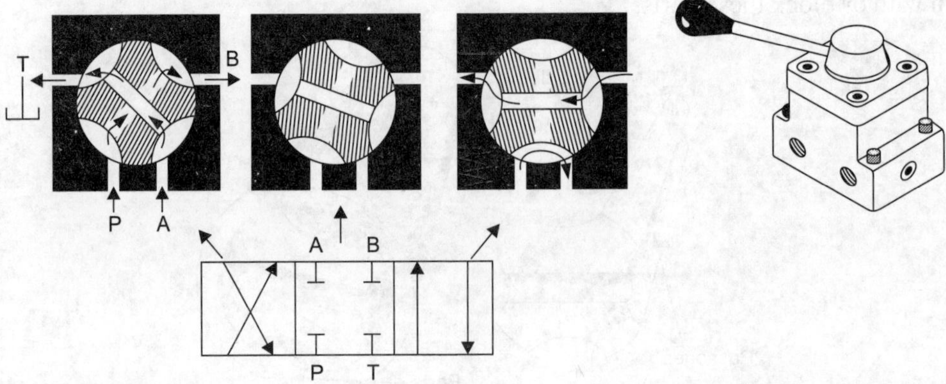

Figure 5.15 *Three positions of a 4/3 rotary spool directional control valve*

Table 5.4 shows schematically the different positions of the core and sleeve for the middle positions of a 4/3 direction control valve.

Table 5.4 *Different positions of core and sleeve for various mid-positions of a 4/3 DCV*

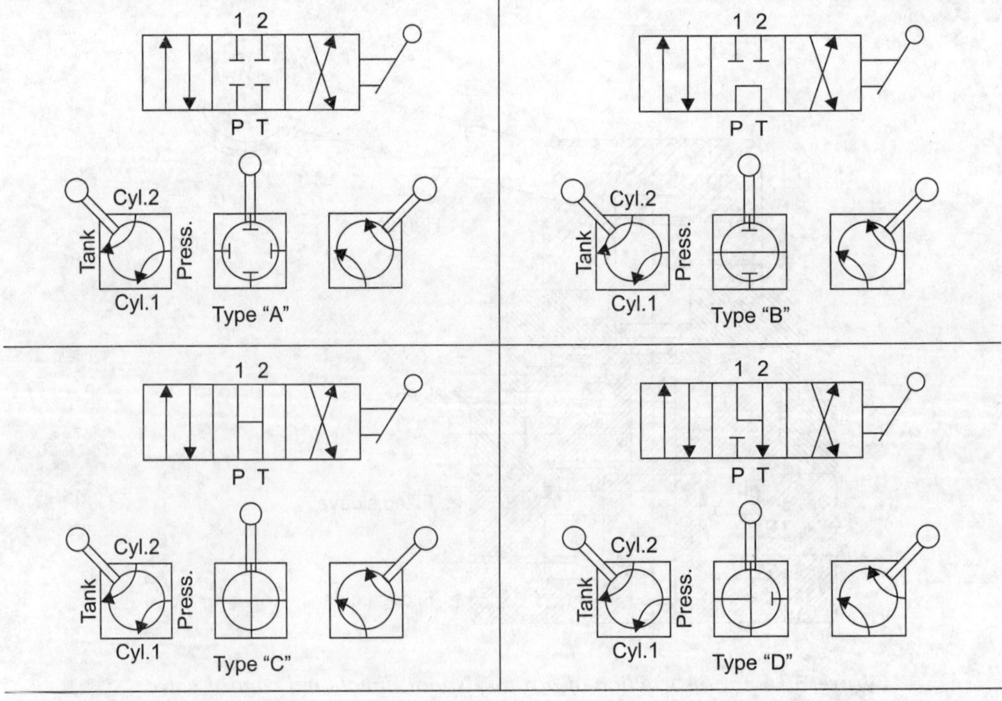

5.2.1.4 DCV classification according to valve operation

Two-way valves turn air on and off in applications such as shop air, branch air lines and supply systems for tools, motors and similar equipment. Two-way valves can also be used as vent valves.

(A) Two-way valves

Normally closed two-way valves block inlet flow in the unactuated position. When actuated, inlet flow is allowed or resumed, as shown in Fig. 5.16. Note that valve symbols show the unactuated position in the lowest block.

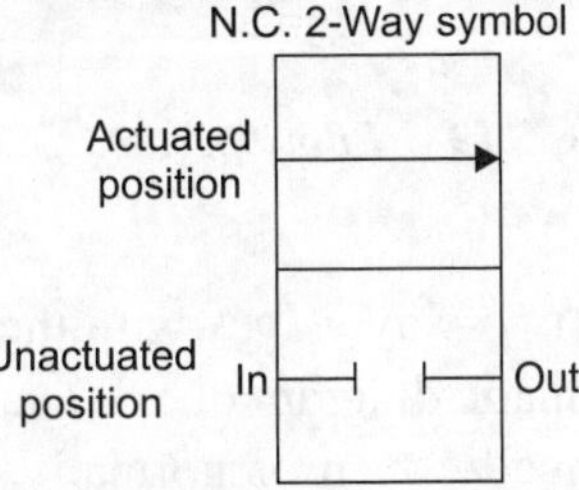

Figure 5.16 *Normally closed 2/2 valve*

Normally open two-way valves allow inlet flow when unactuated and block flow when actuated (Fig. 5.17). Normally open valves can be used in safety applications where a normally blocked cooling flow starts when a failure event actuates the valve.

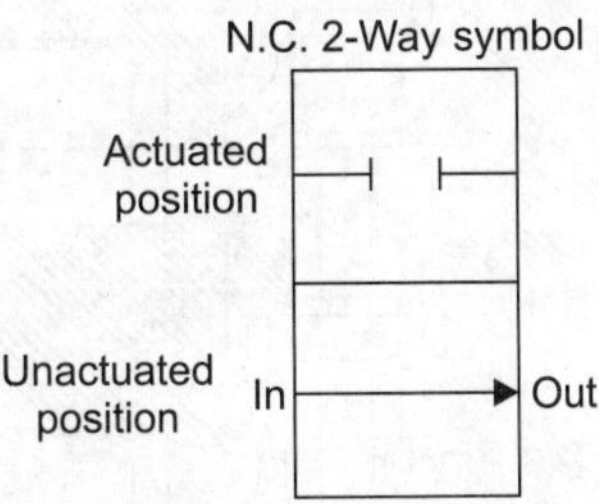

Figure 5.17 *Normally open 2/2 valve*

In spring return valves, a spring returns the valve to its normal unactuated position (Fig. 5.18).

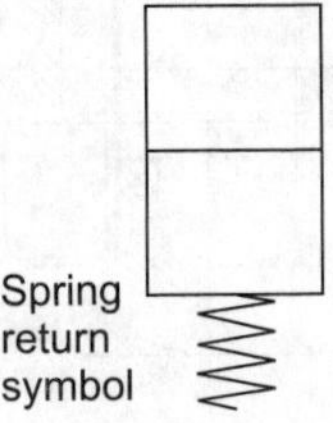

Figure 5.18 *Spring return 2/2 valve*

In a pilot operated configuration when the valve is actuated, the flow path shown in the upper block replaces the flow path in the lower block (Fig. 5.19).

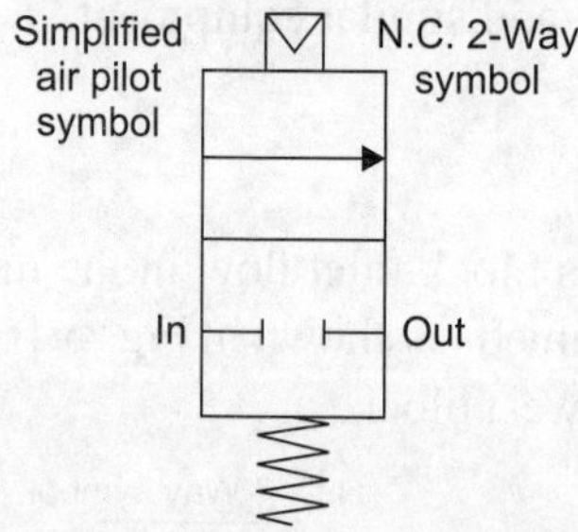

Figure 5.19 *Pilot operated 2/2 valve*

(B) Three-way valves

Three-way valves are similar to two-way valves with the addition of a third port that exhausts downstream air. The schematic diagram of a 3/2 valve is shown in Fig. 5.20. Three-way valves are available in normally open or normally closed operating configurations. These valves are used to control single acting or spring return cylinders and loads, which must be alternately pressurised and exhausted. The valves can also pilot other air operated valves.

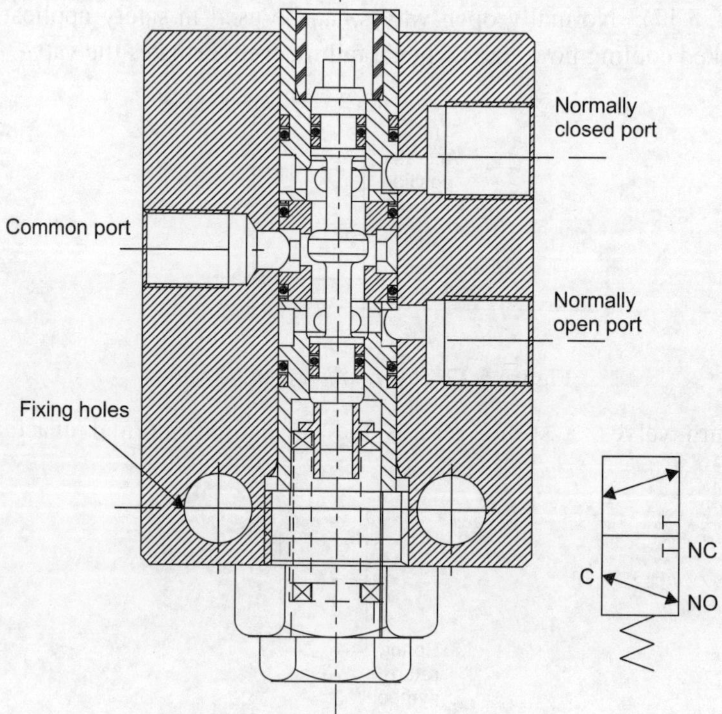

Figure 5.20 *Pilot operated 3/2 valve*

(i) Normally closed three way valves When unactuated, inlet flow is blocked, and the outlet port is connected to the exhaust port. When actuated the inlet port is connected to the outer port, and the exhaust port is blocked (Fig. 5.21). The construction details of a 3/2 valve (normally closed) is given in Fig. 5.22.

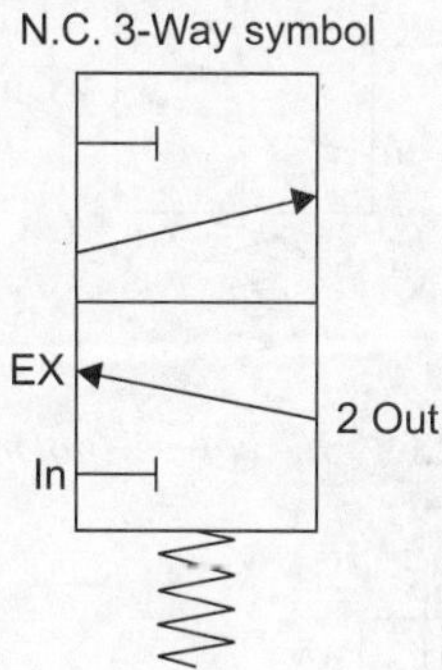

Figure 5.21 *3/2 valve (normally closed)*

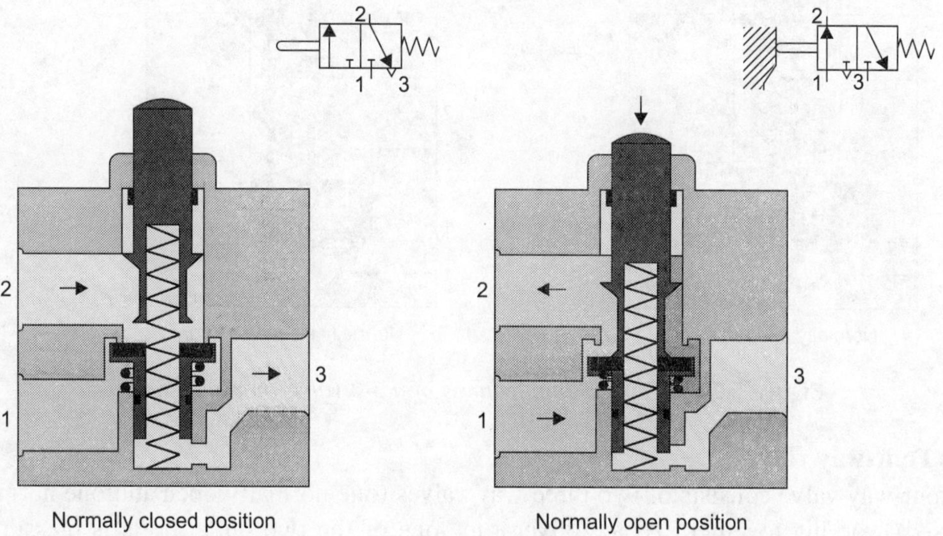

Figure 5.22 *Construction features of a 3/2 valve (normally closed)*

(ii) Normally open three-way valves When unactuated, inlet flows to the outlet and the exhaust port is blocked. When actuated, inlet flow is blocked and the outlet port is connected to the exhaust port (Fig. 5.23). The construction details of a 3/2 valve (normally open) are given in Fig. 5.24.

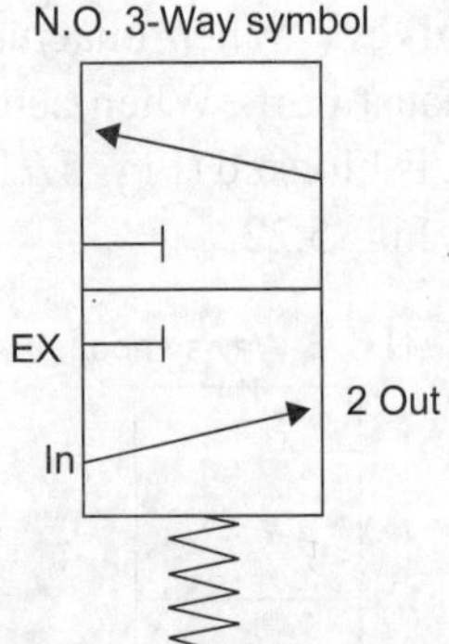

Figure 5.23 *3/2 valves (normally open)*

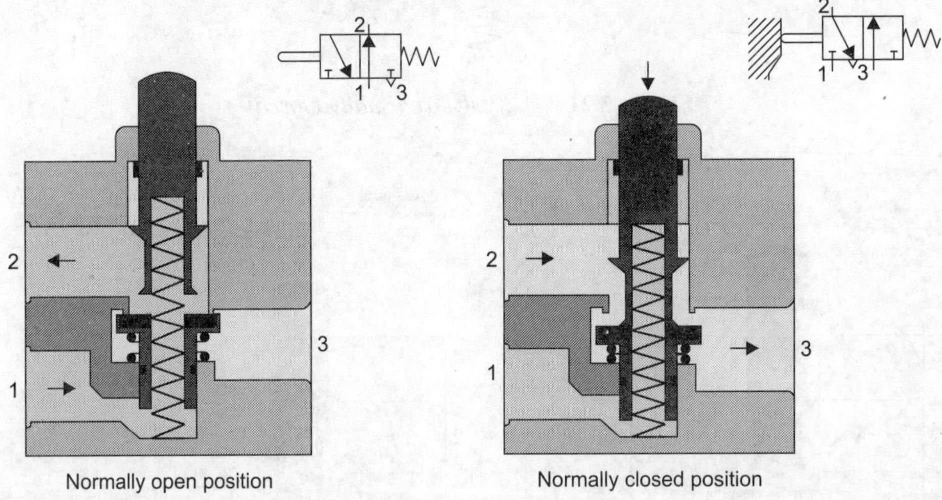

Figure 5.24 *Construction features of a 3/2 valve (normally open)*

(C) Four-way valves

A four-way valve consists of two three-way valves (one normally open and one normally closed) working together. At any given time one of the two outlet ports is pressurised and the other is venting. These valves are used to operate double acting air cylinders and control bidirectional air motors. A four-way valve can also simultaneously control two single acting cylinders.

Figure 5.25 shows the detailed construction features of a 4/2 pilot operated spool valve. There are four ports: cylinder 1, cylinder 2, normally open and normally closed. Figure 5.26 shows the construction features of a 4/2 stem operated valve. When the valve is not actuated, inlet and cylinder 1 ports are connected and so are outlet and cylinder 2. When the valve is actuated, outlet and cylinder 1 are connected as are inlet and cylinder 2, as shown Fig. 5.26.

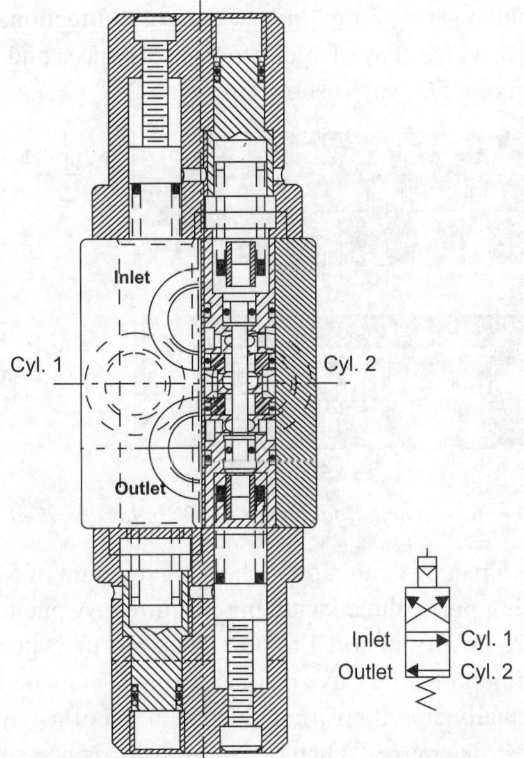

Figure 5.25 *Construction features of a 4/2 pilot operated spool valve*

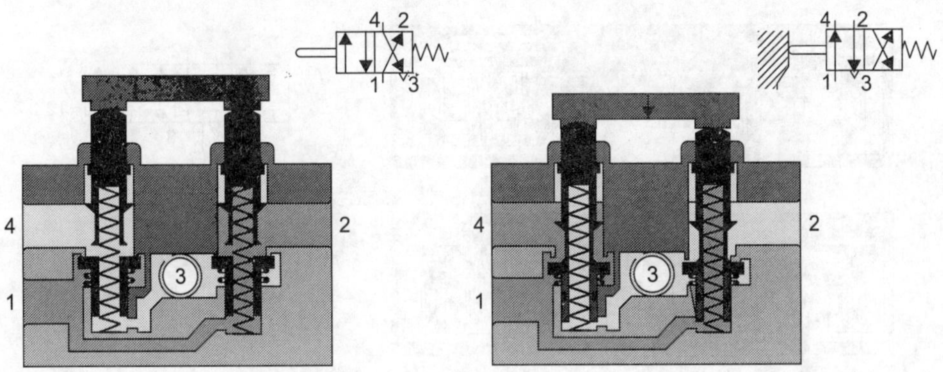

Figure 5.26 *Construction feature of 4/2 pilot operated spool valve*

(D) Five-way valves

A five-way valve has a memory function. A short impulse signal is sufficient for actuation. A pneumatic signal applied to pilot port 12 causes flow from port 1 to port 2. A pneumatic signal applied to pilot port 14 causes flow from port 1 to port 4. If both signals are applied simultaneously to pilot ports, the signal that arrives first is dominant. Five-way valves are used to operate double acting air cylinders, control bidirectional air motors and in air circuitry. These types of valves have large actuation distances and short actuation forces. The schematic diagram of a 5/2 way is shown in Fig. 5.27.

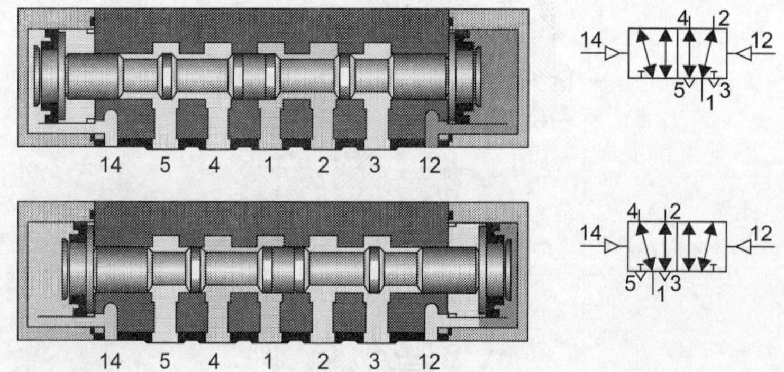

Figure 5.27 *Construction features of a 5/2 pilot operated spool valve*

The five-way valve has a memory function. Schematic diagram of 5/2 way is shown in Fig. 5.28. It has five working ports, three switching positions. A pneumatic signal applied to pilot port 12 (14) causes flow from port 1 to port 2 (4). Spring is present in Fig. 5.28. If no signal applied to the pilot ports, the valve piston is centred in its mid position by the return spring. In mid and normally closed position, the piston rod of a cylinder can be stopped at any position within its stroke range. When the midposition is pressurised, the piston rod of the cylinder moves outward with reduced force (). When the midposition is exhausted, the piston rod of the cylinder can be moved freely to any position by external force.

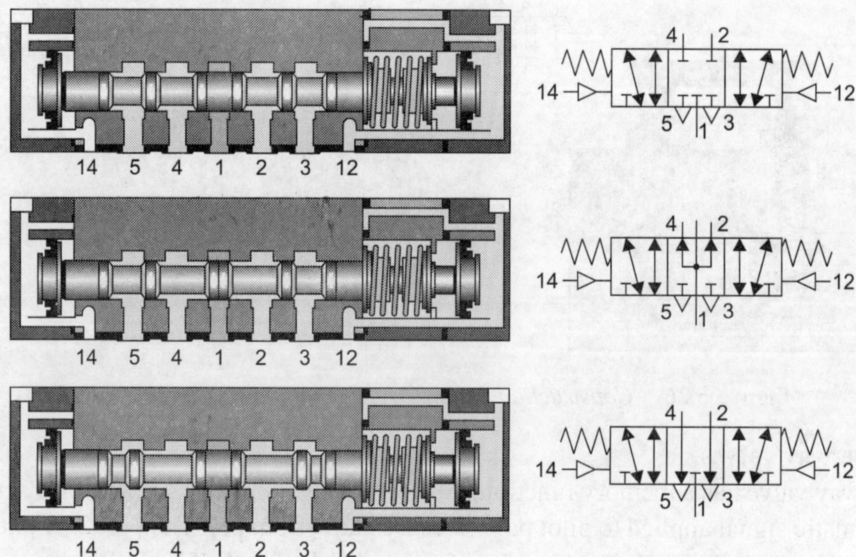

Figure 5.28 *Construction feature of 5/2 pilot and spring operated spool valve*

5.2.1.5 *Methods of actuation*

The methods of actuation of pneumatic directional control valves depend upon the requirements of the task. The types of actuation may be:

- Manual
- Mechanical
- Pneumatic
- Electrical
- Combined

The symbols of the actuation methods are detailed in DIN ISO 1219. When applied to a directional control valve, consideration must be given to methods of both the initial and the return actuation of the valve. Normally these are two separate methods, and each is shown on the sides of the position boxes of the symbol (see Table 5.5). There may also be additional methods of actuation such as manual overrides, which are indicated separately.

Table 5.5 *Methods of actuation*

Type of actuation	Type of control	Symbol
Manual	General	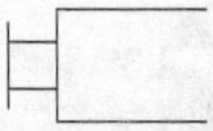
	Pushbutton	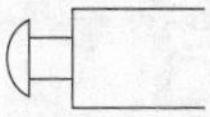
	Detent lever operated	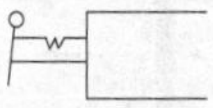
	Foot pedal	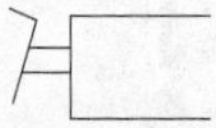
Mechanical	Spring return	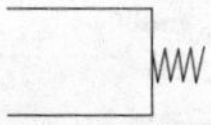
	Spring centred	
	Roller operated	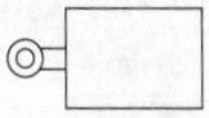
	Idle roller	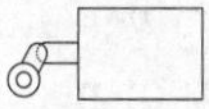

contd...

Table 5.5 *...contd*

Pneumatic	Direct	
	Indirect, pilot operated	
Electrical	Single solenoid	
	Double solenoid	
Combined	Double solenoid with pilot operated	

Some examples are given in Figures 5.29 through 5.32.

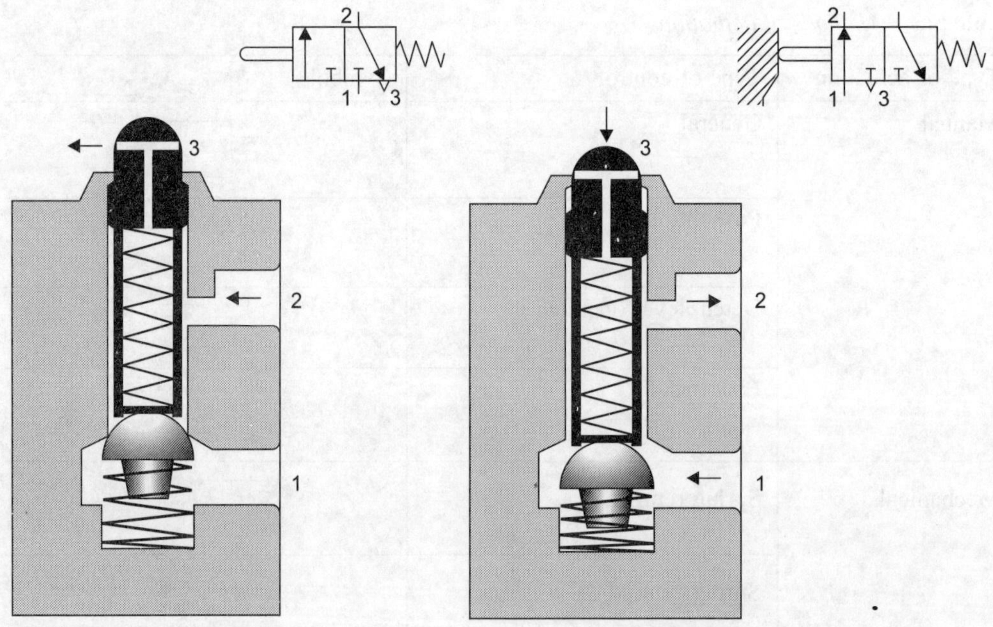

Figure 5.29 *Stem actuated 3/2 ball valve*

The two-way valves are shut-off valves with one inlet port and one outlet port, and can control the flow of neutral gases and low to medium viscosity liquids (Fig. 5.30 shows a normally closed valve).

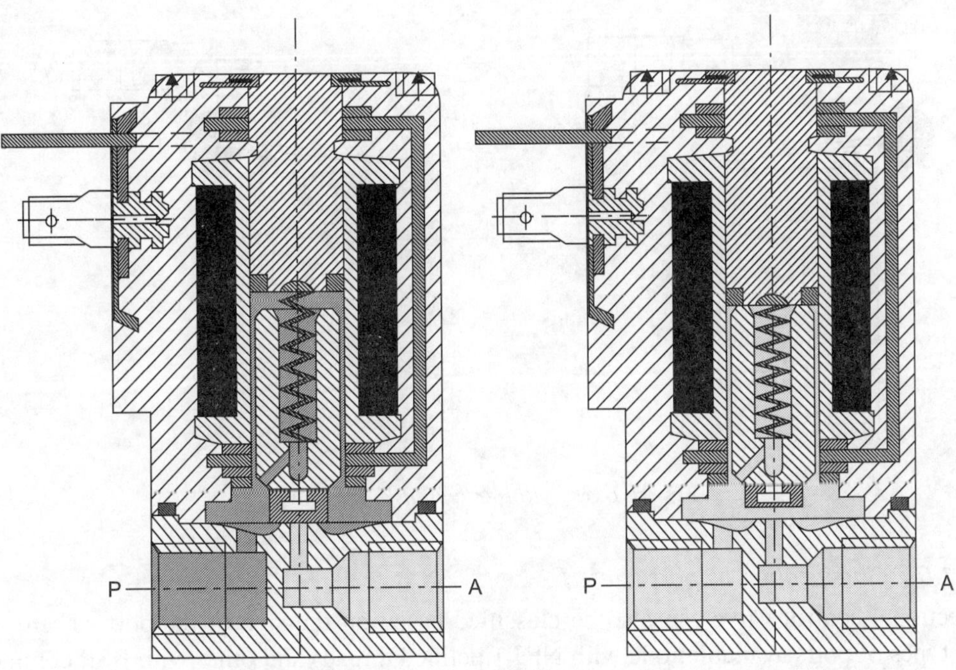

Figure 5.30 *Solenoid actuated 3/2 valve*

Closed position When de-energised, the armature spring assisted by the fluid pressure, holds the valve seal on the valve seat to shut off the flow.

Open position When energised, the armature and seal are pulled into the solenoid coil to open the valve.

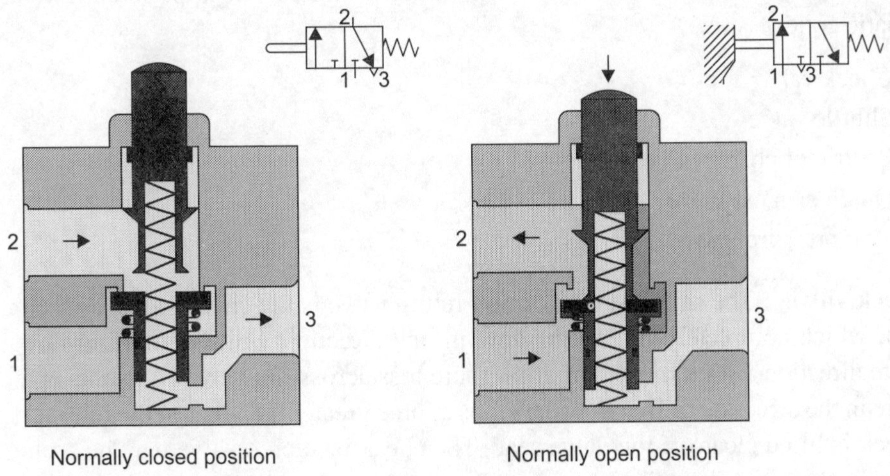

Figure 5.31 *Steam actuated 3/2 disk valve*

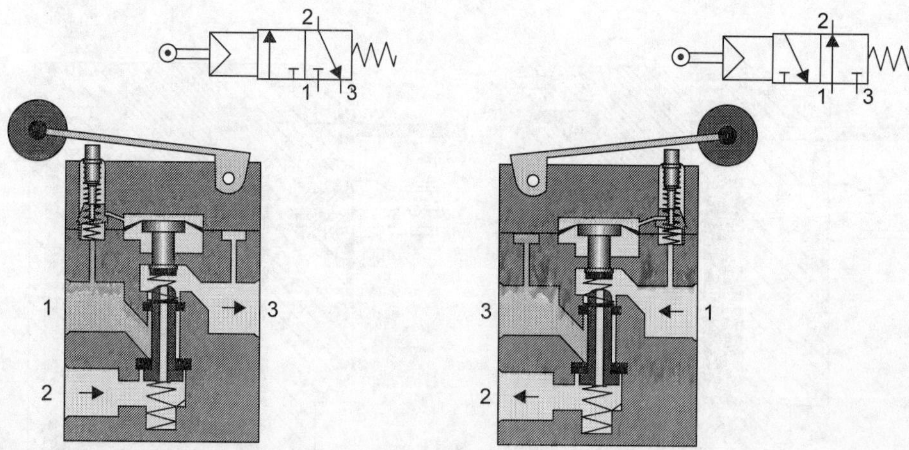

Figure 5.32 *Roller actuated 3/2 valve*

5.2.1.6 *Classification by port size*

Directional control valve can also be classified based on the size of the port. There are two types of ports available; one with NPT (inclined thread) and other with BSP common knowledge (parallel thread). Valves can also be mounted directly on the manifold.

5.2.2 Non-return valves

Non-return valves permit air flow in one direction only, the other direction through the valve is blocked completely. Usually the valves are designed so that the check valve is additionally loaded by the downstream air pressure, thus preventing the return.

Among the various types of non-return valves available, those preferentially used in pneumatic controls are:

(i) Check valve
(ii) Shuttle valve
(iii) Restrictor check valve
(iv) Quick exhaust valve
(v) Two pressure valve

(i) Check valve The simplest type of non-return valve is the check valve, as seen in Fig. 5.33(a), which completely blocks air flow in one direction while permitting flow in the opposite direction with a minimum of pressure loss across the valve. As soon as the inlet pressure in the direction of free flow develops a force greater than that of the internal spring, the check is lifted clear off the valve seat. The check in such a valve may be a plug, ball, plate or diaphragm. Figure 5.33(b) shows the cartridge type check valve.

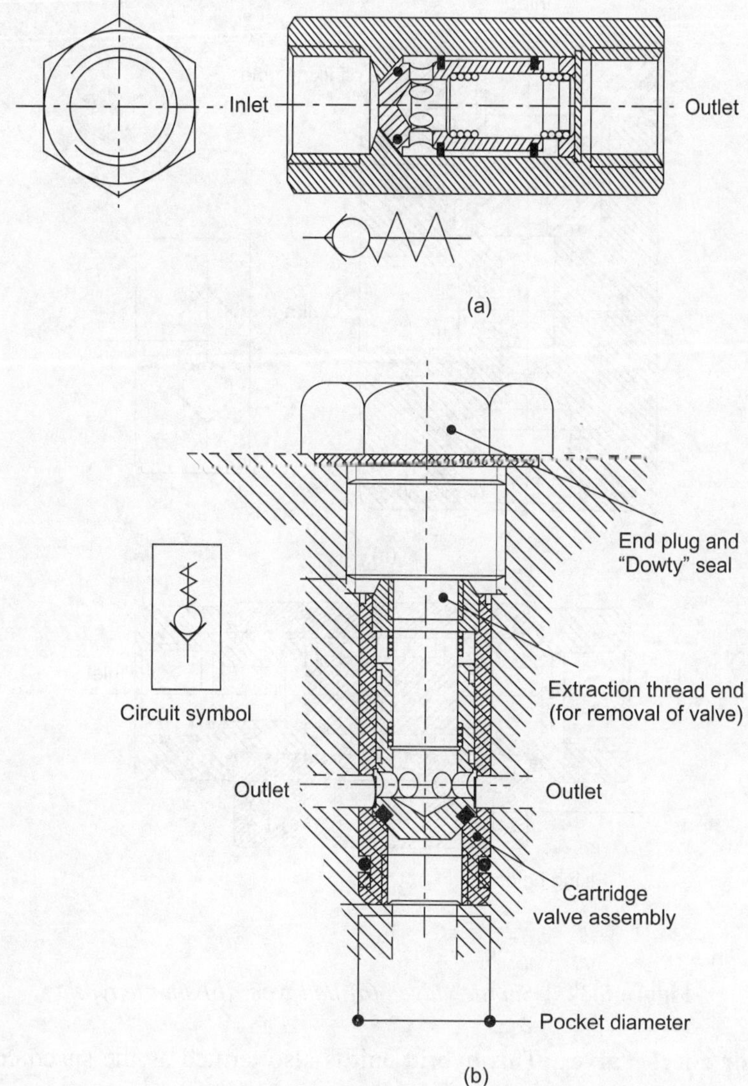

Figure 5.33 *Check valve: (a) plug type, (b) cartridge type*

(ii) Shuttle valve The main purpose of the shuttle valve is to isolate the normal system from an alternate or emergency system. It is small and simple; yet, it is a very important component. A shuttle valve is also known as a double control valve or double check valve, and has two inlets and one outlet. At any given time, flow is shut off in the direction of whichever inlet is unloaded and is open from the loaded inlet to the outlet (Fig. 5.34). A shuttle valve may be installed, for example, when a power unit (cylinder) or a control unit (valve) is to be actuated from two points, which may be remote from one other.

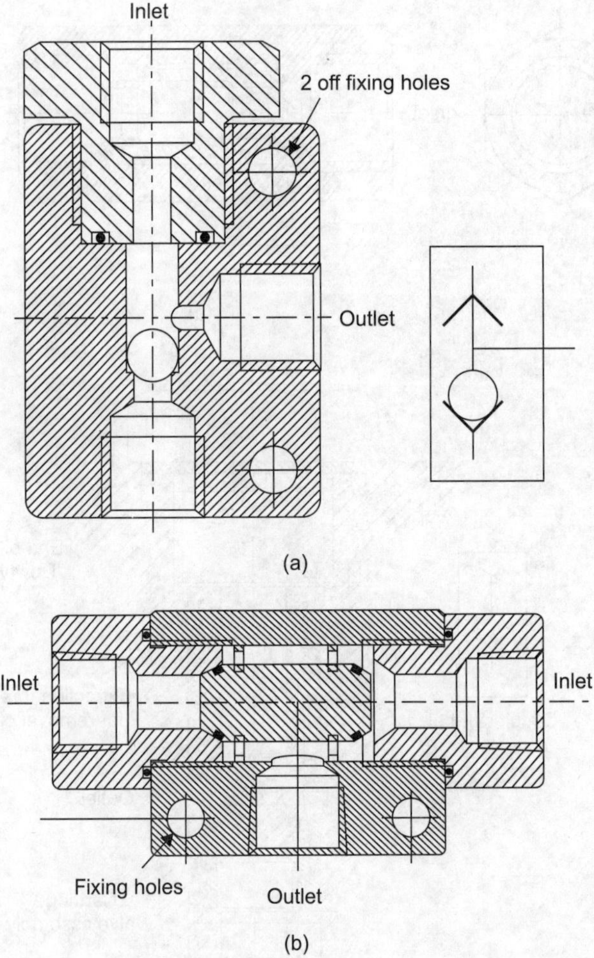

Figure 5.34 *Shuttle valve: (a) ball type, (b) shuttle type*

(iii) Restrictor check valve This hybrid unit is also termed as the speed control valve for pneumatic applications. They are used as flow control valves in pneumatics due to their throttling ability. Incorporating check functionality also makes them non–return valves and this is how they are generally classified.

Usually the throttle of a restrictor check valve is adjustable to permit the regulation of air flow through the valve. The throttling action is effective only in one direction of flow, while in the other direction, free flow is provided through the check (Fig. 5.35). When restrictor check valves are used to control the speed of pneumatic cylinders, a distinction is made between throttling supply air and throttling exhaust air.

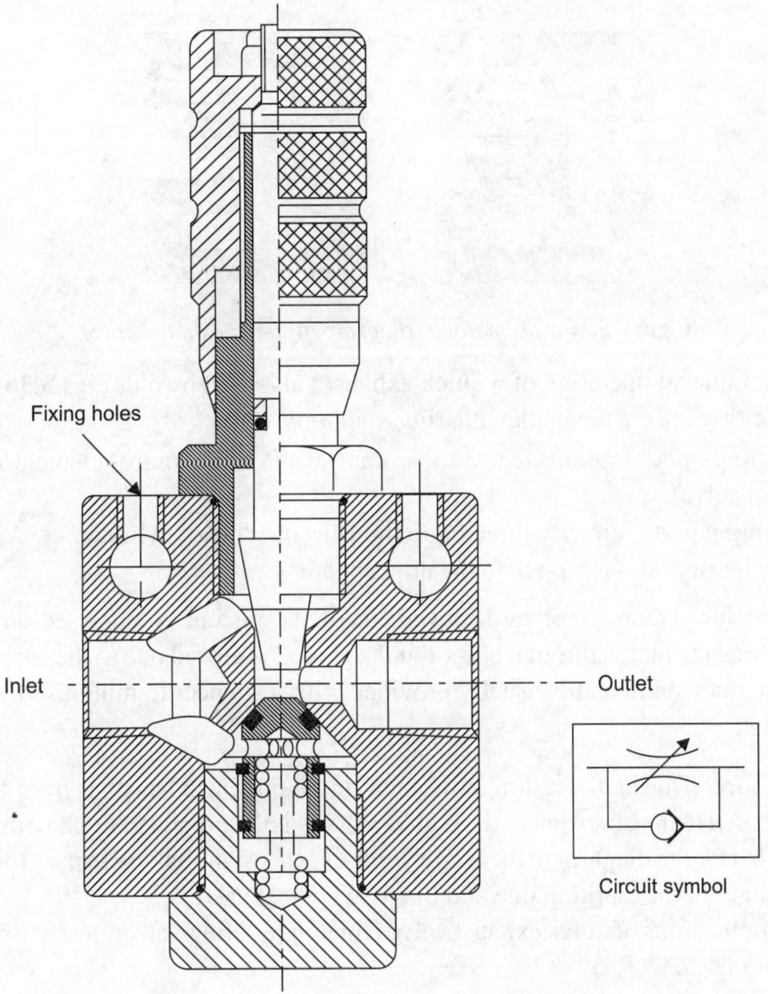

Figure 5.35 *Functional diagram of a restrictor check valve*

(iv) Quick exhaust valves A quick exhaust valve is a typical shuttle valve and is used to exhaust cylinder air quickly to the atmosphere. A schematic diagram of a quick exhaust valve is shown in Fig. 5.36. In many applications, especially those with single acting cylinders, it is common practice to increase the piston speed during the retraction of the cylinder to reduce the cycle time. The higher speed of the piston is made possible by reducing the resistance to the flow of exhausting air during the motion of the cylinder. The resistance can be reduced by expelling the exhaust air to the atmosphere faster by using a quick exhaust valve.

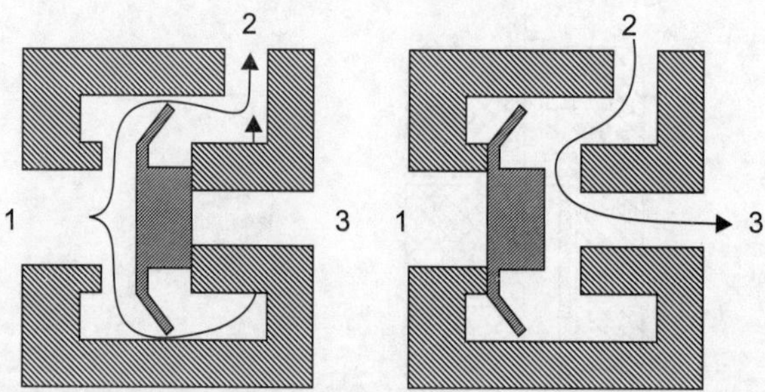

Figure 5.36 *Functional diagram of quick exhaust valve*

The construction and operation of a quick exhaust valve is shown in Fig. 5.36. It consists of a movable disc (also called a flexible ring) and three ports:

- The supply port 1 connected to the output of the final control element (directional control valve).
- The output port 2 directly fitted on to the working port of cylinder.
- The exhaust port 3 left open to the atmosphere.

Forward motion During forward movement of the piston, compressed air is directly admitted behind the piston through ports 1 and 2. Port 3 is closed due to the supply pressure acting on the diaphragm and is usually provided with a silencer to minimise the noise due to the exhaust.

Return motion During the return movement of the piston, exhaust air from the cylinder is directly vented to the atmosphere through port 3 (usually larger and fitted with a silencer). Port 2 is sealed by the diaphragm there by avoiding exhaust air having to pass through long, narrow passages in the working line and the final control valve.

Typical applications of quick exhaust valves for single acting and double acting cylinders are shown in Fig. 5.37.

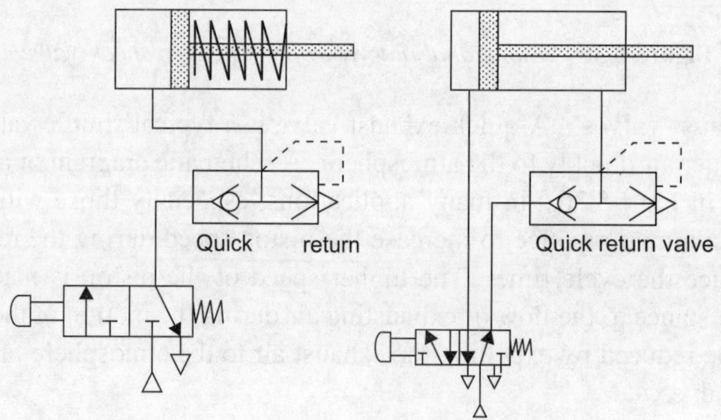

Figure 5.37 *Application of a quick exhaust valve*

(v) Two pressure valve This valve is a pneumatic AND valve. It is also a derivative of the non-return valve. A two pressure valve requires two pressurised inputs to allow an output from itself. The cross sectional views of a two pressure valve in two positions are given in Fig. 5.38. As seen in the figure, this valve has two inputs, 12 and 14, and one output 2. If the compressed air is applied to either input 12 or 14 alone, the spool moves to block the flow, and no signal appears at output 2. If signals are applied to both the inputs 12 and 14, the compressed air flows through the valve, and the signal appears at output 2.

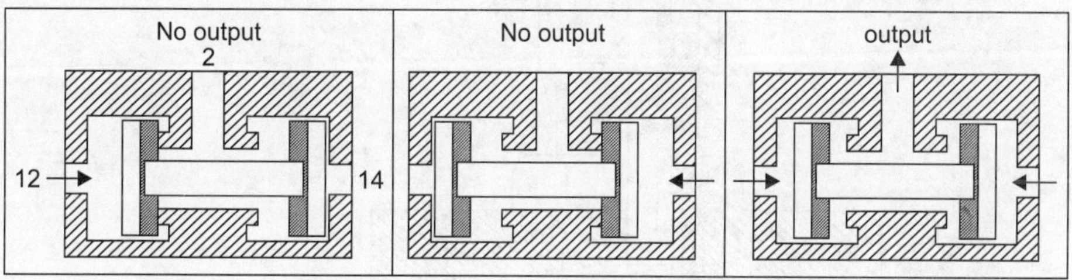

Figure 5.38 *Two pressure valve*

5.2.3 Flow control valves

The function of a flow control valve is self-evident from its name; it regulates the rate of air flow. The control action is limited to the air flow passing through the valve when it is open, maintaining a fixed volume per unit of time. Figure 5.39(a) shows a manifold type variable restrictor flow control valve. Figure 5.39(b) shows an inline type variable restrictor flow control valve. Figure 5.40 shows another design of the flow control valve, in which the flow can be set by turning the knob.

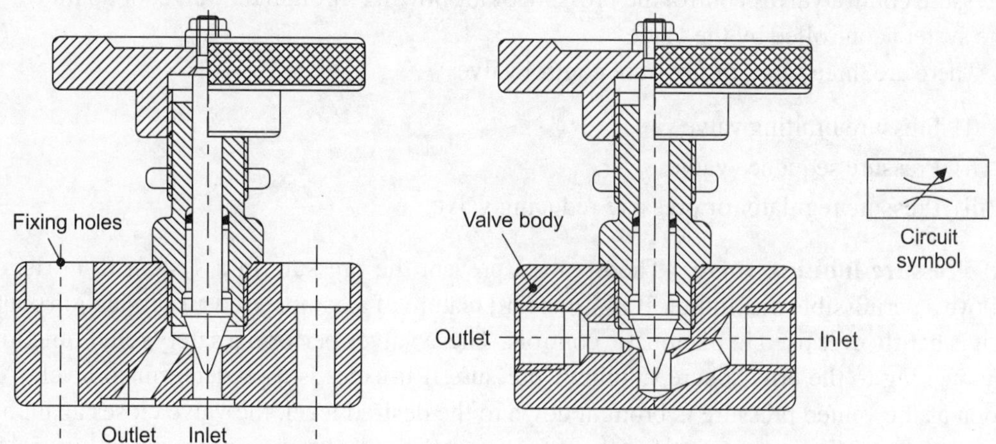

Figure 5.39 *Flow control valve: (a) manifold, (b) inline*

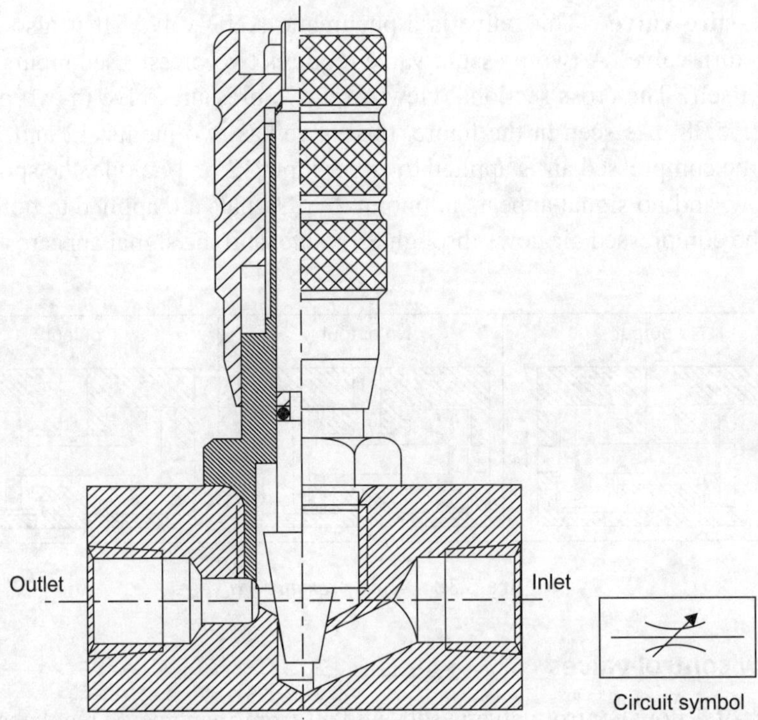

Figure 5.40 *Flow control valve (adjustable)*

5.2.4 Pressure control valve

Compared with hydraulic systems, few pressure control valves are used in pneumatics. Pressure control valves control the pressure of air flowing through the valve, or confined in the system controlled by the valve.

There are three types of pressure control valves:

(i) Pressure limiting valve
(ii) Pressure sequence valve
(iii) Pressure regulator or pressure reducing valve

(i) Pressure limiting valve These valves prevent the pressure in a system from rising above a permissible maximum. It is a standard feature of a compressed air production plant but is hardly ever used in pneumatic controls. These valves perform a safety relief function by opening to the atmosphere to relieve pressure if it exceeds a predetermined level. As soon as the vented pressure is brought down to the desired level, the valve closes again by spring force. The construction features of a pressure limiting valve are shown in Fig. 5.41.

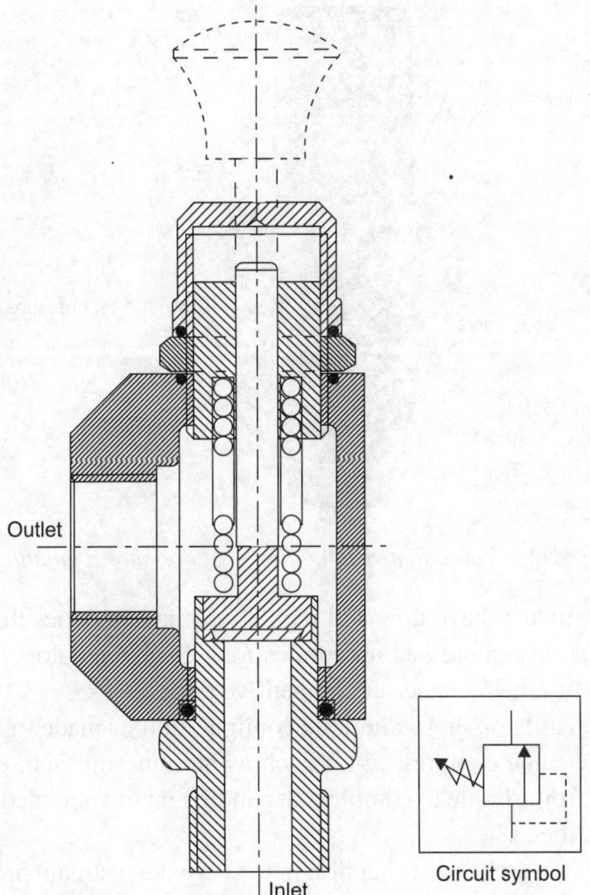

Figure 5.41 *Pressure limiting valve*

(ii) Pressure sequence valve The function of the sequence valve is very similar to that of a pressure limiting valve. It is however used for a different purpose. The outlet of the pressure sequence valve remains closed until the pressure upstream of it builds up to a predetermined value. Only then does the valve open to permit air from the inlet to flow out. The sequence valve is incorporated into pneumatic controls where a minimum pressure level is required for operation to be initiated. They are also used in systems with priority air consumers, where lower priority consumers are supplied with air only after adequate pressure is assured.

(iii) Pressure reducing valve or regulator Pressure regulators maintain a constant output pressure regardless of variations in input pressure or output flow. Regulators contain loading, sensing, actuating, and control components. They can be broadly classified according to their use in the industry- general purpose, special purpose or precision. Three dimensional view of a pressure reducing valve is shown in Fig. 5.42.

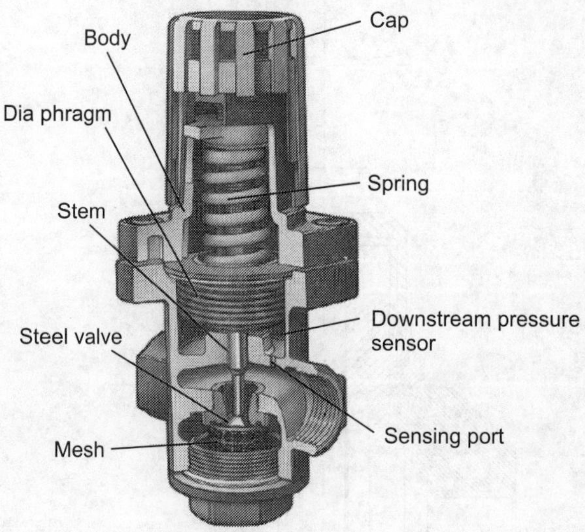

Figure 5.42 *Three dimensional figure of pressure regulating valve*

General-purpose regulators have flow and regulation characteristics that meet the requirements of most industrial compressed-air applications. Such regulators provide long service life and relative ease of maintenance at competitive prices.

Special-purpose regulators have a unique configuration or made of special material for use with fluids other than compressed air such as helium, nitrogen, etc. Regulator construction can range from simple to complex, depending on the intended application and the performance requirements.

Regulators use either a piston or diaphragm to sense downstream pressure. Diaphragms are generally more sensitive to pressure changes Pistons, on the other hand, are generally more rugged and provide a larger effective sensing area in a given size regulator.

When selecting a pressure regulator, the important factors to consider are:

1. Line pressure
2. Minimum and maximum regulated pressure required (maximum flow)
3. Pipe size
4. Regulator adjustment frequency
5. Degree of pressure precision required
6. Accessories or options such as guages and panel mounting
7. Environmental or fluid conditions that could be incompatible with materials used in the regulator

5.3 Valve Specification

The following parameters are considered when selecting a valve:

1. Maximum operating pressure and flow rate required
2. Minimum and maximum operating temperatures
3. Rate of cycling
4. Response time
5. Method of operation
6. Method of mounting

5.4 Valve Sizing

Valve size is an indication of the size of the end connections; however the flow capacity of the valve is the more important criterion. Since gases are compressible and their density changes with pressure, their flow analysis is more complex. The variation in flow based on the pressure drop is also a factor to consider during the valve size determination process.

5.4.1 Valve sizing of flow control valves

An orifice meter measures discharge using the pressure difference between varying cross-sections. In low pressure drop conditions, flow through the outlet opening is restricted when outlet pressure is more than half the inlet pressure (Fig. 5.43a). The flow increases as the outlet pressure decreases.

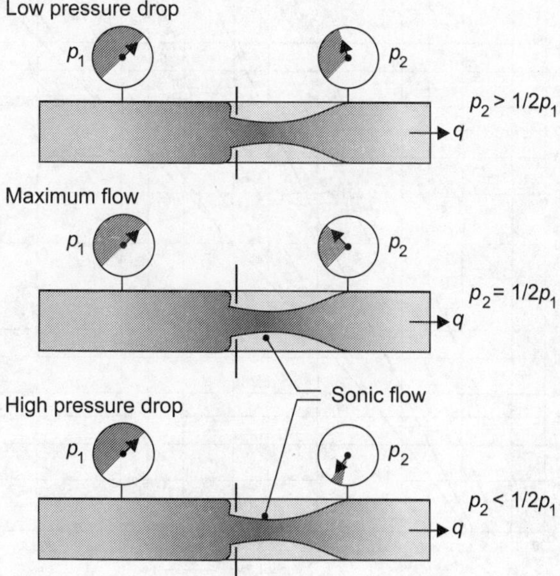

Figure 5.43 *Valve sizing: (a) low pressure drop, (b) choked condition, (c) high pressure drop*

The maximum flow rate cannot exceed the speed of sound and is also known as the choked or critical flow (Fig. 5.43b). This maximum is reached when the outlet pressure decreases to half the inlet pressure. A further decrease in the outlet pressure does not increase the flow at all. High pressure drop flow only depends on the inlet pressure; the outlet pressure is not a factor (Fig. 5.43c).

(i) Equation for low pressure drop

Equation for the low pressure drop is given by:

$$q = N_1 C_v p_1 \left(1 - \frac{2\Delta p}{3p_1}\right) \sqrt{\frac{\Delta p}{p_1 G_g T_1}}$$

where C_v = flow coefficient
q = flow rate, LPM
p_1 = inlet pressure, bar
p_2 = outlet pressure, bar
Δp = pressure drop, bar
G_g = specific gravity of gas = 1 for air
N_1 = constants = 14.42 (supplied by valve manufacturer)

Nomogram for low pressure drop is given in Fig. 5.44. Note that this chart is for $C_v = 1$.

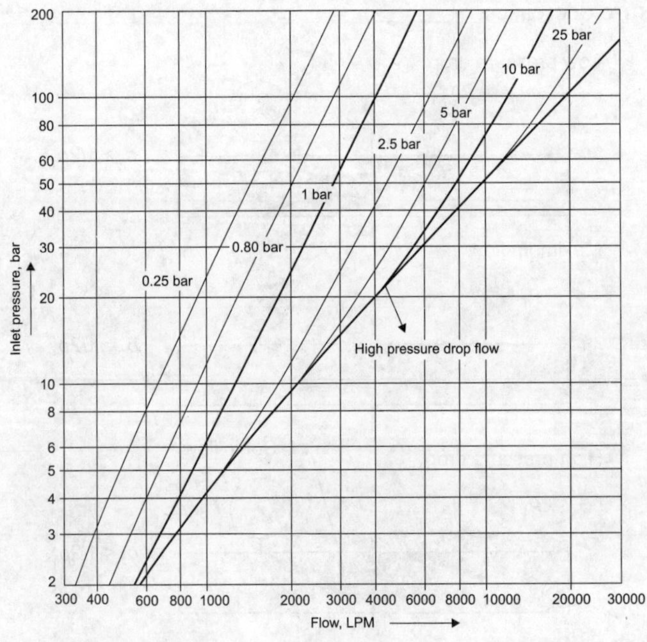

Figure 5.44 *Nomogram for low pressure drop*

(ii) Equation for high pressure drop The equation for the high pressure drop is simple because it depends only on inlet pressure and temperature.

Equation for the high pressure drop is given by:

$$q = 0.471\, N_2 C_v p_1 \sqrt{\frac{1}{G_g T_1}}.$$

where N_2 = constant = 6950 (supplied by the valve manufacturer), T_1 = inlet temperature (K), other variables being the same as defined earlier.

Nomograms for high pressure drops are shown in Figures 5.45 and 5.46.

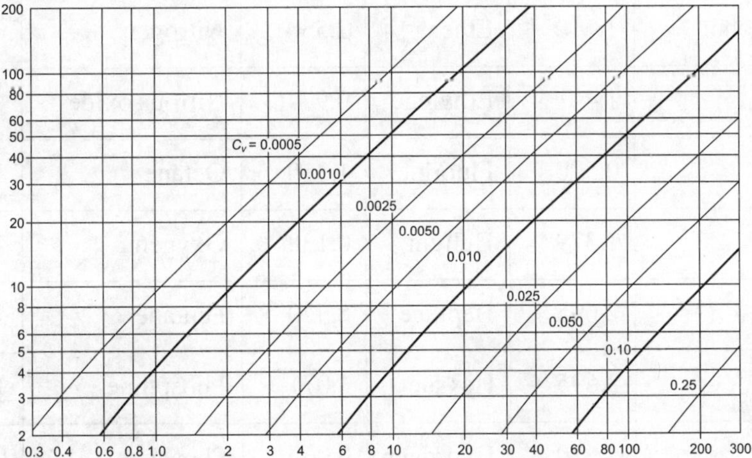

Figure 5.45 *Nomogram for high pressure drop*

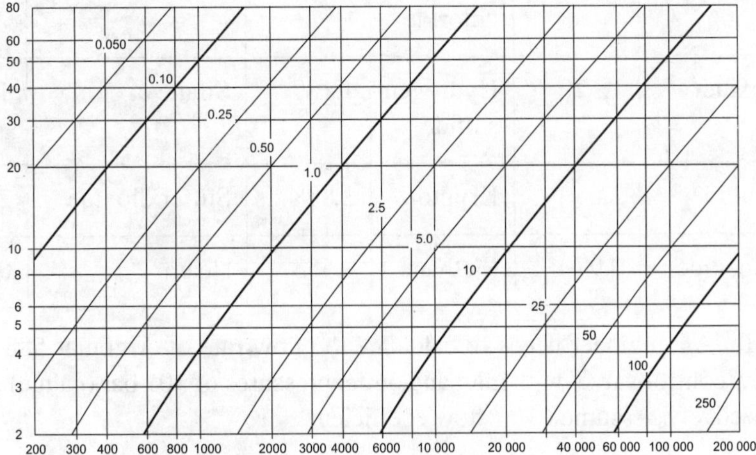

Figure 5.46 *Nomogram for high pressure drop*

Table 5.6 *Specific gravity of commonly used gas in industries*

Gas	Specific gravity	Gas	Specific gravity	Gas	Specific gravity
Acetylene	0.905	Chlorine	2.480	Methyl bromide	3.270
Air	1	Diborane	0.950	Methyl chloride	1.740
Ethyl alcohol	1.590	Ethane	1.040	Naphtalene	4.430
Methyl alcohol	1.110	Ether	2.550	Nitrogen	0.967
Allene	1.40	Ethylene	0.974	Nitrous oxide	1.530
Ammonia	0.590	Fluorine	1.310	Octane	3.943
Argon	1.379	Helium	0.138	Oxygen	1.105
Arsine	2.695	Heptane	5.520	Pentane	2.491
Benzene	2.695	Hexane	2.071	Phosphine	1.183
Bromine	5.520	Hydrogen	0.069	Phosgene	0.420
Butane	2.071	Hydrogen chloride	1.270	Propane	1.547
Carbon dioxide	1.520	Hydrogen sulfide	1.189	Silane	1.114
Carbon disulfide	2.630	Krypton	2.898	Sulfer dioxide	2.210
Carbon monoxide	0.970	Methane	0.555	steam	0.620

Example 5.1 An engineer needs to calculate the flow rate of Argon at 588°K, flowing through a flow control valve, with inlet and outlet pressures of 101 bar (gauge) and 100 bar (gauge) respectively. Assume C_v = flow coefficient = 1

Solution

Note: Use the absolute pressure for calculating the critical pressure ratio.

$$\text{Check } p_2 > \frac{p_1}{2}, \; 101 > \frac{102}{2}$$

Since the condition for low pressure drop is satisfied, we can use the nomogram (Fig. 5.47) for the low pressure drop.

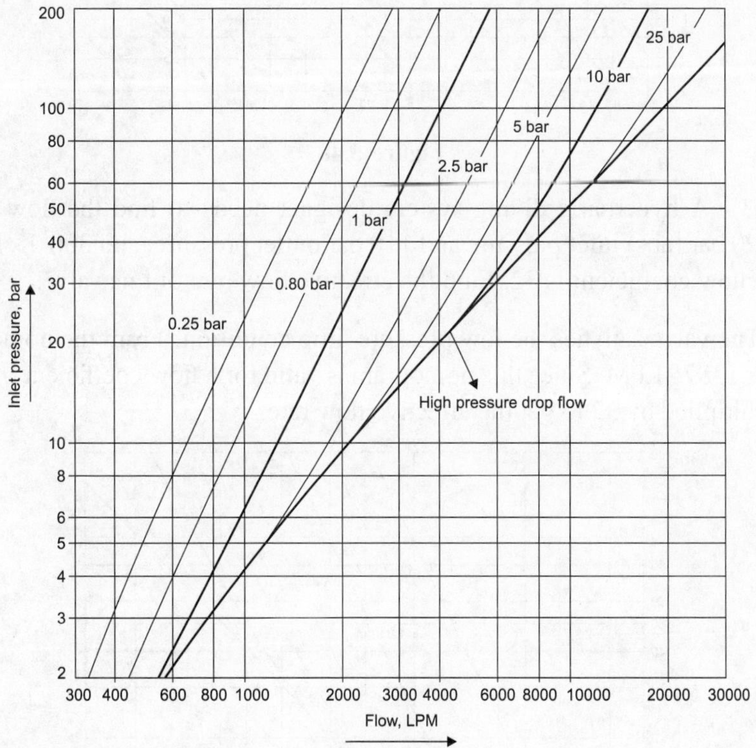

Figure 5.47

From the nomogram, flow $q = 4000$ LPM.

Example 5.2 A semiconductor process engineer in a wafer fabrication unit is replacing a diaphragm shut-off valve in a computer chip manufacturing facility. The process uses phosphine (PH_3) gas at 588°K. The gas runs in a 6.35 mm tube line, supplying an inlet pressure of 20 bar to the shut-off valve. He selects a diaphragm valve of 50 LPM. Verify whether his selection is correct. Assume flow coefficient to be 0.01.

Solution Since the outlet pressure is unknown, a high pressure drop condition can be assumed.

From the nomogram (Fig. 5.48), the flow is 40 LPM, Therefore his selection is correct.

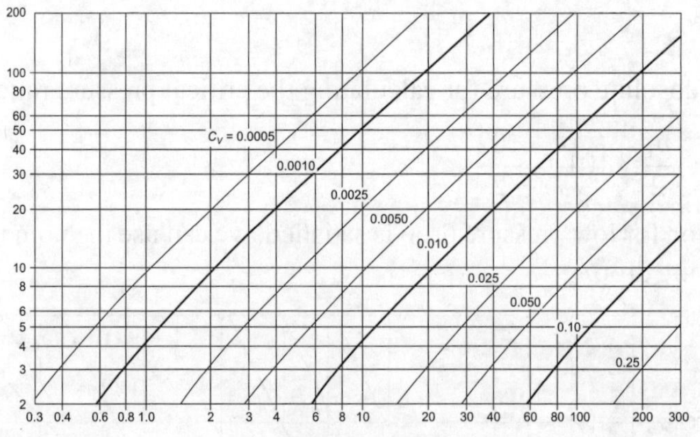

Figure 5.48

Example 5.3 A hyperbaric diving system designer needs to find the flow rate for air flowing at 120 bar (abs) inlet pressure and 110 bar outlet pressure, and 300°K. He selects a valve whose flow coefficient is 32. Find the effective flow rate of this valve.

Solution The valve satisfies the low pressure drop condition. From the nomogram (Fig. 5.49), flow is 13875 LPM. Since this nomogram is valid for a flow coefficient of 1, the flow should be multiplied by 32 to get the effective flow rate.

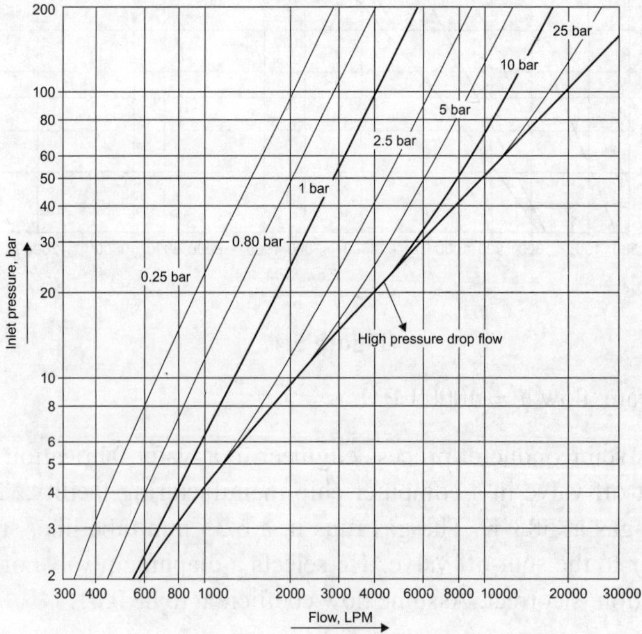

Figure 5.49

Therefore effective flow rate = 13875 × 32 = 444000 LPM.

5.4.2 Sizing of direction control valves

The choice of a solenoid valve determines the regulation and performance of the system. In order to select the most appropriate choice of solenoid valve, several parameters have to be established. The calculation method based on the flow coefficient C_v, has proved highly is the most practical since it can be based on:

- Required flow rate
- Flow resistance
- Type of fluid and its relative viscosity
- Specific gravity and temperature

(i) Solenoid valves for gases

The equation for the low pressure drop $\Delta p \leq \frac{1}{2}p_1$ is given by:

$$q = N_1 C_v \sqrt{\frac{\Delta p \times p_2}{G_g T_1}},$$

where, C_v is the flow coefficient; q the flow rate in (N m/hour); p_1 and p_2 the outlet pressures in (bar); Δp the pressure drop; G_g the specific gravity of gas; T_1 the inlet temperature in (°K) and

N_1 a constant $=$ 514 (supplied by valve manufacturer).

When the low pressure drop $\Delta p > \frac{1}{2}p_1$,

constant $N_1 = 757$ (supplied by the manufacturer).

The equation for low pressure drop $\Delta p \leq \frac{1}{2}p_1$ is given by

$$q = N_3\, C_v \sqrt{\frac{\Delta p}{v_{\text{steam}}}},$$

where, C_v is the flow coefficient; q the flow rate in (kg/hour); Δp the pressure drop; v_{steam} the specific volume of steam in (m^{3}kg); and

$N_3 =$ a constant $=$ 31.7 (supplied by the manufacturer).

Example 5.4 Size the directional control valve, using the nomogram given in Fig. 5.50.
Given air (G_g = 1.3); inlet temperature = 20°C; $C_v = 0.6$; inlet pressure $=$ 12 bar; and pressure drop $=$ 3 bar.

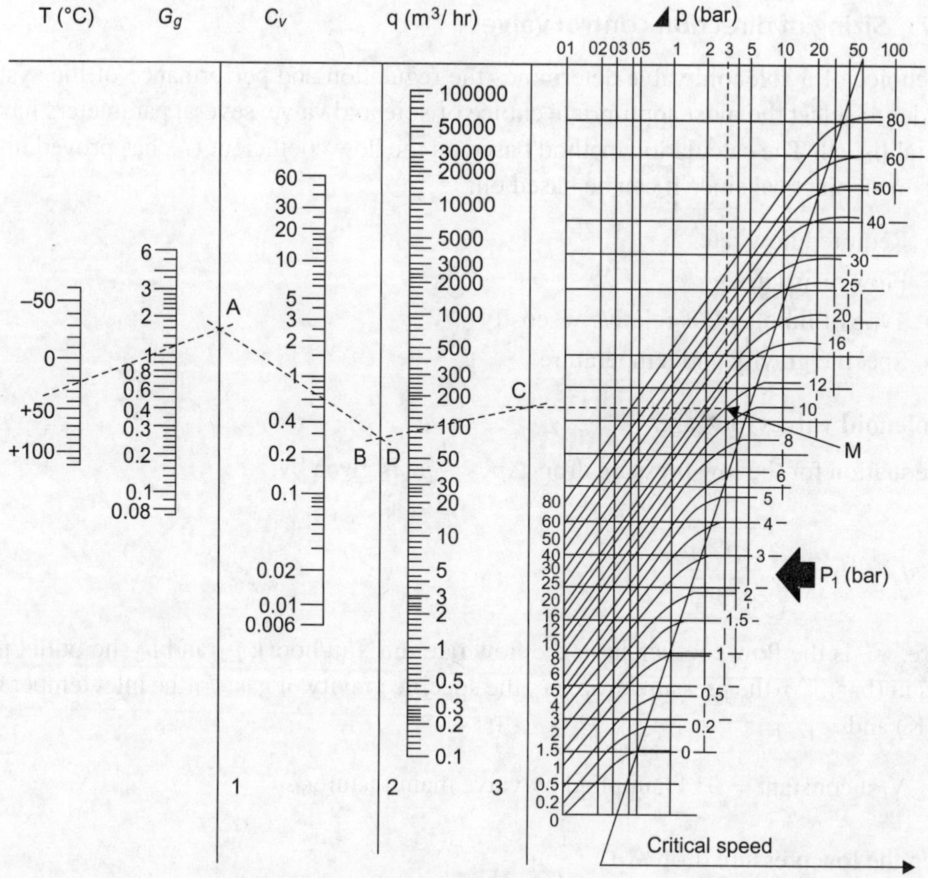

Figure 5.50

Solution Locate 20°C on the temperature scale and 1.3 on G_g axis, extend to meet axis 1 at A.

Locate 0.6 on C_v axis and join this point and A. Extend to meet axis 2 at B.

Locate pressure drop of 3 bar on Δp axis. Let it cut the inlet pressure line at M. Draw a horizontal line to cut the axis at C. Join C and D, it cuts Q axis at D. Read the Q value at point D = 80 Nm³/hour.

Equation for the low pressure drop $\Delta p \leq \frac{1}{2}p_1$ is given by:

$$q = N_1 C_v \sqrt{\frac{\Delta p \times p_2}{G_g T_1}}$$

$$q = 514 \times 0.6 \sqrt{\frac{3 \times 9}{1.3 \times (20 + 273)}}$$

$q = 82$ N m³/hour, which is almost close to nomogram value.

(ii) Solenoid valves for steam

The equation for low pressure drop $\Delta p > \frac{1}{2}p_1$ is given by:

$$q = N_4 C_v \sqrt{\frac{p_1}{v_{\text{steam}}}},$$

where p_1 is the outlet pressures in (bar), and

$N_4 = a$ constant $= 22.4$ (supplied by the manufacturer).

The other variables are the same as defined earlier.

Example 5.5 Size the directional control valve using the nomogram given in Fig. 5.51. For dry saturated steam, $C_v = 0.53$; outlet pressure $= 5.5$ bar and pressure drop is 2 bar.

Solution

Equation for the low pressure drop $\Delta p \leq \frac{1}{2}p_1$ is given by:

$$q = N_3 C_v \sqrt{\frac{\Delta p}{v_{\text{steam}}}}.$$

$C_v =$ flow coefficient $= 0.53$; $p_1 = 7$ bar; $p_2 = 5$ bar; $\Delta p = 2$ bar

$v_{\text{steam}} =$ specific volume of steam $= 0.35$ m^3/kg at the rate of 5.5 bar

$N_3 = 31.7$ (supplied by valve manufacturer)

$$q = 31.7 \times 0.5\sqrt{\frac{1.6}{0.35}}.$$

Solving, we get:

$q = 36.88$ kg/hour.

Solution using a nomogram is shown in Fig. 5.51. From the nomogram: $q = 36$ kg/hour, which is close to the theoretical value.

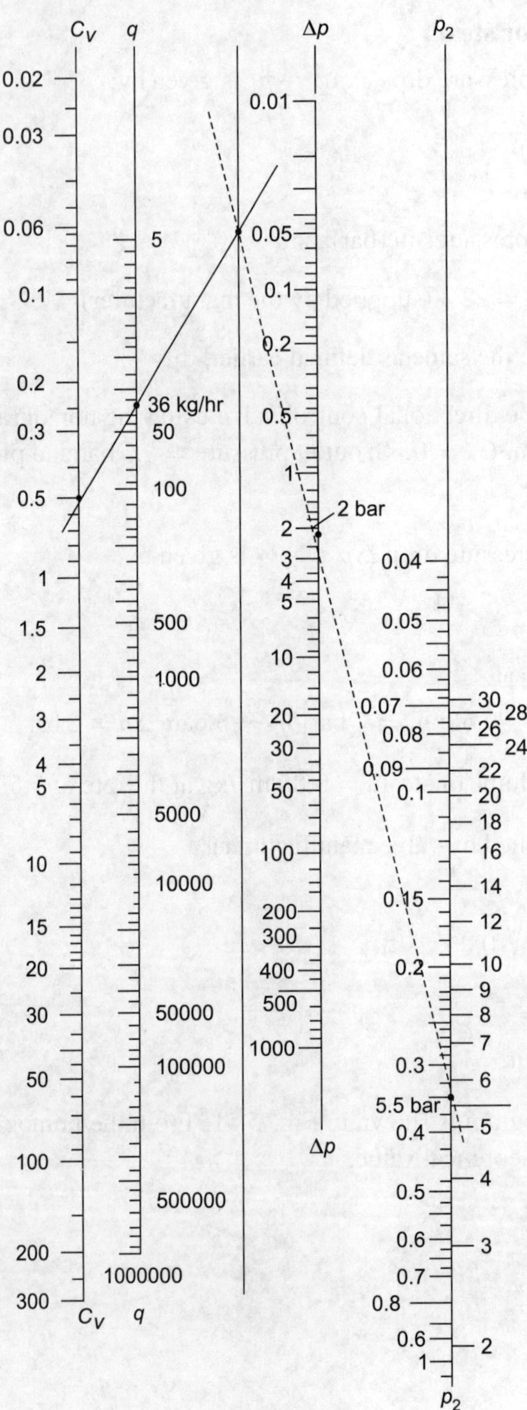

C_V
q
Δp
p_2
0.02
0.03
0.06
0.1
0.2
0.3
0.5
1
1.5
2
3
4
5
10
15
20
30
50
100
200
300
5
36 kg/hr
50
100
500
1000
5000
10000
50000
100000
500000
1000000
0.01
0.05
0.1
0.2
0.5
1
2 bar
2
3
4
5
10
20
30
50
100
200
300
400
500
1000
0.04
0.05
0.06
0.07
0.08
0.09
0.1
0.15
0.2
0.3
5.5 bar
0.4
0.5
0.6
0.7
0.8
0.6
1
30
28
26
24
22
20
18
16
14
12
10
9
8
7
6
5
4
3
2

Figure 5.51

Summary

1. A valve is defined as any device by which the flow of fluid may be started, stopped or otherwise regulated by a movable part that opens or obstructs passage. As applied in pneumatic power systems, valves are used for controlling the flow, the pressure and the direction of the fluid flow.
2. Valves are classified according to their use: flow control, pressure control, and directional control. Some valves have multiple functions that fall into more than one classification.
3. Directional control valves are designed to direct the flow of fluid at the desired time, to the point in a fluid power system where it will do work. The driving of a ram back and forth in its cylinder is an example of when a directional control valve is used.
4. Directional control valves for hydraulic and pneumatic systems are similar in design and operation, with one major difference. The return port of a hydraulic valve is ported through a return line to the reservoir; the corresponding port of a pneumatic valve (the exhaust port), is usually vented to the atmosphere.
5. Directional control valves may be operated by differences in pressure acting on opposite sides of the valve element, or they may be positioned manually, mechanically or electrically. Often two or more methods of operating the same valve are used in different phases of its action.
6. Directional control valves may be classified in several ways: by the type of control, the number of ports in the valve housing, and the specific function of the valve. The most common method is by the type of valve element used in the construction of the valve. The most common types of valve elements are the ball, cone or sleeve, poppet, rotary spool and sliding spool. The basic operating principles of the poppet, rotary spool, and sliding spool valve elements are discussed in this chapter.
7. Check valves are used in fluid systems to permit flow in one direction and to prevent flow in the other direction. They are classified as one-way directional control valves. The check valve may be installed independently in a line to allow flow in one direction only, or it may be used as an integral part of globe, sequence and pressure-reducing valves.
8. Check valves are available in various designs. They are opened by the force of fluid in motion flowing in one direction, and are closed by fluid attempting to flow in the opposite direction. The force of gravity or the action of a spring aids in closing the valve.
9. The main purpose of the shuttle valve is to isolate the normal system from an alternate or emergency system. It is small and simple; yet, it is a very important component.
10. The term two-way indicates that the valve contains and controls two functional flow control ports-an inlet and an outlet. Three-way valves contain a pressure port, a cylinder port, and a return or exhaust port. The three-way directional control valve

is designed to operate an actuating unit in one direction; it permits either the load on the actuating unit or a spring to return the unit to its original position. Most actuating devices require system pressure for operation in either direction. The four-way directional control valve, which contains four ports, is used to control the operation of such devices. The four-way valve is also used in some systems to control the operation of other valves. It is one of the most widely used directional control valves in fluid power systems.

11. A quick exhaust valve is a typical shuttle valve and is used to exhaust cylinder air quickly to the atmosphere.
12. Pressure limiting valves prevent the pressure in a system from rising above a permissible maximum. The functions of a pressure limiting and a pressure sequence valve are similar. A pressure regulator maintains constant pressure in the out stream.

EXERCISES

Fill in the blanks

1. Valves are defined as devices to control, or regulate the commencement and — of air.
2. On resetting valves, e.g., those equipped with a return spring, the — position is the position assumed by the moving parts of the valve when it is connected but not actuated.
3. Direct control of a valve means that valve operates directly by means of an actuating element without any — elements being operated.
4. —— automatically limit flow to a single direction at the point where they are installed in a compressed air line.
5. In a pressure regulating valve the inlet pressure is —— than the outlet pressure.
6. Poppet valves give —— stroke and fewer moving parts give minimum wear.
7. The quick exhaust valve is used to exhaust the cylinder air quickly to ——.
8. Compared with hydraulic systems, — varieties of pressure control valves are used in pneumatics.
9. As per ISO 1219-1:2006 designation, number 12 or 14 indicate —— ports.
10. In a rotary spool directional control valve, the rotating part is called — and the stationary part is called —.
11. The two most common basic flow control devices used in a pneumatic system are fixed-sized orifices and — valves.
12. The meter — flow control circuit is the preferred method to use for controlling the operating speed of cylinders in pneumatic circuits.

State whether true or false

1. Restrictor check valves are non-return valves which are also employed as flow control valves.
2. Quick exhaust valves are designed to decrease the position speed in the cylinder.
3. Spring force set on a pressure limiting valve or a sequence valve corresponds to the minimum permissible or minimum desired pressure of the controlled fluid.
4. Ball seat valves ensure perfect sealing at all times in pneumatic circuits.
5. Two-way valves are used where a pure control function is required, i.e., when downstream equipment does not need exhausting to the atmosphere via this valve.
6. The pressure sequence valve holds the working pressure largely constant.
7. A two-way pressure valve is the pneumatic OR valve.
8. A shuttle valve is the pneumatic OR valve.
9. Pressure loss in a manifold type valve is more than in inline type valves.
10. A suspended disc type valve has very long actuation movement.

Review questions

1. List the functions of a pneumatic valve.
2. Classify the various pneumatic valves.
3. Classify the various direction control valve(DCV).
4. How do 2/2 valves differ from 4/2 pneumatic direction control valves?
5. What are the advantages of a poppet valve over a ball valve?
6. Explain the working of a 3/2 direction control valve with a neat sketch.
7. Explain the working of a 5/3 direction control valve with a neat sketch.
8. Mention some applications of a 4/3 direction control valve.
9. Differentiate between rotary valve and spool type valve.
10. How are non-return valves classified?
11. Mention some applications of non-return valves.
12. With the help of a neat sketch, explain the construction and working of quick exhaust valves.
13. How are pressure control valves classified?
14. Explain the difference between pressure limiting valves and sequence valves.
15. Explain the working of pressure limiting valves.
16. Compare and contrast two-way valves and shuttle valves. Mention their applications.

Problems

1. A process engineer needs to calculate the flow rate of Fluorine at 605°K, flowing through a flow control valve, with inlet and outlet pressures of 99 bar (gauge) and 96.5 bar (gauge) respectively. Assume C_v = flow coefficient = 1.1 .

2. A semiconductor process engineer in a wafer fabrication unit is replacing a diaphragm shut-off valve in a computer chip manufacturing facility. The process uses phosphine (PH_3) gas at 550°K. The gas runs in a 7 mm tube line, supplying an inlet pressure of 30 bar to the shut-off valve. He selects a diaphragm valve of 55 LPM. Verify whether his selection is correct. Assume flow coefficient to be 0.01.
3. A hyperbaric diving system designer needs to find the flow rate for air flowing at 100 bar (abs) inlet pressure and 105 outlet pressure, and 300°K. He selects a valve whose flow coefficient is 10. Find the effective flow rate of this valve.
4. Size the following directional control valve using the nomogram method.

   Gas ($G_g = 1.3$); inlet temperature = 50°C; $C_v = 0.61$; inlet pressure = 12 bar; pressure drop = 3 bar
5. Using nomogram, size the following directional control valve.

   $C_v = 0.52$ = for dry saturated steam; outlet pressure = 5 bar; pressure drop is = 1.8 bar.

Answers

Fill in the blanks

1. direction 2. neutral 3. intermediate 4. Non-return valves 5. higher

6. shorter 7. atmosphere 8. few 9. pilot 10. core/sleeve

11. needle 12. out

True or false

1. True 2. False 3. False 4. False 5. True

6. True 7. False 8. True 9. False 10. False

Problems

1. 6820 LPM 2. 60 LPM 3.90000 LPM 4. 82 N m^3/hour 5. 36.87 kg/hour

6 | SINGLE ACTUATOR CIRCUITS

Upon completion of this chapter, the reader should be able to

- Differentiate between pneumatic circuits and pneumatic circuit diagrams
- Understand the basic rules used in the design of pneumatic circuits
- Explain memory, delay, AND, OR and NOT functions
- Explain direct and indirect control of single acting cylinders
- Explain direct and indirect control of double acting cylinders
- Differentiate between throttling of supply and exhaust air
- Study methods of checking end positions of a cylinder
- Design pressure and time dependent circuits

6.1 Pneumatic Circuits and Pneumatic Circuit Diagrams

Pneumatic control systems can be designed in the form of pneumatic circuits. A pneumatic circuit is composed of pneumatic components such as cylinders, directional control valves, flow control valves, pressure regulators, signal processing elements such as shuttle valves, two pressure valves etc. Pneumatic circuits have the following functions:

1. To control the entry and exit of compressed air in cylinders
2. To use one valve to control another valve
3. To control actuators or any other pneumatic devices

A pneumatic circuit diagram uses pneumatic symbols to describe its design, in accordance to some basic rules. To be able to design pneumatic circuits, a basic knowledge of simple pneumatic circuits is necessary as it helps in designing more complicated circuits involving multiple cylinders.

6.2 Single Acting Cylinder Control

6.2.1 Direct control of a single acting cylinder

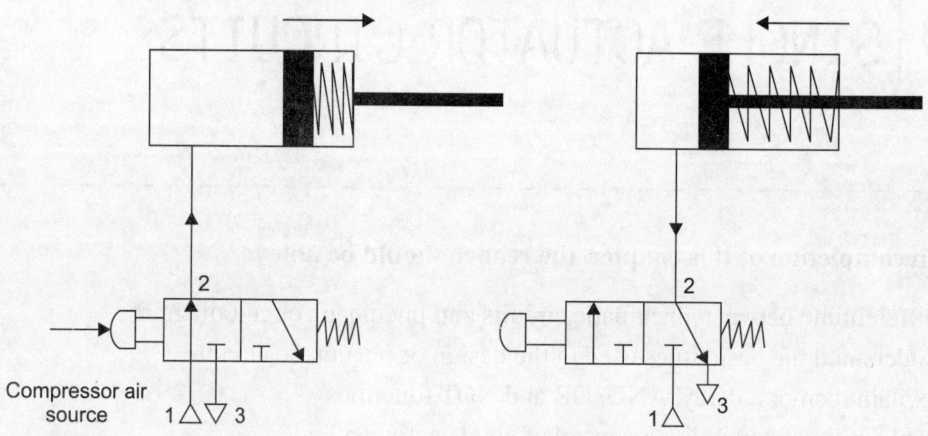

Figure 6.1 *Direct control of a single acting cylinder*

Pneumatic cylinders can be directly controlled by the actuation of final directional control valves (Fig. 6.1). These valves can be controlled manually or electrically. This type of circuit can be used for small cylinders, as well as cylinders operating at low speeds with low flow rate requirements. When the directional control valve is actuated by a push button, the valve switches over to the open position, allowing the working medium into the cylinder volume. This results in the forward motion of the piston. When the push button is released, the reset spring of the valve restores the valve to the initial closed position. The cylinder space is connected to the exhaust port and the piston retracts, either due to spring action or due to supply pressure applied from the other port.

Example 6.1 A small single acting cylinder is to extend and clamp a workpiece when a push button is pressed. As long as the push button is activated, the cylinder should remain in the clamped position. If the push button is released, the clamp is to retract. Use an additional start button. The schematic diagram of the setup is shown in Fig. 6.2.

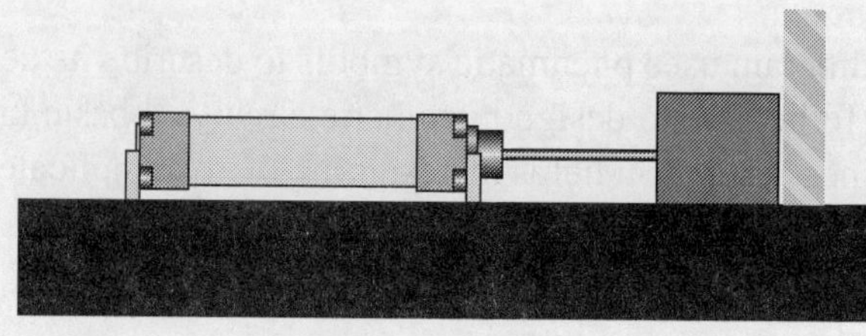

Figure 6.2

Solution The control valve used for the single acting cylinder is a 3/2 valve. In this case, since the cylinder is of small capacity, the operation can be directly controlled by a push button 3/2 directional control valve with a spring return. Valve 1 is usually referred to as final control valve.

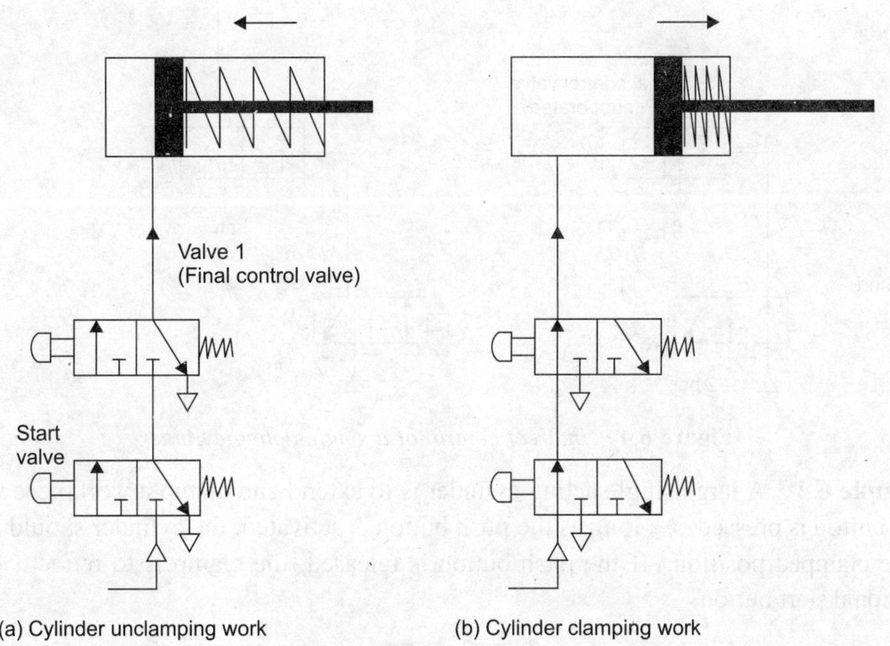

Figure 6.3

When the start button and valve 1 are operated, the cylinder moves forward to clamp the workpiece. When the start button and valve 1 are released, the cylinder comes back to the retracted position as shown in Fig. 6.3.

6.2.2 Indirect control of single acting cylinder

This type of circuit (Fig. 6.4) is suitable for large single cylinders as well as cylinders operating at high speeds. The final pilot control valve is actuated by a normally closed 3/2 push button operated valve and can handle large quantities of air. When the push button is pressed, the 3/2 NC valve generates a pilot signal (12) to the final valve and this connects the working medium to the piston side of the cylinder, thereby advancing it. When the push button is released, pilot air from final valve is vented to the atmosphere through 3/2 NC DCV.

The signal pressure required is around 1–1.5 bar. The working pressure passing through the final control valve depends on the force requirement, which is around 4–6 bar. Indirect control permits processing of input signals. Single piloted valves are rarely used in applications where the piston has to retract immediately upon receiving the set pilot signal.

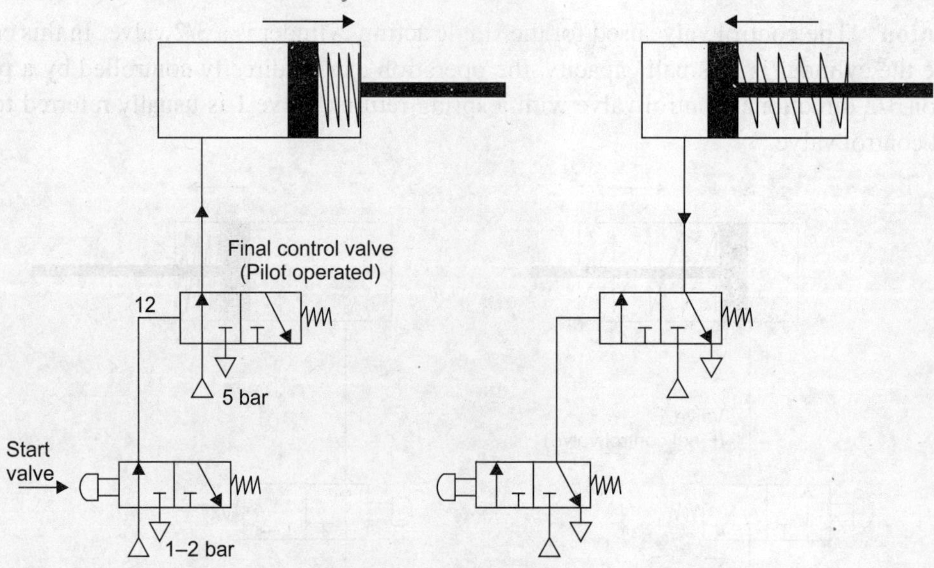

Figure 6.4 *Indirect control of a single acting cylinder*

Example 6.2 A large single acting cylinder is to extend and clamp a workpiece when a push button is pressed. As long as the push button is activated, the cylinder should remain in the clamped position. If the push button is released, the clamp is to retract. Use an additional start button.

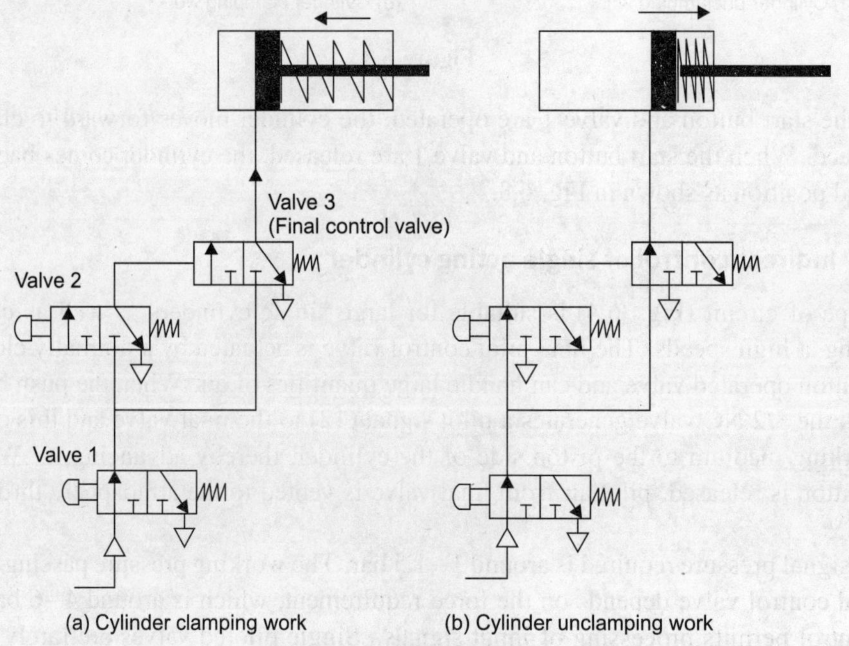

Figure 6.5

The control valve used for a single acting cylinder is the 3/2 valve. In this case, since the cylinder is of large capacity, indirect control should be used for the operation as shown in Fig. 6.5. Valve 3 is usually called the final control valve.

Valve 2 is a small capacity valve which controls the large capacity valve 3. When valve 2 is unactuated, the cylinder is in the retracted condition. When valve 2 is actuated the cylinder is in the extended position to clamp the workpiece.

6.2.3 Control of a single acting cylinder using an "OR" valve

A shuttle valve is also known as a double control valve or a double check valve. This valve is also called an OR valve. A shuttle valve has two inlets and one outlet (Fig. 6.6). At any given time, the flow is shut off in the direction of whichever inlet is unloaded, and is open from the loaded inlet to the outlet. A shuttle valve may be installed when the cylinder or valve is to be actuated from two points remote from each other.

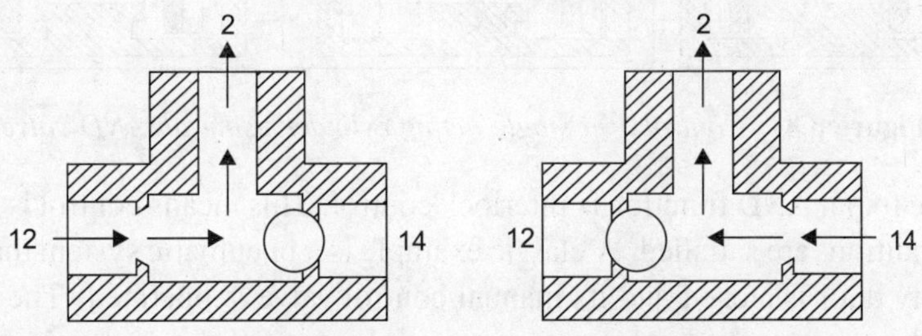

Figure 6.6 *Shuttle valve (OR valve)*

The single acting cylinder in Fig. 6.7 can be operated by two different circuits; e.g., manual operation in combination with automatic circuit signals. Either control valve ① or roller operated control valve ② can independently operate the cylinder, demonstrating the OR functionality in the circuit in Fig. 6.7.

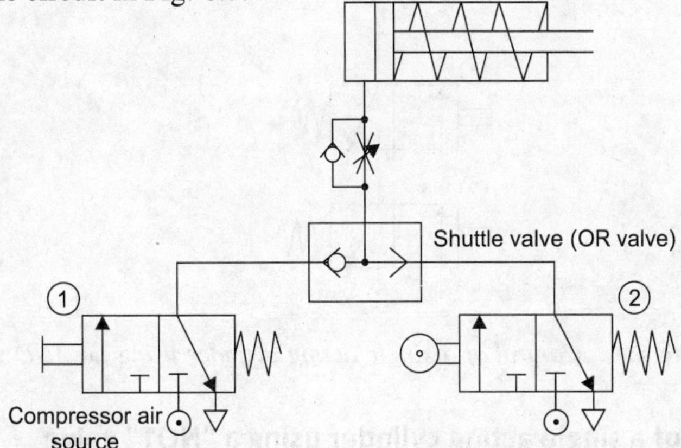

Figure 6.7 *Control of a single acting cylinder using OR valve*

6.2.4 Control of single acting cylinder using an "AND" valve

This pneumatic AND valve is a derivative of a non-return valve. A two-way pressure valve requires two pressurised inputs to allow an output from itself. The cross-sectional views of two-way pressure valves in two positions are given in Fig. 6.8. As shown in the Fig. 6.8, this valve has two inputs 12 and 14, and one output 2. If compressed air is applied to either of the inputs 12 or 14, the spool moves to block the flow, and no signal appears at output 2. If signals are applied to both the inputs 12 and 14, compressed air flows through the valve, and the signal appears at output 2.

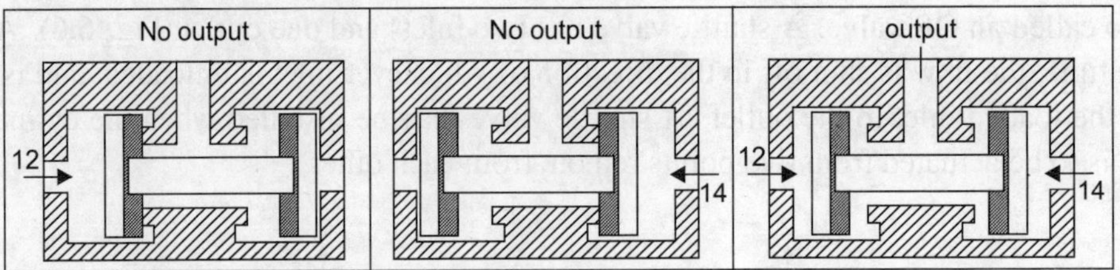

Figure 6.8 *Control of a single acting cylinder using an AND valve*

Another name for an AND function is interlock control. This means control is possible only when two conditions are satisfied. A classic example is a pneumatic system that works only when its safety door is closed and its manual control valve is operated. The flow passage only opens when both control valves are operated. Figure 6.9 shows the circuit diagram of an AND function circuit. The cylinder works only when both valve ① and ② are operated.

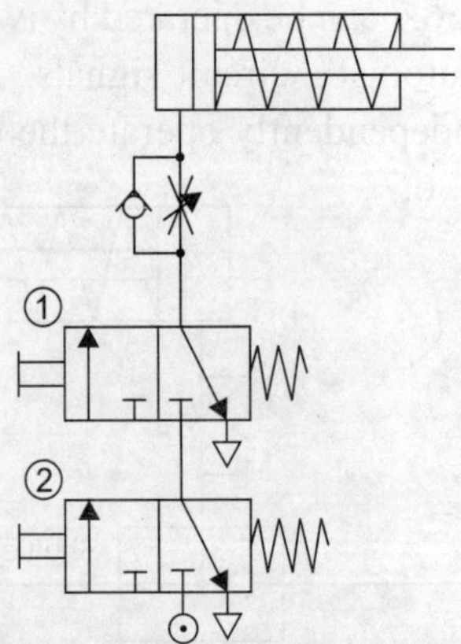

Figure 6.9 *Control of a single acting cylinder using an AND valve*

6.2.5 Control of a single acting cylinder using a "NOT" valve

Another name for a NOT function is inverse control. In order to hold or lock an operating conveyor or a similar machine, the cylinder must be locked until a signal to cancel the lock is received. Therefore, the signal to cancel the lock should be operated by a normally open control valve. To cancel the lock, however, the same signal must also cancel locks on other devices, like the indication signal ③. Figure 6.10 shows how a normally closed control valve ① can be used to cut off the normally open control valve ② to achieve the goal of changing the signal.

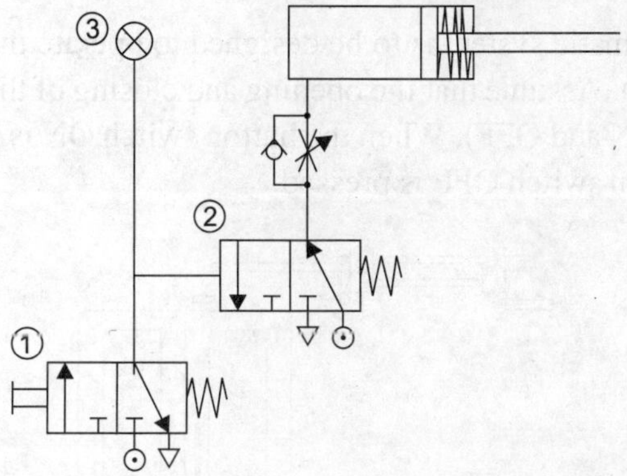

Figure 6.10 *Control of a single acting cylinder using a NOT valve*

6.3 Direct Control of a Double Acting Cylinder

The only difference between a single acting cylinder and a double acting cylinder is that the latter uses a 5/2 directional control valve instead of a 3/2 directional control valve (Fig. 6.11). Usually when a double acting cylinder is not being operated, outlet B and inlet P are connected. In this circuit, when the operation button is manually pushed, the double acting cylinder moves back and forth once.

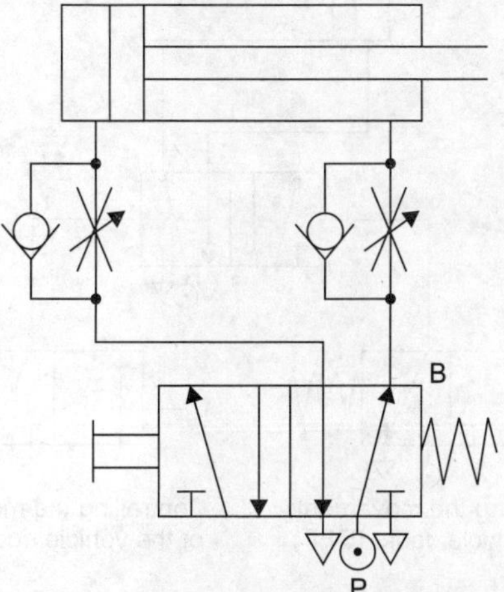

Figure 6.11 *Direct control of a double acting cylinder*

In order to control the speed in both directions, flow control valves are connected to the inlets on both sides of the cylinder. The direction of the flow control valve is opposite to that of the release of air by the flow control valve in a single acting cylinder. Compared to the throttle inlet, the flow control valve is tougher and more stable. Connecting the circuit in this way allows sufficient air pressure and energy at the input to drive the piston.

Example 6.3 A pneumatic system is to be designed to operate the doors of public transport vehicles (Fig. 6.12). Assume that the opening and closing of the doors is controlled by two button switches (ON and OFF). When the button switch ON is pressed the door opens; it closes when the button switch OFF is pressed.

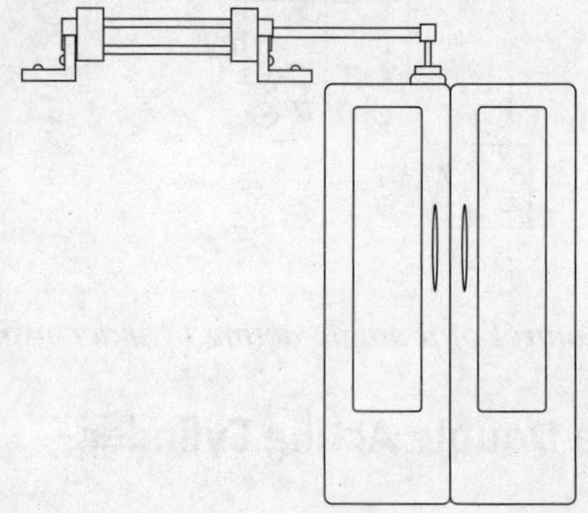

Figure 6.12 *Pneumatic system to control vehicle door*

Solution The solution is given in Fig. 6.13, which is self-explanatory.

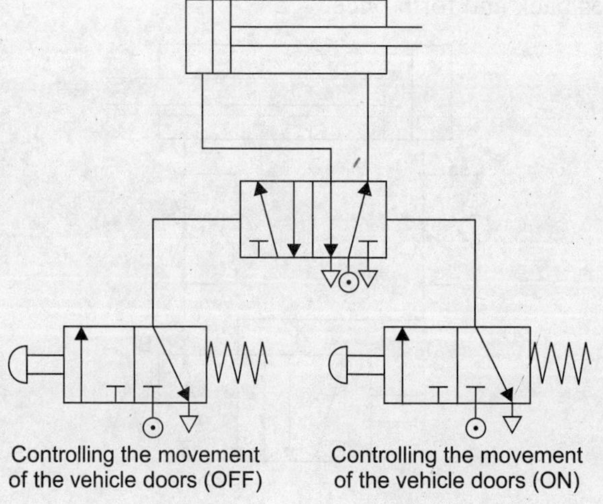

Figure 6.13 *Pneumatic circuit to control the vehicle door*

6.3.1 Indirect control of double acting cylinder using a memory valve

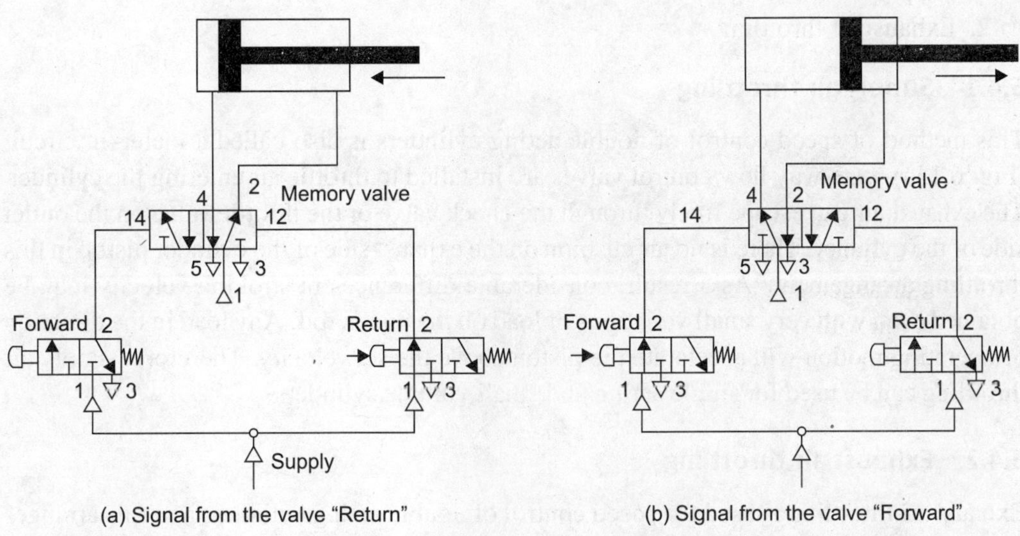

Figure 6.14 *Indirect control of a double acting cylinder using a memory valve*

When the 3/2 valve meant for forward motion (Fig. 6.14b) is pressed, the 5/2 memory valve switches over as a result of the signal applied to its pilot port 14. The piston travels out and remains in the forward end position. The double piloted valve is called a memory valve because at this point even if the push button meant for forward motion is released, the final 5/2 control valve remains in the actuated state as both the pilot ports of the 5/2 valves are exposed to atmospheric pressure and the piston remains in the forward end position.

When the 3/2 way valve meant for return motion (Fig. 6.14a) is pressed, the 5/2 way valve switches back to the initial position as a result of the signal applied to its pilot port 12. The piston now returns to its initial position at the rear end and remains there. Now even if the push button for return is released, the state of the cylinder does not change.

The circuit is called a memory circuit because it uses a 5/2 double pilot memory valve. The 5/2 valve remembers the last signal applied in terms of the position of the spool in the absence of reset springs, thus memorising or storing the pneumatic signal. Double piloted 4/2 valves also can be used as memory valves.

6.4 Supply and Exhaust Air Throttling

It is always necessary to reduce the speed of a cylinder from the maximum (based on the size of the chosen final control valve) to the nominal speed depending on the application. Speed control of pneumatic cylinders can be conveniently achieved by regulating the flow rate of supply or exhaust air. The volume flow rate of air can be controlled by using one-way or two-way flow control valves.

There are two types of throttling circuits for double acting cylinders:

1. Supply air throttling
2. Exhaust air throttling

6.4.1 Supply air throttling

This method of speed control of double acting cylinders is also called a meter–in circuit (Fig. 6.15a). One-way flow control valves are installed to throttle air entering the cylinder. The exhaust air can escape freely through the check valve or the throttle valve on the outlet side of the cylinder. There is no air cushion on the exhaust side of the cylinder piston in this throttling arrangement. As a result, considerable differences in stroking velocity may be obtained even with very small variations of load on the piston rod. Any load in the direction of operating motion will accelerate the piston above the set velocity. Therefore, supply air throttling can be used for single acting and small volume cylinders.

6.4.2 Exhaust air throttling

Exhaust air throttling is used for speed control of double acting cylinders and this arrangement is also called a meter-out circuit (Fig. 6.15b). Throttle relief valves are installed between the cylinder and the main valve so that the exhaust air leaving the cylinder is throttled in both directions of motion of the cylinder. The supply air can pass freely through the corresponding check valves in both cases. The piston is loaded between two cushions of air while the cylinder is in motion to ensure smooth movement. The first cushioning effect is from the supply air entering the cylinder through the check valve, and the second cushioning effect is from the exhaust air leaving the cylinder through the throttle valve at a slower rate. Exhaust air throttling is commonly used in pneumatic control.

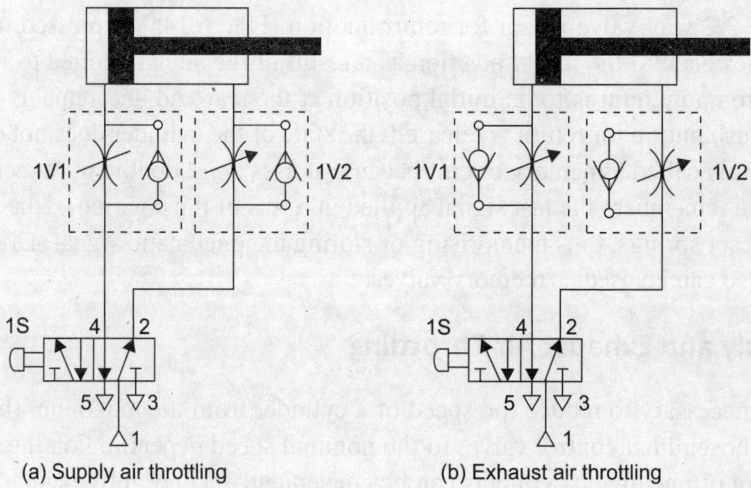

Figure 6.15 *Throttling circuits*

6.5 Methods of Checking the End Position of a Cylinder

The following methods are commonly used to locate the end position of the piston in a cylinder:

1. Mechanically operated limit switches
   (i) Roller lever type (ii) Idle return roller type
2. Reed sensors (normally used in cylinders with magnetically coupled slides)
   (i) With electrical output (ii) With pneumatic output
3. Electrical proximity switches
4. Pneumatic signal generators

This section only discusses the use of limit switches to find the end position of a piston in a cylinder.

6.5.1 Use of limit switches

S1 and S2 are the limit switches corresponding to the home and the extended positions respectively. Although they are located in the path of the movement of the piston rod, the normal practice is to represent the symbol of the limit switches on either side of the 3/2 control valve without the signals connected to the pilot ports of the valve (in Fig. 6.16, pilot signal actuations are shown for clarity). The roller lever limit switches are essentially 3/2 ball seat or disc seat valves handling pneumatic signals. These are available in direct actuation and internal pilot actuation versions. Idle return roller limit switches are used for actuation only in one direction, and are used in signal elimination devices in case of signal overlap.

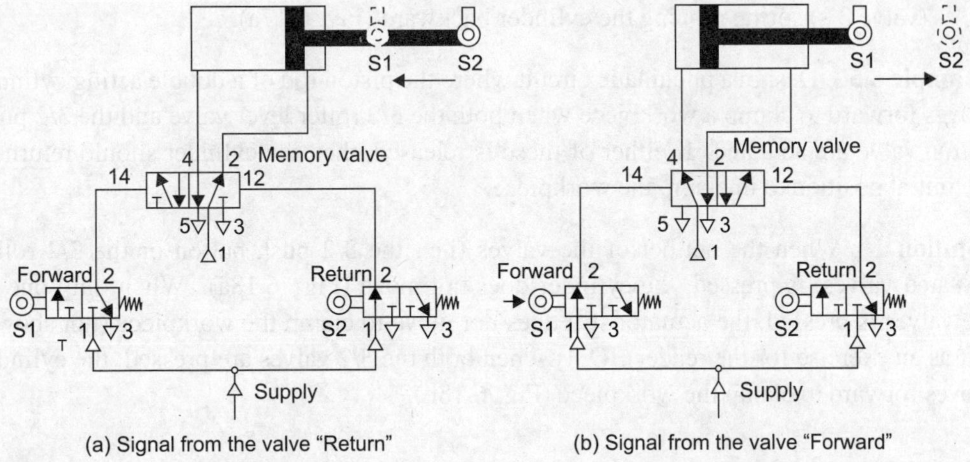

Figure 6.16 *Limit switches in pneumatic circuits*

Example 4: A double acting cylinder is to advance when a push button is operated. Upon the release of the push button, the cylinder is to be retracted. The cylinder is 300 mm in diameter and consumes a relatively large volume of air.

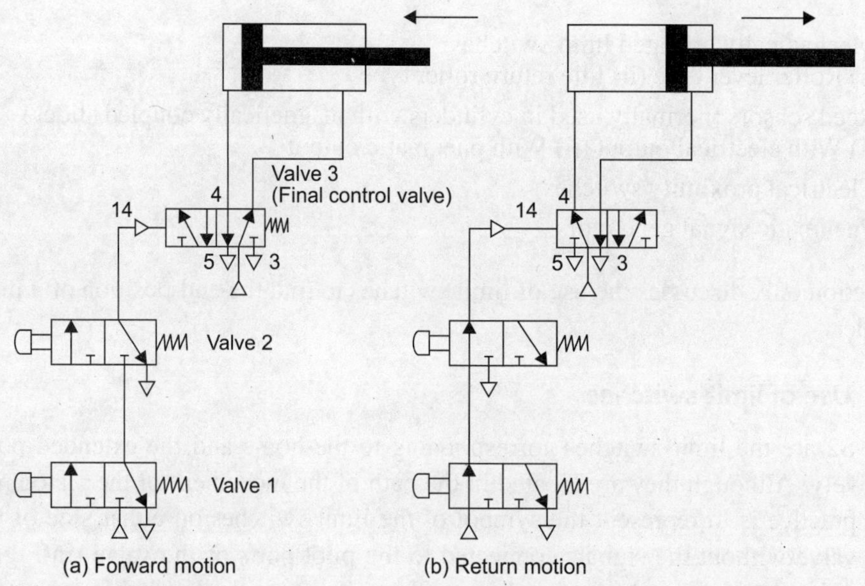

(a) Forward motion (b) Return motion

Figure 6.17

When valve 2 is actuated, it sends a pilot signal to the larger valve 3 changing its position and moving the cylinder forward (Fig. 6.17b).

When valve 2 is unactuated, there is no pilot signal at valve 3 and its position is changed by its (valve 3's) spring moving the cylinder backward (Fig. 6.17a).

Example 6.5 Design a pneumatic circuit where the piston rod of a double acting cylinder moves forward to clamp a workpiece when both the 3/2 roller lever valve and the 3/2 push button valve are actuated. If either of these is released, then the cylinder should return to the initial position to unclamp the workpiece.

Solution 1 When the neither of the valves (i.e., the 3/2 push button or the 3/2 roller actuated valves) is pressed, the cylinder does not move (Fig. 6.18a). When only one of the valves is pressed, the actuator still does not move to clamp the workpiece (not shown, left as an exercise for the reader). Only when both the 3/2 valves are pressed, the cylinder moves forward to clamp the workpiece (Fig. 6.18b).

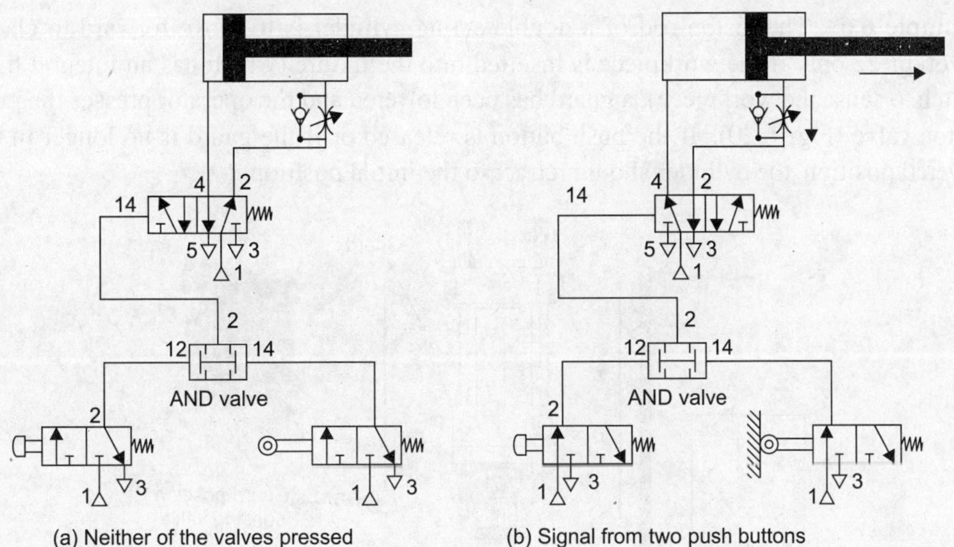

Figure 6.18

Solution 2 An alternative solution is to use the 3/2 roller actuated valve and the 3/2 push button valve in series to achieve AND functionality.

When only one of the valves (i.e., 3/2 push button or 3/2 roller actuated) is pressed, the cylinder does not move (Fig. 6.19a). When neither valve is pressed, the actuator does not move to forward position to clamp the workpiece (not shown and left as an exercise for the reader). Only when both the 3/2 valves are pressed, does the cylinder move forward to clamp the workpiece (Fig. 6.19b).

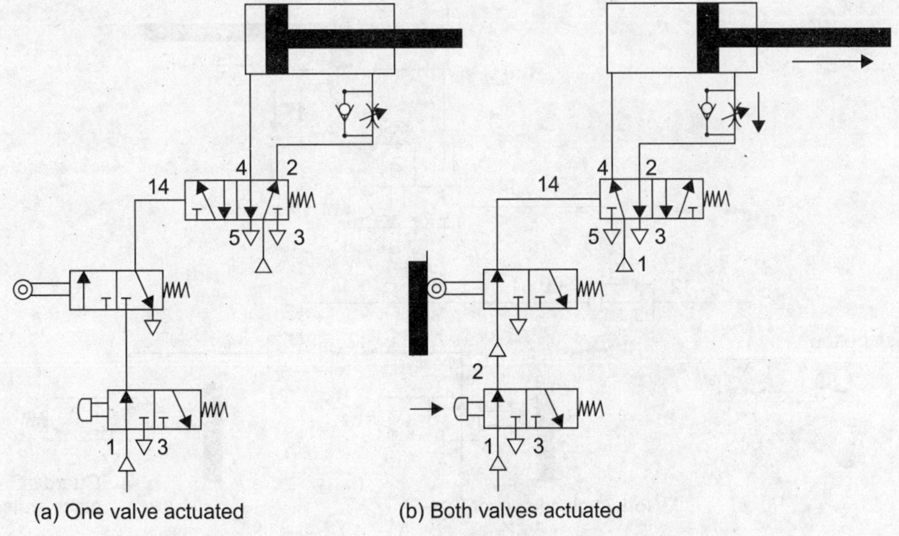

Figure 6.19

Example 6.6 The piston rod of a double acting cylinder is to move forward to clamp a workpiece only if the workpiece is inserted into the fixture (which has an integral limit switch to sense the workpiece), a guard has been lowered and the operator presses the push button valve (Fig. 6.20). If the push button is released or if the guard is no longer in the lowered position, the cylinder should retract to the initial position.

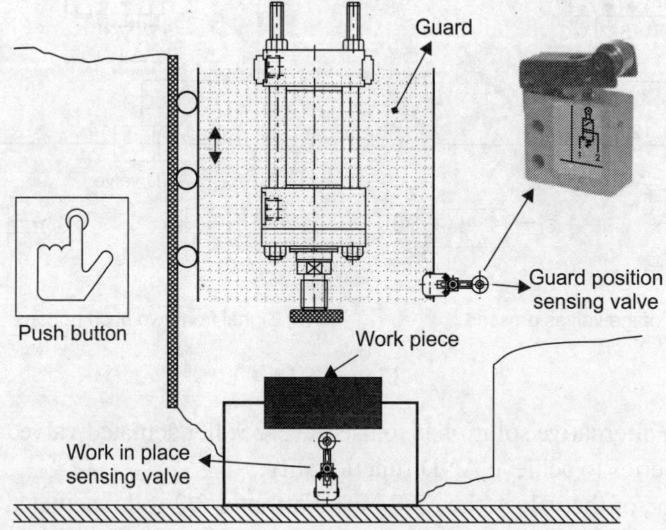

Figure 6.20

Solution is illustrated in Fig. 6.21, which is self-explanatory.

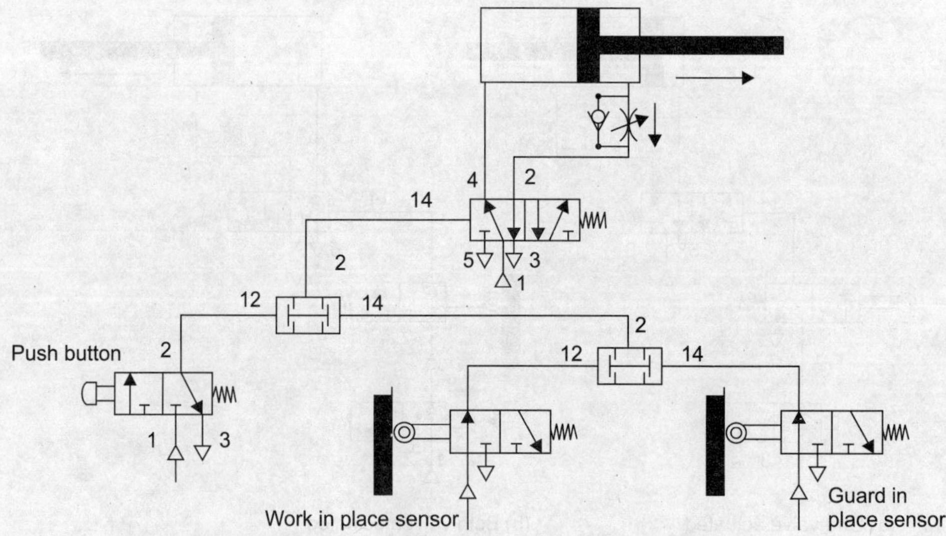

Figure 6.21

Example 6.7 Design a pneumatic circuit where a double acting cylinder is to extend and punch a workpiece when either of the push button valves are pressed. The cylinder is to retract if both buttons are released.

The input 12 and 14 of the shuttle valve 1V1 are connected to the working connections of valves 1S1 and 1S2. When one of the push buttons is operated, the corresponding valve (1S1 or 1S2) is actuated and a signal is generated at the 12 or 14 side of the shuttle valve. The OR condition is fulfilled and the signal passes through the shuttle valve and is emitted at port 2. The signal pressure is prevented from escaping via the exhaust of the unactuated valve by closing the line in the shuttle valve. The signal affects the switching of the control element 1V2. The piston side of the cylinder is pressurised and the piston rod advances (Fig. 6.22b).

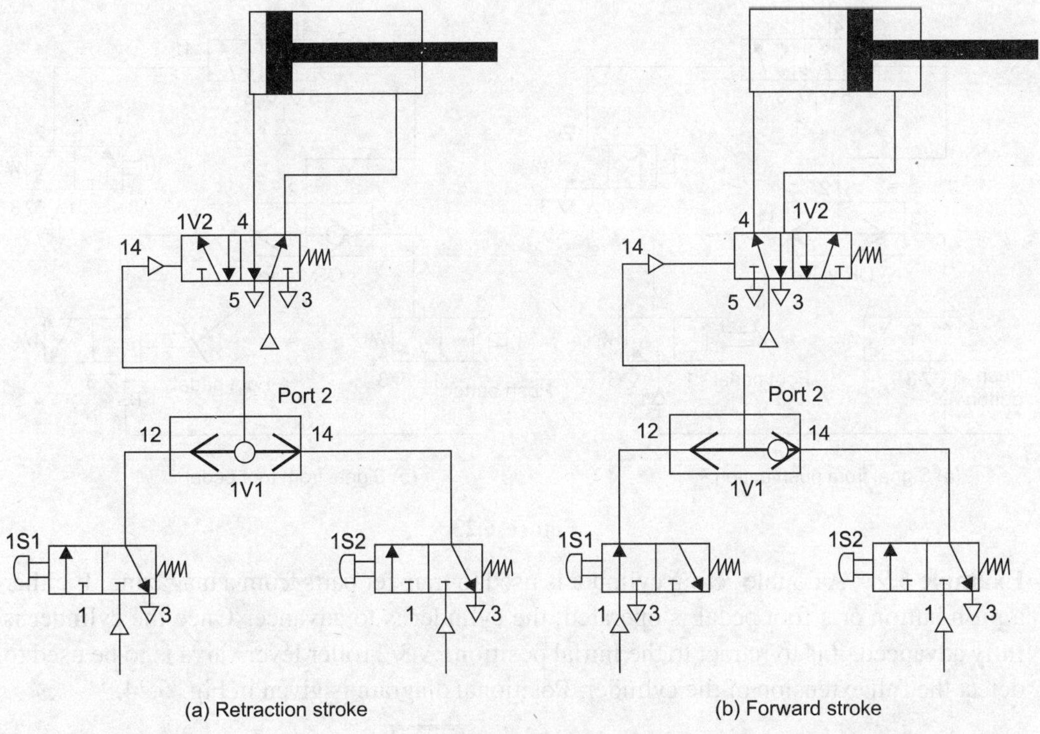

Figure 6.22

When the actuated push button is released, the signal pressure is relieved via the corresponding valve (1S1 or 1S2) and the control element is switched back to its initial position. The pressure now building up on the piston rod side ensures the retraction of the piston rod (Fig. 6.22a).

A similar scenario occurs when the other valve (i.e., 1S2 instead of IS1) is pressed and is left as an exercise to the reader to draw the circuit.

Example 6.8 A double acting cylinder is to extend and punch a workpiece when either a push button valve or a foot-operated pedal is pressed. The cylinder is to retract when a second push button is pressed. Develop a pneumatic circuit to implement the given control task using OR logic.

The two positions of the circuit for the control task are shown in Fig. 6.23. The double acting cylinder extends as a result of the application of a signal from either the push button valve or the foot-pedal valve. This is illustrated in self-explanatory figures 6.23a and 6.23b.

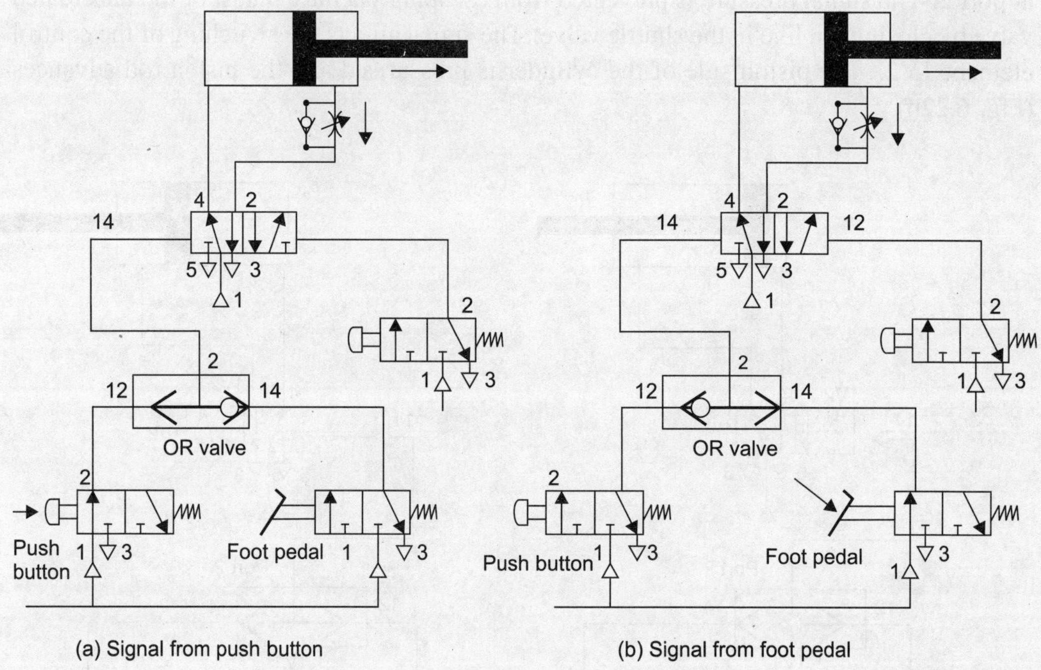

Figure 6.23

Example 6.9 A double acting cylinder is used to transfer parts from a magazine. If either a push button or a foot pedal is operated, the cylinder is to advance. Once the cylinder is fully advanced, it is to retract to the initial position. A 3/2 roller lever valve is to be used to detect the full extension of the cylinder. Positional diagram is given in Fig. 6.24.

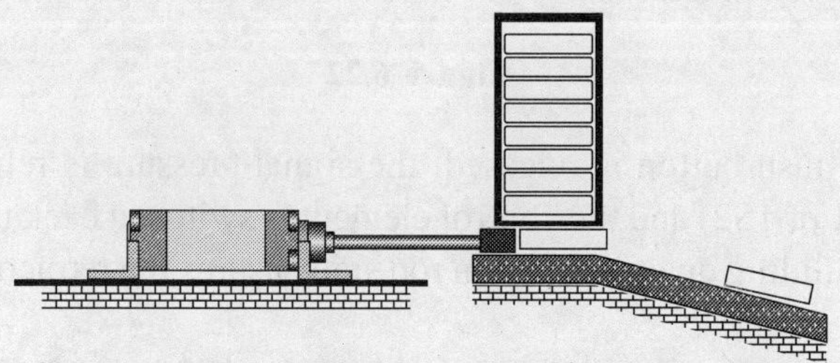

Figure 6.24

The solution is shown in Fig. 6.25, which is self-explanatory.

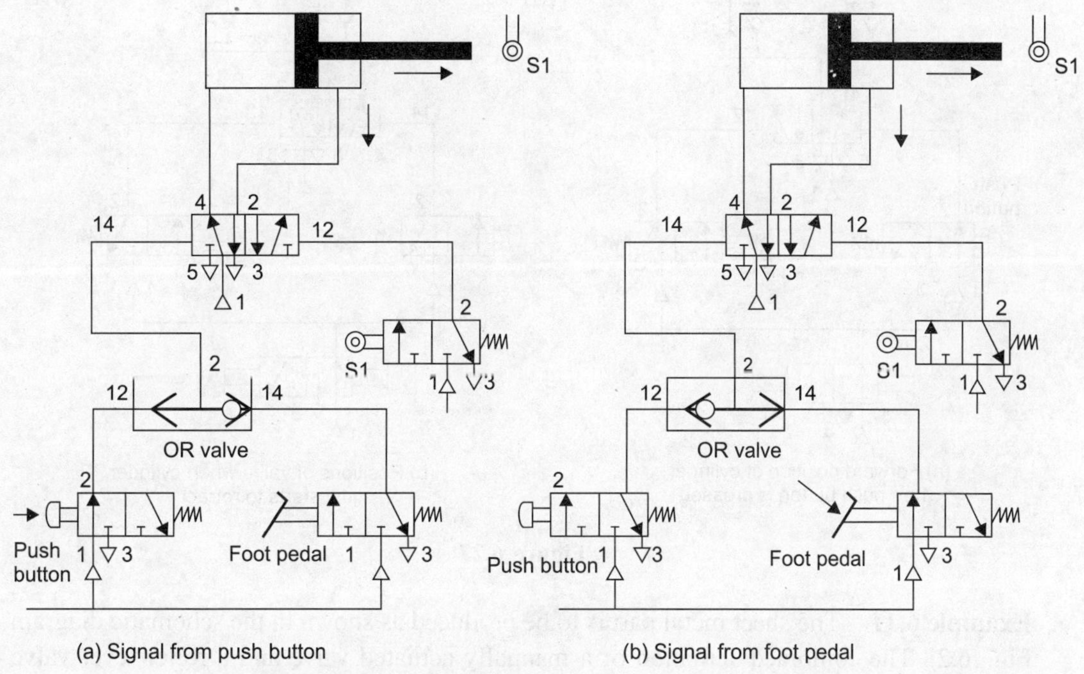

Figure 6.25

Example 6.10 A double acting cylinder is used to transfer parts from a magazine. The cylinder is to fully advance when a push button is operated and then retract automatically. Full extension is detected by a roller lever valve. The cylinder is to continue forward even if the push button is released before full extension is reached. The speed of the cylinder is to be adjustable in both directions of motion. Positional diagram is given in Fig. 6.26.

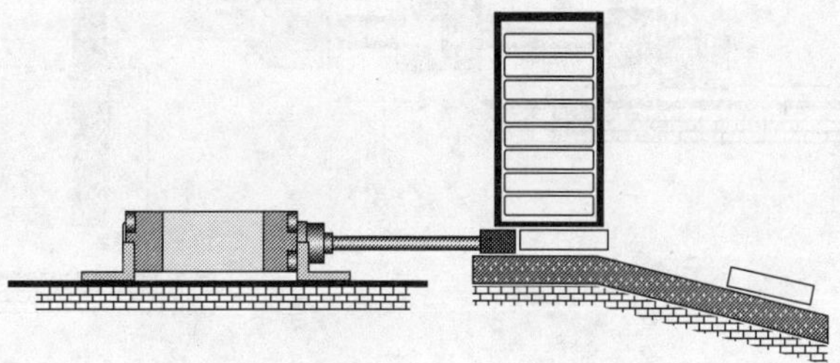

Figure 6.26

The solution is shown in Fig. 6.27, which is self-explanatory.

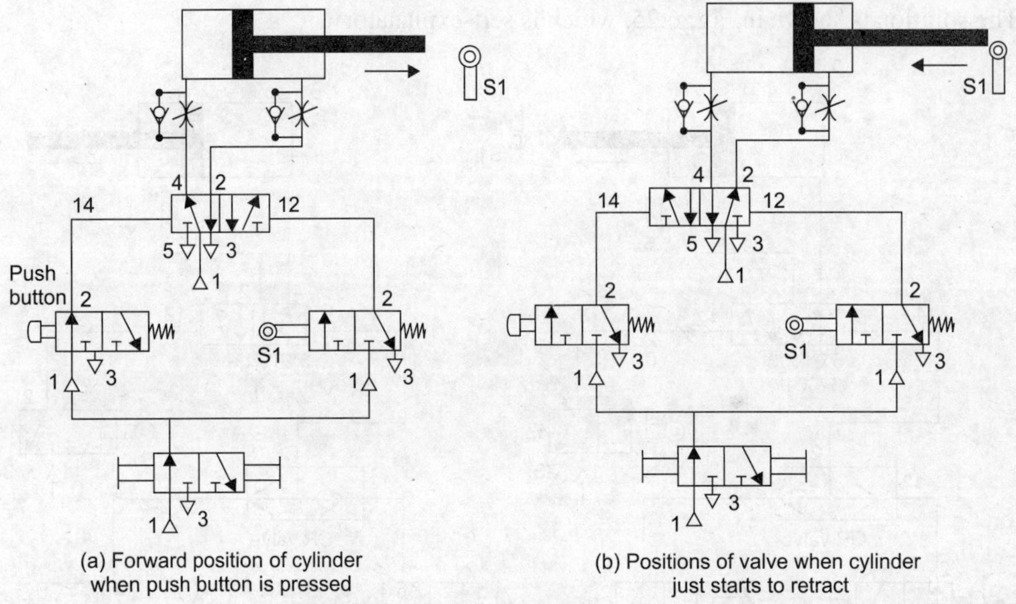

Figure 6.27

Example 6.11 The sheet metal part is to be produced as shown in the schematic diagram Fig. 6.28 The combined actuation of a manually actuated valve and a roller lever valve advances a forming tool on an edge-folding device. The forming tool is driven by a double acting cylinder. For rapid forward travel, the circuit utilises a quick exhaust valve. The retracting speed is to be adjustable. If either of the two valves is released, the tool returns to its initial position. The double acting cylinder shown below is used to bend the sheet metal component. The circuit needs a two-hand safety feature.

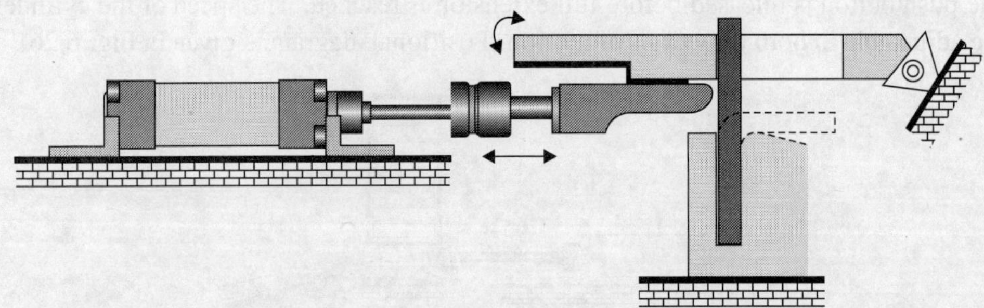

Figure 6.28

Rapid forward stroke: When two push buttons are pressed simultaneously, the 5/2 valve is shifted as shown in the figure and cylinder moves forward. Since rapid forward movement is required, a quick return valve is used at the exit port of the cylinder (Fig. 6. 29a).

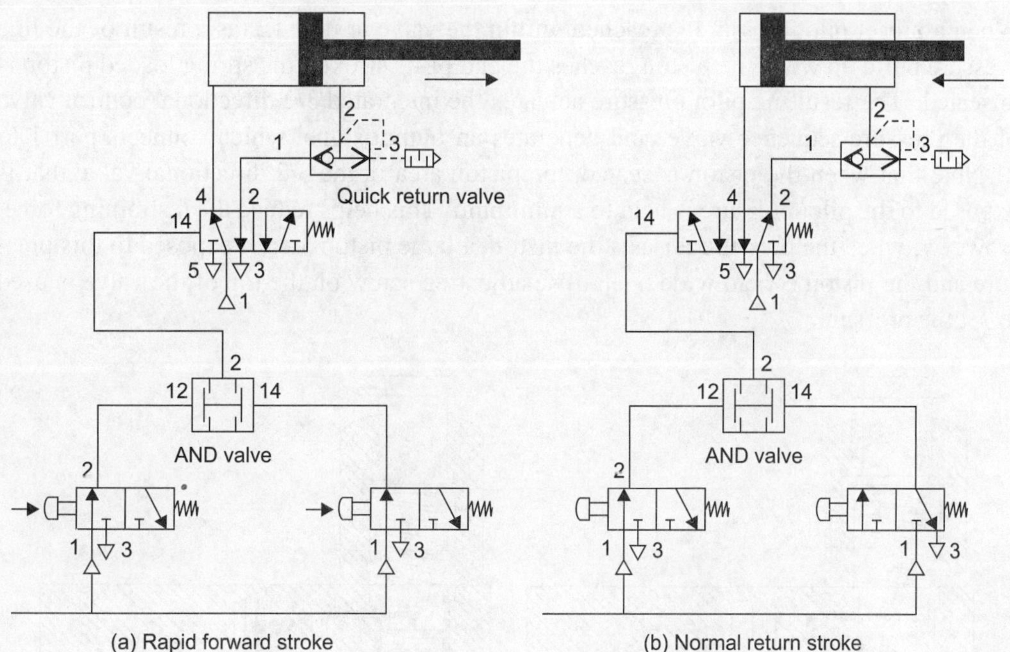

Figure 6.29

Normal return stroke: When both push buttons are released or only one button is pressed, then there is no supply to the 5/2 valve and the cylinder returns with normal speed (Fig. 6.29 b).

6.6 Pressure Dependent Controls

A pressure sequence valve is essentially a switch-on or switch-off valve. A sequence valve generates a pneumatic signal if the sensing pressure (signal input) is more than the desired preset pressure. This generated output signal is used as a set or reset signal to the final control valve to produce forward or return motion in the cylinder. It is used in applications such as bonding cylinders, clamping cylinders, etc. to ensure the desired minimum pressure in the cylinder. This is a combination valve with two sections; one is a 3/2 directional control and the other a pressure control valve.

A sensing pressure signal is introduced at port 12. A spring loaded cap/screw knob is used to manually adjust the pressure. The knob is rotated clockwise for a higher pressure setting, and anti-clockwise for a lower pressure setting. The right section is a 3/2 NC directional control valve and is pilot operated using a pressure signal from the left section.

The cross-sectional views and symbols of a pressure sequence valve are shown in Fig. 6.30. This valve consists of a 3/2 directional control valve and a pressure control valve. The principle of operation of a sequence valve is that the outlet of the sequence valve remains closed until the pressure upstream of it builds up to a predetermined value.

When a preset pilot pressure is reached within the valve at port 12 (as a result of the line pressure build up when the piston reaches the end of its stroke), the spring loaded piston is unseated. The resulting pilot pressure actuates the integrated 3/2 directional control valve of the pressure sequence valve, and generates an output signal which connects port 1 to 2. Note that when the piston is seated, the piston area in the 3/2 directional valve that is exposed to the pilot pressure is held to a minimum. This helps reduce the restraining force. However, when the pressure unseats the piston, a large piston area is exposed to this pressure and the piston is held wide open. The adjusting screw on the top of the valve is used to set the pressure.

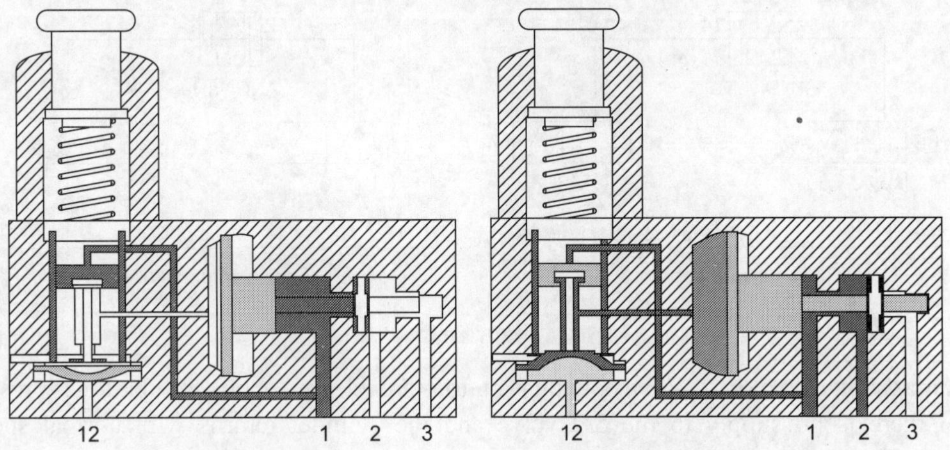

Figure 6.30 *Pressure dependent control valve*

Example 6.12 A double acting cylinder is used to press together glued components (the estimated pressure is around 4 bar). Upon pressing a push button, the clamping cylinder is to extend and trip the roller valve. Once the fully extended position of the cylinder has been reached and sufficient clamping force has been developed, the cylinder is to retract to the initial position. Develop the pneumatic circuit using a pressure sequence valve.

The two positions (cylinder extending and fully extended) of the pneumatic circuit for the control task are shown in Fig. 6.31. The pressure in the sequence valve is set to the working pressure of 4 bar, and the signal input to the pressure sequence valve is tapped from the power line from port 4 of valve 1.1 of the cylinder to gauge the pressure on the piston side of the cylinder. As shown in Fig. 6.31, valve 1.2 initiates the forward motion of the cylinder. While the cylinder is moving forward, the pressure in the power line from port 4 of valve 1.1 to the cylinder does not build up to the working pressure. It is only after the cylinder is fully extended that the pressure in the line builds up to reach the level required for the gluing task. When the preset pressure in the sequence valve is reached, the integrated 3/2 directional control valve is actuated, generating an output signal. This signal is used to reset the final control element 1.1, thus causing the return motion of the cylinder.

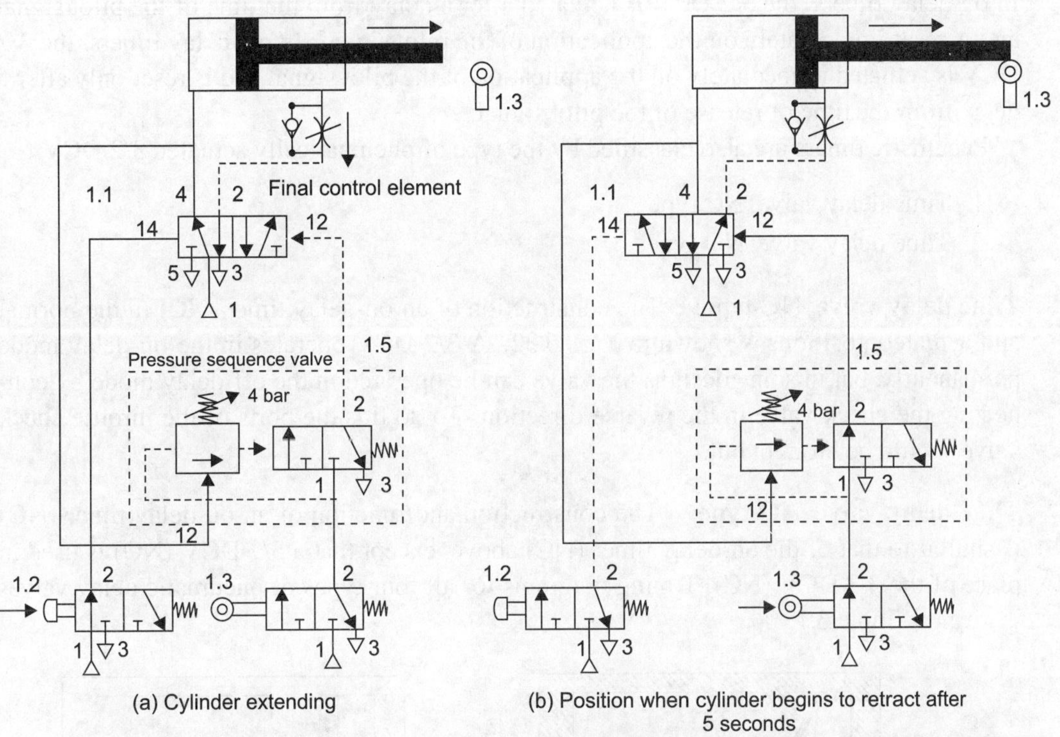

(a) Cylinder extending

(b) Position when cylinder begins to retract after 5 seconds

Figure 6.31 *Pressure dependent control valve*

6.7 Time Dependent Controls

Pneumatic timers are used to create time delay of signals in pilot operated circuits and are available as normally closed and normally open timers. Pneumatic timers are usually delay timers; delayed signals are commonly used in applications involving bonding of two pieces. Normally open pneumatic timers are also used in signal elimination and in safety devices in two-hand blocks.

Time delay valve is a combination of a pneumatically actuated 3/2 direction control valve, an air reservoir and a throttle relief valve. The time delay function is obtained by controlling the air flow rate to or from the reservoir, by using the throttle valve. Adjustment of the throttle valve permits fine control of the time delay between minimum and maximum times. In pneumatic time delay valves, typical time delays in the range 5-30 seconds are possible. The time delay can be extended with the addition of an external reservoir. Pneumatic timers can be classified as:

1. On-delay timer
2. Off-delay timer

In on-delay timers, the 3/2 DCV is actuated after a delay from the time of the pilot signal and is reset immediately on the application of the pilot signal. In off-delay timers, the 3/2 DCV is actuated immediately on the application of the pilot signal and is reset only after a delay from the time of release of the pilot signal.

Pneumatic timers are also classified by the type of pneumatically actuated 3/2 DCV:

1. Time delay valve, NC type
2. Time delay valve, NO type

Time delay valve, NC type: The construction of an on-delay timer (NC) in the normal and actuated positions is shown in Fig. 6.32. A 3/2 DCV operates in the on-delay mode permanently, but in some designs the valve can be operated in the off-delay mode by connecting the check valve in the reverse direction. To do this the ports of the throttle check valve should be brought out.

Time delay valve, NO type: The construction and function of an on-delay timer (NO) is similar to that of the on-delay timer (NC) above, except that a 3/2 DCV (NO) is used in place of the 3/2 DCV (NC). Timing diagrams for all four types of pneumatic delay valves is given in Table 6.1.

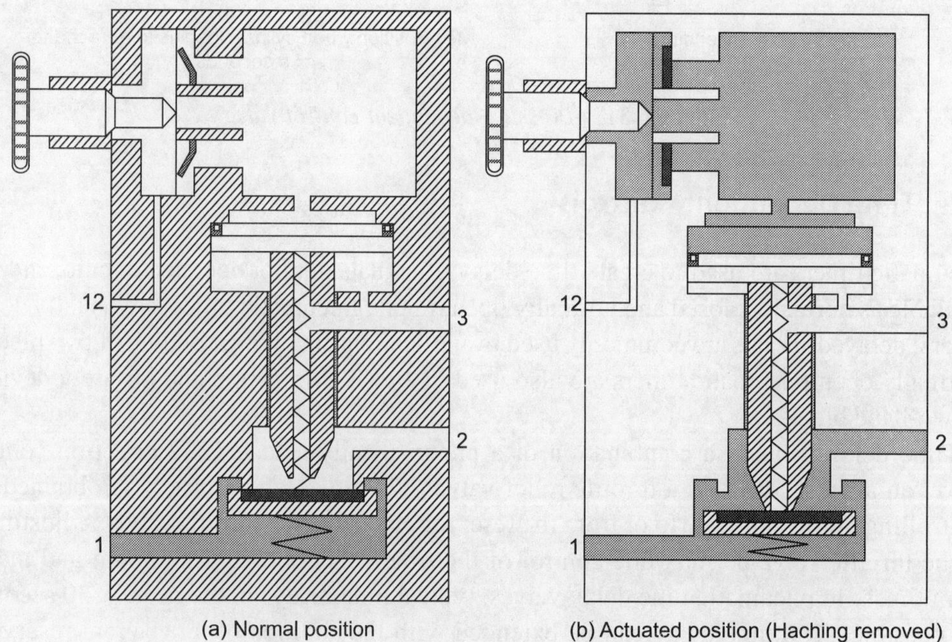

(a) Normal position (b) Actuated position (Haching removed)

Figure 6.32

Table 6.1 *Timing diagrams for all four types of pneumatic delay valves*

Timing diagram	Symbol

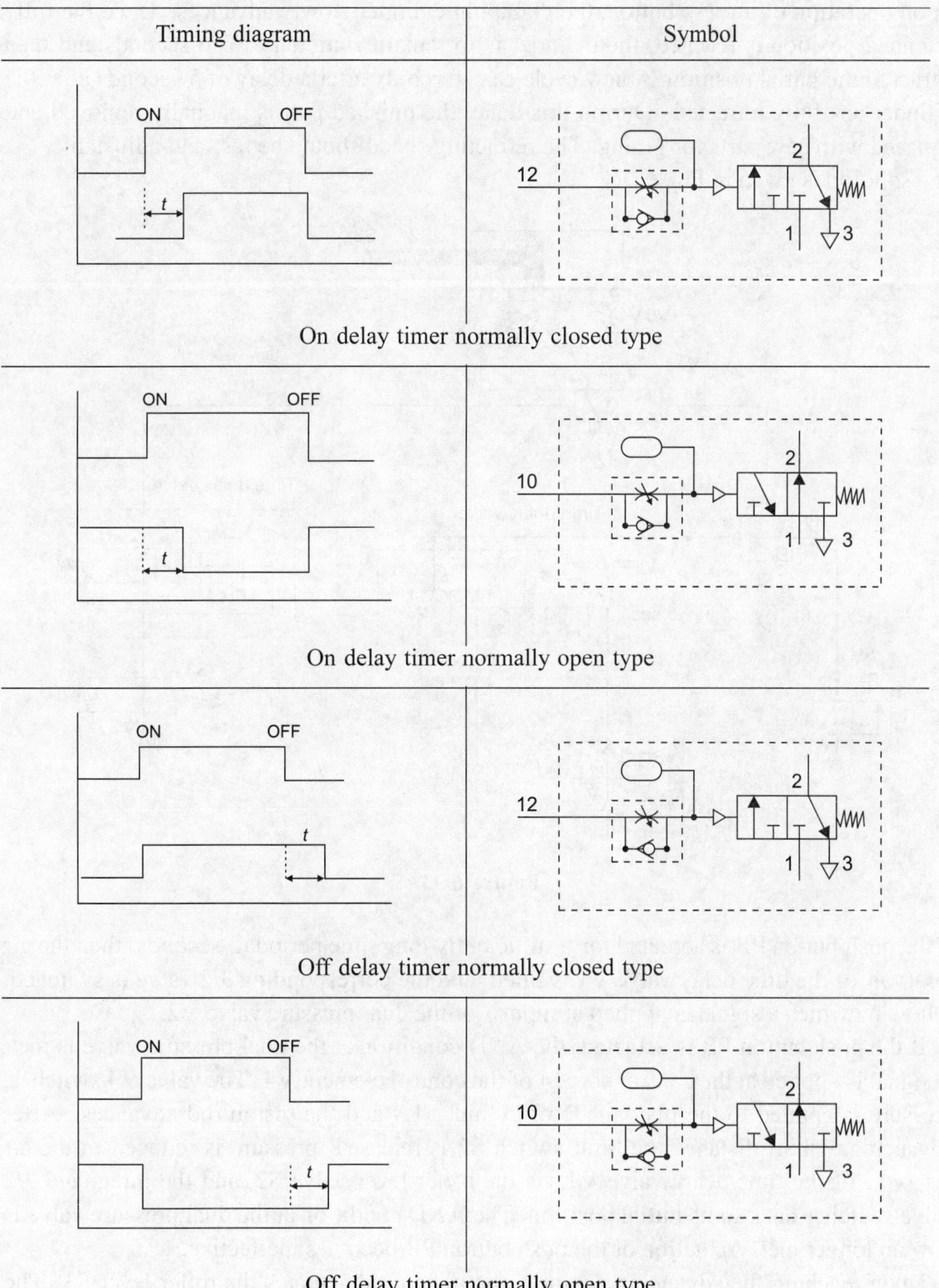

Example 6.13 A double acting cylinder is used to press together glued components. Upon operation of a press button, the clamping cylinder slowly advances. Once the fully extended position is reached, the cylinder is to remain stationary for 6 seconds and then retract to the initial position. A new cycle can start only after a delay of 5 seconds after the cylinder has fully retracted. During this delay the finished part is manually removed and replaced with new parts for gluing. The retracting speed should be fast, but adjustable. The solution is given in Fig. 6.33.

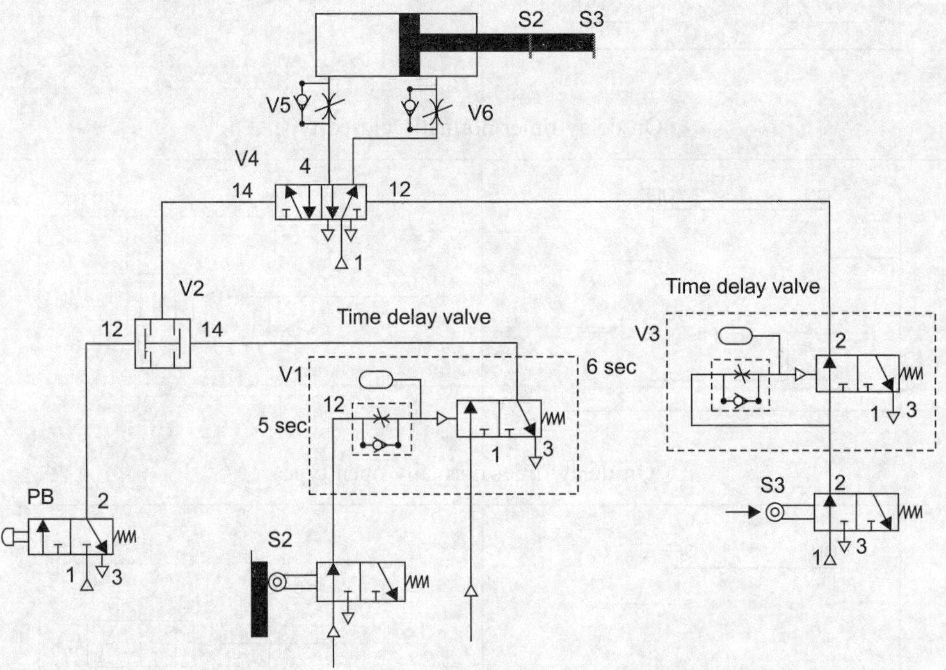

Figure 6.33

If the push button PB is actuated for a sufficiently long time period (5 seconds) then the air reservoir of the time delay valve V1 is filled, and the corresponding 3/2 valve is switched, following which a signal is applied at input 1 of the dual pressure valve V2.

If the push button PB is actuated, the AND condition at the dual pressure valve is met. A signal is applied at the control port 14 of the control element V4. The valve V4 switches, pressure is applied to the piston side of cylinder 1A and the piston rod advances. After advancing a short distance the limit switch S2 is released, pressure is reduced in the air reservoir of the time delay valve V1 via the roller lever valve S2, and the integrated 3/2 valve switches back to its initial position. The AND condition at the dual pressure valve is now no longer met. Actuation of the push button PB becomes ineffective.

Upon reaching the advancing position, the piston rod actuates the roller lever S3. The pressure line to the time delay valve V3 is now released and pressure in the air reservoir is increased. The rate of the pressure increase is adjustable via the integrated flow control valve. When the switching pressure has been reached, the integrated 3/2 valve switches and a signal is applied at the control port 12 of the final control element V4. The valve V4 reverses and the piston rod retracts. Upon release of the limit switch S3, the time delay valve V3 switches to its initial position again.

The limit switch S2 is actuated when the piston rod reaches its initial position, the pressure in the air reservoir of the time delay valve V1 starts to increase until the switching pressure has been reached after 5 seconds, causing the integrated 3/2 valve to switch. The initial status of the system is now reached again and a new cycle can be started. The piston rod speed is set at the restrictors of the one way flow control valves V5 and V6.

Example 6.14 Pins are to be fed to the next processing station using a double acting cylinder. The arrangement is shown in Fig. 6.34. The pins are fed continuously by the to and fro motion of the cylinder, triggered by the start push button PB1. Pressing the stop push button PB2 stops feeding. Forward and return speeds should be adjustable independently. Develop a pneumatic circuit to implement the control.

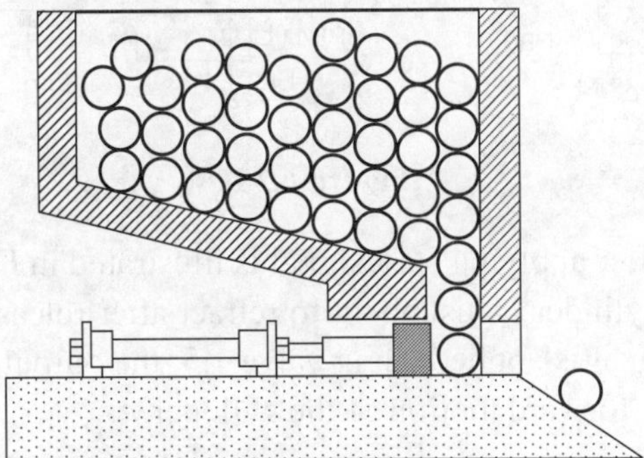

Figure 6.34

Solution is given Figure 6.35

A cylinder can perform a continuous to and fro motion with the use of two sensors 1.2 and 1.3 placed at the extreme positions of the cylinder piston range of motion. The sensor 1.2 (limit switch) generates a signal when the cylinder is in the retracted position, and the sensor 1.3 generates a signal when the cylinder is in the extended position. The signal from 1.2 is applied to port 14 of valve 1.1 and the signal from the sensor 1.3 is applied to port 12 of valve 1.1 to trigger automatic forward and return motion of the cylinder.

To incorporate the necessary start and stop control, a start push button and a stop push button can be used to control the air supply to pilot port 14 of the final control element 1.1 using a 3/2 (or 5/2 with port 2 blocked) double pilot DC valve.

When a momentary input signal is given by the "start push button", the double pilot DC generates a continuous output signal. This output signal goes off only when a momentary input signal is given by the stop push button.

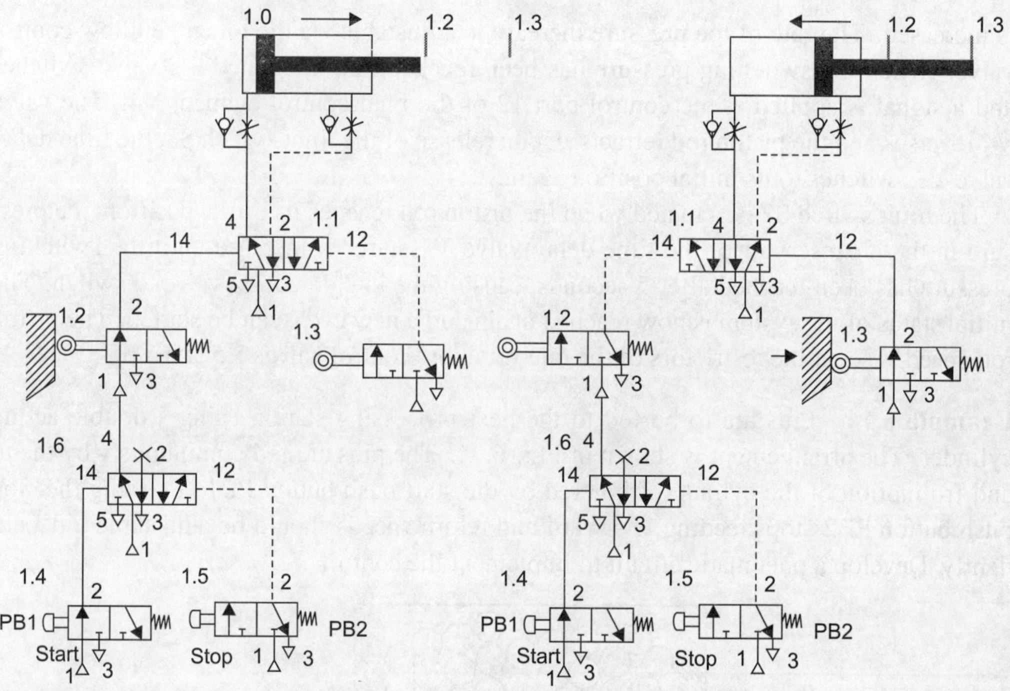

Figure 6.35

The state of the circuit after applying start signal is illustrated in Fig. 6.35(a), and the state of the circuit when the cylinder is just about to retract after releasing roller valve1.3 (Fig. 6.35 b). When a stop signal is applied using valve 1.5, the output signal of valve 1.6 goes off, thus stopping further forward motion of the cylinder.

Example 6.15 A double acting cylinder is used to press glued components together for half a minute. The arrangement is shown in Fig. 6.36. A push button PB1 is used to control the forward stroke of the cylinder. After pressing the components together for half a minute, the cylinder has to retract automatically. The return stroke must occur even if the start button is still actuated. A new start signal may only take effect after the initial position of the piston is reached and the push button is released. Develop a pneumatic circuit to implement the control task.

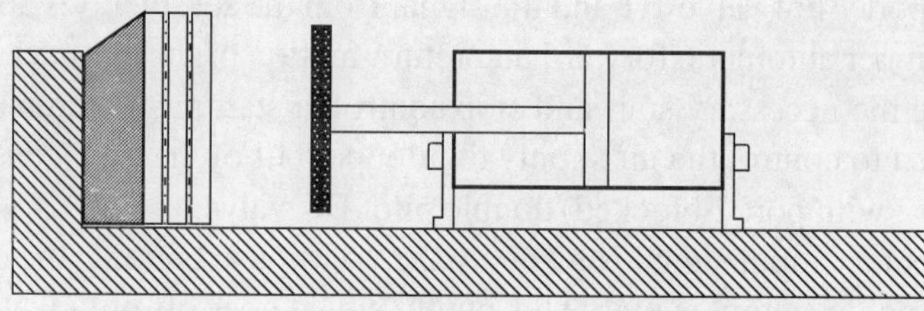

Figure 6.36

The solution is given in Fig. 6.37, which is self explanatory.

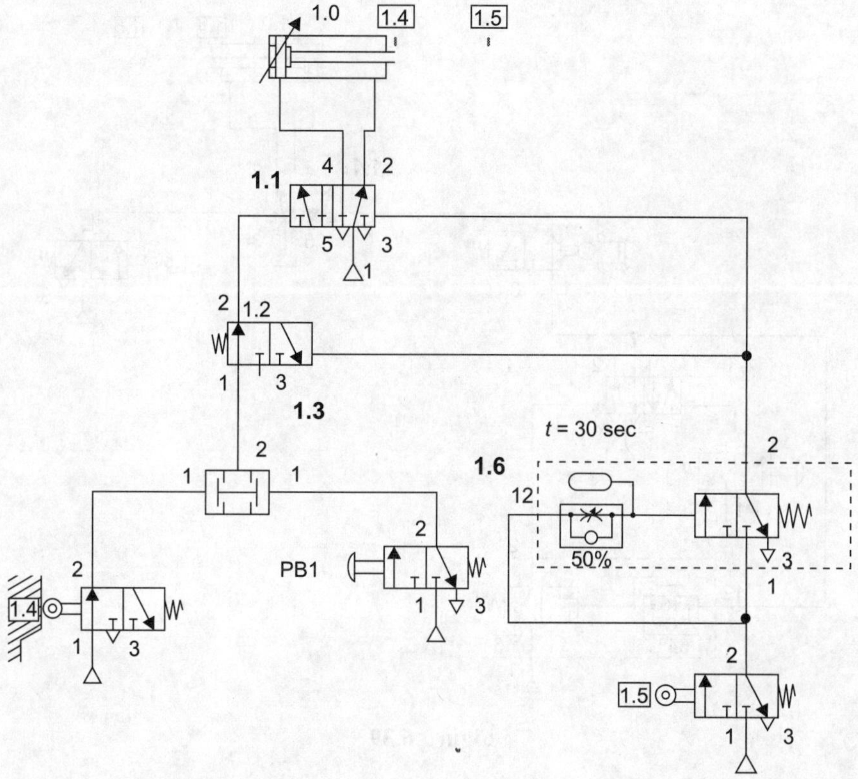

Figure 6.37

Example 6.16 Figure 6.38 shows the arrangement for quenching heat treated valves for automobile engines. A double acting cylinder is used to move the container filled with valves up and down in the bath a few times. The operator provides a start signal manually. The valves heat treatment is turned off automatically after a preset time. Develop a pneumatic control circuit to implement the control task.

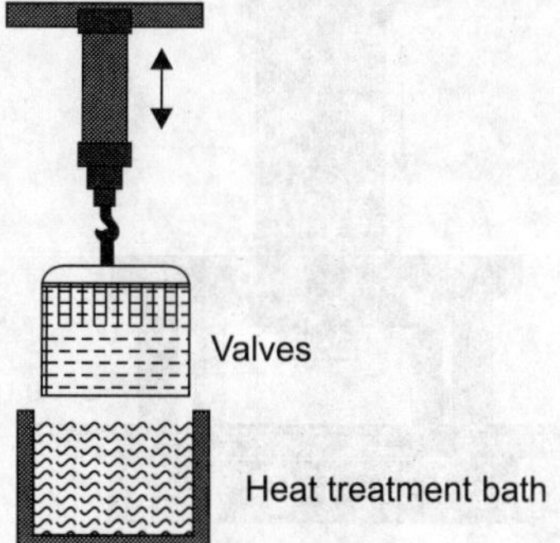

Figure 6.38

The solution is shown in Fig. 6.39.

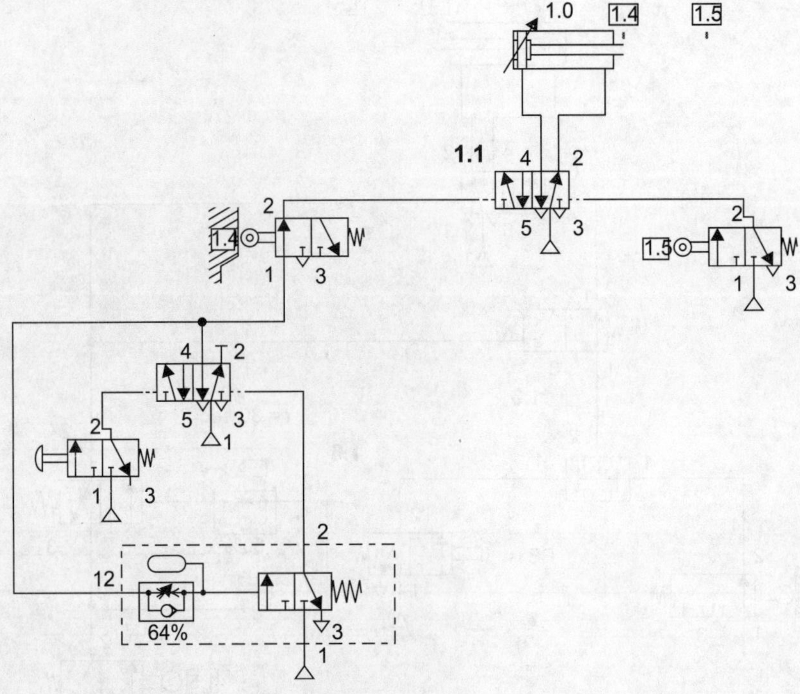

Figure 6.39

Example 6.17 A double acting cylinder is used for embossing slide rules as shown in Fig. 6.40. The cylinder extends only when two push buttons are pressed simultaneously within one second to ensure the complete safety of the operators. The cylinder should retract immediately if any of the push buttons is released. Develop a pneumatic circuit to implement the control task.

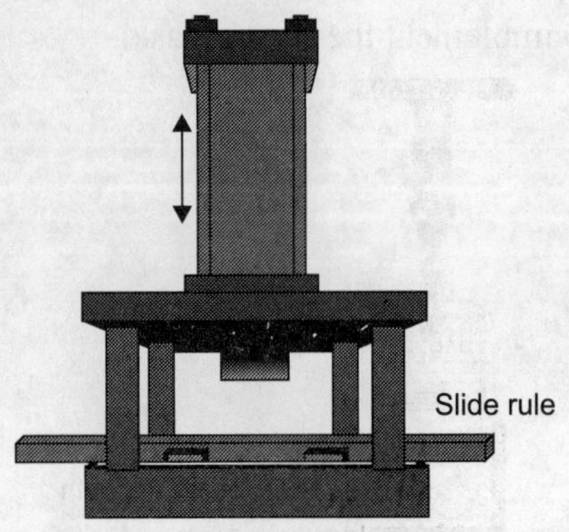

Figure 6.40

The solution is shown in Fig. 6.41.

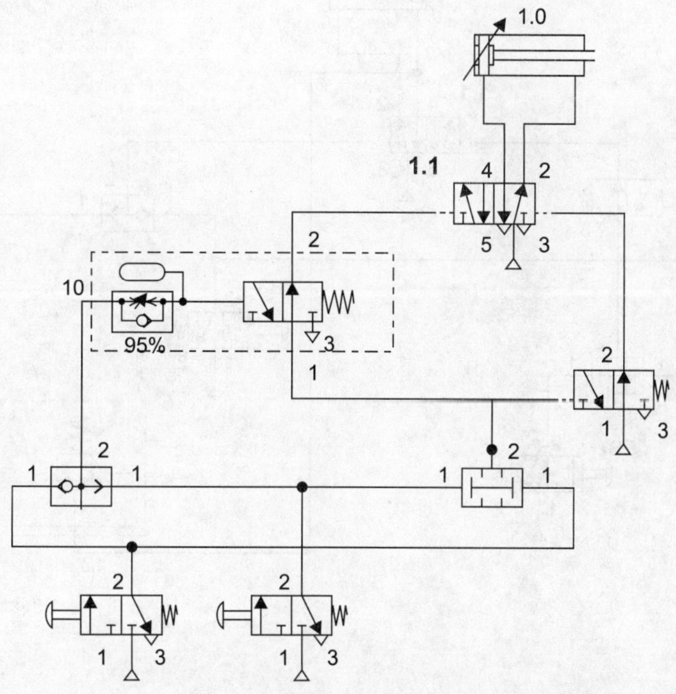

Figure 6.41

Example 6.18 A pneumatically controlled press with a stamping die shown in Fig. 6.42 is used to produce badges from a thin metal sheet. A double acting cylinder is to extend when the push buttons PB1 and PB2 are pressed simultaneously. The cylinder is to retract automatically after reaching the forward end position and attaining a preset pressure to get consistent quality. The cylinder should retract immediately if the emergency push button PB3 is pressed. Develop a pneumatic control circuit to implement the control task.

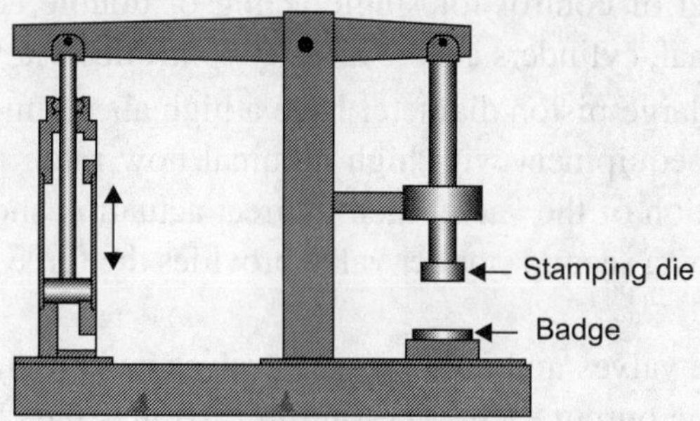

Figure 6.42

The solution is shown in Fig. 6.43.

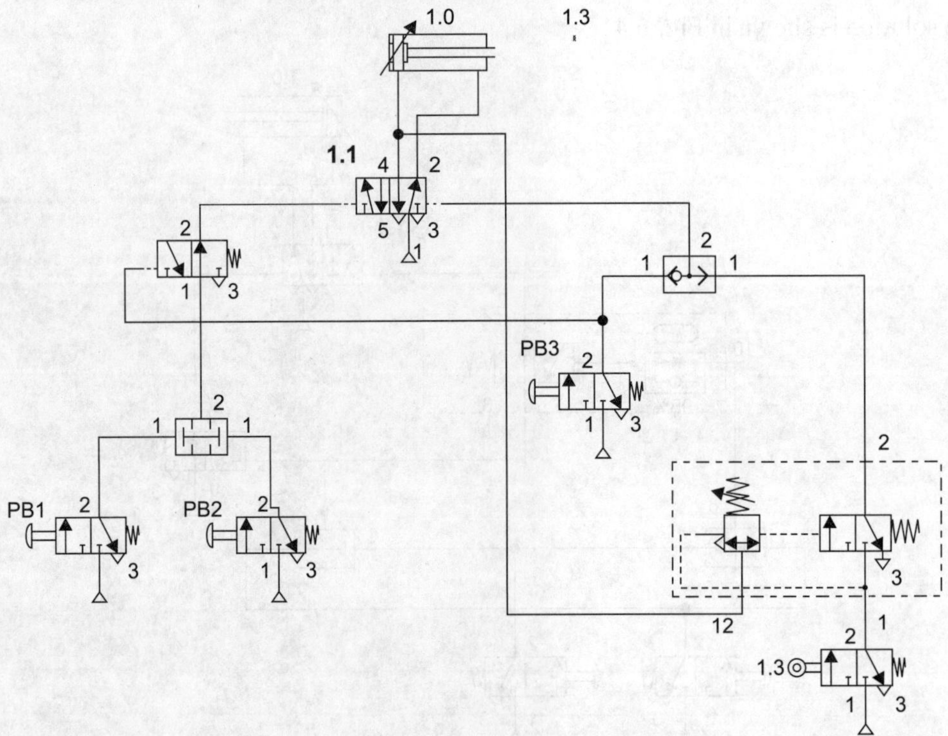

Figure 6.43

Summary

1. Pneumatic control systems can be designed in the form of pneumatic circuits. A pneumatic circuit is assembled from various pneumatic components, such as cylinders, directional control valves, flow control valves, pressure regulators, signal processing elements such as shuttle valves, two pressure valves, etc. A pneumatic circuit diagram uses pneumatic symbols to describe its design.
2. The simplest level of control for single acting or double acting cylinders involve direct control signal; cylinders can be directly controlled via valves.
3. Cylinders with a large piston diameter have a high air requirement and must be actuated by control equipment with high nominal flow rate. If the force is too high for manual actuation of the valve, then indirect actuation should be used where the signal generated by a second smaller valve provides the force necessary to switch the control element.
4. Pneumatic shuttle valves and dual pressure valves have logic functions. Both have two inputs and one output each. The shuttle valve has the characteristics of an OR function; where at least one of the two inputs is required to generate an output for the valve. In the case of dual pressure valve, the characteristics are that of an AND function, where both inputs are required to generate an output.

5. 4/2 and 5/2 double piloted valves possess a memory function; the valve retains its last switched position until an opposing signal is received. This characteristic is independent of the time period for which the signal is applied to the switching device.
6. A quick exhaust valve is a typical shuttle valve and is used to exhaust the cylinder air quickly to the atmosphere. For rapid movement of the cylinders, circuits utilise the quick exhaust valve. If the speed has to be adjustable, then flow control valves are used in either one or both sides of a double acting cylinder, depending upon the requirement.
7. In pneumatic application, it is the flow that controls the speed. Flow control valves are used in pneumatic circuits to control the rate of flow of compressed air from one part of the circuit to the other. Their function is to regulate the amount of compressed air passing through a valve by means of a metering orifice. Accurate flow control is difficult to achieve in pneumatics systems if there is variation in load.
8. A throttle relief valve permits the full flow of air in one direction of motion and reduced flow of air in the other. Throttle relief valves can be installed for speed control in a double acting cylinder in two ways: supply air throttling and exhaust air throttling.
9. In many pneumatic systems, deliberate time delays are required between specific operations and can be achieved using pneumatic time delay valves. A pneumatic time delay valve is used to generate or cancel an output signal after a specified interval of time from the application or removal of a pilot signal. Pneumatic time delay can be implemented with either on-delay or off-delay timers.
10. The pressure sequence valve is a combination of two valves: an adjustable relief valve and a 3/2 direction control valve. Both these valves are enveloped together.

EXERCISES

Fill in the Blanks

1. For relatively small single acting and double acting cylinders —— type of control is preferred.
2. A single acting pneumatic cylinder or a non-reversing air motor can be controlled by — way valve as the signal output unit. When a four-way control valve is used one outlet port is ——.
3. Double acting cylinders can be controlled by 2/3 way valve or ——— way valve corrected .
4. A throttle relief valve or throttle check valve is used to control the ——— in pneumatic circuits.
5. In supply air throttling, air ———— the cylinder is throttled.
6. A shuttle valve or a double check valve can be used as logic — valve.

7. In ——— pneumatic timer, the 3/2 DC valve is actuated after a delay with reference to the application of the pilot signal and is reset immediately after the release of the pilot signal.
8. The pressure sequence valve is a combination of two valves: an adjustable pressure ——— valve and a 3/2 direction control valve.
9. As per ISO 5599, pilot lines ports are designated using 12 or ———.
10. The most important threads used on pneumatic valves are metric thread (M), National Pipe Thread (NPT), Unified fine thread (UNF) and ——— (BSP~G)

State whether true or false

1. Direct control is preferred for large single acting or double acting cylinders.
2. Supply air throttling is used only for the speed control of single acting cylinders and small volume cylinders.
3. Exhaust air throttling is practically used in all double acting cylinders due to the double cushioning effect.
4. Double piloted 4/2 and 5/2 valves can be used as memory valves in pneumatic circuit designs.
5. There is no possibility of signal conflict in memory valves.
6. Two pressure valves can be used as a NOT valve in pneumatic circuit design.
7. Return speed of double acting cylinders can be increased by means of a quick exhaust valve.
8. In off-delay timers, the 3/2 DC valve is actuated immediately on the application of the pilot signal and is reset only after a delay from the time of the release of the pilot signal.
9. There are four possible combinations of pneumatic delay timer using NO and NC controls.
10. ISO 5599 systems recommend alphanumeric port designation for direction control valves.

Review questions

1. Where are three-way and four-way valves used?
2. List the basic five rules that are important in design of pneumatic circuits.
3. State three types of signal processing elements used in pneumatic control.
4. Explain the working of a NOT valve and provide one application of this valve.
5. Mention some applications of AND and OR valves.
6. What are the functions of pneumatic timer delay valves?
7. How are timer delay valves classified?
8. Differentiate between on-delay and off-delay timers with help of symbols.

9. Explain the construction and working of a pressure sequence valve with the help of a neat sketch.
10. Give two applications of pressure sequence valves and time delay valves.
11. Compare and contrast two-way valves and shuttle valves. Mention their application.
12. Differentiate between supply air throttling and exhaust air throttling.
13. Why is exhaust air throttling used for speed control in double acting cylinders.
14. What is the advantage of an internal pilot valve in pneumatic valves?
15. State the difference between a final control element and a signal element.

Problems

1. A double acting cylinder is to extended when a push button is operated. Upon release of the push button, the cylinder is to retract (Fig. 6.44). The cylinder is of small bore (25 mm diameter) requiring a small flow ratc to opcrate at the correct speed.

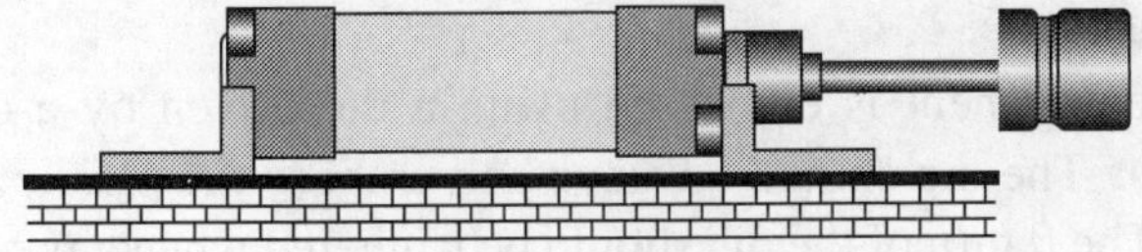

Figure 6.44

2. Study the following circuit and label all the components. Explain what happens if 1S1 or 1S2 is pressed. How is the cylinder retracted?

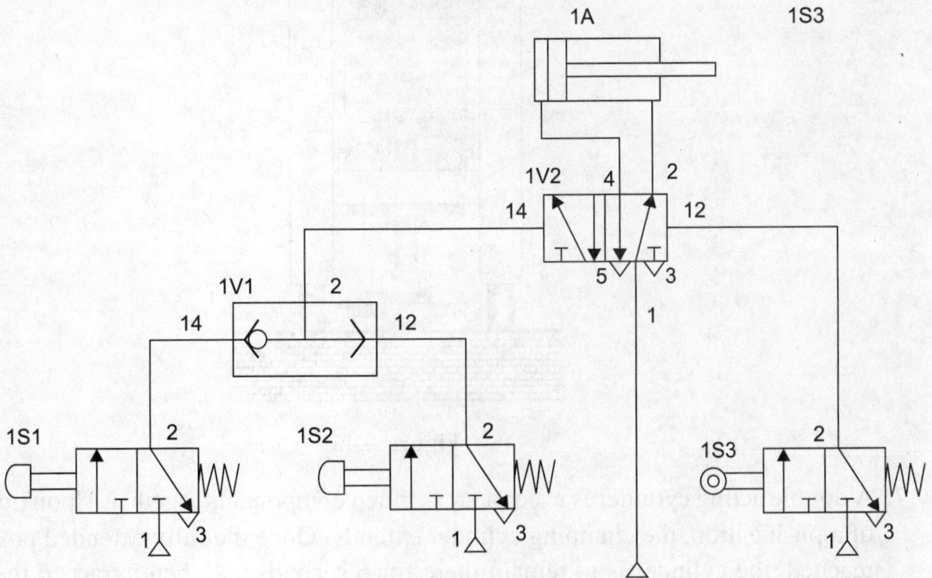

Figure 6.45

3. A double acting cylinder is to be advanced using three hand operated valves which are positioned at different locations from the cylinder. If any of the three push buttons are pressed, the cylinder must extend. Cylinder must be in the retracted condition if none of the push buttons are pressed.
4. A double acting cylinder is to extend and clamp a workpiece when two push button valves are pressed simultaneously. For safety reasons, these push buttons valves are installed in such a way that both valves cannot be operated with one hand; i.e., both hands must be used to operate these two valves. The cylinder is to retract when a one or two push buttons is released. Develop a pneumatic circuit to implement the given control task using AND logic.
5. A currency coin is embossed using a die driven by a double acting cylinder (Fig. 6.46). The die should advance and emboss the coin when a push button is operated. The return of the die is triggered when a preset pressure is reached. The embossing pressure should be adjustable so that different coins can be embossed on the same machine.
6. A plastic component is embossed using a die driven by a double acting cylinder (Fig. 6.46). The die should advance and emboss the plastic when a push button is operated. The return of the die should be triggered when a preset pressure is reached. The embossing pressure should be adjustable.

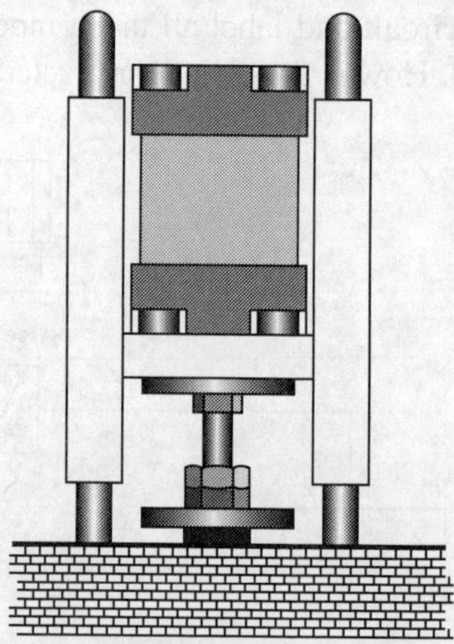

Figure 6.46

7. A double acting cylinder is used to press glued components together. Upon operation of a push button, the clamping cylinder extends. Once the fully extended position is reached, the cylinder is to remain there for 6 seconds and then retract to the initial position. The cylinder retraction is to be adjustable. A new cycle can start only after the cylinder has fully retracted.

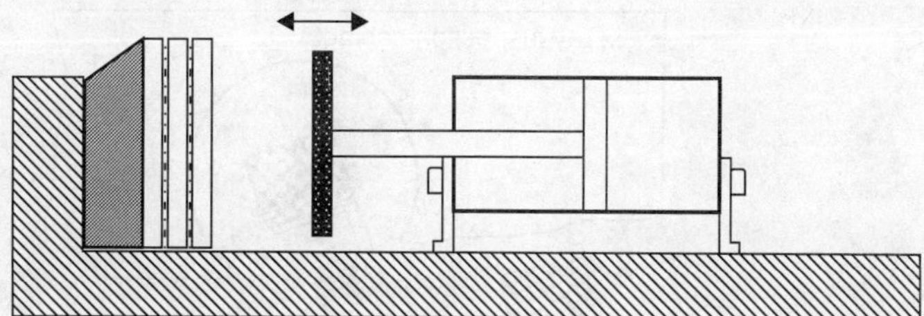

Figure 6.47

8. A single acting cylinder is used to sort disoriented metal stampings. If the metal stampings are disoriented, they are pushed to a second conveyor (Fig. 6.48). If a push button is pressed, a single acting cylinder must actuate to push stampings to the second conveyor which is inclined downwards; the cylinder has to retract if the push button is released. The forward motion of the cylinder takes 0.5 seconds. A pressure gauge is to be fitted before and after the one–way flow control valve. Design a pneumatic circuit.

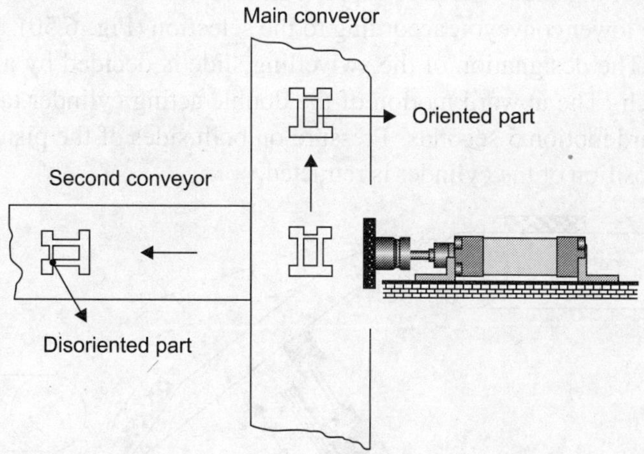

Figure 6.48

9. In a parcel delivery room of a central post office, a parcel separating device feeds post from a sloping conveyor to an X-ray checking unit before delivery (Fig. 6.49). Operating a push button causes very rapid retraction of the single acting actuator, and the piston rod advances. Forward motion time is 10 seconds. A pressure gauge has to fitted before and aftĕr the flow control valve.

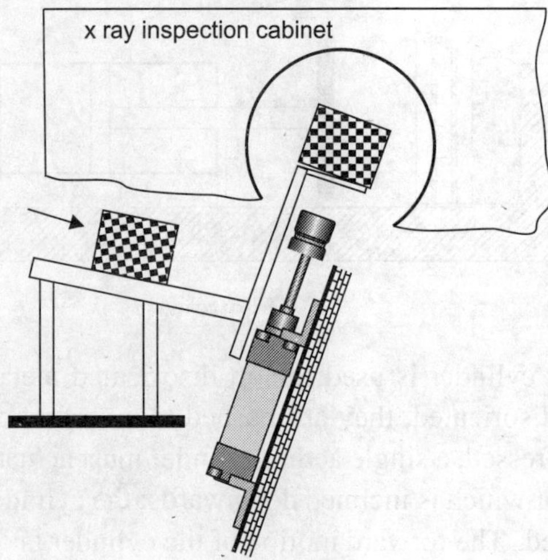

Figure 6.49

10. The swivelling slide is used to supply soft drinks packets either to an upper conveyor or to the lower conveyor, according to the selection (Fig. 6.50). Design a pneumatic circuit. The designation of the swivelling slide is decided by a valve with a selector switch. The upward motion of the double acting cylinder takes 10 seconds, the downward motion 5 seconds. Pressure on both sides of the piston is indicated. The initial position of the cylinder is retracted.

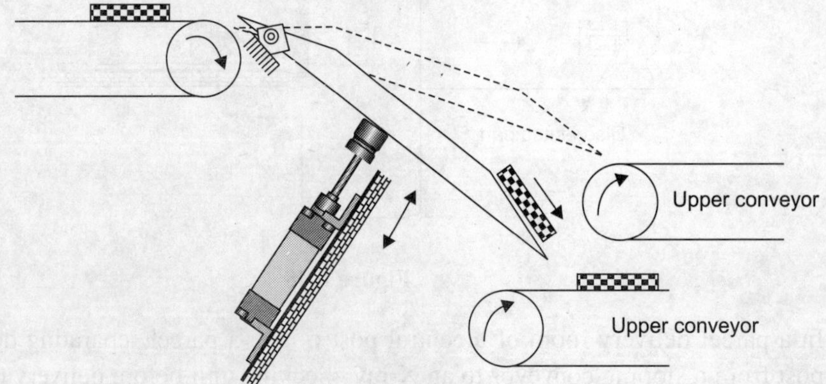

Figure 6.50

11. A double acting cylinder guides cylinder pins towards a measuring device. The pins are separated by means of a continuous to and fro movement. The oscillating motion can be started by means of a valve with selector switch (Fig. 6.51). The duration of the forward stroke of the cylinder is to be 6 seconds. The return stroke is 4 seconds. The cylinder is to remain in the forward end position for 10 seconds, resulting in a cycle time of 20 seconds.

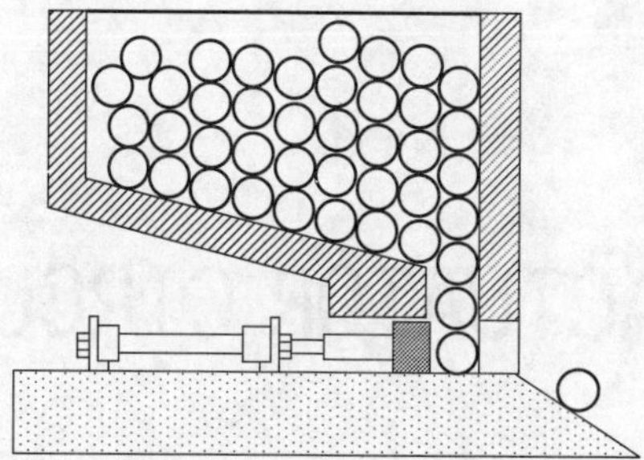

Figure 6.51

12. After the liquid paint colours have been poured together into a bucket, they are mixed in by a vibrating machine (Fig. 6.52). When a push button has been pressed, the extended cylinder (1A) retracts completely and executes a to and fro movement in the rear stroke range. The oscillation is limited to the retracted end position by a roller lever valve as well as a second roller lever valve in the central position. The frequency of oscillation is adjustable within limits by setting a pressure regulator controlling the amount of air supply. Set an operating pressure of 4 bar. After a specified interval, the oscillator is switched off. The double acting cylinder extends completely and actuates the third roller lever valve. Set a vibration time of 5 seconds.

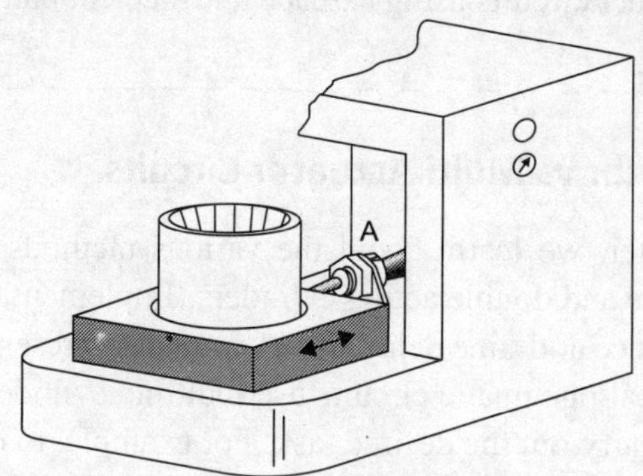

Figure 6.52

Answers

Fill in the blanks

1. direct 2. three / plugged 3. four/five 4. flow(speed) 5. entering
6. OR 7. On -delay 8. Relief 9. 14 10. British standard pipe thread (BSP-G)

True or false

1. False 2. True 3. True 4. True 5. False 6. False 7. True 8. True 9. True 10. False

7 MULTI ACTUATOR CIRCUITS – PART I

Upon completion of this chapter, the reader should be able to

- Differentiate between single and multi actuator circuits
- List the methods available for pneumatic circuit design
- Explain signal conflict in double piloted memory valves
- Draw motion and displacement diagrams for multi cylinders
- List and explain signal elimination methods
- Explain the working of idle roller valves and their use
- Design multi cylinder circuits using cascade and step counter methods

7.1 Single Actuator vs Multi Actuator Circuits

In the previous chapter, we learnt about the various methods to control single actuator circuits for both single and double acting cylinders. Implementation of logic gates and the use of pressure sequence and time delay valves were also presented.

Most of the practical pneumatic circuits use multiple cylinders. They are operated in a specific sequence to carry out the desired task. For example, to drill a wooden component, it first needs to be clamped and then drilled. The cylinder can be unclamped only after the drill is withdrawn from the workpiece. The sequence of movements of the cylinders that clamp and those that carry the drill is important. This sequence is determined by the actuation of the appropriate final control valves (e.g., DCVs). The position of the cylinders is gauged by sensors like limit switches and roller or cam operated valves.

Multi cylinder pneumatic circuits can be designed in various ways. There is no universal circuit design method that suits all types of circuits. Some methods are commonly used for compound circuits but would be too expensive for simple circuits. There are five common methods used in engineering and they are given below:

- Classic or intuitive method
- Cascade method

- Step counter method
- Karnaugh–Veitch (KV) method
- Combinational circuit design

In this chapter, the classic, cascade and step counter methods are discussed. Chapter 8 deals with Boolean algebra, KV mapping and combinational circuit methods. Since double piloted 4/2 and 5/2 directional control valves are susceptible to signal conflicts, the more systematic methods like cascading and step counter can be used to eliminate signal conflict when these DCVs are used and are better choices than the intuitive methods.

7.2 Classic or Intuitive Method

In the intuitive method, circuit design is done intuitively based on the knowledge of pneumatics. In general, the steps are:

- Writing down the sequence of tasks and drawing the motion diagrams
- Drawing the cylinders and control valves
- Completing the circuits intuitively

7.2.1 Coordinated and sequential motion control

In majority of pneumatic applications, more than one cylinder is used. The movements of these cylinders are coordinated according to the required sequence. Sensors (usually limit switches) are used to verify the position of the cylinder and determine the subsequent actuation of the final control element. The activation of the limit switches of the component cylinders provides the "set" or "reset" signals to the final control valves for further control of the cylinders and their movements. A motion diagram is used to determine the location of the limit switches.

7.2.2 Demonstration of the classic method

In order to develop control circuitry for multi cylinder applications, it is necessary to draw the motion diagram to understand the sequence of actuation of various signal input switches (limit switches and sensors). The motion diagram represents the possible position of the cylinder—whether extended or retracted in a particular sequence.

Example 7.1 In a press shop, a stamping operation is to be performed using a stamping machine. The workpiece has to be first clamped under the stamping station. Then the stamping tool gets into position and performs the stamping operation. The workpiece must be unclamped only after the operation is completed.

Step 1: Write the statement of the problem.
A is the clamping cylinder and B is the stamping cylinder, as shown in Fig. 7.1. First, cylinder A extends and brings the workpiece under the stamping station where cylinder

B is located. Cylinder B then extends and stamps the job. Cylinder A returns only after cylinder B has retracted fully.

Step 2: Draw the positional layout (Fig. 7.1).

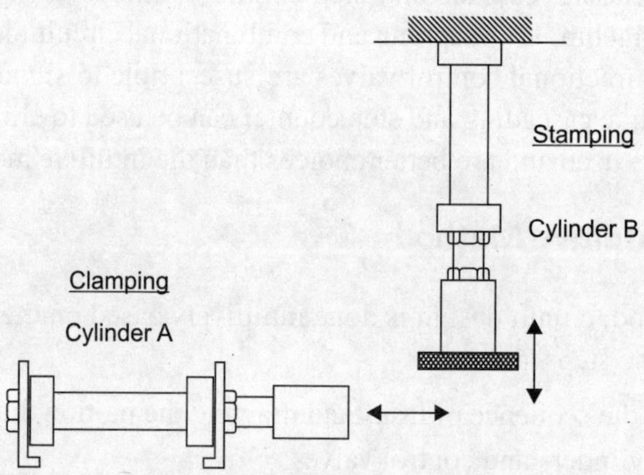

Figure 7.1 *Positional layout*

Step 3: Represent the control task using notational form.

Cylinder A advancing is designated as A+
Cylinder A retracting is designated as A–
Cylinder B advancing is designated as B+
Cylinder B retracting is designated as B–

The sequence for clamping and stamping is A+B+B–A–

Step 4: Draw the displacement–step diagram (Fig. 7.2).

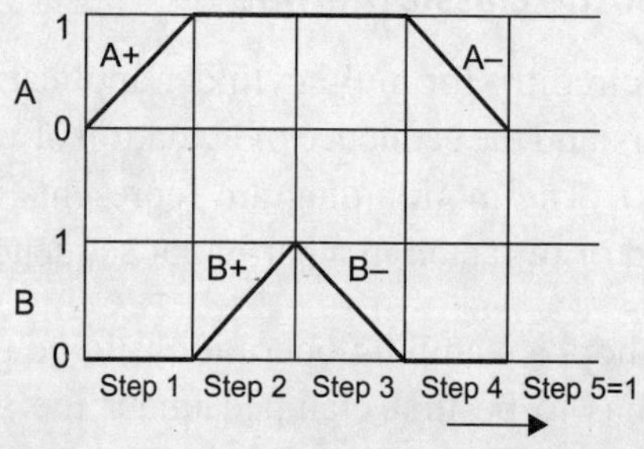

Figure 7.2 *Displacement–step diagram*

Step 5: Draw the displacement–time diagram (Fig. 7.3).

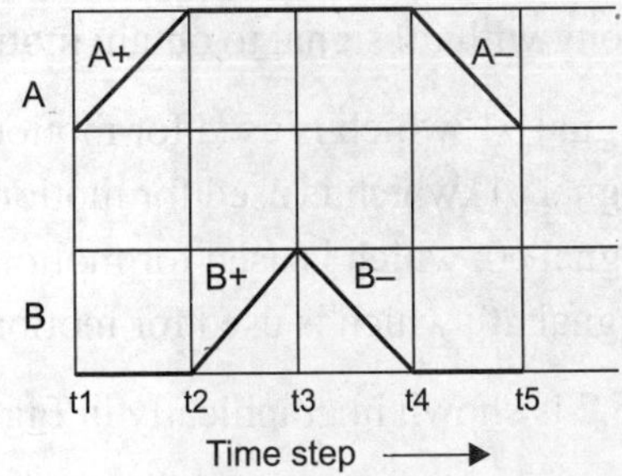

Figure 7.3 *Displacement–time diagram*

Step 6: Analyse and draw the pneumatic circuit.
Step 6.1: Analyse input and output signals.

Input signals

Cylinder A: Limit switch at home position $a0$
Limit switch at home position $a1$
Cylinder B: Limit switch at home position $b0$
Limit switch at home position $b1$

Output signal

Forward motion of cylinder A (A+)
Return motion of cylinder A (A−)
Forward motion of cylinder B(B+)
Return motion of cylinder B(B−)

Step 6.2: Use the displacement–time/step diagrams to link input and output signals (Fig. 7.4).

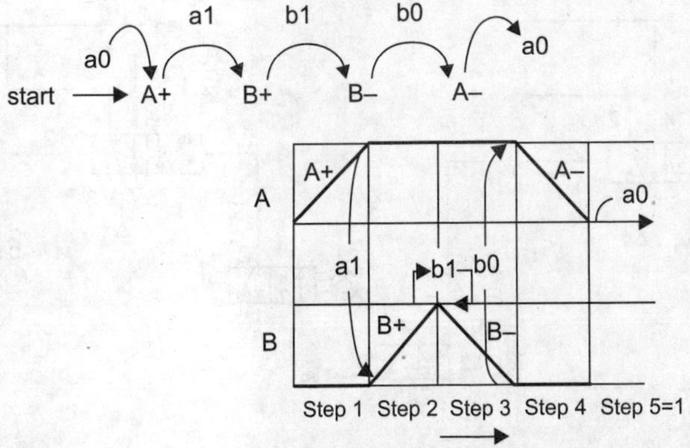

Figure 7.4 *Input–output signal flow*

The start signal is required along with $a0$ signal to obtain motion A+.

1. A+ generates sensor signal $a1$, which is used for motion B+.
2. B+ generates sensor signal $b1$, which is used for motion B−.
3. B− generates sensor signal $b0$, which is used for motion A−.
4. A− generates sensor signal $a0$, which is used for motion A+.

The information above Step 6.2 is shown in graphically in Fig. 7.4.

Step 7: Draw the power circuit (Fig. 7.5).

(i) Draw the cylinders A(1.0) and B(2.0).
(ii) Draw the DCVs 1.1 and 2.1 in the unactuated state.
(iii) Mark the limit switch positions for cylinders A(1.0) and B(2.0).

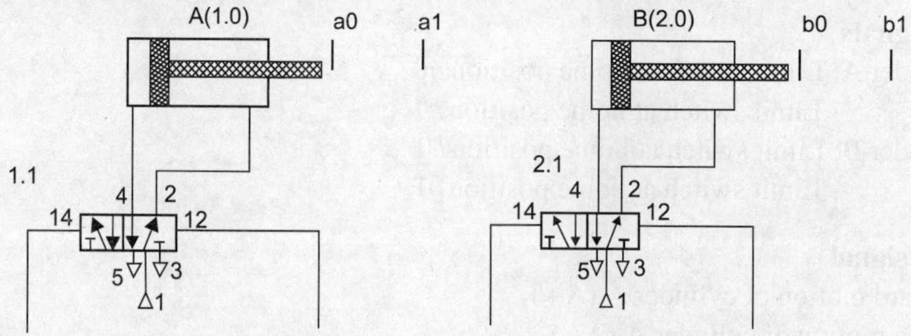

Figure 7.5 *Power circuit diagram*

Step 8: Draw the control circuit (Fig. 7.6).

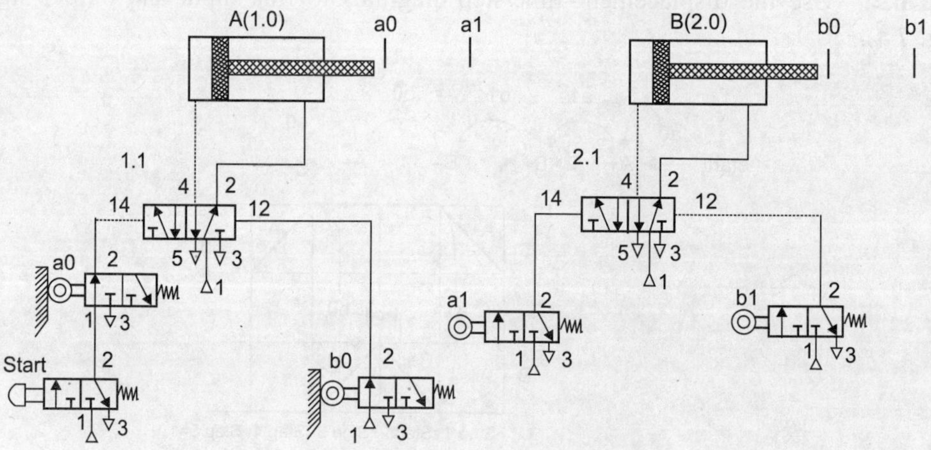

Figure 7.6 *Control circuit diagram*

Step 9: Analysis of the pneumatic circuit:

1. When the start button is pressed, the signal appears at port 14 of valve 1.1 through the limit switch signal $a0$.
2. Check for and confirm the presence of the signal at the other end (port 12) of valve 1.1 (because $b0$ is also pressed). This results in a signal conflict and valve 1.1 is unable to move (Fig. 7.7).
3. Assume for now that $b0$ is somehow disengaged so that valve 1.1 can switch over and consequently cylinder A can extend when the start button is pressed (Fig. 7.8).
4. When cylinder A extends fully, it generates a limit switch signal $a1$, which is applied to port 14 of the valve 2.1.
5. Check for the presence of the signal at the other end (port 12) of valve 2.1. There is no signal and hence no signal conflict.
6. Signal applied to port 14 of valve 2.1 causes the shifting of DCV 2.1 and cylinder B extends.
7. When cylinder B extends fully, it generates a limit switch signal $b1$, which is applied to port 12 of valve 2.1.
8. Checking for the presence of the signal at the other end of valve 2.1 (port 14) confirms that the signal is present there as well (because $a1$ is also pressed). This results in a signal conflict and valve 2.1 is unable to move (Fig. 7.9).
9. Assume for now that $a1$ is somehow disengaged so that valve 2.1 can switch over and consequently cylinder B can retract (Fig. 7.10).

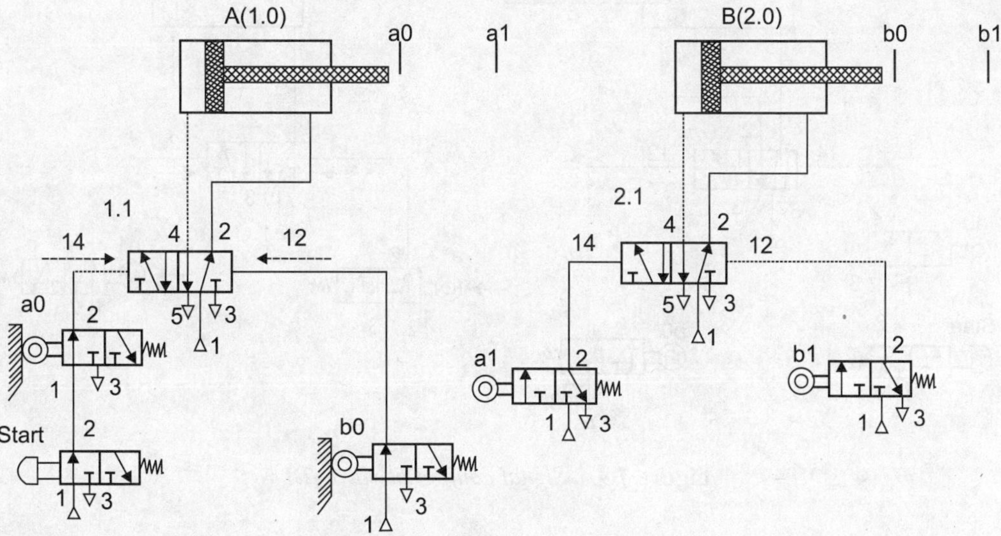

Figure 7.7 *Signal conflict at valve 1.1*

10. When cylinder B is fully retracted, it generates a limit switch signal $b0$, which is applied to port 12 of valve 1.1 (Fig. 7.11).
11. Check for a signal at the other end (port 14) of valve 1.1; there is no signal present and hence no conflict. So valve 1.1 can switch over and cylinder A can retract.

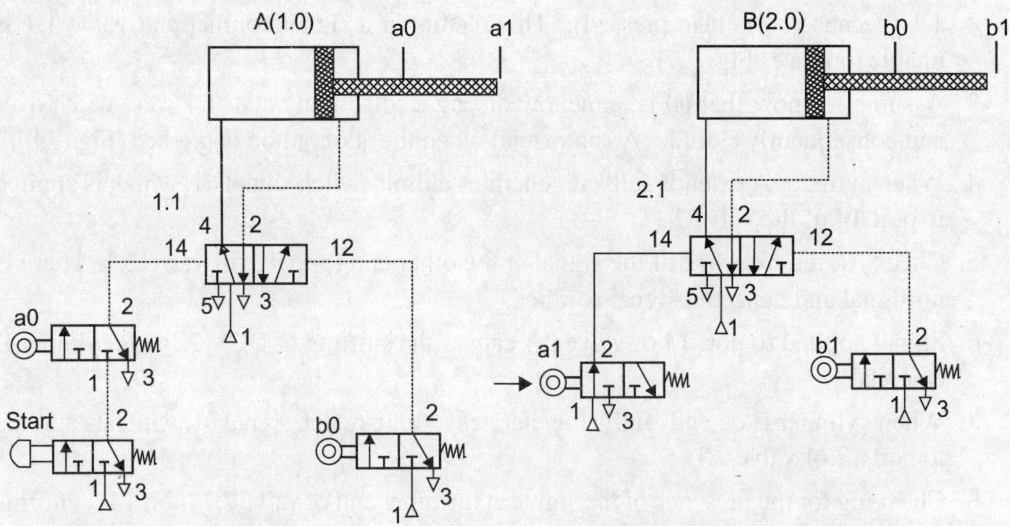

Figure 7.8 *Pneumatic circuit after cylinders A and B are extended* $(A+B+)$

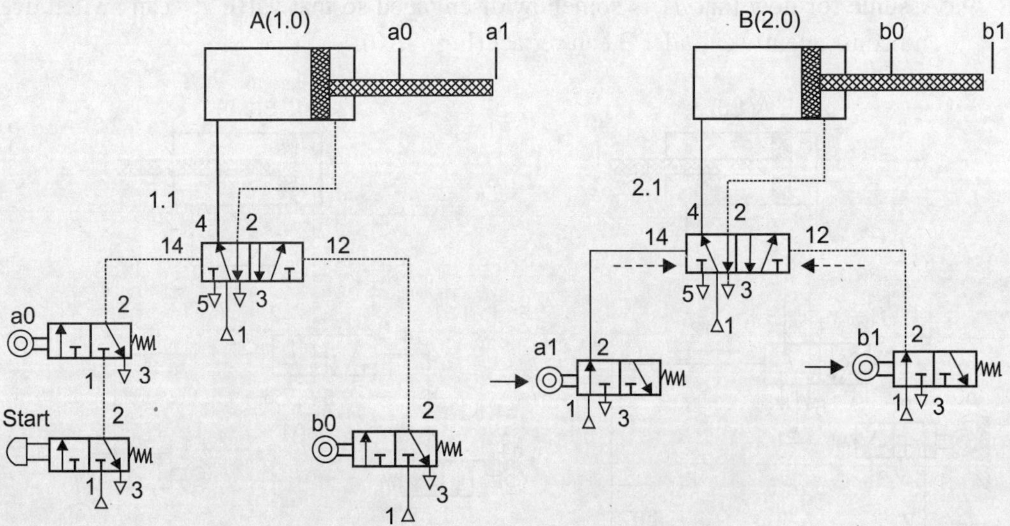

Figure 7.9 *Signal conflict at valve 2.1*

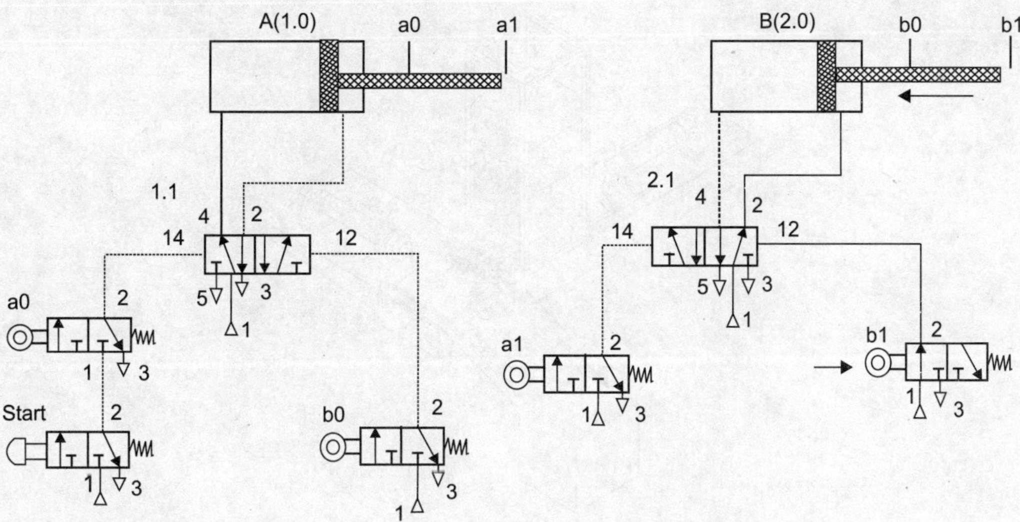

Figure 7.10 *Position when cylinder B is reversing (B−)*

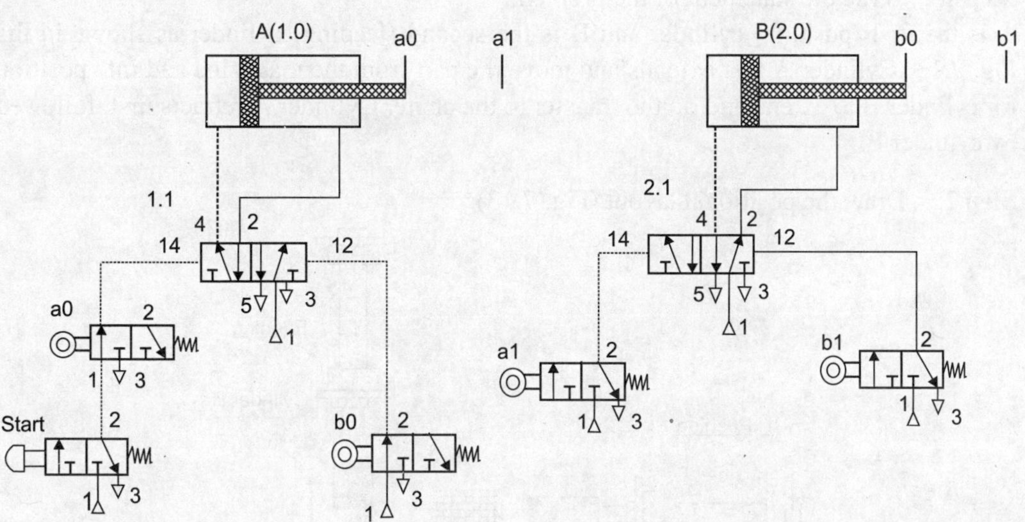

Figure 7.11 *Position when cylinder A has retracted fully (A−)*

Example 7.2 Two cylinders are used to transfer parts from a magazine on to a chute (Fig. 7.12). When a push button is pressed, the first cylinder extends, pushing the part from the magazine and positioning it in preparation for transfer by the second cylinder on to the out-feed chute. Once the part is transferred, the first cylinder retracts followed by the second. Confirmation of all extended and retracted positions are required.

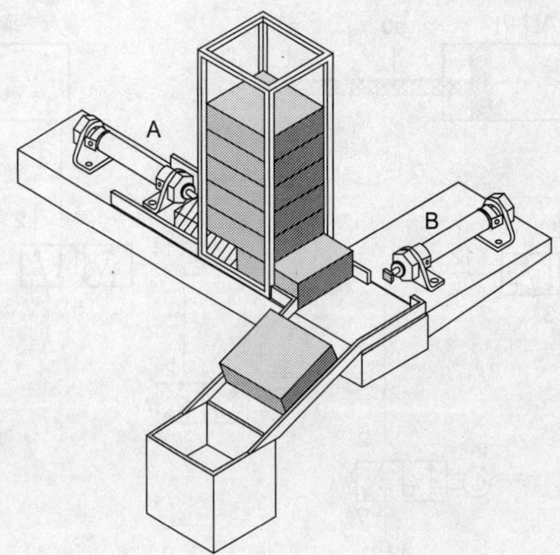

Figure 7.12 *Positional diagram*

Step 1: Write the statement of the problem.

A is the first (pushing) cylinder and B is the second (feeding) cylinder as shown in the Fig. 7.13. Cylinder A first extends and move the part from the magazine and into position for cylinder B to extend and do the transfer to the chute. Cylinder A retracts first, followed by cylinder B.

Step 2: Draw the positional layout (Fig. 7.13).

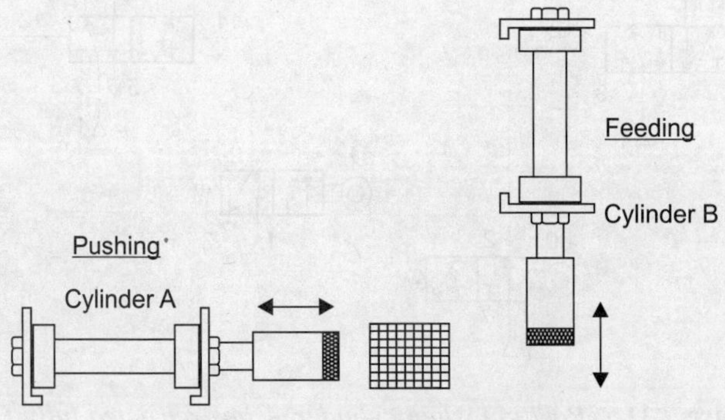

Figure 7.13 *Positional diagram*

Step 3: Represent the control task using notational form.

Cylinder A advancing is designated as A+

Cylinder A retracting is designated as A–

Cylinder B advancing is designated as B+

Cylinder B retracting is designated as B–

The sequence for the push and transfer operation is A+B+A–B–.

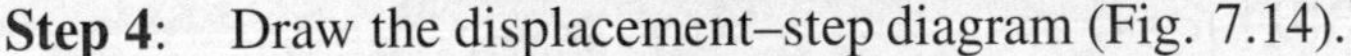

Step 4: Draw the displacement–step diagram (Fig. 7.14).

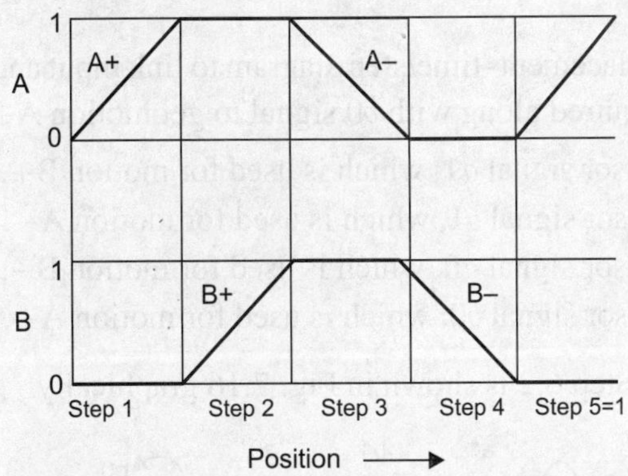

Figure 7.14 *Displacement–step diagram*

Step 5: Draw the displacement–time diagram (Fig. 7.15).

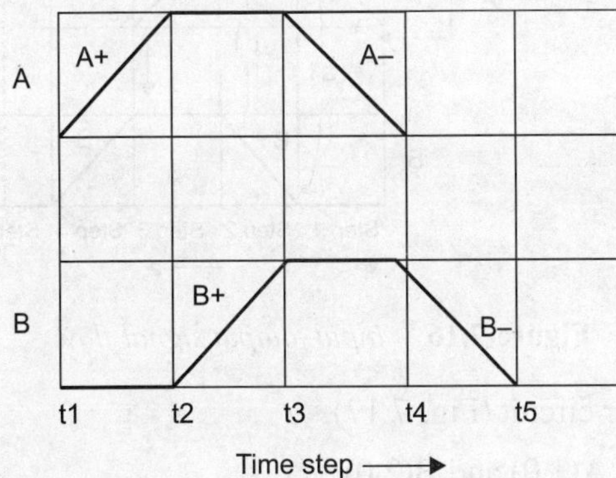

Figure 7.15 *Displacement–time diagram*

Step 6: Analyse and draw the pneumatic circuit.
Step 6.1: Analyse input and output signals.

Input signals

Cylinder A: Limit switch at home position $a0$
Limit switch at home position $a1$

Cylinder B: Limit switch at home position $b0$
Limit switch at home position $b1$

Output signal

Forward motion of cylinder A(A+)
Return motion of cylinder A(A–)

Forward motion of cylinder B(B+)
Return motion of cylinder B(B−)

Step 6.2: Use the displacement–time/step diagram to link input and output signals (Fig. 7.16). The start signal is required along with $b0$ signal to get motion A+.

1. A+ generates sensor signal $a1$, which is used for motion B+.
2. B+ generates sensor signal $b1$, which is used for motion A−.
3. A− generates sensor signal $a0$, which is used for motion B−.
4. B− generates sensor signal $b0$, which is used for motion A+.

The information above Step 6.2 is shown in Fig. 7.16 graphically.

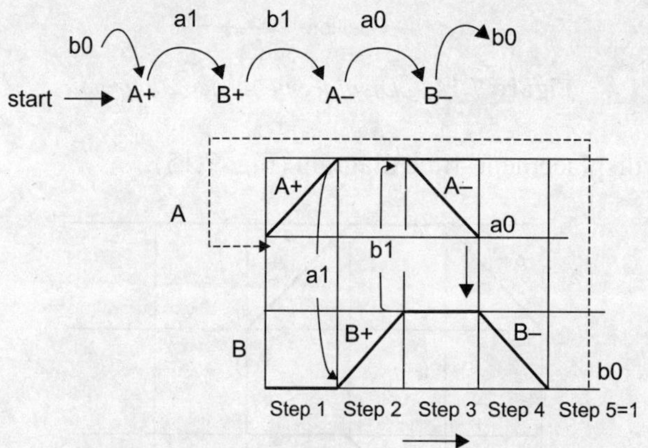

Figure 7.16 *Input–output signal flow*

Step 7: Draw the power circuit (Fig. 7.17).

(i) Draw the cylinders A(1.0) and B(2.0).
(ii) Draw the DCVs 1.1 and 2.1 in the unactuated state.
(iii) Mark the limit switch positions for cylinders A(1.0) and B(2.0).

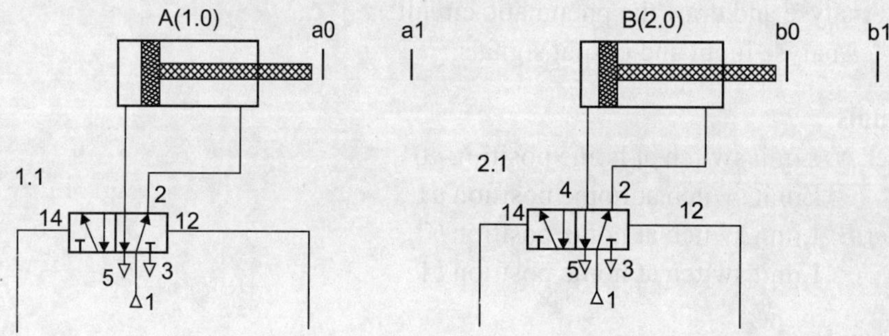

Figure 7.17 *Power circuit diagram*

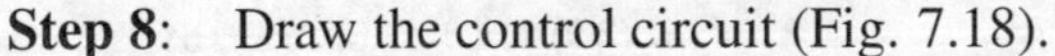

Step 8: Draw the control circuit (Fig. 7.18).

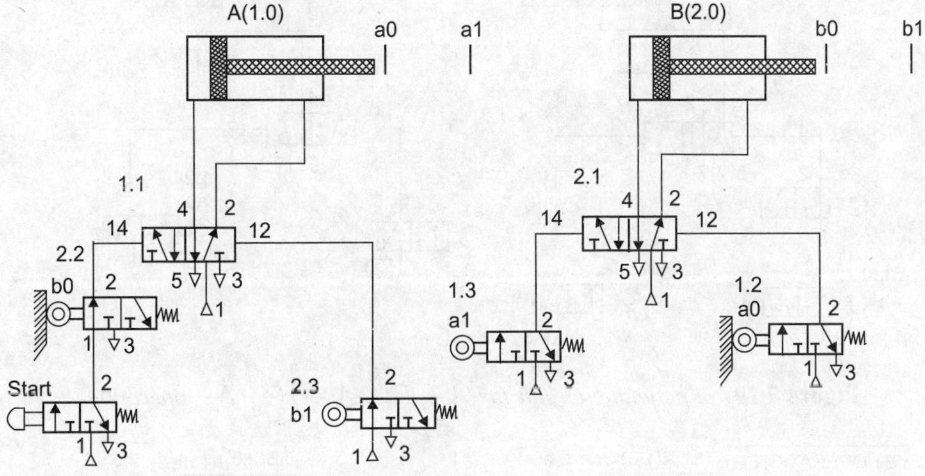

Figure 7.18

Step 9: Analysis of the pneumatic circuit:

1. When the start button is pressed, the signal appears at port 14 of valve 1.1 through the limit switch signal $b0$.
2. Check for the presence of the signal at the other end (12) of valve 1.1 and verify that no signal is present at port there (because $b1$ is not pressed). There is no signal conflict and valve 1.1 is able to move and cylinder A advances to the forward position.
3. When cylinder A extends fully, it generates a limit switch signal $a1$, which is applied to port 14 of valve 2.1. Cylinder B advances to the forward position.
4. Check for the presence of the signal at the other end (port 12) of valve 2.1 and verify that no signal is present there (because $a0$ is not pressed, A is already in extended position now) and so there is no signal conflict.
5. A signal applied to port 14 of valve 2.1 causes the shifting of DCV 2.1 and cylinder B extends.
6. When cylinder B extends fully, it generates a limit switch signal $b1$, which is applied to port 12 of valve 1.1. Cylinder A returns and ao is pressed. There is no signal conflict, as $a0$ and $a1$ are mutually exclusive signals.
7. When cylinder A is fully retracted, it generates a limit switch signal $a0$, which is applied to port 12 of valve 2.1. Cylinder B retracts.

All five operations are shown in Figures 7.19 to 7.23.

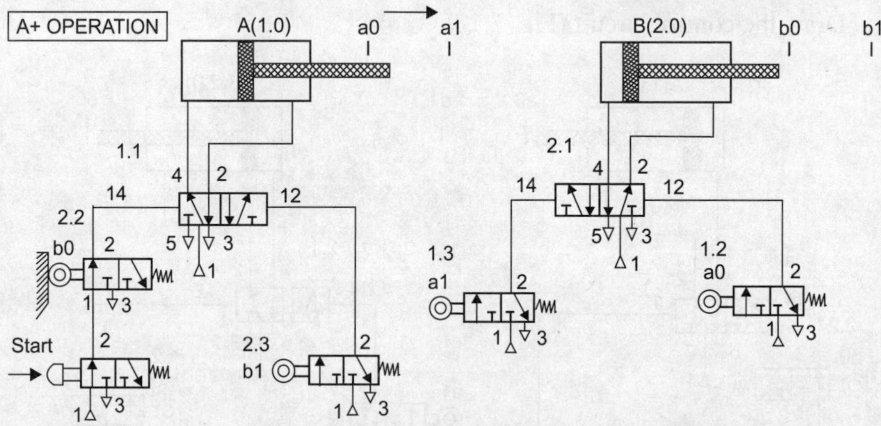

Figure 7.19 *Pneumatic circuit for* $A + B + A - B-$ *(A+ operation)*

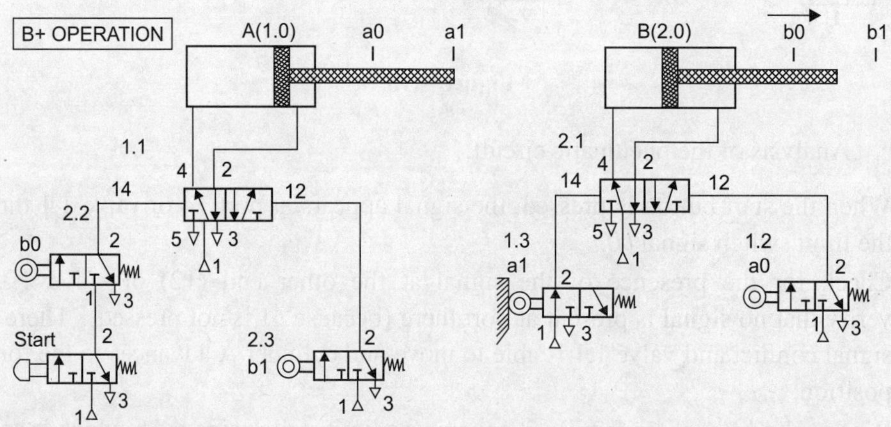

Figure 7.20 *Pneumatic circuit for* $A + B + A - B-$ *(B+ operation)*

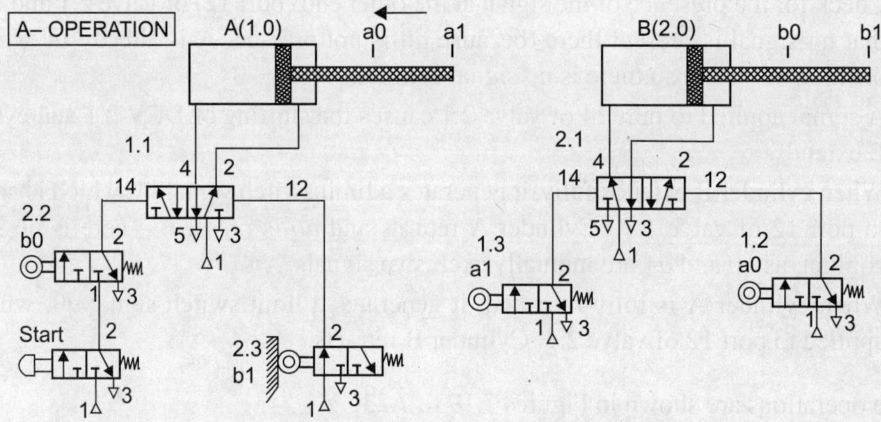

Figure 7.21 *Pneumatic circuit for* $A + B + A - B-$ *(A– operation)*

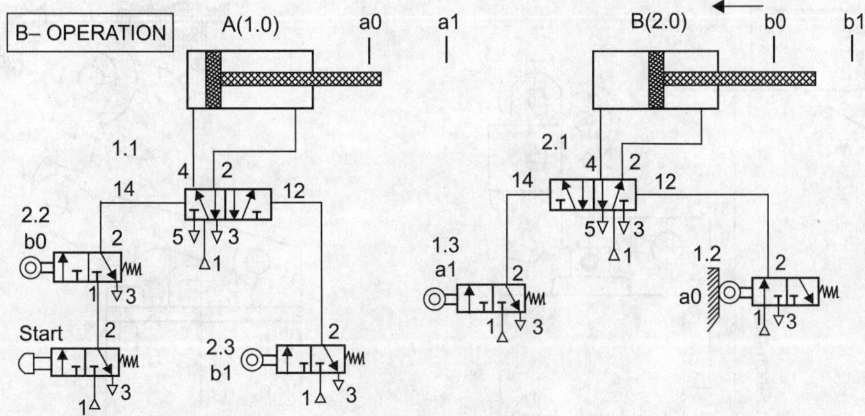

Figure 7.22 *Pneumatic circuit for $A + B + A - B-$ (B– operation)*

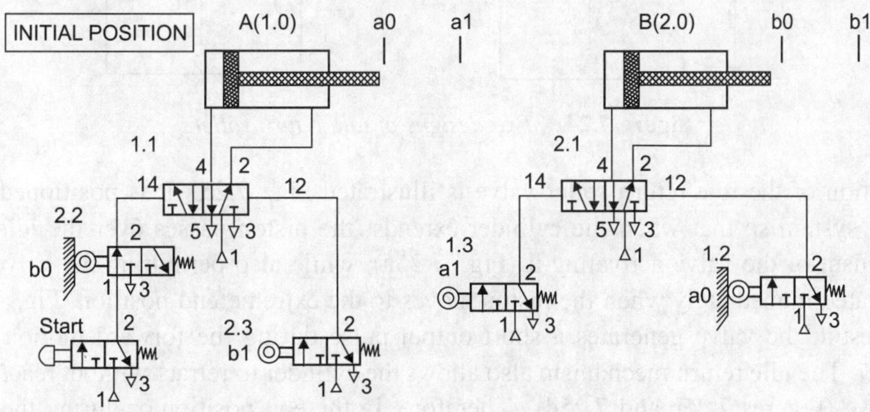

Figure 7.23 *Pneumatic circuit for $A + B + A - B-$ (initial position)*

7.2.3 Elimination of signal conflict

Various methods are used to solve the problem of signal conflicts in multi cylinder circuits.

a) Idle return roller
b) Reversing valves (memory valves)
c) Modules consisting of combination of valves

The cascading method uses reversing valves (also known group changing valves) and the step counter method uses modular valves. Both methods are discussed in subsequent sections of this chapter.

7.2.3.1 Use of idle return rollers

An idle return roller valve consists of a 3/2 DCV fitted with an idle return roller mechanism. The two designs of an idle roller are shown in Fig. 7.24.

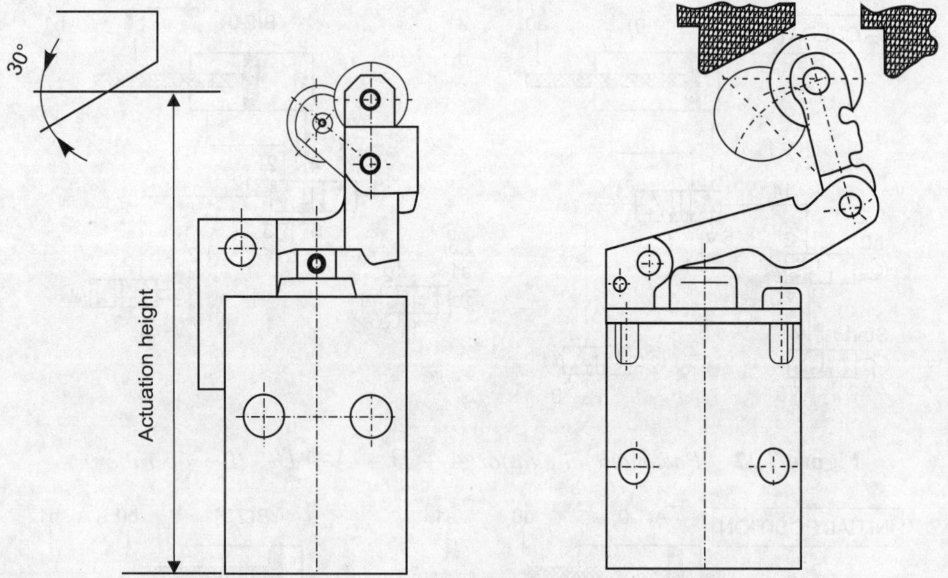

Figure 7.24 *Two designs of idle return rollers*

The action of the idle return roller valve is illustrated Fig. 7.25. It is positioned in the control system so that when the cylinder extends, the piston passes over the idle roller mechanism of the valve activating it (Fig. 7.25a), while also permitting the valve to be deactivated immediately when the piston moves to the extreme end position (Fig. 7.25b). As a result, the valve generates a short output pulse during the forward motion of the cylinder. The idle return mechanism also allows the cylinder to retract without reactivating the valve (Figures 7.25c and 7.25d). Therefore, in the end position or during the return motion of the piston, the valve does not get actuated and no output signal is produced. For the idle return roller valve to generate a short output pulse during the return motion of the cylinder, the valve is positioned in the direction opposite to that in the earlier sequence of forward motion of the cylinder.

The problem of signal conflict caused by roller valves $b1$ and $a1$ in the previous sequence problem is eliminated by replacing them with idle return rollers.

Drawbacks of idle return rollers:

1. This method is not reliable since the end position cannot be sensed accurately.
2. The control system is slow.

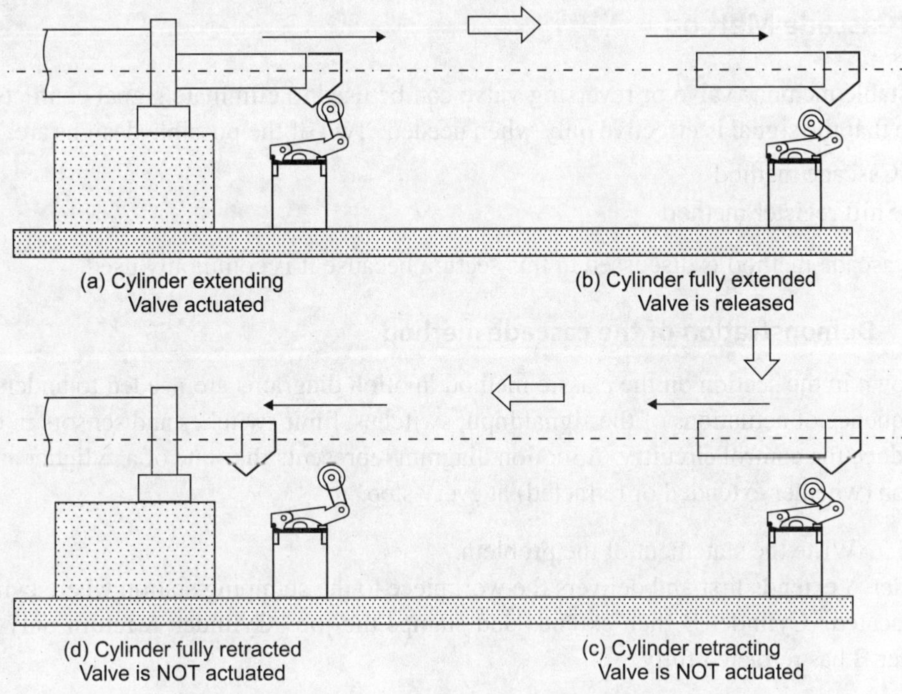

Figure 7.25 *Action of idle return rollers*

Example 7.3 Develop a pneumatic circuit for the sequence A+ B+ B− A−. Avoid signal conflict by using an idle return roller.

Figure 7.26 shows the circuit for a control sequence A+ B+ B− A− using idle return rollers at positions $b0$ and $a1$ to prevent signal conflict.

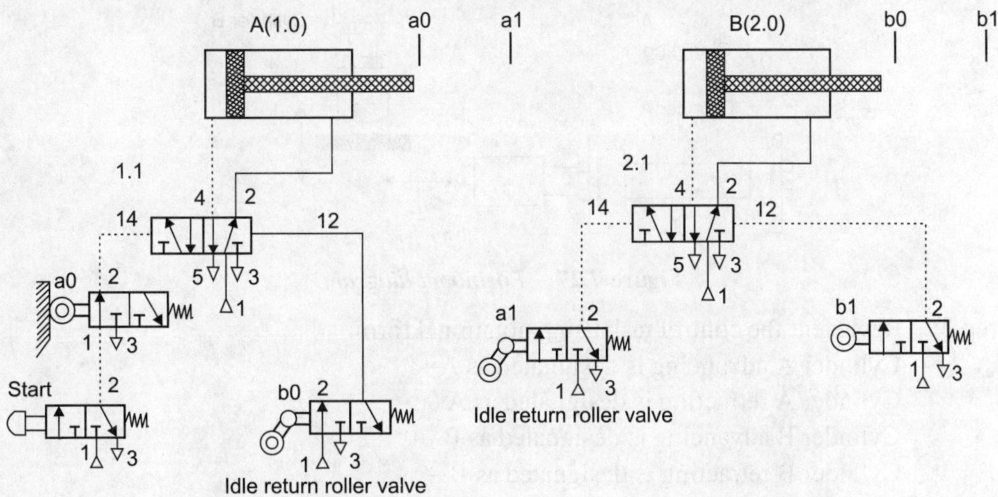

Figure 7.26 *Pneumatic circuits for* A + B + B − A− *(with idle return roller)*

7.3 Cascade Method

A bi-stable memory valve or reversing valve can be used to eliminate signal conflicts and ensure that the signal is effective only when needed. Two of the possible designs are:

(i) Cascade method
(ii) Shift register method

Only cascade method is discussed in this section because it is commonly used.

7.3.1 Demonstration of the cascade method

As shown in the section on the classic method, motion diagrams are needed to understand the sequence of actuations of the signal input switches, limit switches and sensors in order to produce the control circuitry. A motion diagram represents the state of a cylinder and its position (whether extended or retracted) at every step.

Step 1: Write the statement of the problem.
Cylinder A extends first and delivers the workpiece to the stamping station where cylinder B is located. Cylinder B then extends and stamps the job. Cylinder A return only after cylinder B has retracted fully.

Step 2: Draw the positional layout (Fig. 7.27).

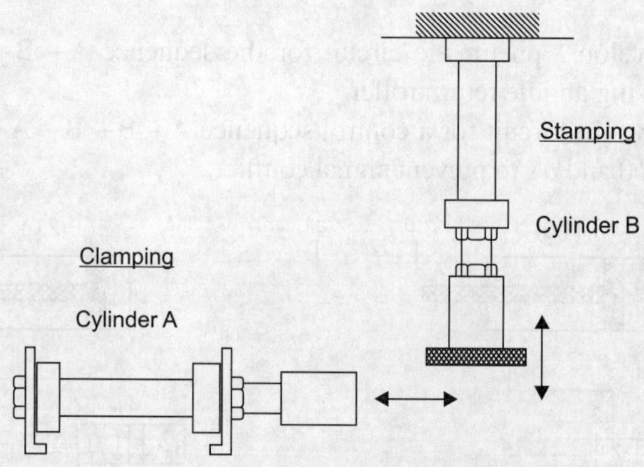

Figure 7.27 *Positional diagram*

Step 3: Represent the control task using notational form
Cylinder A advancing is designated as A+
Cylinder A retracting is designated as A−
Cylinder B advancing is designated as B+
Cylinder B retracting is designated as B−

The sequence for clamping and stamping is A+B+B−A−.

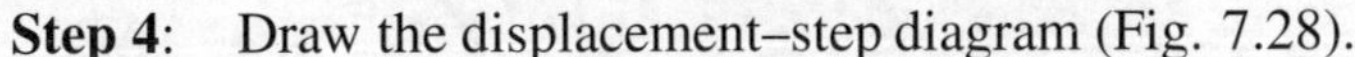

Step 4: Draw the displacement–step diagram (Fig. 7.28).

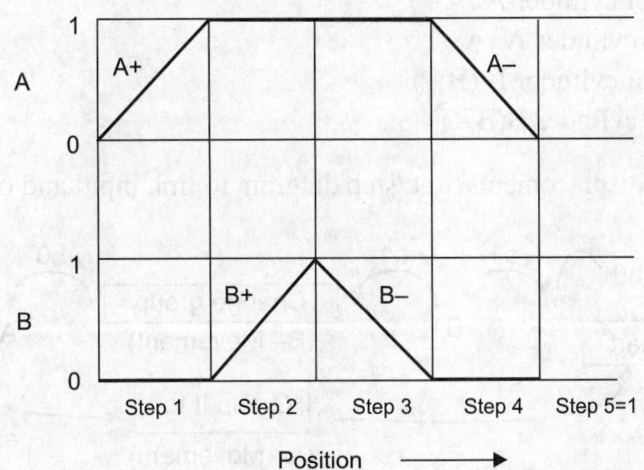

Figure 7.28 *Displacement–step diagram*

Step 5: Draw the displacement–time diagram (Fig. 7.29).

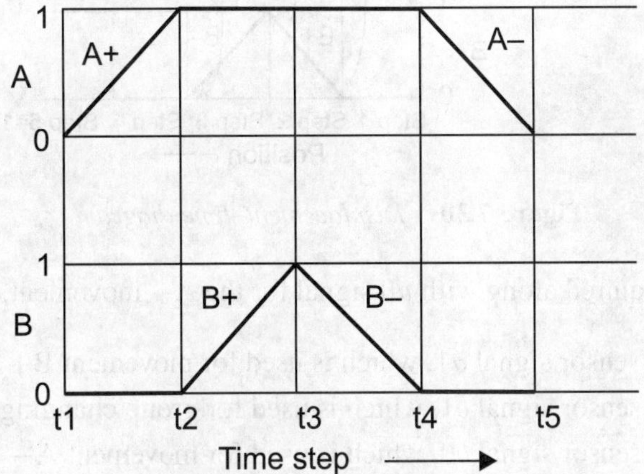

Figure 7.29 *Displacement–time diagram*

Step 6: Analyse and draw the pneumatic circuit.

Step 6.1: Analyse input and output signals.

Input signals

Cylinder A: Limit switch at home position $a0$

Limit switch at home position $a1$

Cylinder B: Limit switch at home position $b0$

Limit switch at home position $b1$

Output signal

Forward motion of cylinder A (A+)
Return motion of cylinder A (A–)
Forward motion of cylinder B (B+)
Return motion of cylinder B(B–)

Step 6.2: Use the displacement–time/step diagram to link input and output signals (T).

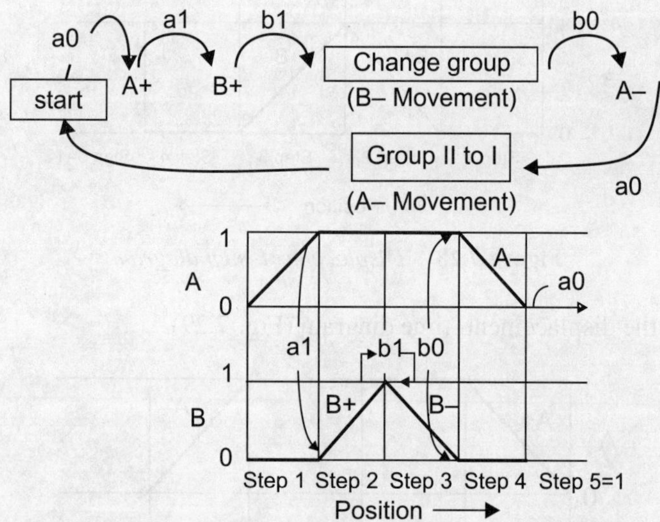

Figure 7.30 *Displacement–time diagram*

The start signal is required along with $a0$ signal for the A+ movement.

1. A+ generates sensor signal $a1$, which is used for movement B+
2. B+ generates sensor signal $b1$, which is used for group changing
3. B– generates sensor signal $b0$, which is used for movement A–
4. A– generates sensor signal $a0$, which is used for group changing

The information (step 6.2) is shown below graphically.

Step 7: Draw the power circuit.

(i) Divide the given circuits into groups. Grouping should be done so that there is no signal conflict. Do not put A+ and A– in the same group. Similarly B+ and B- should be in different groups to avoid signal conflict.

In this example, (A+ B+ B– A–) can be grouped as:

A+ B+	*B– A–*
Group 1	Group 2

(ii) The number of group changing valves is one less than the number of groups. In this example, we have 2 groups. So we need one group changing valve.

Connect the group changing valve as shown in Fig. 7.31. From the figure it is clear that when the control signals I and II are applied to the group changing valve, the air (power) supply changes from Group 1(G1) to Group 2 (G2).

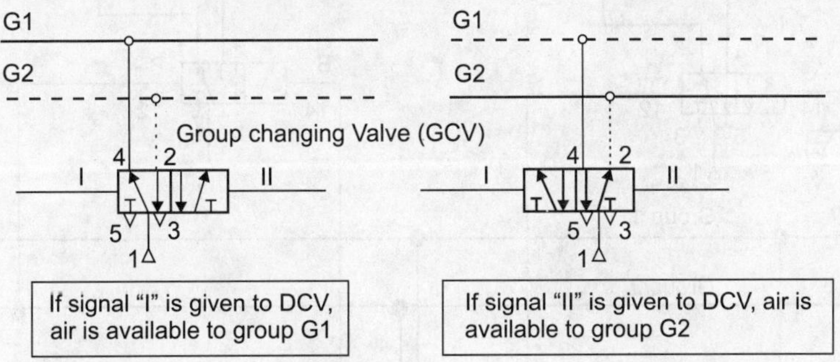

Figure 7.31

(iii) Arrange the limit switch and start button as shown in Fig. 7.32.

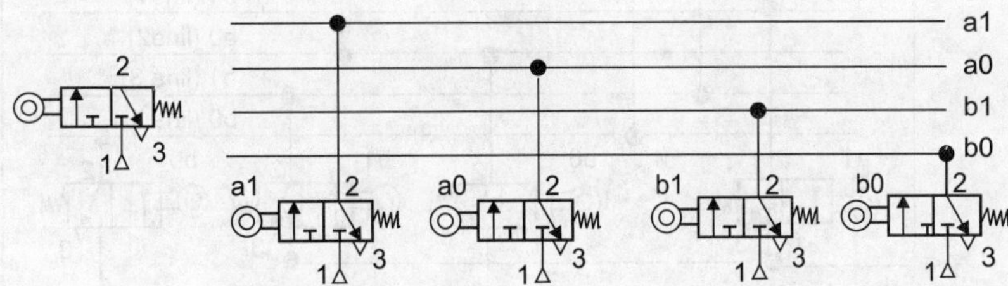

Figure 7.32

(iv) Draw the power circuit (Fig. 7.33).

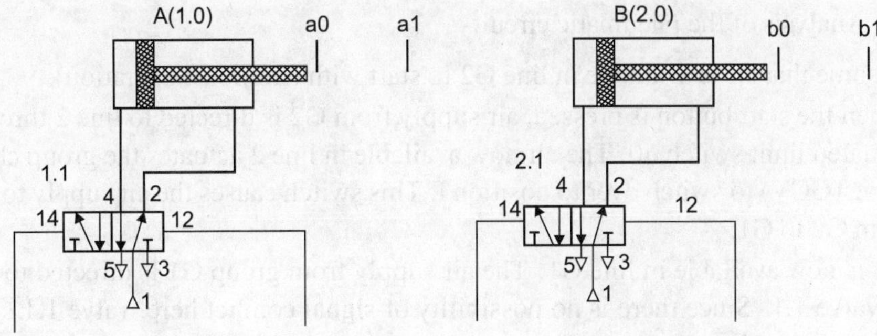

Figure 7.33

Step 8: Draw the control circuit (Fig. 7.34).

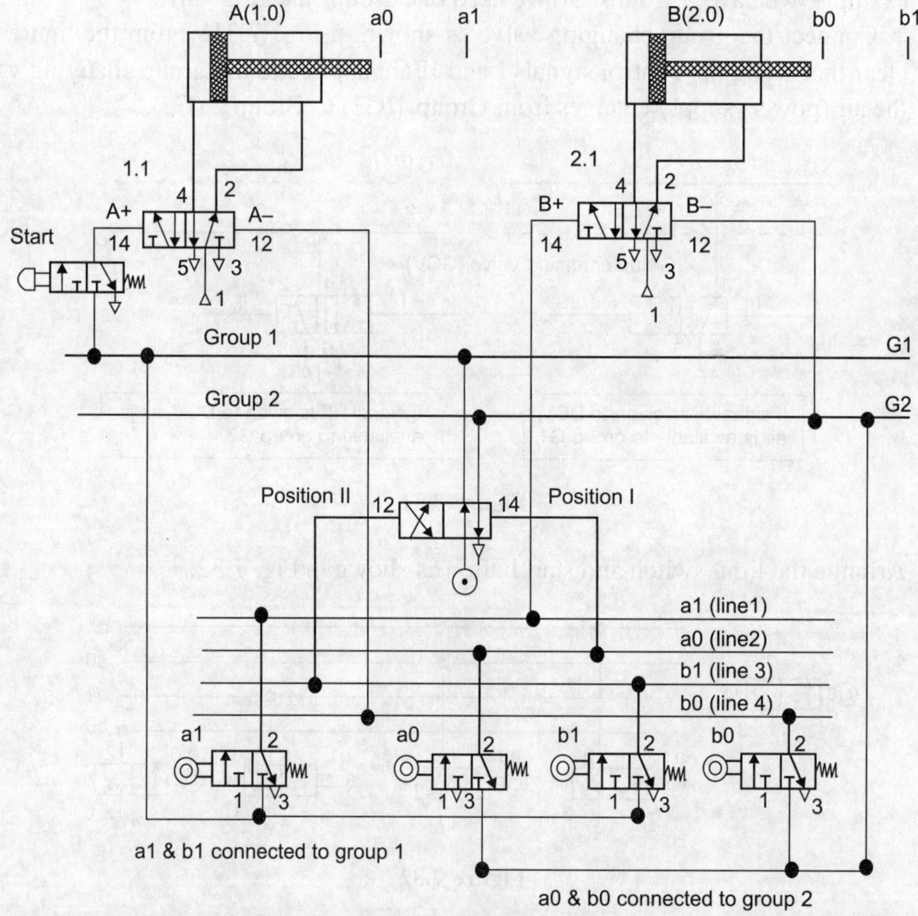

Figure 7.34 *Pneumatic circuits for A+ B+ B− A−*

Step 9: Analysis of the pneumatic circuit

1. Assume that air is available in line G2 to start with (from last operation).
2. When the start button is pressed, air supply from G2 is directed to line 2 through the actuated limit switch $a0$. The air now available in line 2 actuates the group changing valve (GCV) to switch over to position I. This switch causes the air supply to change from G2 to G1.
3. Air is now available in line G1. The air supply from group G1 is directed to port 14 of valve 1.1. Since there is no possibility of signal conflict here, valve 1.1 switches over causing the movement A+.
4. Sensor $a1$ is actuated as a result of movement A+, allowing the air supply from group G1 to reach to line 1 through $a1$. The air now available reaches port 14 of valve 2.1.

As there is no possibility of signal conflict, valve 2.1 switches over, causing the movement B+ automatically.

5. Sensor b1 is actuated as a result of the movement B+, allowing air supply in line 3. Air from line 3 reaches port 12 of the group changing valve (also called reversing valve). As a result, the GCV switches over, causing the group supply to change from G1 to G2.
6. Air is now available in G2 and acts on port 12 of valve 2.1. As there is no possibility of signal conflict, valve 2.1 switches over, causing movement B− automatically.
7. The sensor is actuated as a result of movement B−. Air is now available in line 4 and it reaches port 12 of valve 1.1.With no possibility of signal conflict, valve 2.1 switches over, causing the movement A− automatically.

The cascade system provides a straightforward method of designing any sequential circuit. The following are important points to note:

a) **Preset**: The system must be set to the last group for start-up
b) **Pressure drop**: Since the air supply is cascaded, large circuits can suffer from large pressure drops.
c) **Cost**: Additional reversing valves and other hardware add to the expense.

Example 7.4 Using a transfer station, parts are to be transferred from a vertical magazine onto a chute (Fig. 7.35). The parts are pushed out of the magazine by cylinder A and then transferred onto the chute by cylinder B. The piston rod of cylinder A may be extended only after cylinder B has retracted. The cycle is to start when the start button is pressed. Limit switches are used to confirm the cylinder position. Also, cylinder A retracts and cylinder B extends at the same time.

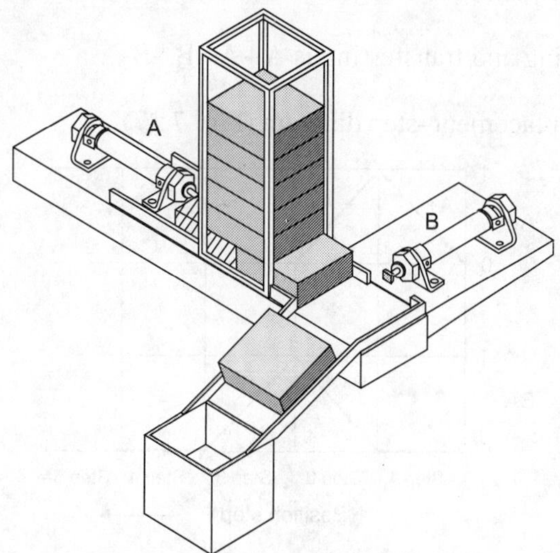

Figure 7.35 *Positional diagram*

Step 1: Write the statement of the problem.

Cylinder A extends first to transfer the part from the chute to the next station. Cylinder B then extends to push the part and at the same time A retracts and then retracts. Cylinder A extends again only after cylinder B has retracted.

Step 2: Draw the positional layout (Fig. 7.36).

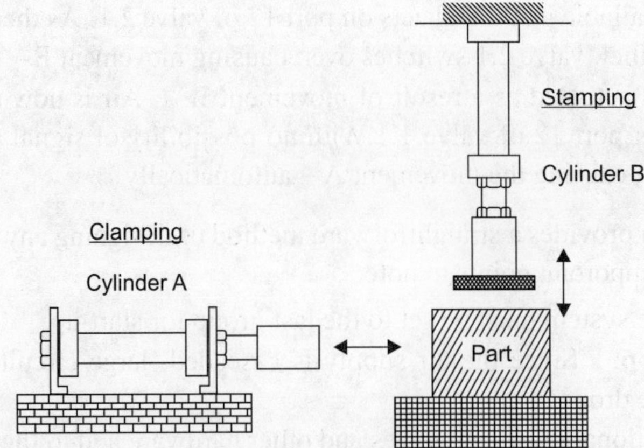

Figure 7.36 *Positional layout*

Step 3: Represent the control task using notational form.
Cylinder A advancing is designated as A+
Cylinder A retracting is designated as A−
Cylinder B advancing is designated as B+
Cylinder B retracting is designated as B−

The sequence for pushing and transferring is A+A−B+B−.

Step 4: Draw the displacement–step diagram (Fig. 7.37).

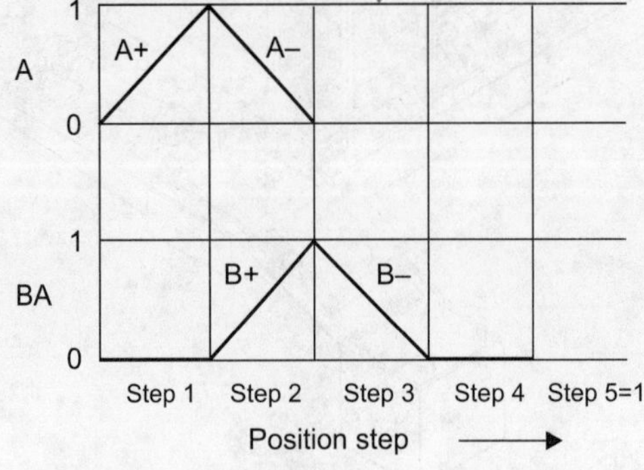

Figure 7.37 *Displacement–step diagram*

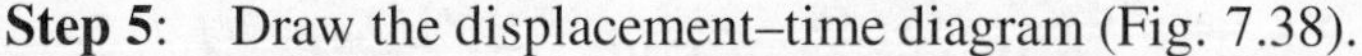

Step 5: Draw the displacement–time diagram (Fig. 7.38).

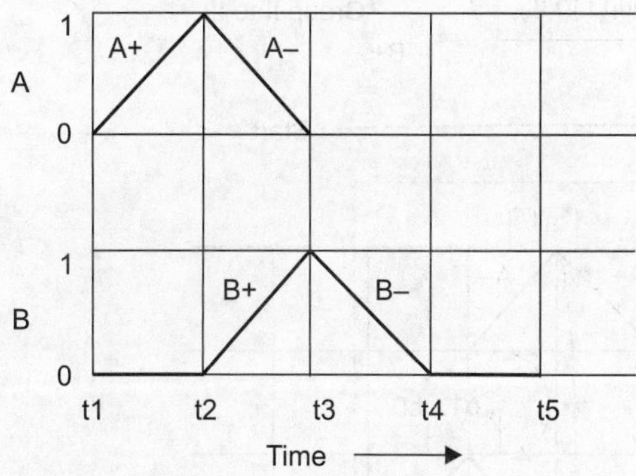

Figure 7.38 *Displacement–time diagram*

Step 6: Analyse and draw pneumatic circuit.
Step 6.1: Analyse input and output signals.

Input signals

Cylinder A: Limit switch at home position $a0$
Limit switch at home position $a1$
Cylinder B: Limit switch at home position $b0$
Limit switch at home position $b1$

Output signal

Forward motion of cylinder A(A+)
Return motion of cylinder A (A−)
Forward motion of cylinder B (B+)
Return motion of cylinder B (B−)

Step 6.2: Use the displacement time/step diagram to link input and output signals (Fig. 7.39).

The start signal is required along with $a0$ signal for the movement A+.

1. A+ generates sensor signal $a1$, which is used for group changing.
2. (i) A− generates sensor signal $a0$, which is used for B+.
   (ii) B+ generates sensor signal $a0$ and $b1$, which are used for group changing.
3. B− generates sensor signal $b0$, which is used for group changing.

The information above (step 6.2) is shown graphically in Fig. 7.39.

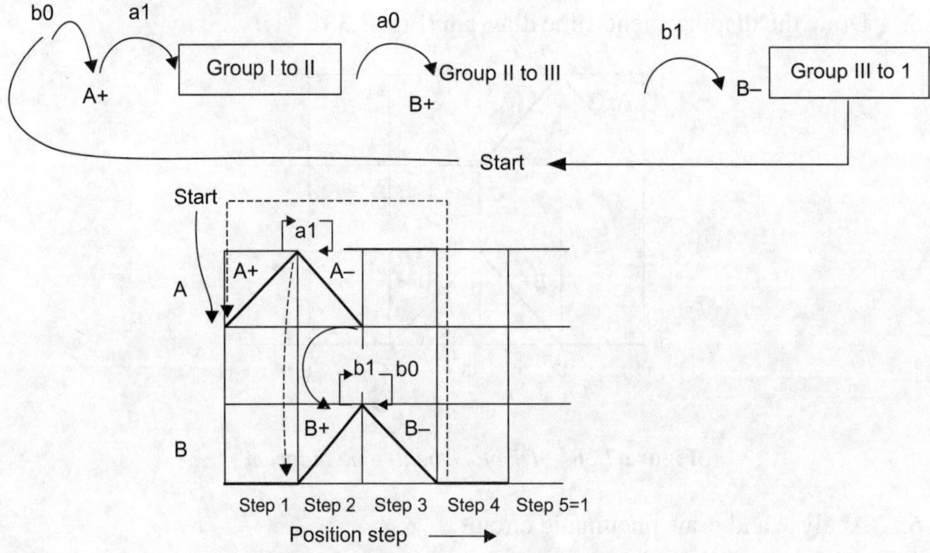

Figure 7.39 *Input–output signal flow*

Step 7: Draw the power circuit (Fig. 7.40).

(i) Divide the given circuits into groups. Grouping should be done so that there is no signal conflict. Do not put A+ and A– in the same group. Similarly B+ and B– should be in different groups to avoid signal conflict.

In our example, A+ A– B+B– can be grouped as shown.

(ii) The number of group changing valves is one less than the number of groups. In our example, we have three groups. So we need two group changing valves.

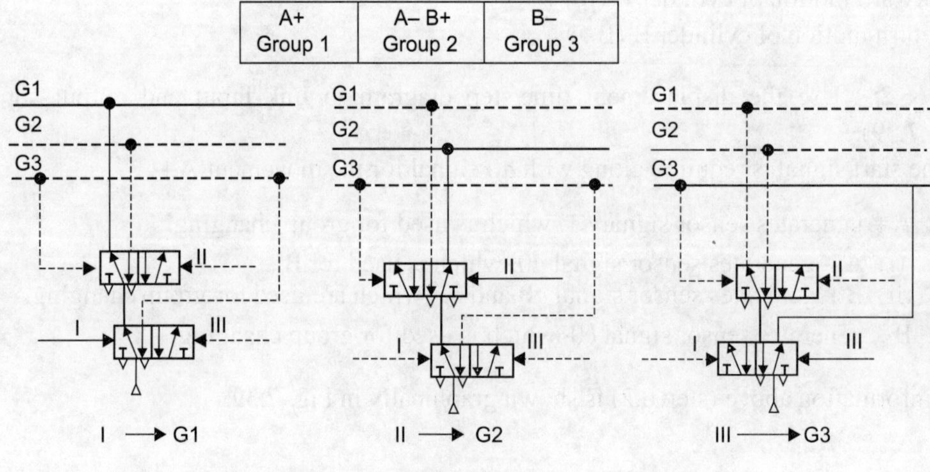

Figure 7.40

(iii) Arrange the limit switch as shown below (Fig. 7.41).

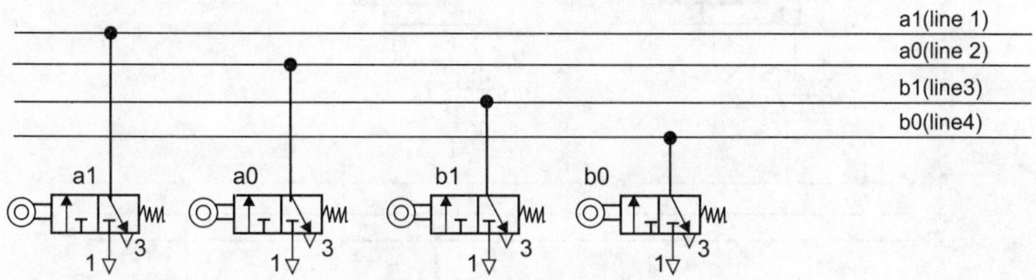

Figure 7.41

(iv) Draw the power circuit (Fig. 7.42).

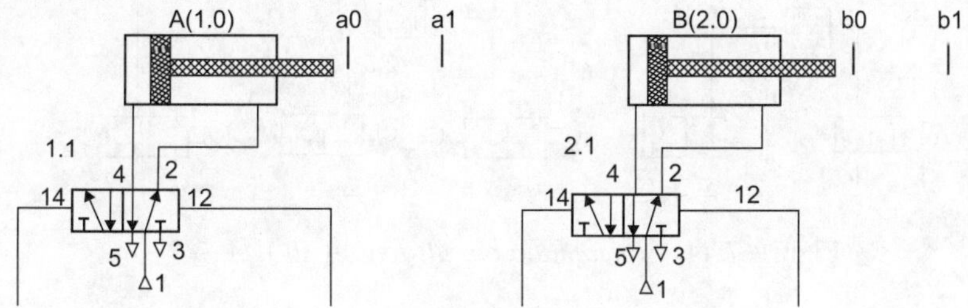

Figure 7.42

Step 8: Draw the control circuit (Fig. 7.43 and Fig. 7.44).

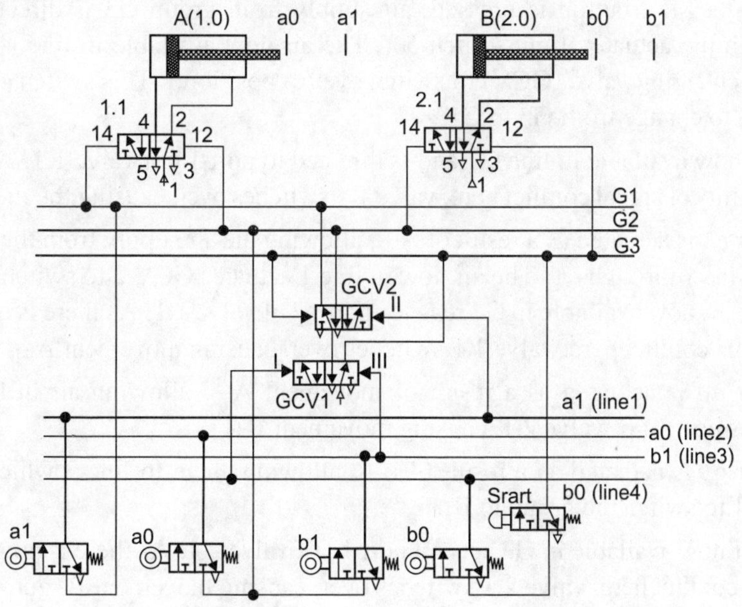

Figure 7.43 *Pneumatic circuit for A+ A− B+ B−*

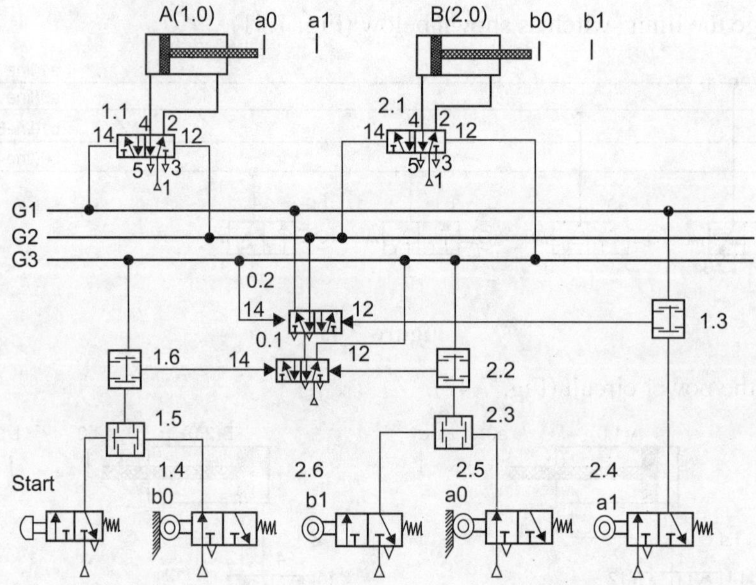

Figure 7.44 *Pneumatic circuit for A+ B+ A− B−*

Step 9: Analysis of pneumatic circuit

1. Assume that air is available in line G3 to start with (from last operation).
2. When the start button is pressed, air supply from group G3 is directed to line 4 through the actuated limit switch $b0$. The air now available in line 4 actuates the group changing valve (GCV) to switch over to position I. This switch causes the air supply to change from G3 to G1.
3. Air is now available in line G1 and is directed to port 14 of valve 1.1. As there is no possibility of signal conflict here, valve 1.1 switches over, causing the movement A+.
4. Sensor $a1$ is actuated as a result of A+, allowing the air supply from the group G1 to reach line 1 through $a1$. The air now in line 1 actuates GCV 2 to switch the group to II. The air now available in G2 reaches port 12 of valve 1.1. As there is no possibility of signal conflict here, valve 1.1 switches over, causing movement A−.
5. Sensor $a0$ is actuated as a result of movement A−, allowing air in line 2 which reaches port 14 of valve 2.1, causing movement B+.
6. Sensor $b1$ is actuated as a result of B+, allowing air in to line 3 which causes the GCV 1 to switch the group to III.
7. The air now available in G3 acts on port 12 of valve 2.1. As there is no possibility of signal conflict here, valve 2.1 switches over, causing movement B- automatically.
8. Sensor $b0$ is actuated as a result of B-action. The air now available in line 4 actuates GCV1 to switch the group to I.

9. The air now available in line G1 reaches port 14 of valve 1.1 causing movement A+ and the cycle continues.

The analysis of Fig. 7.44 is lift as an exercise for the reader.

7.4 Step Counter Method

A step counter is a digital modular counter constructed from stepping units which serve as sequence steppers for sequential circuit. The basic stepping unit shown in Fig. 7.45 is built from a signal output memory valve and a pre-switched AND valve with two inputs. Achieving an output signal from the stepping unit requires:

1. A signal input at line 1 which serves as a preparation signal from the preceding stepping unit
2. A confirmation signal at line 2 which confirms the completion of the cylinder movement by the preceding stepping unit
3. Line 4 resets the memory valve of the preceding stepping unit.
4. Line 5 serves as a preparation signal to prepare the next stepping unit
5. Line 6 is the reset signal from the next stepping unit.
6. Line 3 switches the DCV.

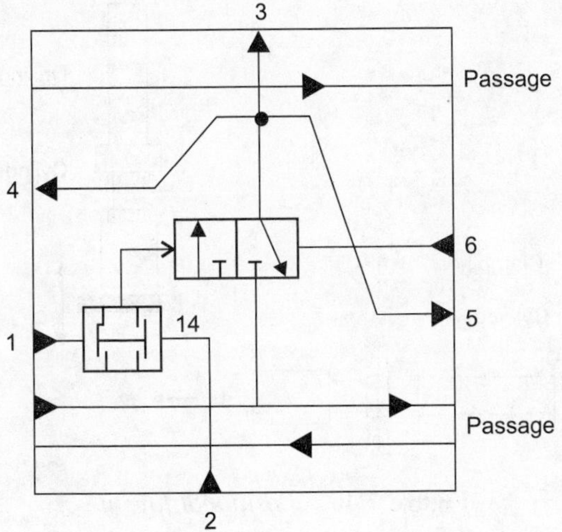

Figure 7.45 *Step counter*

Steps involved

Step 1: Draw the position–step diagram used to represent the movement of drive elements. The X axis represents steps; the sequence of operation is divided into a number of steps (starting at 1). The Y axis represents the position (starting at 0). The functional lines are drawn thick and they determine the position of the driving unit. Any change of position of a member has to start or stop at a corner of the square.

Step 2: From the position–step diagram, the number of stepping units are decided.

Step 3: Draw the cylinders and their memory valves (directional control valves). Connect the memories to the cylinders.

Step 4: Draw the step counters.

Step 5: Draw the position sequence valves.

Step 6: Connect the stepping units and valves.

Example 7.5 Consider an automatic drilling machine. The complete cycle is as follows: cylinder A extends to clamp the workpiece, then cylinder B extends to drill a hole and then retracts. Cylinder A then retracts to unclamp the workpiece. Design a control circuit applying the step counter method. The circuit is provided with a start valve to avoid continuous cycles.

Step 1: Write the statement of the problem

First cylinder A extends and delivers the workpiece under the drilling station where cylinder B is located. Cylinder B then extends and drills the workpiece. Cylinder A can return only when cylinder B has retracted fully.

Step 2: Draw the positional layout (Fig. 7.46).

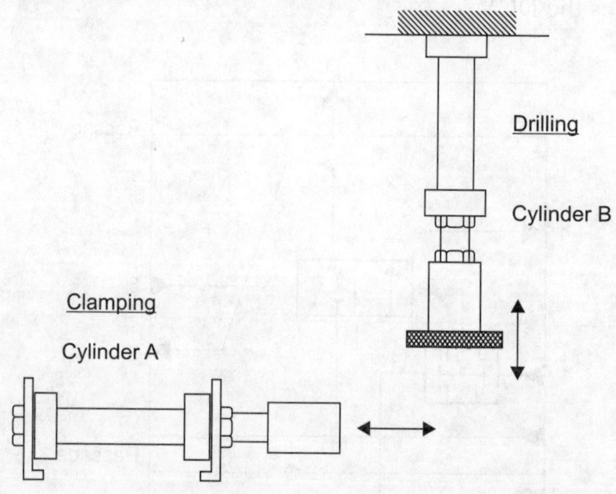

Figure 7.46 *Positional layout*

Step 3: Represent the control task using notational form.

Cylinder A advancing is designated as A+

Cylinder A retracting is designated as A−

Cylinder B advancing is designated as B+

Cylinder B retracting is designated as B−

The sequence for positioning and drilling is A+B+B−A−

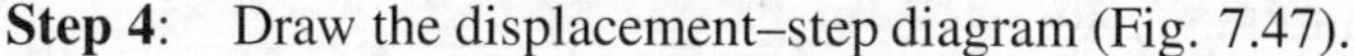

Step 4: Draw the displacement–step diagram (Fig. 7.47).

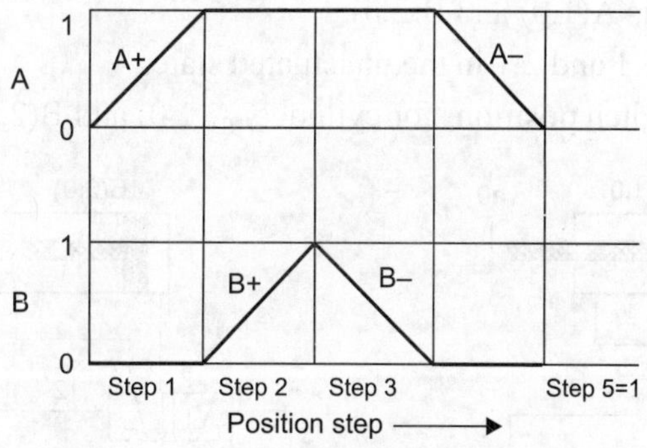

Figure 7.47 *Displacement–step diagram*

Step 5: Draw the displacement–time diagram (Fig. 7.48).

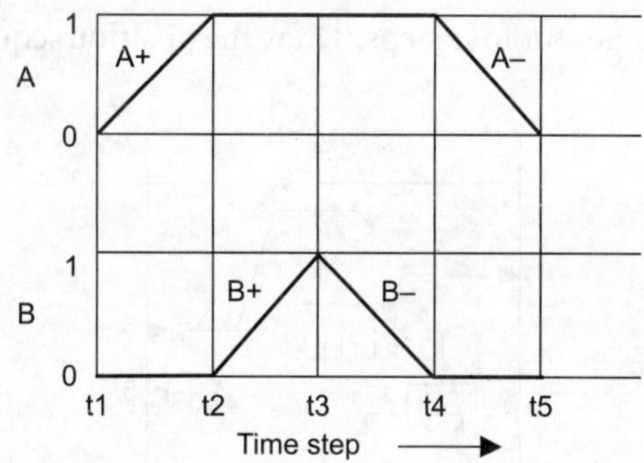

Figure 7.48 *Displacement–time diagram*

Step 6: Analyse and draw the pneumatic circuit.
Analyse input and output signals.

Input signals

Cylinder A: Limit switch at home position $a0$
Limit switch at home position $a1$
Cylinder B: Limit switch at home position $b0$
Limit switch at home position $b1$

Output signal

Forward motion of cylinder A (A+)
Return motion of cylinder A (A−)
Forward motion of cylinder B (B+)
Return motion of cylinder B (B−)

Step 7: Draw the power circuit (Fig. 7.49).

(i) Draw the cylinders A(1.0) and B(2.0).
(ii) Draw the DCVs 1.1 and 2.1 in the unactuated state.
(iii) Mark the limit switch positions for cylinders A(1.0) and B(2.0).

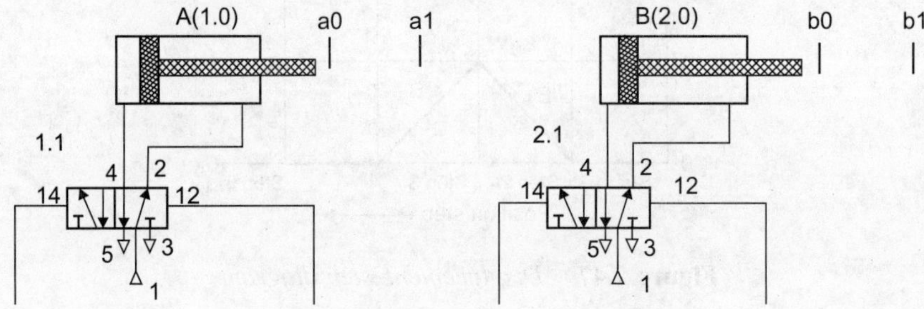

Figure 7.49 *Control circuit*

Step 8: Draw a step counter for four steps. Draw the position sequence valves (Fig. 7.50).

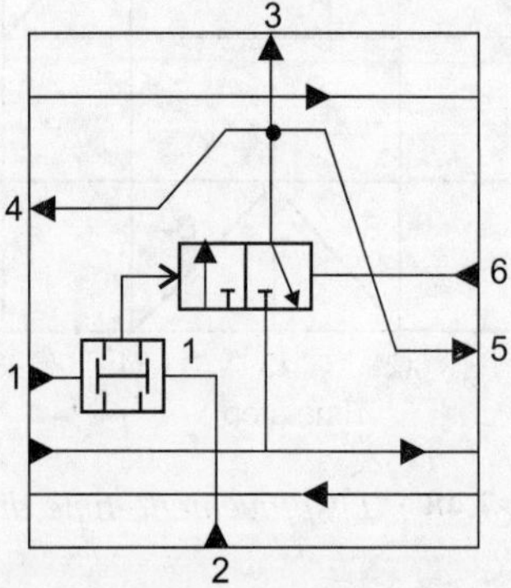

Figure 7.50 *Step counter*

1. The confirmation signal for step 1 originates from the position sequence valve controlling the completion of the previous step. This is step 4 of the previous cycle and so this is at the retracted end of cylinder A. Therefore connect $a0$ to step 1.
2. This first step serves a triple purpose. It signals cylinder A to extend by providing the A+ signal, it gives a preparation signal to step 2 and it sends a reset signal to the previous step (step 4).
3. The confirming signal for step 2 originates from the position sequence valve controlling the completion of the previous step (step 1) therefore connect $a1$ to step2.
4. The second step serves a triple function. It signals cylinder B to extend by providing the B+ signal, it gives a preparation signal to step 3 and it sends a reset signal to the previous step (step 1).

5. Continue in the same manner throughout the entire circuit. Label the circuit as shown in Fig. 7.51.

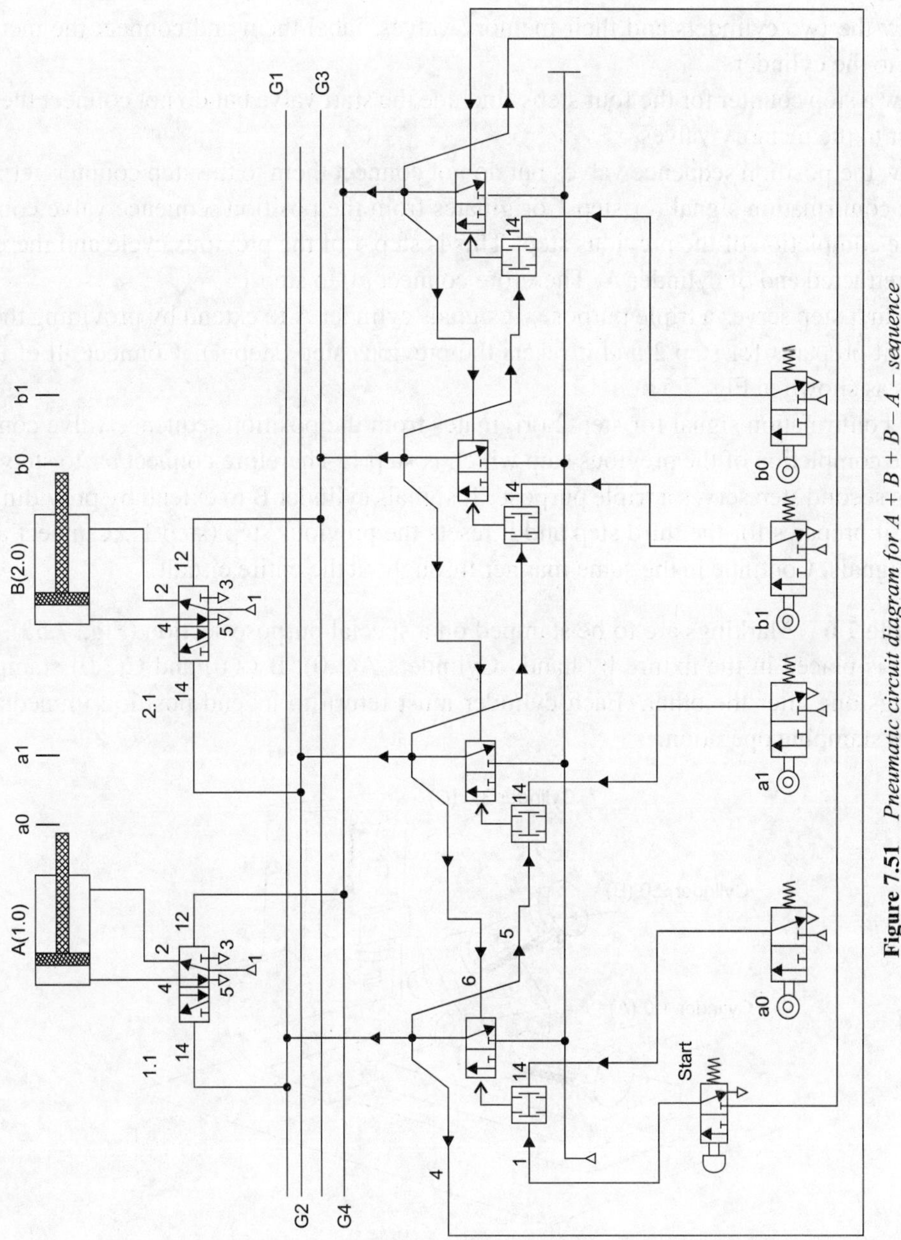

Figure 7.51 *Pneumatic circuit diagram for A + B + B – A – sequence*

Design considerations From the traverse-time diagram, it is evident that the step counter requires four stepping units. The last point represents step 1 of the next cycle, so is not counted.

Draw the two cylinders and their memory valves, label them and connect the memory valves to the cylinders.

Draw a step counter for the four steps. Include the start valve but do not connect the step counter to the memory valves.

Draw the position sequence valves but do not connect them to the step counter yet.

The confirmation signal for step 1 originates from the position sequence valve controlling the completion of the previous step. This is step 4 of the previous cycle and therefore at the retracted end of cylinder A. Therefore connect $a0$ to step 1.

The first step serves a triple purpose. It signals cylinder A to extend by providing the A1 signal, it prepares for step 2 and it resets the previous step (step 4). Connect all of these signals as shown in Fig. 7.51.

The confirmation signal for step 2 originates from the position sequence valve controlling the completion of the previous step which is step 1. Therefore connect $a1$ to step 2.

This second step serves a triple purpose. It signals cylinder B to extend by providing B1 signal, it prepares for the third step and it resets the previous step (step 1). Connect all of these signals. Continue in the same manner throughout the entire circuit.

Example 7.6 Markings are to be stamped on a special purpose section (Fig. 7.53). The section is placed in the fixture by hand. Cylinders A(1.0), B (2.0) and C(3.0) stamp the markings one after the other. Each cylinder must return to its end position immediately after its stamping operation.

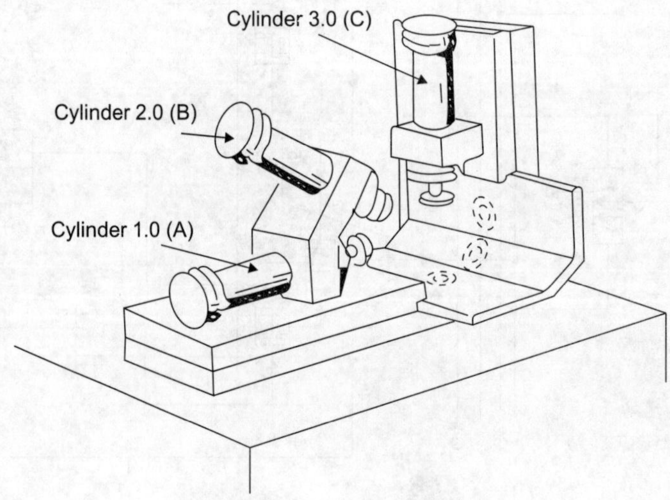

Figure 7.52 *Positional diagram*

Step 1: Write the statement of the problem.

Cylinder A extends first, stamps the marking and returns. Cylinder B extends next, stamps the marking and returns. Similarly, cylinder C extends last, stamps the marking and returns.

Step 2: Draw the positional layout (Fig. 7.53).

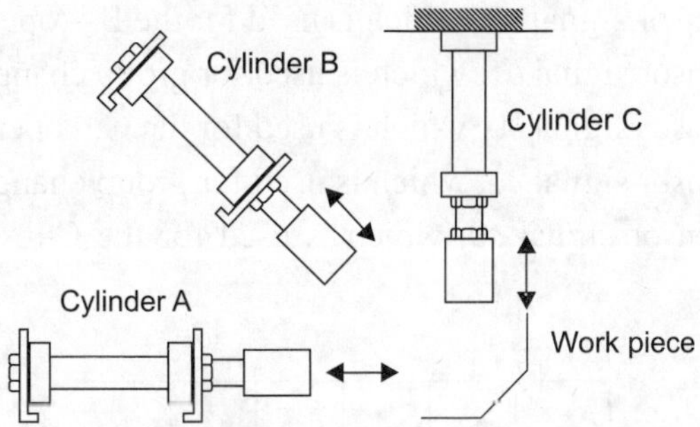

Figure 7.53 *Positional Layout*

Step 3: Represent the control task using notational form.

Cylinder **A** advancing is designated as **A+**
Cylinder A retracting is designated as A–
Cylinder B advancing is designated as B+
Cylinder B retracting is designated as B–
Cylinder C advancing is designated as C+
Cylinder C retracting is designated as C–

The sequence for clamping and stamping is A+A–B+B–C+C–

Step 4: Draw the displacement–step diagram (Fig. 7.54).

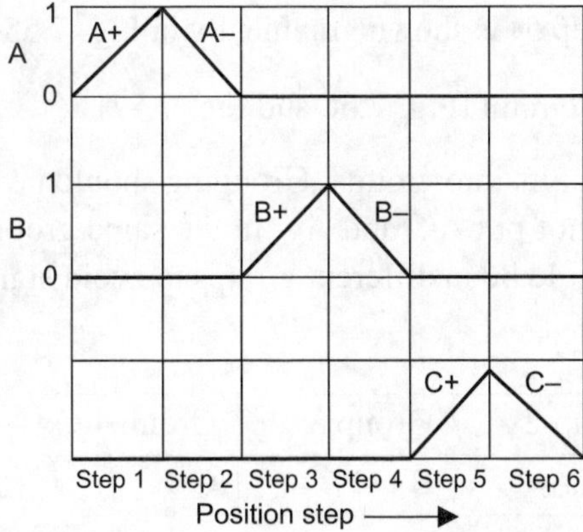

Figure 7.54 *Displacement–step diagram*

The required sequence is A+A–B+B–C+C–

Step 5: Use the displacement–time/step diagram to link input and output signals (Fig. 7.55).

The start signal is required along with $a0$ signal for movement A+.

1. A+ generates sensor signal $a1$, which is used for group changing.
2. A− generates sensor signal $a0$, which is used for the B+ operation.
3. B+ generates sensor signal $b1$, which is used for group changing.
4. B− generates sensor signal $b0$, which is used for the C+ operation.
5. C+ generates sensor signal $c1$, which is used for group changing
6. C− generates sensor signal $c0$, which is used for the C+ operation and the push button signal.

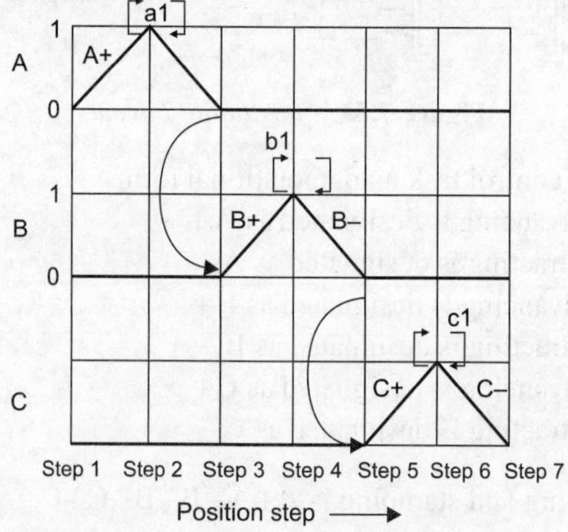

Figure 7.55 *Input and output signal flow*

The information above (step 5) is shown graphically in Fig. 7.55.

Step 6: Draw the power circuit (Fig. 7.56 and Fig. 7.57).

(i) Divide the given circuits into groups. Grouping should be done such that there is no signal conflict. Do not put A+ and A− in the same group. Similarly, B+ and B− and C+ and C− should be in different groups to avoid signal conflict.

A+ A− B+ B− C+ C−

Group 1	Group 2	Group 3	Group 4
A+	A−B+	B+C+	C−

(ii) The number of group changing valve is one less than the number of groups.

There are four groups in this example; therefore three group changing valves are needed. Connect the group changing valve as given in Figs 7.56 and 7.57

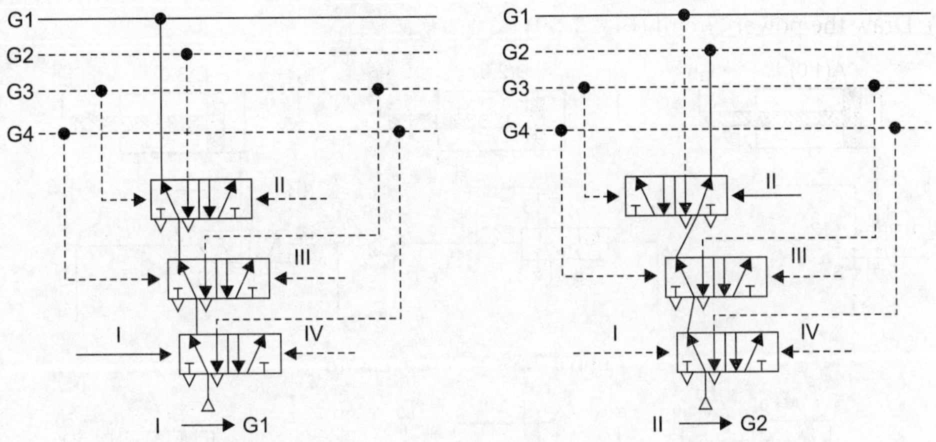

Figure 7.56 *Group changing valves (G1 and G2)*

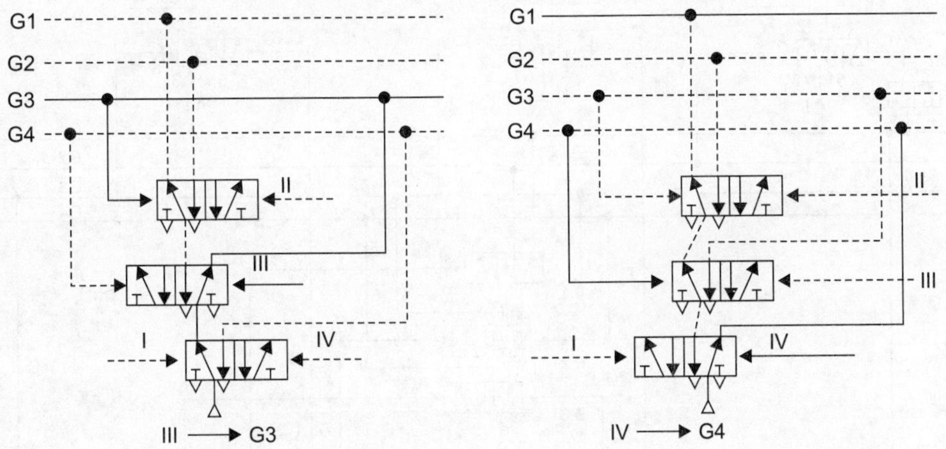

Figure 7.57 *Group changing valves (G3 and G4)*

(iii) Arrange the limit switches and start button as given in Fig. 7.58:

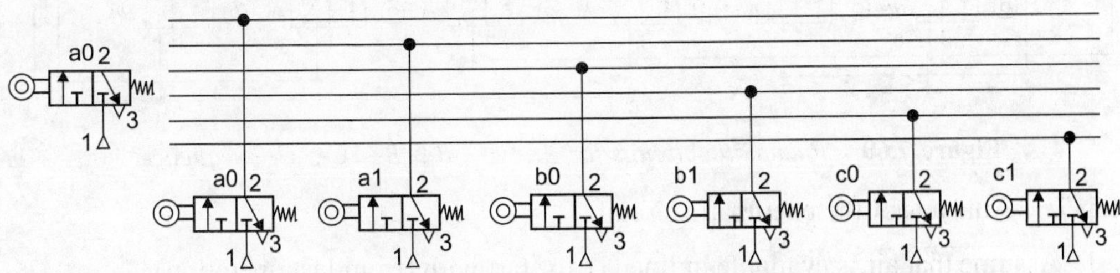

Figure 7.58 *Arrangement of limit switches*

(iv) Draw the power circuit (Fig. 7.59).

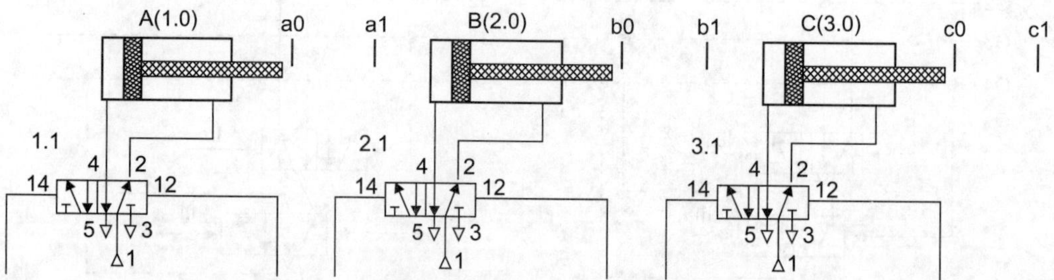

Figure 7.59 *Power circuit*

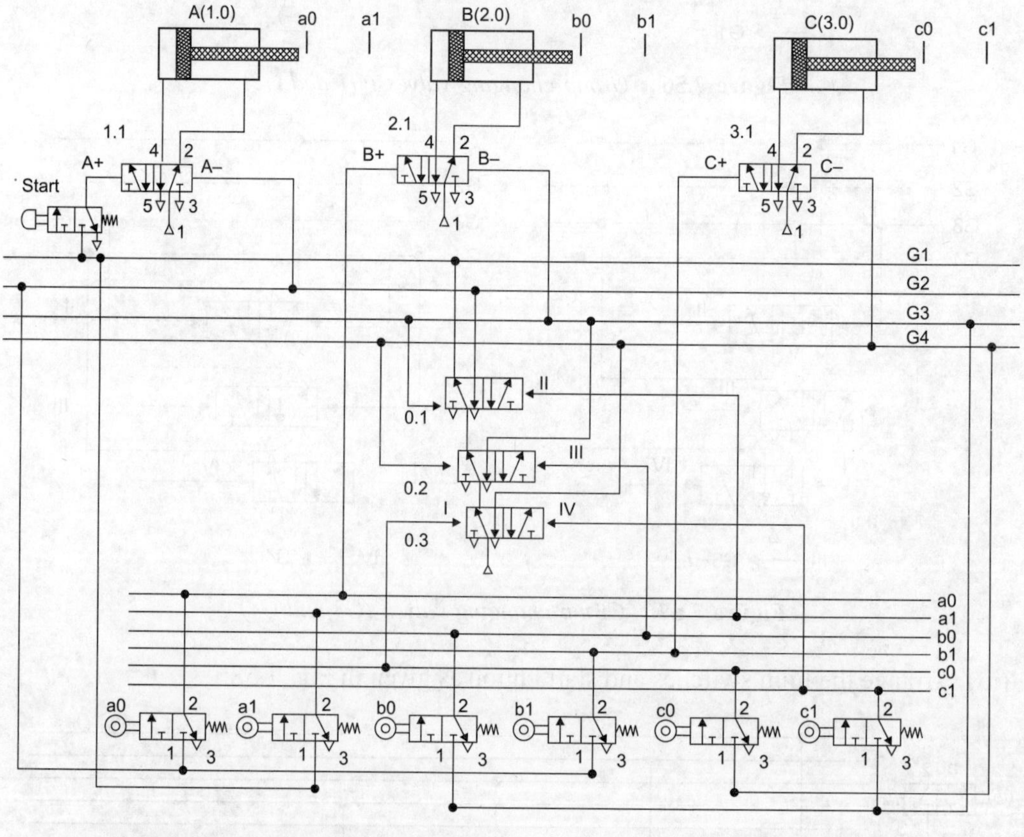

Figure 7.60 *Pneumatic circuits for A+ A− B+ B− C+ C− sequence*

Step 7: Analysis of the pneumatic circuit.

1. Assume that air is available in line G1 to start with (from last operation).
2. When the start button is pressed, air supply from group G1 is directed to port 14 of valve 1.1 (A+) and A extends. When cylinder A extends, it touches limit switch $a1$ and gives an output signal to group changing valve 0.1 to change the group to G2.

3. Air is now available in line G2 and is directed to port 12 of valve 1.1 (A−). As there is no possibility of signal conflict, valve 1.1 switches over causing movement A−.
4. Sensor $a0$ is actuated as a result of movement A−, allowing air supply into line $a0$, which causes cylinder B to extend.
5. Sensor $b1$ is actuated as result of movement B+, $b1$ operates on group changing valve 0.2, causing the group supply to change from G2 to G3.
6. Air is now available in G3 and it acts on port 12 of valve 2.1. As there is no possibility of signal conflict, valve 2.1 switches over, causing movement B-automatically.
7. Sensor $b0$ is actuated as the result of movement B−. As there is no possibility of signal conflict, valve 2.1 switches over, causing movement C+ automatically.
8. Sensor $c1$ is actuated as result of movement C+. It operates on group changing valve 0.3, letting in air into G4 which then acts on port 12 of valve 3.1 and causes movement C−.
9. Sensor $c0$ is actuated as a result of movement C−. $c0$ operates the group changing valve 0.3, letting air into G1 and the cycle repeats.

Example 7.7 Copper tubes for plumbing applications are to be flanged in two stages (Fig. 7.61). The copper tube is placed in a clamping fixture against the stop. When the start signal is given, the copper tube is clamped by cylinder 1.0 (A). The first flanging operation is performed by the first flanging station by cylinder 2.0 (B). The first flanging station travels back. Cylinder C moves the second flanging station into position. Cylinder 2.0 (B) travels out again and performs the second and final flanging operation on the copper tube. Cylinder 2.0 (B) travels back to the initial position. Cylinder 1.0 (A) then unclamps the tube and at the same time cylinder 3.0 (C) returns the first flanging station back into position. Layout drawing is shown below.

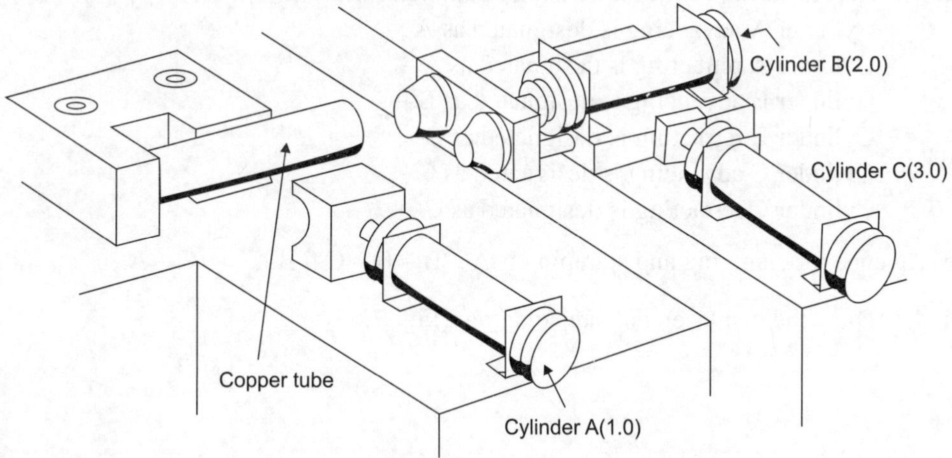

Figure 7.61 *Positional diagram*

Step 1: Write the statement of the problem.

Cylinder A first extends and does the clamping. Cylinder B then extends and completes the first flanging operation and returns. Cylinder C then extends to move the second flanging tool mounted onto the cylinder B slide. B extends to do the second flanging and returns. Then cylinders C and A both return.

Step 2: Draw the positional layout (Fig. 7.62).

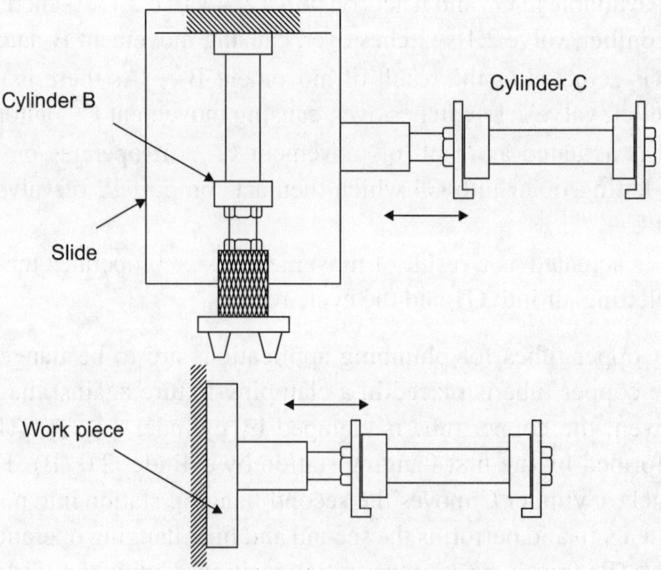

Figure 7.62 *Positional layout*

Step 3: Represent the control task using notational form.

Cylinder A advancing is designated as A+
Cylinder A retracting is designated as A–
Cylinder B advancing is designated as B+
Cylinder B retracting is designated as B–
Cylinder C advancing is designated as C+
Cylinder C retracting is designated as C–

The sequence for clamping and stamping is A+ B+ B– C+ B+ B–(C– A–)

Step 4: Draw the displacement–step diagram (Fig. 7.63).

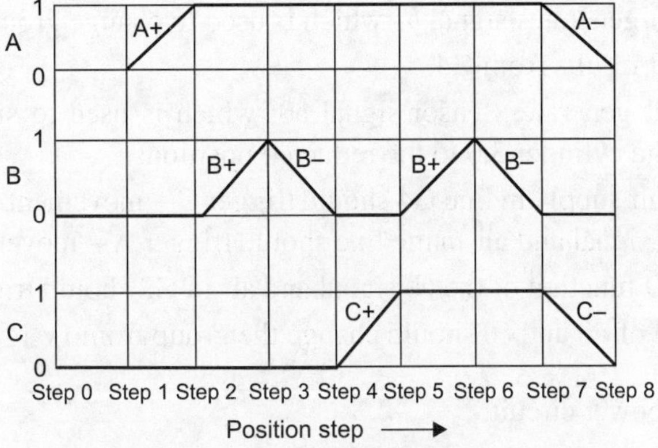

Figure 7.63 *Displacement–step diagram*

The sequence of operation is A+ B+ B− C+ B+ B−(C− A−).

Step 5: Use the displacement–time/step diagram to link input and output signals (Fig. 7.64).

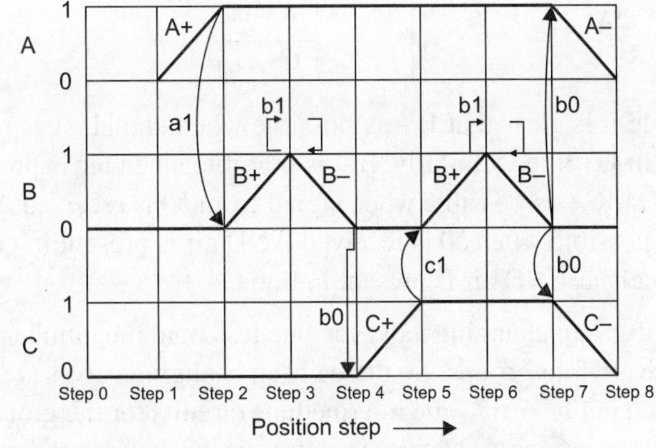

Figure 7.64 *Input and output signal flow*

The start signal is required along with $a0$ signal for the movement A+.

1. A+ generates sensor signal $a1$, which is used for movement B+.
2. B+ generates sensor signal $b1$, which is used to change groups from G1 to G2.
3. B− movement results from G2.
4. B− movement generates sensor signal $b0$, which is used for movement C+.
5. C+ movement generates sensor signal $c1$, which is used to change group from G2 to G3.
6. B+ movement results from G3.

7. B+ movement generates signal $b1$ which is used to change groups from G3 to G4.
8. B− movement results from G4.
9. B− movement generates sensor signal $b0$, which is used to simultaneously bring cylinder 1.0 and cylinder 3.0 to the retracted positions.
10. $b0$ signal and air supply in line G4 should trigger C− movement. Additionally, AND function of $b0$ signal and air in the line should trigger A− movement.
11. Also, the AND function of the $b0$ signal and air in G2 should trigger C+ operation.
12. AND function of $a0$ and $c0$ should change the group from IV to I.

Step 6: Draw the power circuit.

(i) Divide the given circuits into groups. Grouping should be done so that there is no signal conflict. A+ and A− should not be in the same group. Similarly B+, B− and C+, C− should be in different groups to avoid signal conflict.

A+ B+ B− C+ B+ B− C− A−

Group 1	Group 2	Group 3	Group 4	
A+B+	B−C+	B+	B−	A−
				C−

From this table it is clear that B+ is possible when signal $a1$ is received OR when air is present in G3 line. Similarly, B− is possible when air is present in G2 OR air is present in G4. C+ is possible when signal from $b0$ is received AND air is present in G2. C− is possible when $b0$ is received AND air is present in G4. A− is possible when $b0$ is received AND air is present in line G4.

(ii) The number of group changing valves is one less than the number of groups. In this example, there are four groups. So three group changing valves are needed. The final circuit is shown in Fig. 7.65. The intermediate circuits for the group changing valves and the cylinder configuration steps are left as an exercise to the reader.

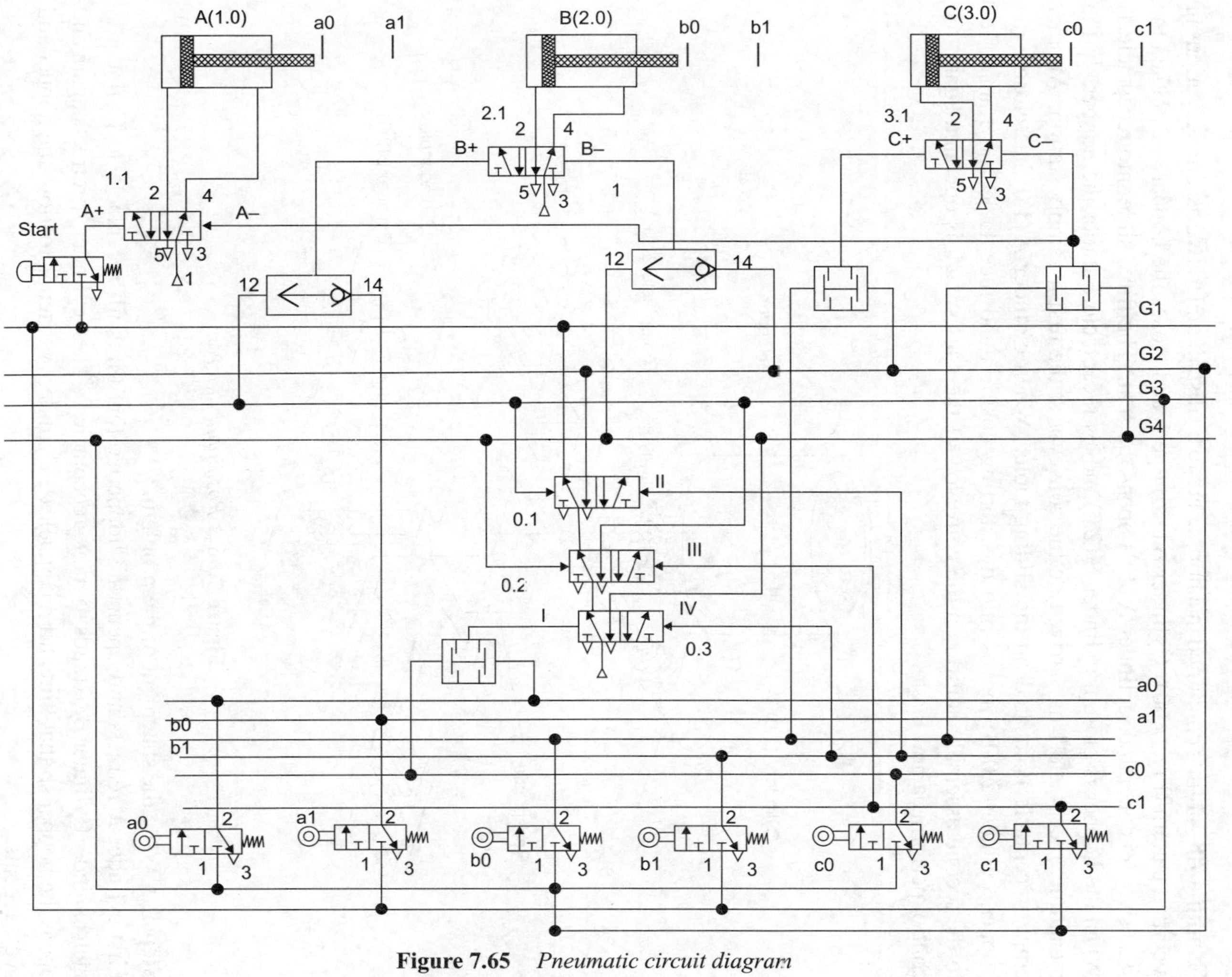

Figure 7.65 *Pneumatic circuit diagram*

Example 7.8 Flints contained in a hopper are to be distributed to two assembly stations in a certain pattern (Fig. 7.66). Cylinder A (1.0) opens and closes the hopper seal. When the start switch is operated, cylinder A (1.0) opens the seal. The flints slide into receptacle 1. When the hopper is closed, cylinder B (2.0) moves receptacle 2 under the hopper. In the meantime, receptacle 1 has arrived on the conveyor at the first assembly station. Another empty receptacle is located on the sliding table. When cylinder A (1.0) has closed, the hopper, cylinder 2.0(B) moves into the starting position and receptacle 2 is transported to the second assembly station on the conveyor. If the start switch is operated again, the control performs a new cycle.

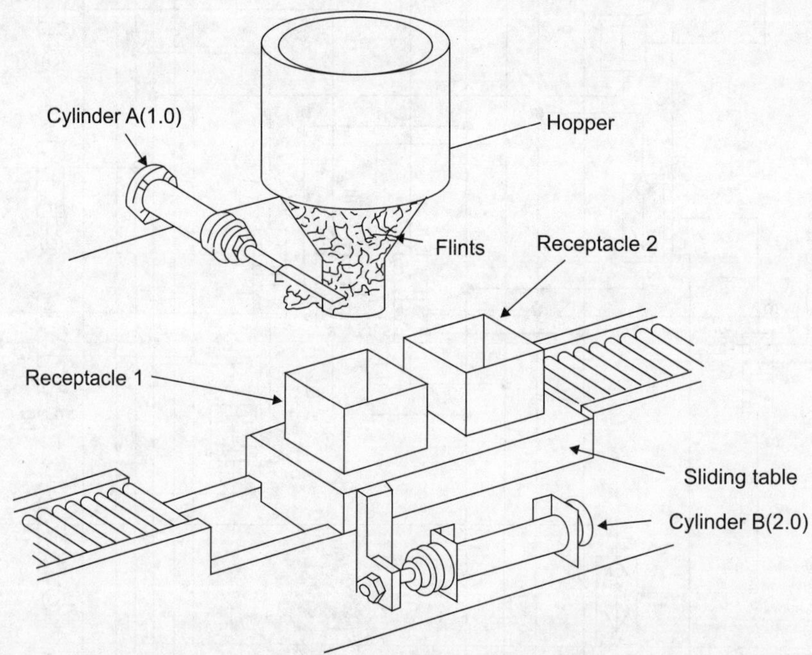

Figure 7.66 *Positional diagram*

Step 1: Write the statement of the problem.
First cylinder A retracts and opens the hopper so that flints fill receptacle 1. Cylinder A then extends. Cylinder B extends to move receptacle 2. Then cylinder A retracts again and opens the hopper so that flints fill receptacle 2. Cylinder A extends to close the hopper and cycle repeats.

Step 2: Draw the positional layout (Fig. 7.67).

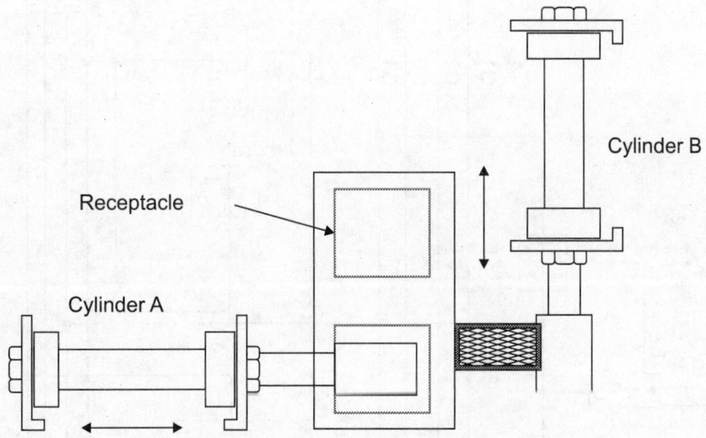

Figure 7.67 *Positional layout*

Step 3: Represent the control task using notational form.
Cylinder A advancing is designated as A+
Cylinder A retracting is designated as A−
Cylinder B advancing is designated as B+
Cylinder B retracting is designated as B−

The sequence for distribution is A−A+B+A−A+B−

Step 4: Draw the displacement–step diagram (Fig. 7.68).

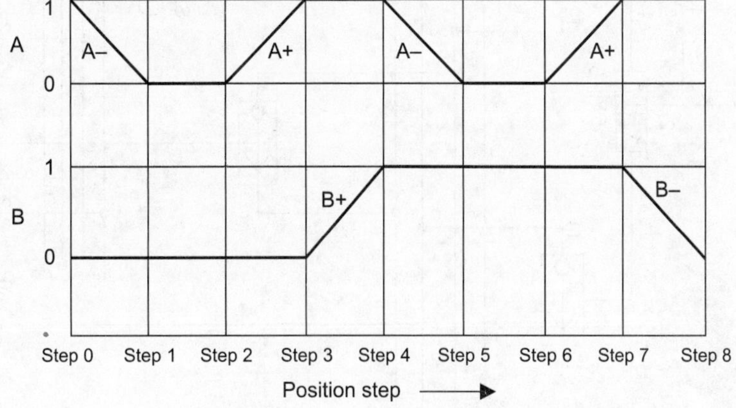

Figure 7.68 *Displacement–step diagram*

The sequence is $A-$ Timer $A + B + A-$ Timer $A + B-$.

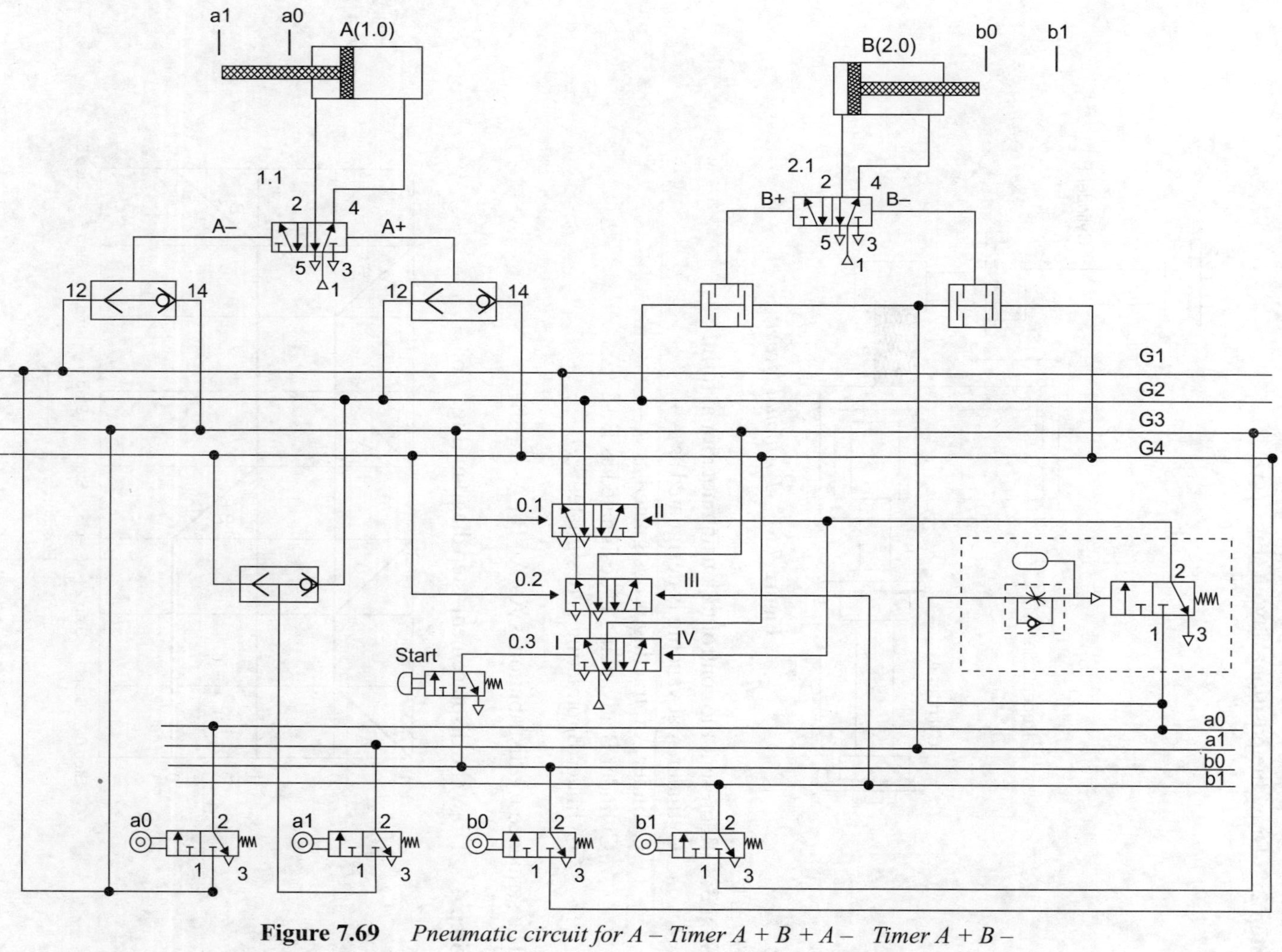

Figure 7.69 *Pneumatic circuit for A – Timer A + B + A – Timer A + B –*

Example 7.9 In an automobile manufacturing unit, components are to be supplied to a processing machine from the discharge rail of a vibrator conveyor as shown in Fig. 7.70. Cylinder 1.0 extends and picks up the component from the discharge rail and then retracts. Cylinder 2.0 extends and shifts the component to the machine. Cylinder 1.0 extends once again, lowering and dropping the component on to the machine. In the end, cylinders 1.0 and 2.0 retract one after another. Develop a pneumatic control circuit to implement the control task. Avoid signal conflicts in the circuit.

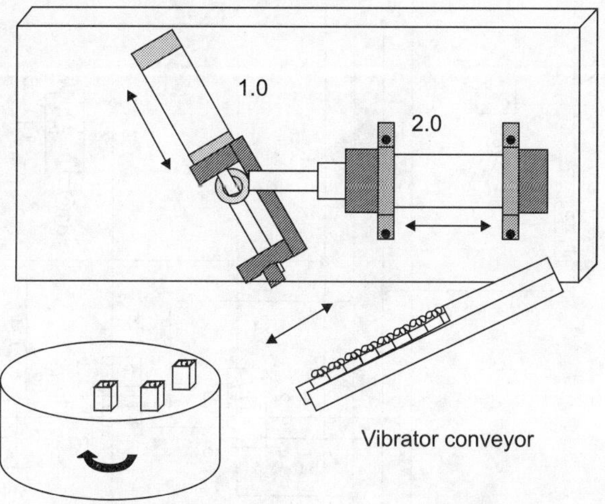

Figure 7.70 *Positional diagram*

Solution The displacement–step diagram is drawn (Fig. 7.71). Remaining steps are left as an exercise to the reader. The final circuit diagram is shown in Fig. 7.72.

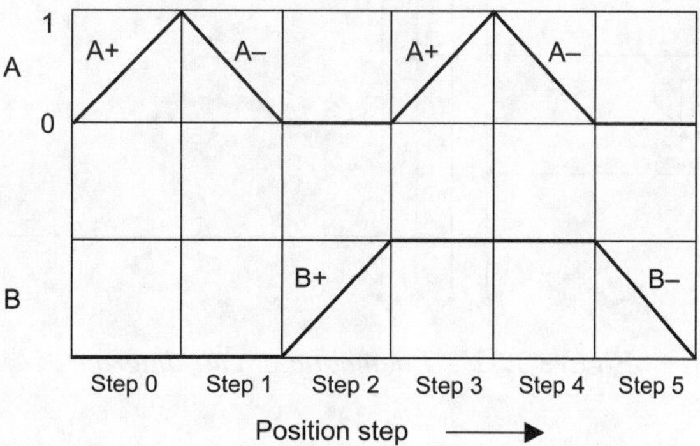

Figure 7.71 *Displacement–step diagram*

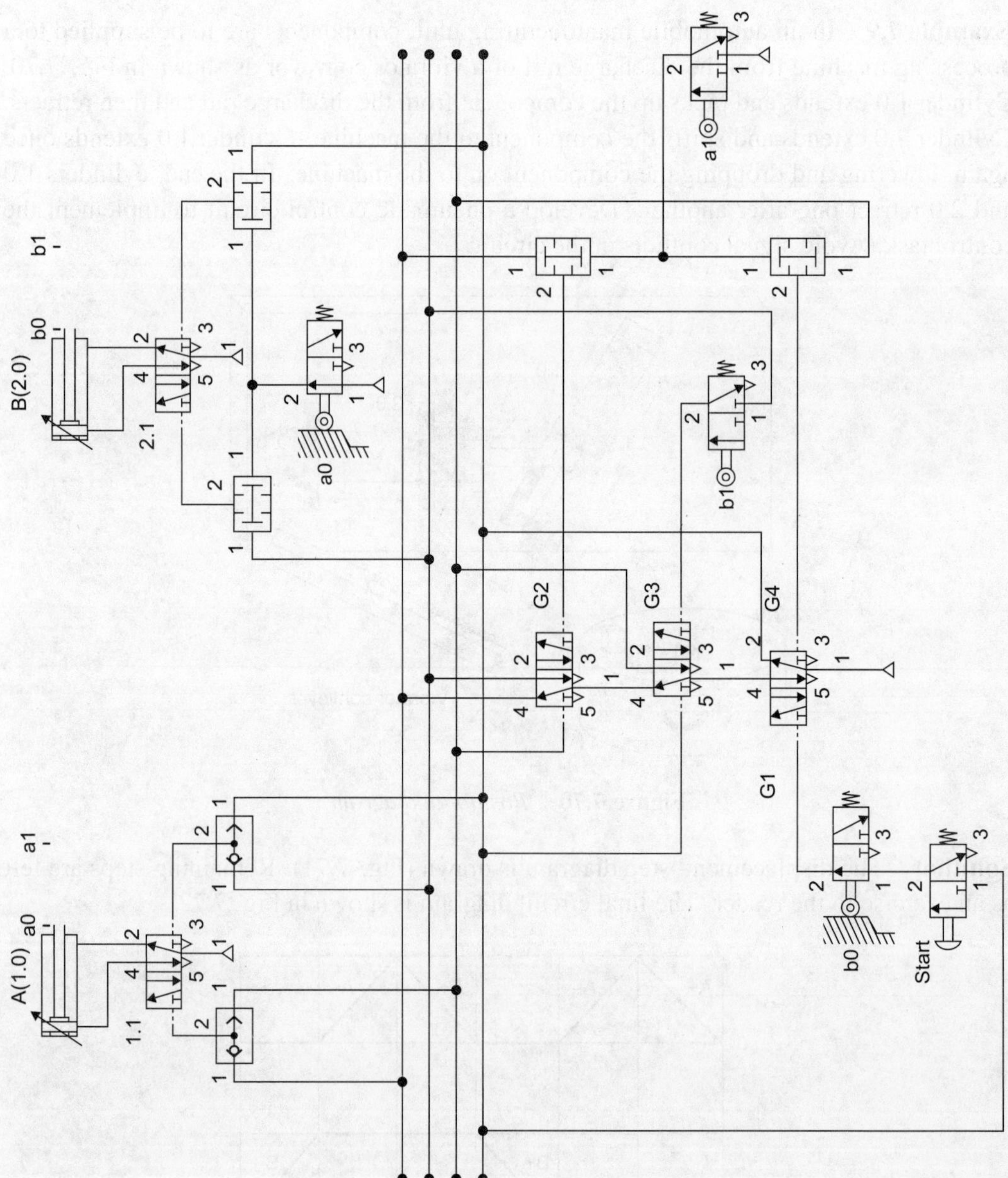

Figure 7.72 *Pneumatic circuit diagram*

Example 7.10 Square shaped workpieces are fed from a gravity magazine to a drilling machine (Fig. 7.73). The horizontal cylinder A (1.0) pushes end pieces out of a gravity feed magazine under the drilling spindle and holds them clamped against the fixed stop. When the required clamping pressure of 4 bar is reached, the drill spindle B (2.0) extends via the feed unit with a hydraulic cushioning cylinder that lowers the drill. Maximum feed force is set at 5 bar pressure. After reaching the depth control stop adjusted by a roller lever valve, the return stroke is commenced. Completion of the return stroke causes the finished end pieces to be ejected by a single acting cylinder C (3.0). After a period of 1 second, the rapid return stroke is commenced. When the ejector cylinder has reached the retracted end position, a fourth roller lever valve is actuated and is used to signal a new cycle. A separate pressure gauge indicates the clamping pressure of the cylinder A (1.0). The control system is set into operation by pressing the start button. To select a continuous cycle, a detent valve is reversed.

Solution The required sequence is A + B + B − A − C + C− and is shown in Fig. 7.74 and the final pneumatic circuit is shown in Fig. 7.75.

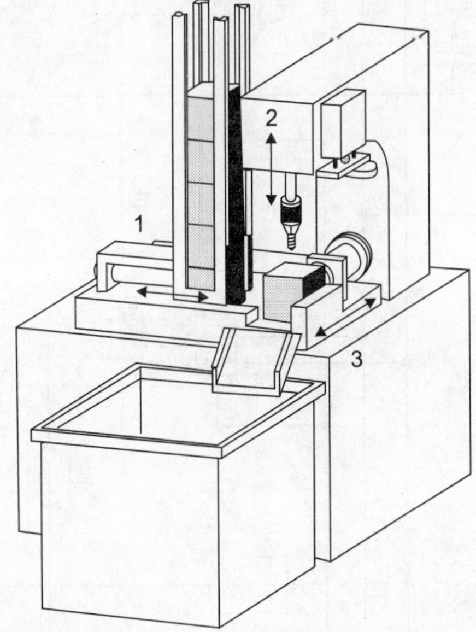

Figure 7.73 *Positional diagram*

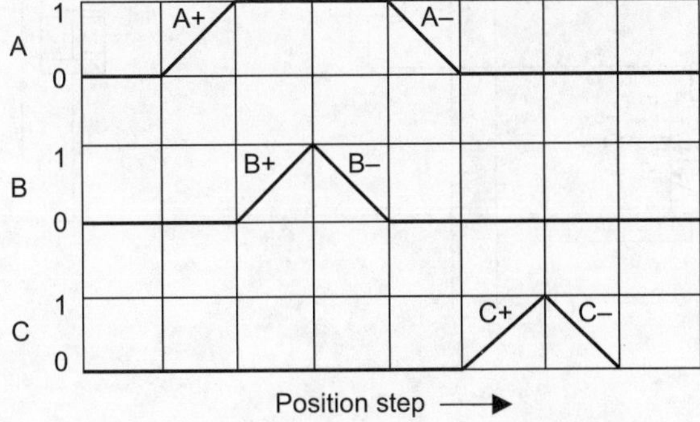

Figure 7.74 *Displacement–step diagram*

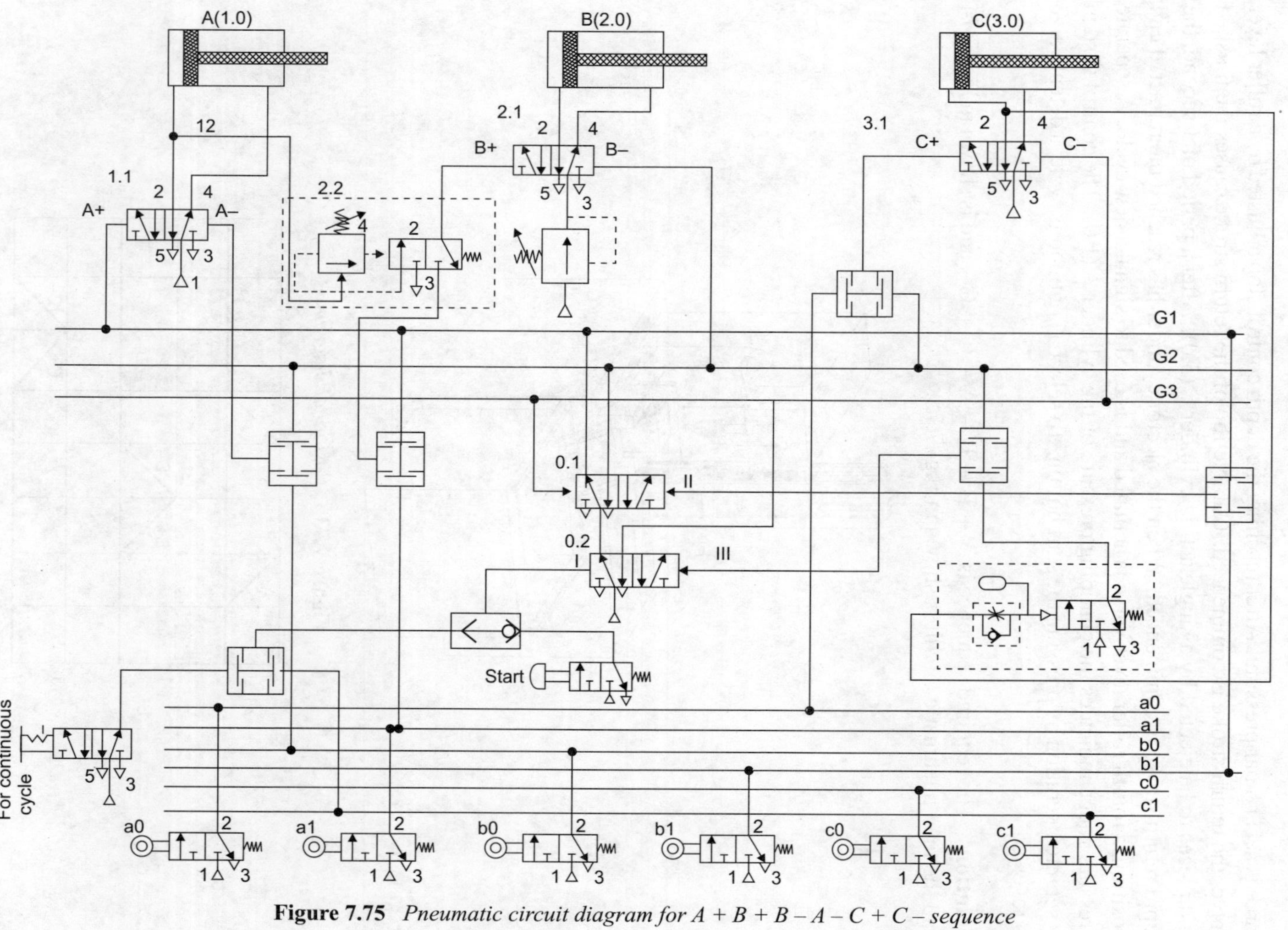

Figure 7.75 *Pneumatic circuit diagram for A + B + B – A – C + C – sequence*

Example 7.11 Metal strips are bent using a bending machine (Fig. 7.76), where they are placed manually in a fixture. Cylinder 1.0 clamps the metal strip when a push button is pressed. Cylinder 2.0 starts the bending operation and retracts, and cylinder 3.0 completes the operation. After cylinder 3.0 returns to its initial position, cylinder 1.0 releases the part.

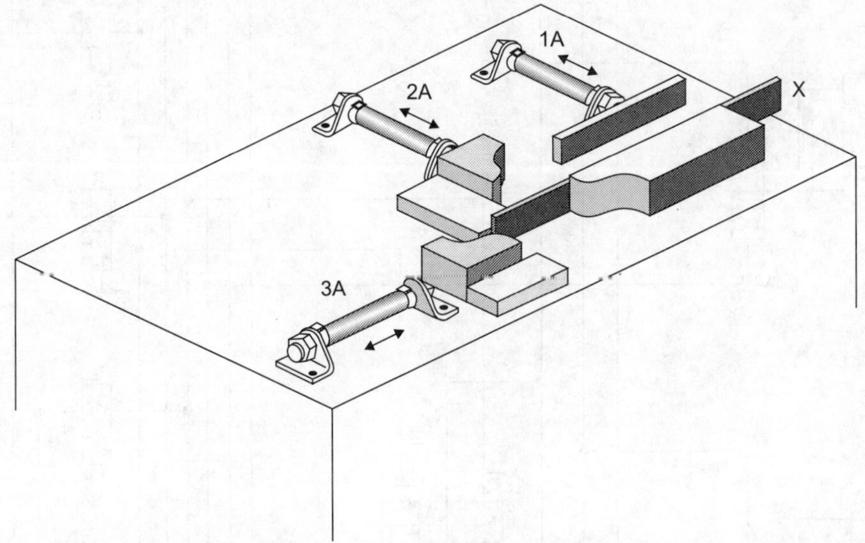

Figure 7.76 *Positional diagram*

Solution The sequence diagram is shown in the Fig. 7.77.

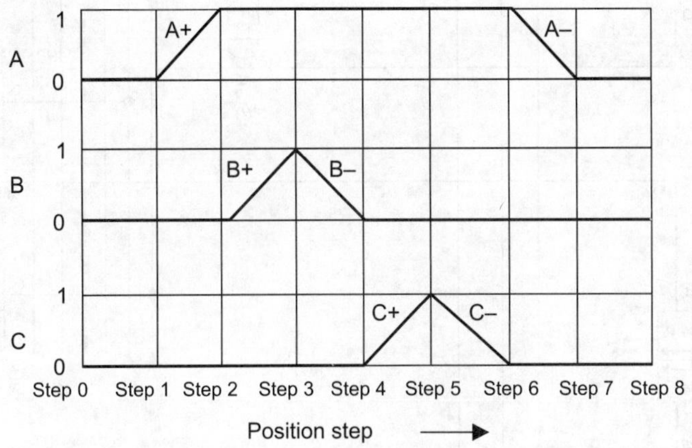

Figure 7.77 *Displacement–step diagram*

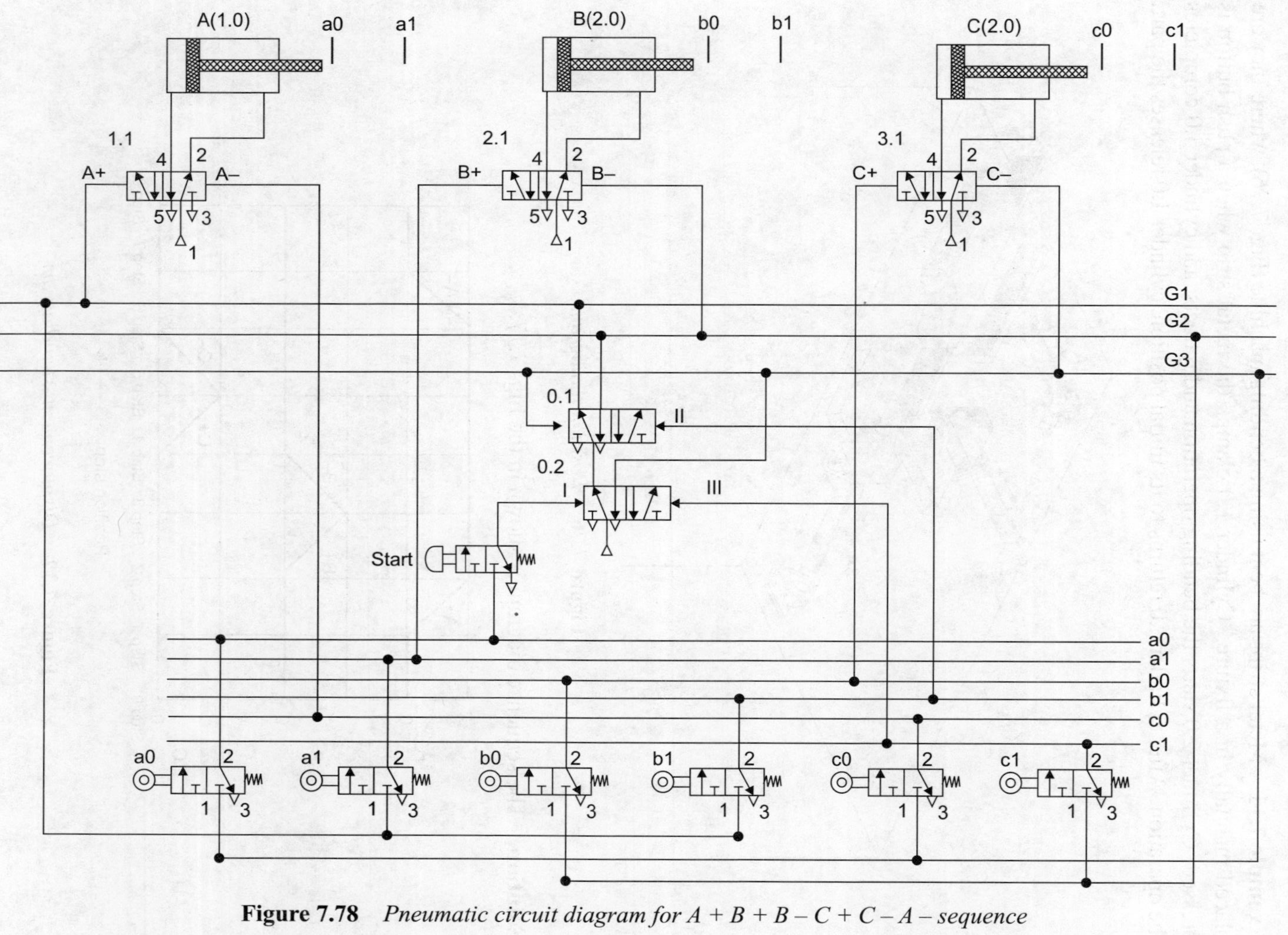

Figure 7.78 *Pneumatic circuit diagram for A + B + B – C + C – A – sequence*

Example 7.12 Rectangular parts are stamped on a special machine (Fig. 7.79). The parts are taken from a gravity feed magazine, pushed into the machine against a stop and clamped by cylinder A (1.0), stamped by cylinder B (2.0) and ejected by the ejector cylinder C (3.0). Design a pneumatic circuit using idle rollers while considering the following auxiliary conditions:

1. The sequence of operations should be performed automatically, with the choice of single or continuous cycle being available.
2. The start signal is provided by a start button.
3. The magazine should be monitored by means of a limit switch. If there is no part left in the magazine, the system must be halted in its starting position and interlocked to prevent restarting.
4. If an emergency stop switch is operated, the piston rods of all the three cylinders should return immediately from wherever they are to their starting position.

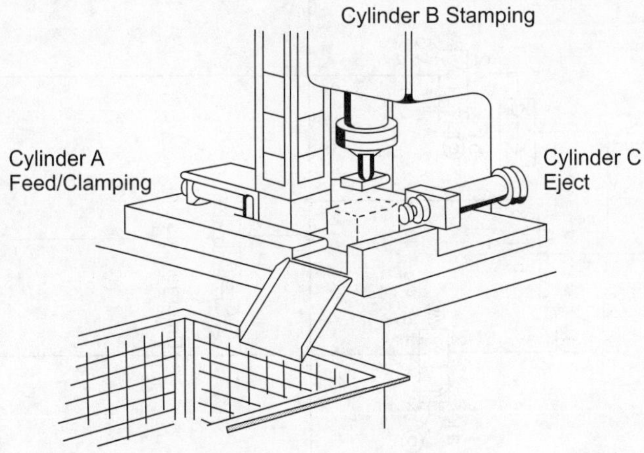

Figure 7.79 *Positional diagram*

Solution The sequence of operation is shown in the displacement–step diagram (Fig. 7.80).

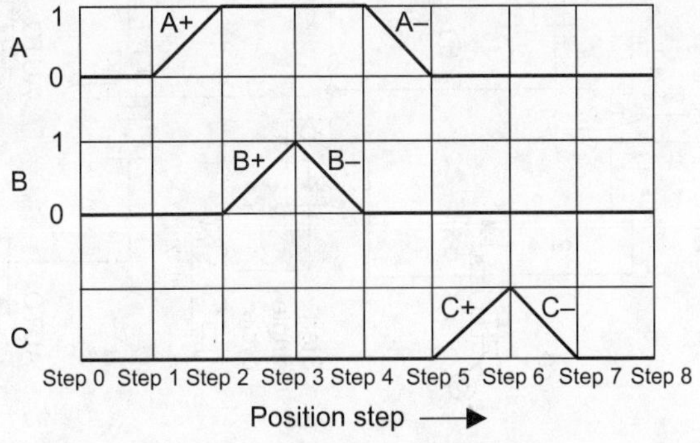

Figure 7.80 *Displacement–step diagram*

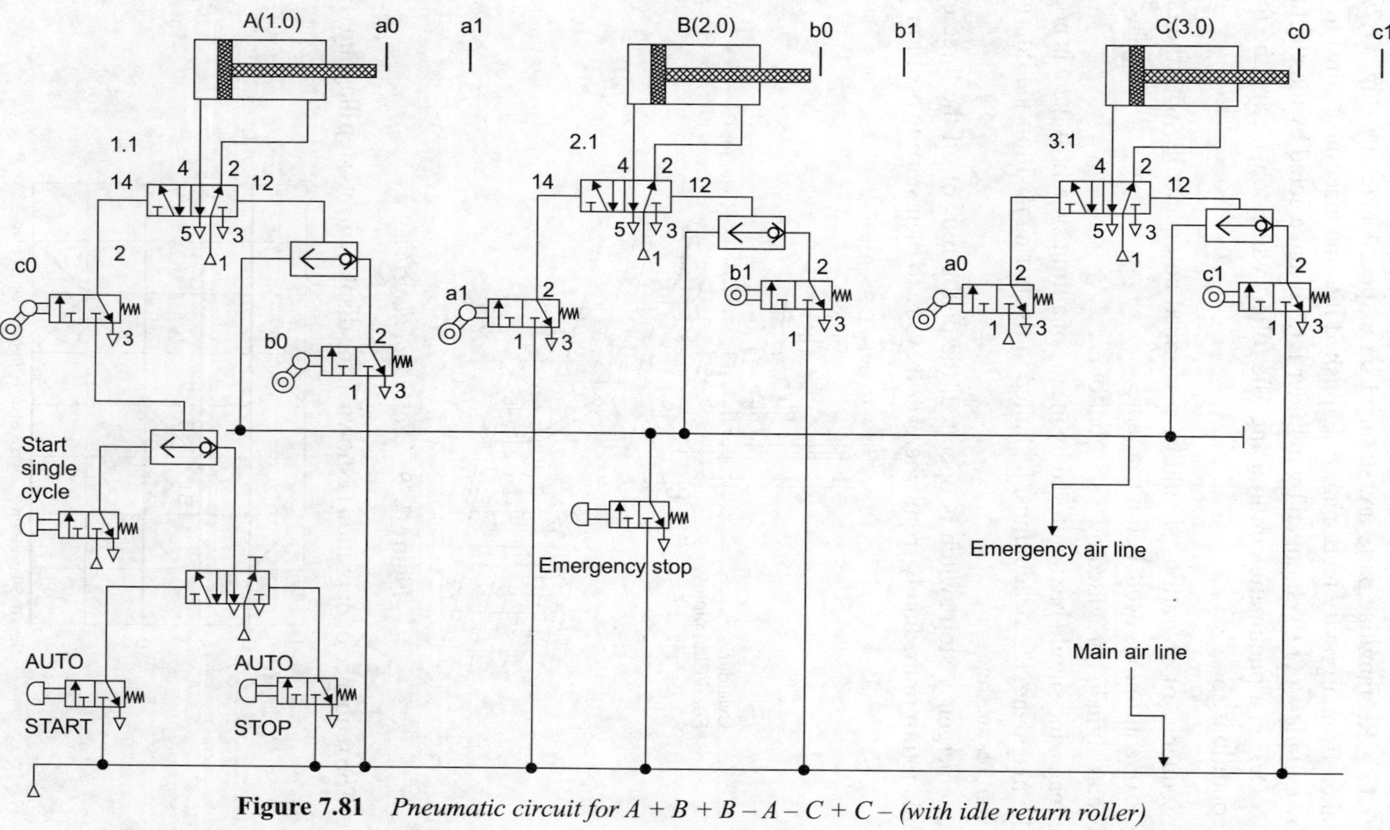

Figure 7.81 *Pneumatic circuit for A + B + B – A – C + C – (with idle return roller)*

Example 7.13 Lubricating oil bottles are pushed against the extended piston rod of the separating cylinder (2.0) by a continuously running conveyor belt (Fig. 7.82). The oil dispensing unit (1.0) closes the feed container in the retracted position. If the push button valve is actuated, the dispensing cylinder (1.0) extends with the exhaust air throttled and then retracts again. Both cylinders' end positions are sensed by roller lever valves. The filling capacity is adjusted by a screw X. The separating cylinder (2.0) then retracts and extends again immediately with the exhaust air throttled. The flow control is to be adjusted in such a way that a new empty oil bottle is located under the filling valve. Two proximity switches signal the cylinder position to the sequencer. If the valve with the selector switch is reset, the motion sequence finishes at the end of the cycle. Use the stepper counter method.

Solution The step diagram is shown in Fig. 7.83 and final pneumatic circuit in Fig. 7.84.

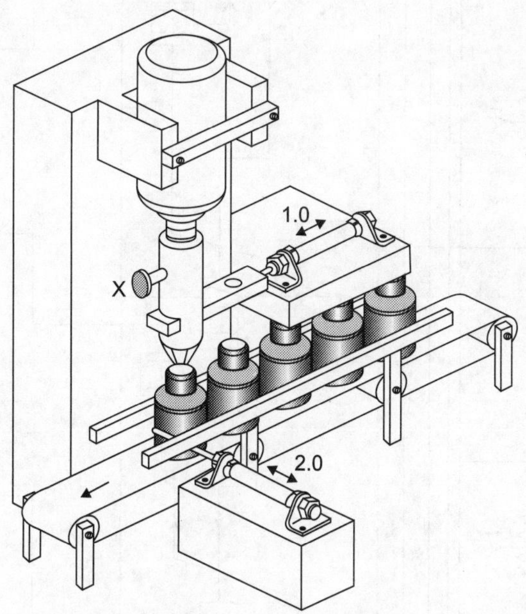

Figure 7.82 *Positional layout*

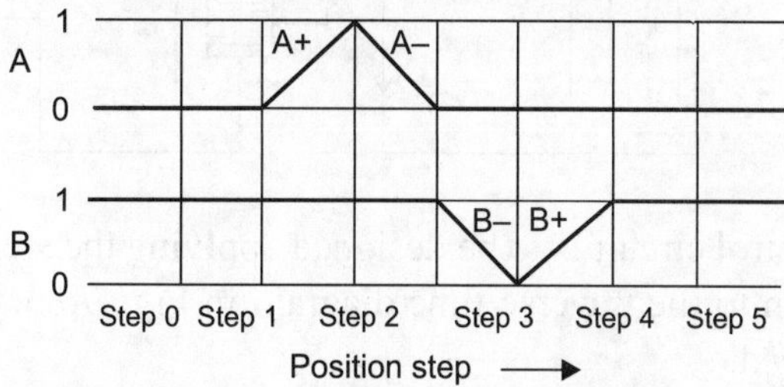

Figure 7.83 *Displacement step diagram*

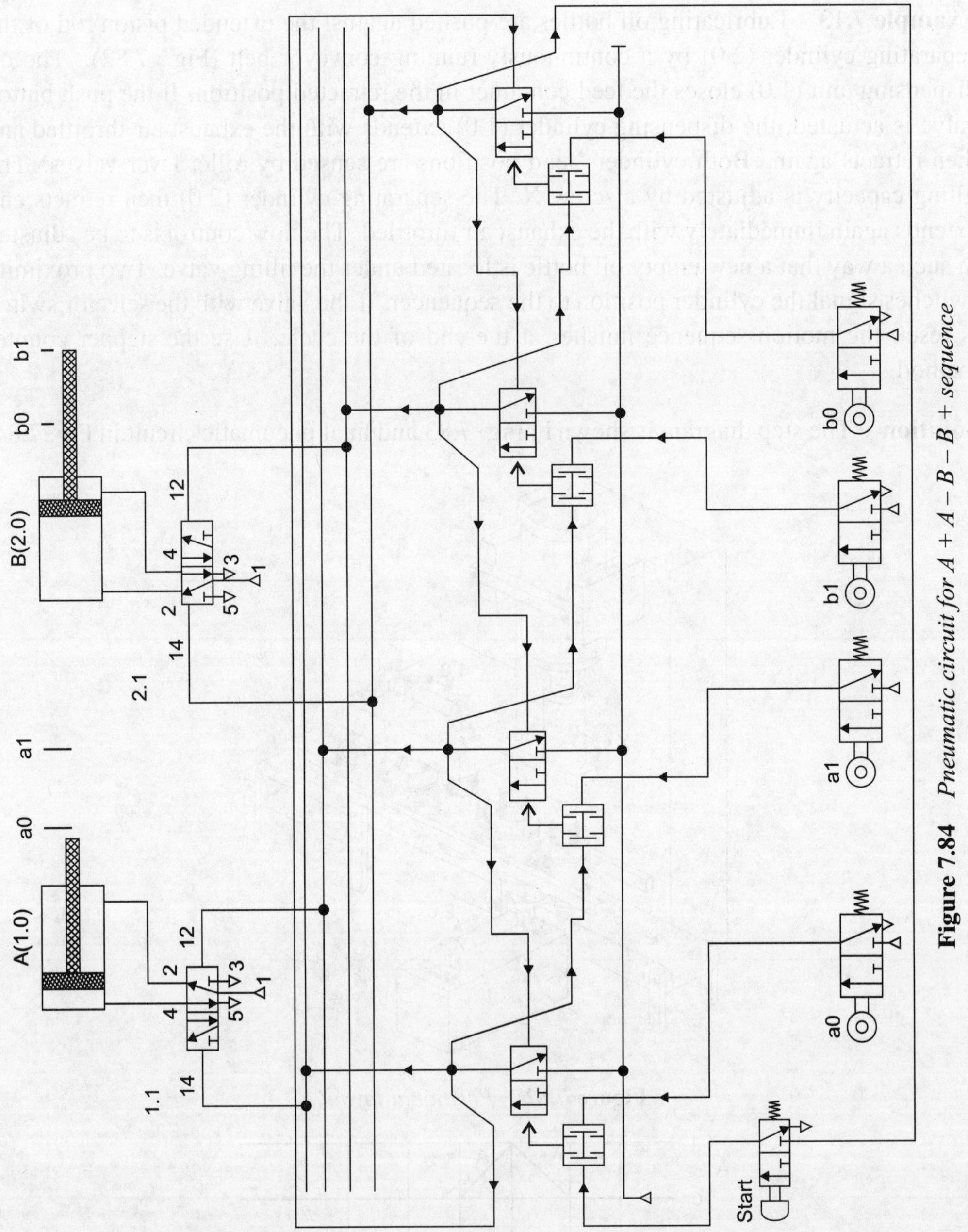

Figure 7.84 *Pneumatic circuit for* $A + A - B - B +$ *sequence*

Example 7.14 A control circuit is to be designed applying the step counter method. The sequence must conform to the traverse time diagram in Fig. 7.85. The cycle starts when the start button is pressed.

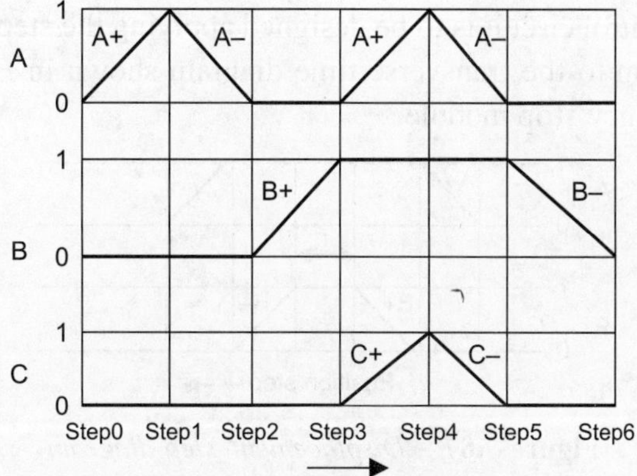

Figure 7.85 *Displacement–step diagram*

Solution The final pneumatic circuit is shown in Fig. 7.86.

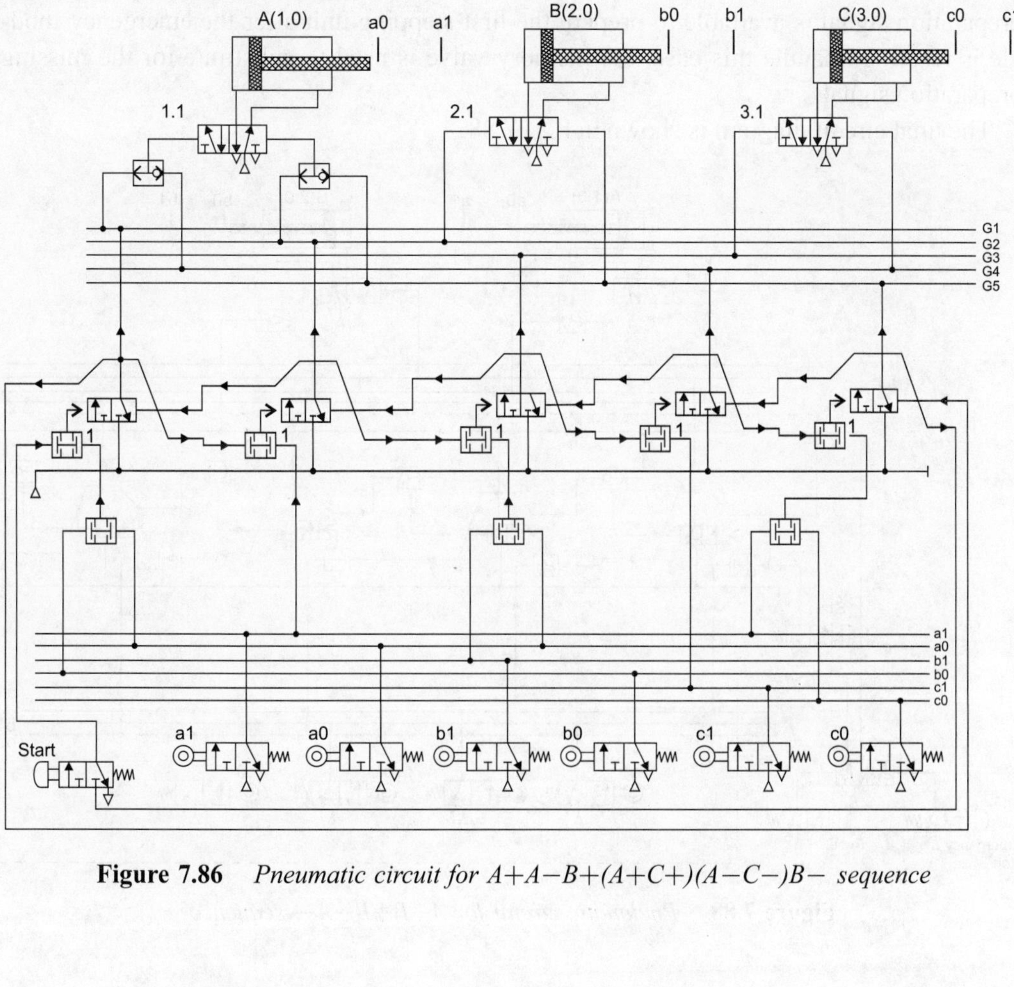

Figure 7.86 *Pneumatic circuit for A+A−B+(A+C+)(A−C−)B− sequence*

Example 7.15 A control circuit is to be designed applying the step-counter method. The sequence must conform to the transverse time diagram shown in Fig. 7.87. The circuit should have an emergency stop module.

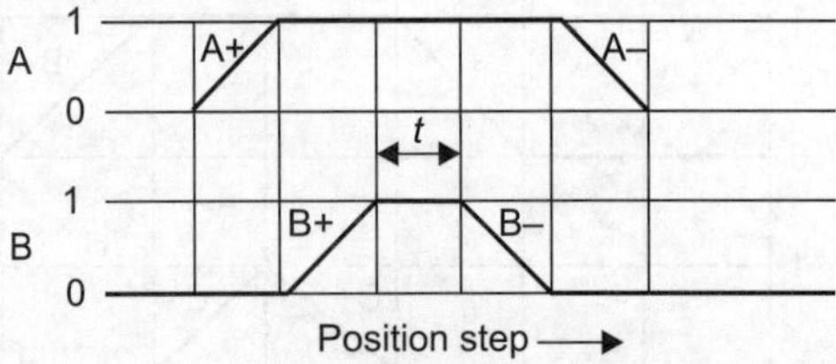

Figure 7.87 *Displacement–step diagram*

Solution In an emergency, the stepping units must be reset to prevent opposing signals acting on the directional control valves. Under normal conditions, the preparation signal for the first stepping unit is provided by the last stepping unit. Under emergency conditions, the circuit could stop anywhere between the first and last stepping units. Therefore, no preparation signal is available to prepare the first stepping unit after the emergency module is reset. To handle this case, the memory valve is used to substitute for the missing preparation signal.

The final circuit diagram is shown in Fig. 7.88.

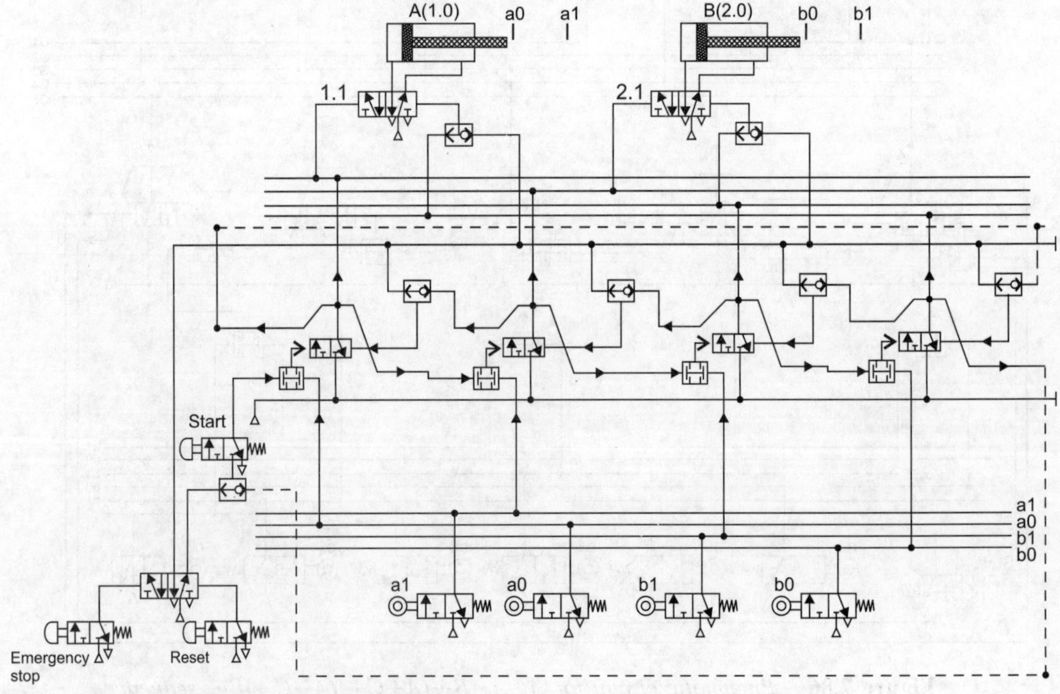

Figure 7.88 *Pneumatic circuit for A+B+B−A− sequence*

Example 7.16 A pipe is to be flanged in two steps in a pneumatic flanging device, as shown in Fig. 7.89. Cylinder 2.0 is in the extended position initially to act as the stop for the advancing pipe. Cylinder 1.0 extends and clamps the pipe, and subsequently, cylinder 2.0 retracts. The flanging cylinder 3.0 extends for the first time for the preliminary flanging operation and then retracts. The tool switchover is performed by cylinder 4.0, and flanging cylinder 3.0 is in action for the second time for the final flanging operation. The complete retraction of the cylinder 3.0 triggers the return motion of clamping cylinder 1.0 and tool switching cylinder 4.0, and the forward motion of position cylinder 2.0. The sequence of operation is shown in the sequence diagram. Develop a pneumatic control circuit to implement the control task.

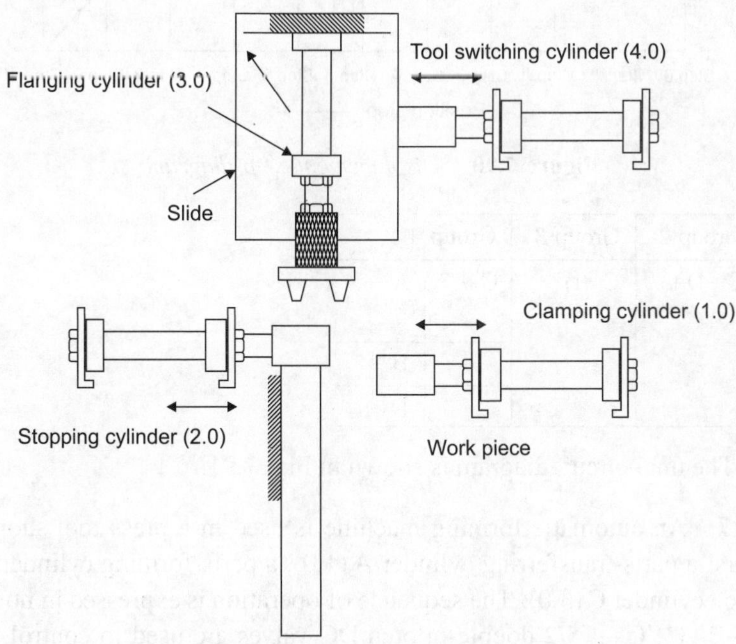

Figure 7.89 *Positional layout*

Step 3: Represent the control task using notational form
Cylinder A advancing is designated as A+
Cylinder A retracting is designated as A−
Cylinder B advancing is designated as B+
Cylinder B retracting is designated as B−
Cylinder C advancing is designated as C+
Cylinder C retracting is designated as C−

The sequence of operations is A+B+B−C+B+B−C−A−.

Step 4: Draw the displacement–step diagram (Fig. 7.90).

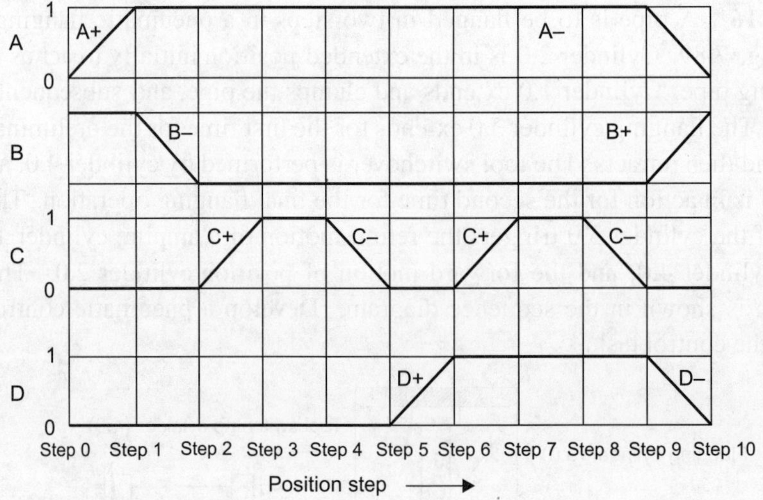

Figure 7.90 *Displacement–step diagram*

Group 1	Group 2	Group 3	Group 4	
A+B− C+(t)	C−D+	C+(t)	C−	A−
				B−
				D−

Grouping The final circuit diagram is shown in Fig. 7.91.

Example 7.17 An automatic forming machine is used in a press tool shop. There are three cylinders: a parts-transferring cylinder A (1.0), a parts-forming cylinder B (2.0), and a parts-ejection cylinder C (3.0). The sequence of operation is expressed in notational form as A+B+A−B−C−C+. 5/2 double-piloted DC valves are used to control the cylinder. Roller valves are mounted at both ends of each cylinder to provide a feedback signal, with an exception for the forward end position of cylinder B for which, instead, a pressure sequence valve is used to verify the extended position of cylinder B. A push button (PB) starts the cycle. Develop a pneumatic control circuit to implement the given control task.

Solution The displacement–step diagram is shown in Fig. 7.92 and the final circuit in Fig. 7.93.

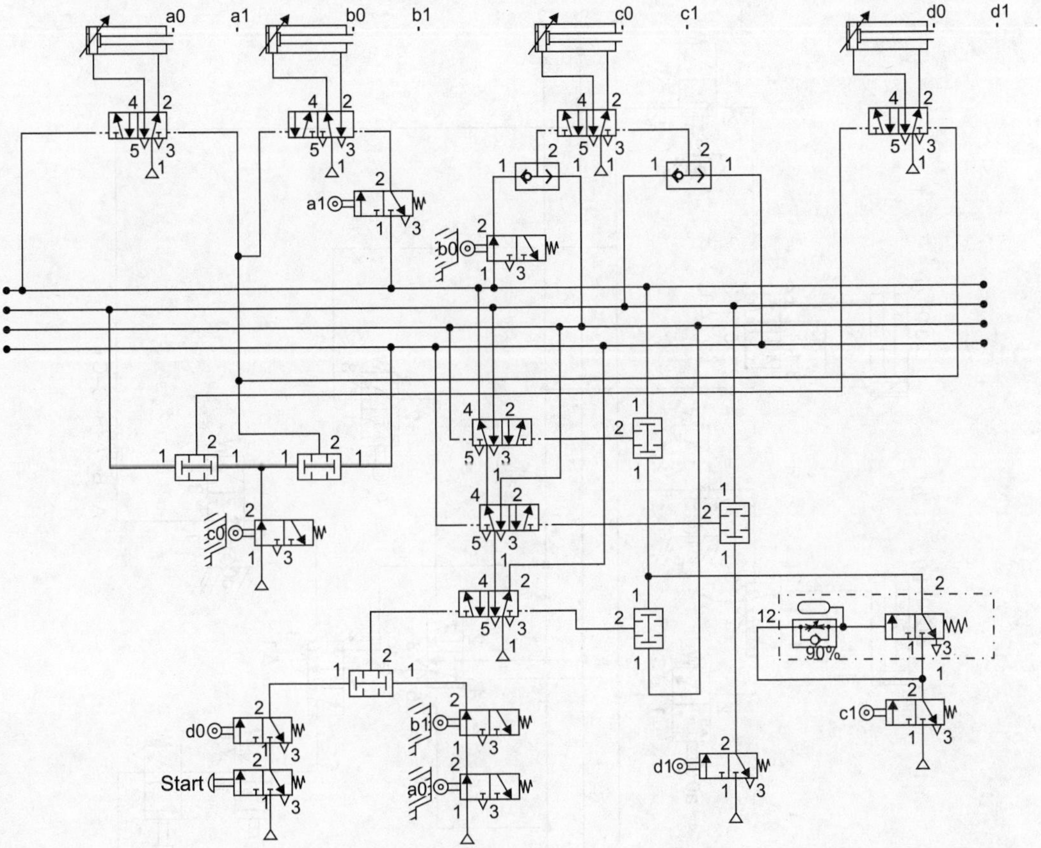

Figure 7.91 *Pneumatic circuit diagram*

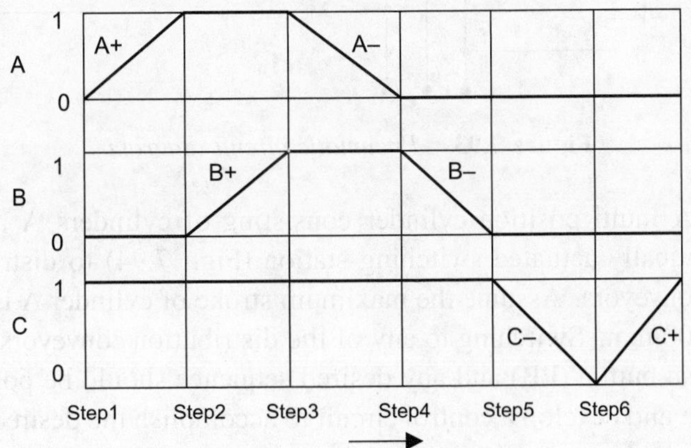

Figure 7.92 *Displacement–step diagram*

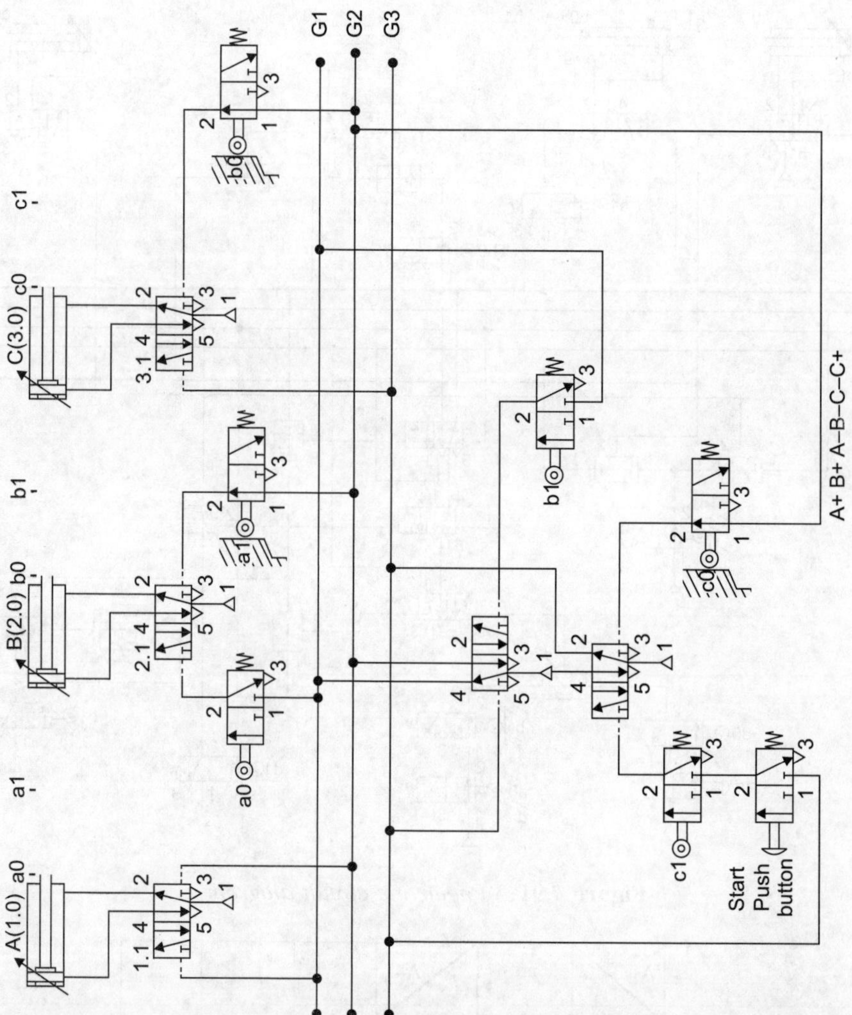

Figure 7.93 *Pneumatic circuit diagram*

Example 7.18 A multi position cylinder consisting of cylinders A and B is used to operate a pneumatically actuated switching station (Fig. 7.94) to distribute workpieces to four different conveyors. Assume the maximum stroke of cylinder A is 50 mm and that of cylinder B is 100 mm. Switching to any of the distribution conveyors is to be initiated by means of a push button (PB) and any desired sequence should be possible. Complete the control scheme and develop a control circuit to accomplish the desired control task.

The final circuit is shown in Fig. 7.95.

Solution

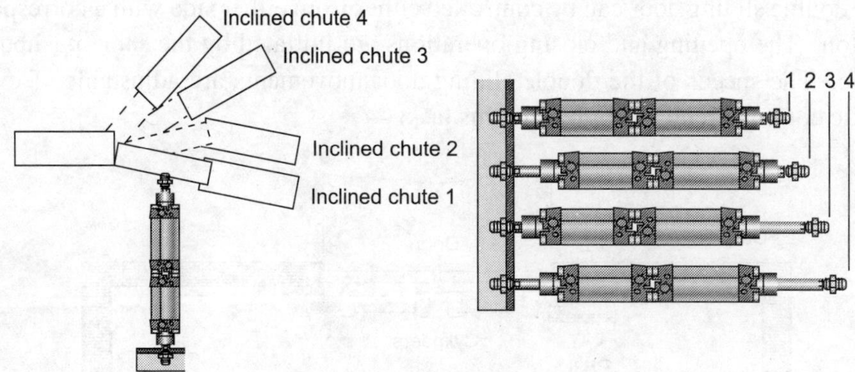

Figure 7.94

Position	Distance	PB	Notation	
1	0 mm	PB1	A−	A+
2	50 mm	PB2	A+	B−
3	100 mm	PB3	A−	B+
4	150	PB4	A+	B+

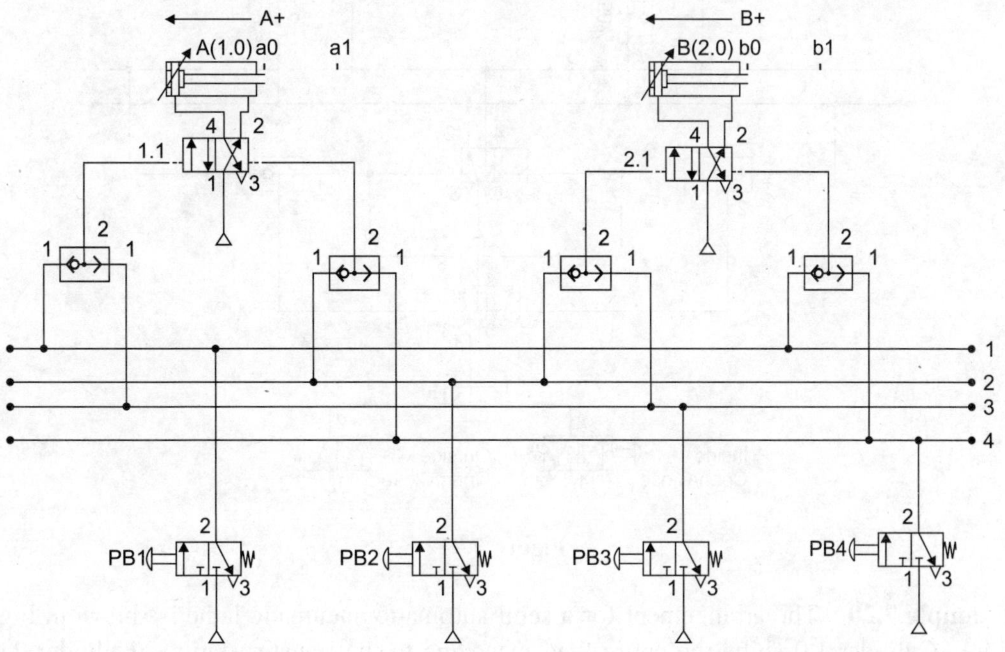

Figure 7.95

Example 7.19 A pneumatically controlled double sliding door is used in a metrology lab. The double sliding door can be controlled either from either side with a corresponding pushbutton. The opening and closing operations are initiated by the same pushbutton in each case. The speeds of the double sliding door movements are adjustable. Develop a pneumatic control circuit to implement this task.

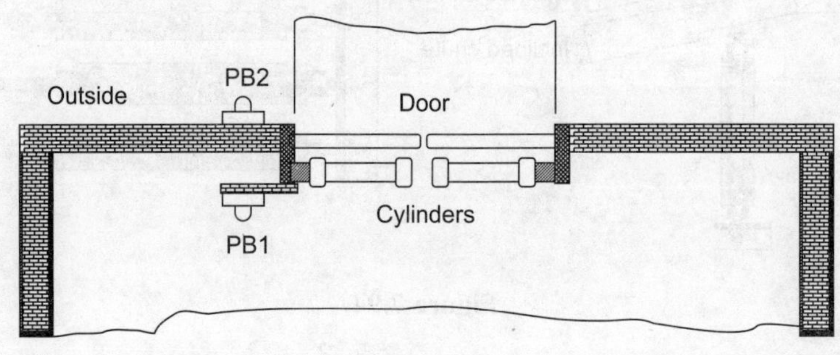

Figure 7.96

Solution Refer to Figs 7.96 for the positional layout and Fig 7.97 for the final circuit.

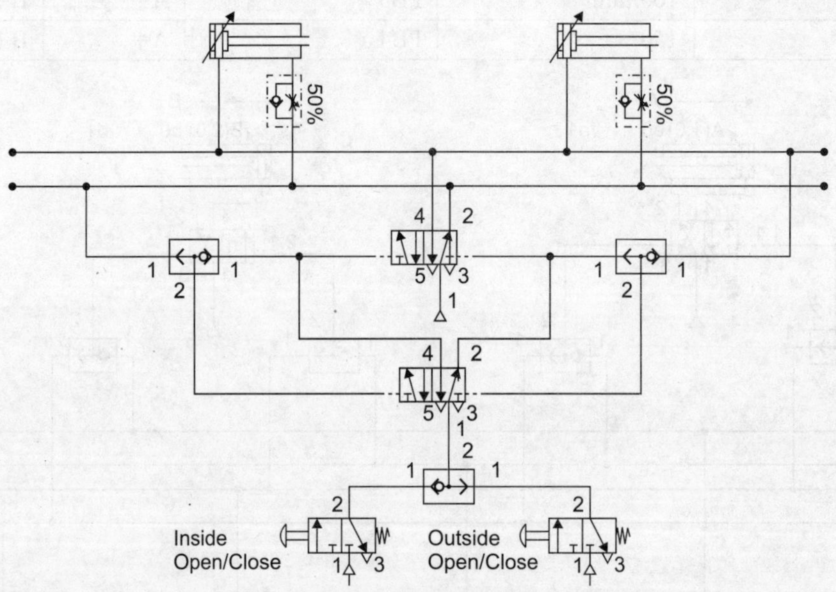

Figure 7.97

Example 7.20 The arrangement for a semi-automatic pneumatic lathe is shown in Fig. 7.99. Cylinder 1.0 shifts the entire feed magazine to the transfer station. Cylinder 2.0 pushes the workpiece into the lathe chuck. Cylinders 1.0 and 2.0 then retract. Cylinder 3.0 moves the slide rest forward and back. Cylinder 4.0 then ejects the workpiece following the release of the lathe chuck. The sequence of operation of these cylinders is shown in the displacement–step diagram (Fig. 7.98). Develop a pneumatic control circuit to implement the given control task: (1) for one cycle and (ii) for continuous sequence of operation.

Solution The final circuit is shown in Fig. 7.100.

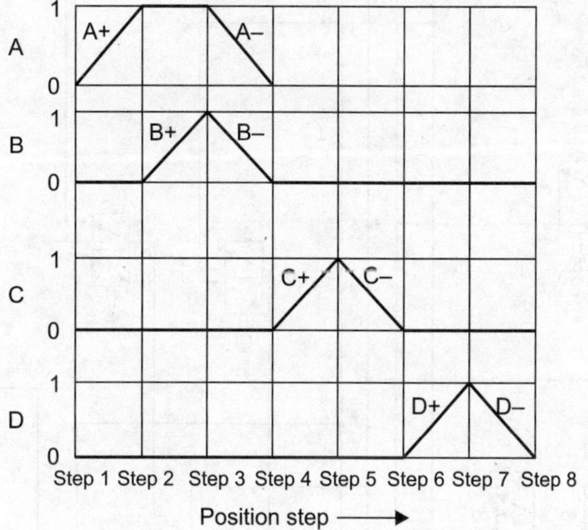

Figure 7.98

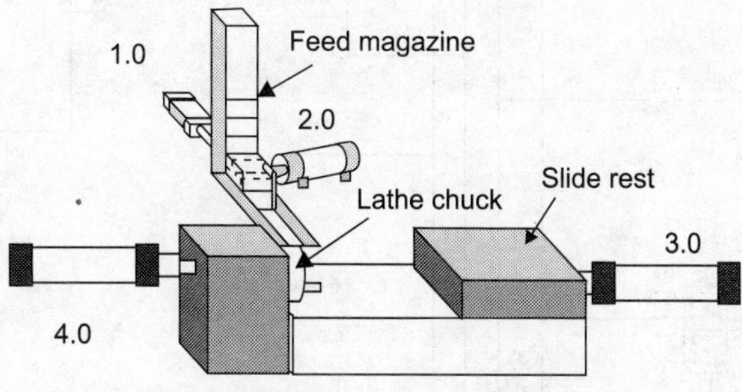

Figure 7.99

Example 7.21 An arrangement for feeding, clamping, processing and ejecting workpieces is shown in Fig. 7.102. Workpieces are taken from the gravity feed magazine and placed into the device by cylinder 1.0. Next, cylinder 2.0 extends and clamps the workpiece. Cylinder 3.0 is used to carry out processing. The processing time is adjustable via a timer. Cylinder 4.0 is used to eject the workpiece. The sequence of operation is shown in the sequence diagram Fig. (7.101). Develop a pneumatic control circuit to implement the control task.

Solution The final circuit is shown in Fig. 7.103.

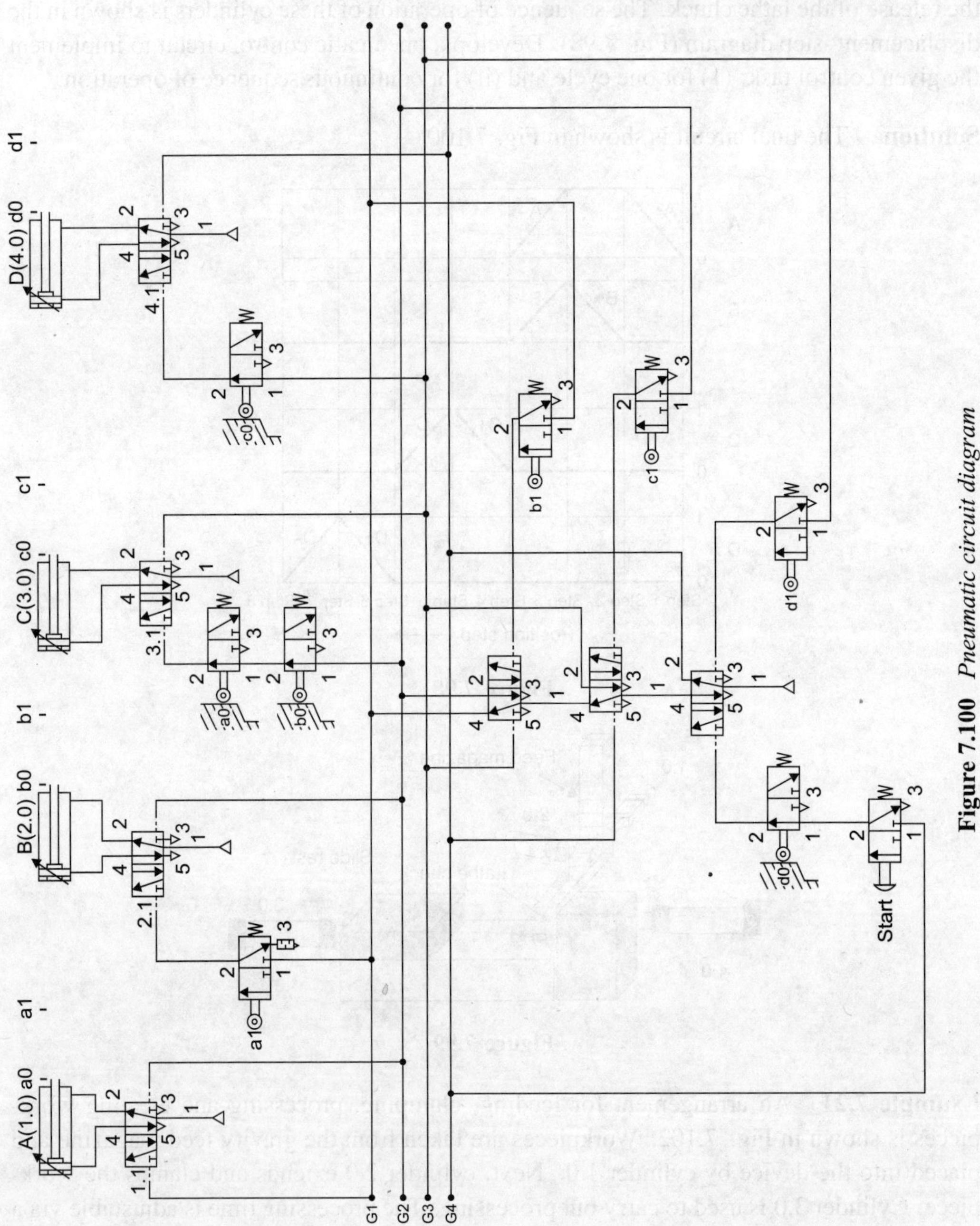

Figure 7.100 *Pneumatic circuit diagram*

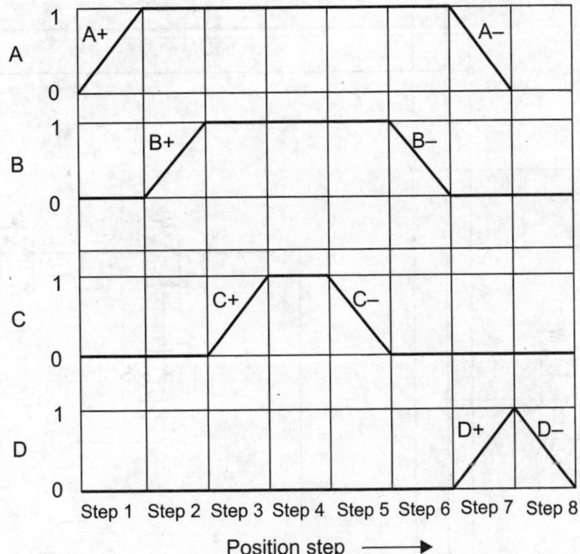

Figure 7.101

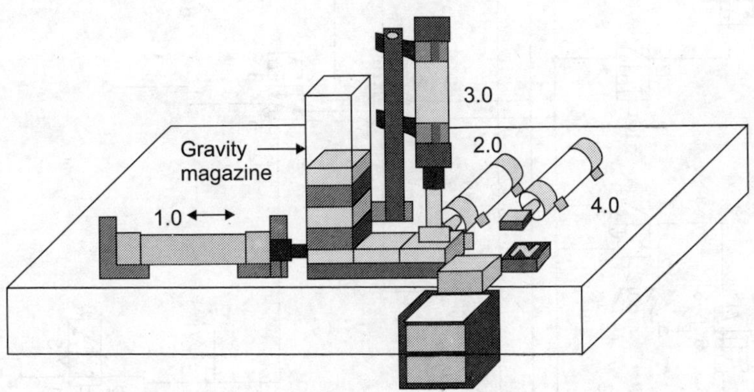

Figure 7.102

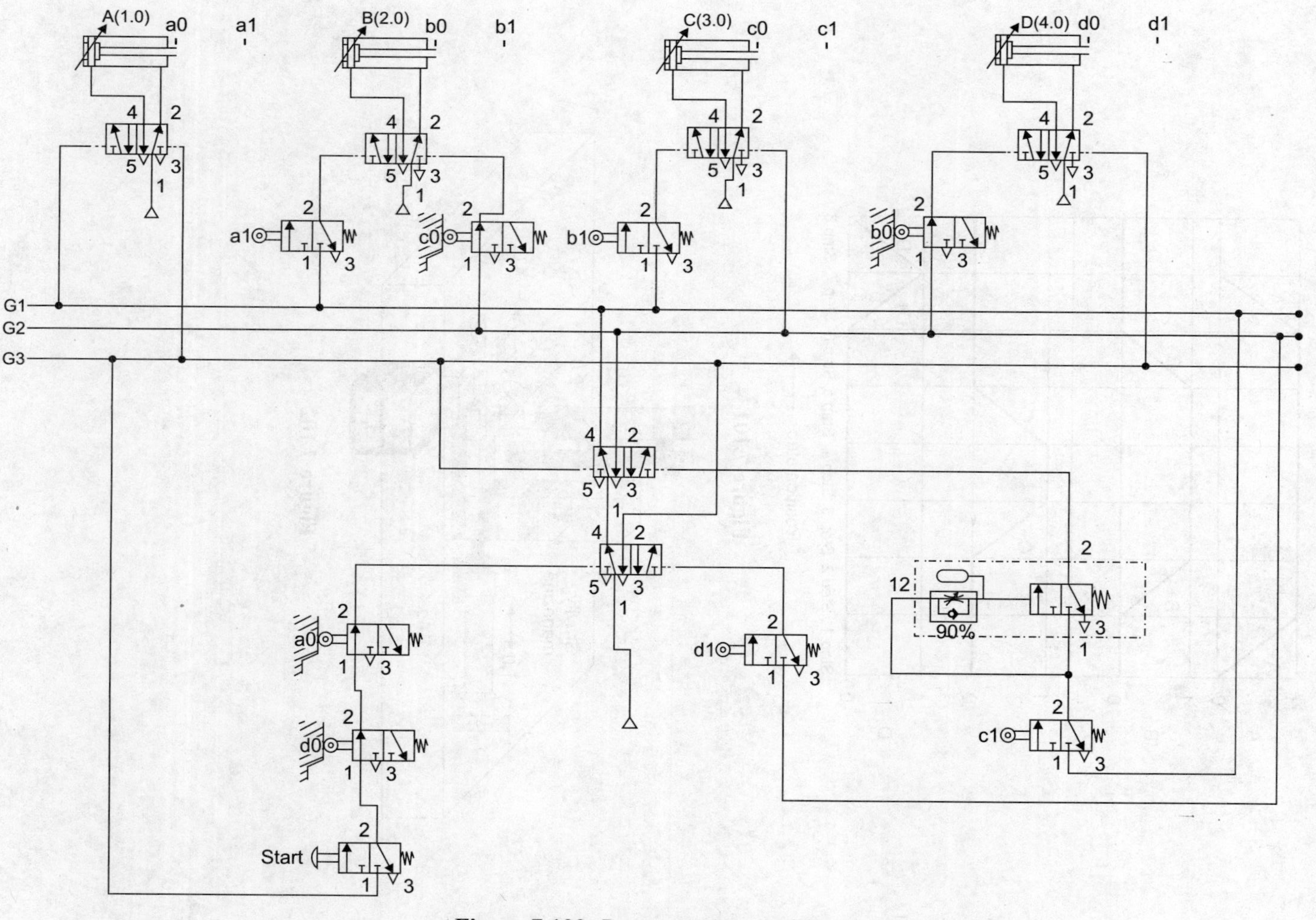

Figure 7.103 *Pneumatic circuit diagram*

Example 7.22 A pneumatic system is to be designed to drill holes as shown in Fig. 7.104. Cylinder 1.0 clamps the workpiece. The feed cylinder comes forward fully and then returns. The feed cycle (B+B–) is repeated thrice. The sequence diagram is shown. Develop a pneumatic control circuit to implement the control task.

Solution The displacement diagram is shown in Fig. 7.104 and the final circuit in Fig. 7.105.

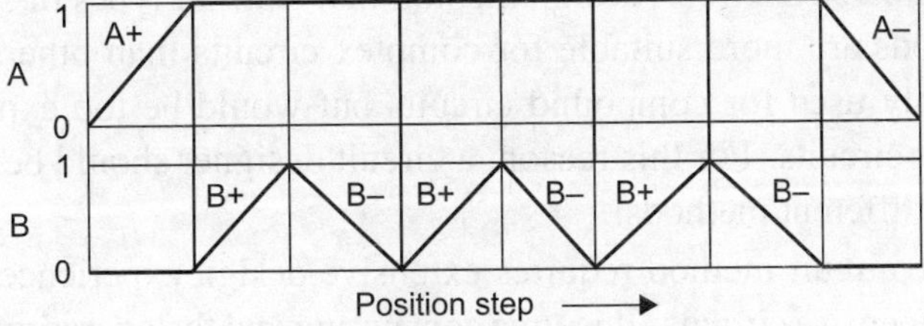

Figure 7.104

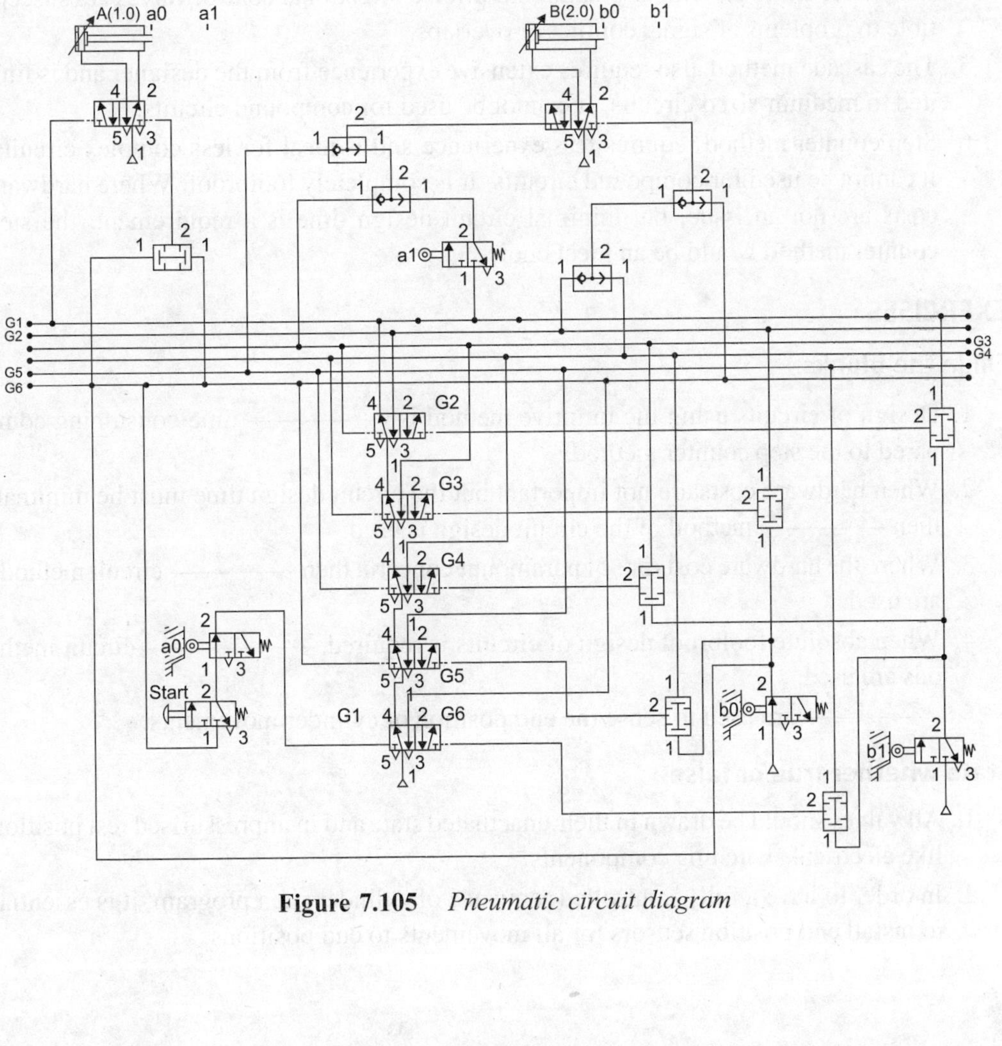

Figure 7.105 *Pneumatic circuit diagram*

Summary

1. Most of the practical pneumatic circuits use multi cylinders. They are operated in specific sequences to carry out the desired tasks. These actuators extend and retract in response to the actuation of directional control valves, which act as final control elements. Sensors are used for judging and verifying the position of the cylinder.
2. There is no universal circuit design method that suits all types of sequential circuits. Some methods are more suitable for complex circuits than others. Other methods are commonly used for compound circuits but would be too expensive for simple two-cylinder circuits. For this reason, a circuit designer should be aware of the pros and cons of different methods.
3. The intuitive circuit method requires extensive design experience and is limited to small non-repetitive circuits. It is time consuming and the operation is not completely fool proof.
4. It is important to remember that double-piloted directional control valves are susceptible to problems of signal conflict or overlaps.
5. The cascade method also requires extensive experience from the designer and is limited to medium-sized circuits. It cannot be used for compound circuits.
6. Step counter method requires less experience and is ideal for less complex circuits. It cannot be used for compound circuits. It is completely foolproof. Where hardware costs are not an issue, but minimal circuit design time is a requirement, the step counter method would be an ideal choice.

EXERCISES

Fill in the blanks

1. Design of circuits using the intuitive method is ________ time consuming compared to the step counter method.
2. When hardware costs are not important but the circuit design time must be minimal, then ________ method of the circuit design is used.
3. Where the hardware costs are of paramount concern, then ________ circuit methods are used.
4. When absolute foolproof design of circuits is required, ________ circuit methods are used.
5. ________ are used to sense the end position of cylinder movements.

State whether true or false

1. All valves should be drawn in their unactuated state and in unpressurised rest position like electrical switching components.
2. In order to have a fully-controlled sequence of cylinders in a program, it is essential to install end position sensors for all movements to end positions.

3. 4/2 or 5/2 double-piloted valves are free from signal overlaps.
4. Step counter method is absolutely free from signal conflicts.
5. Time delay and pressure delay functions cannot be used in multi cylinder circuit design.

Review questions

1. Explain, using án example, how signal conflict occurs .
2. What is the effect of signal conflict in multi actuator circuit designs?
3. What are the different ways of eliminating signal conflict in multi actuator circuit designs?
4. Explain the step–displacement diagram for the sequence A+B+B−A−.
5. List the disadvantages of using idle return rollers for overcoming signal conflicts.
6. Explain the principle of the cascade method with a suitable sequence example.
7. Briefly explain the principles of step counter method with a suitable example.
8. Draw a group changing cascade circuit for two, three, four and five groups.
9. Explain the construction and working of a pressure sequence valve with the help of a neat sketch.
10. What are the different ways to sense the end position and movement of cylinders?

Problems

1. In a press shop, sheet metal strips are continuously fed to a press tool work table and clamped before press tool operation. Figure 7.106 shows the working of the linear indexing for feeding sheet metal strips to the press tool.

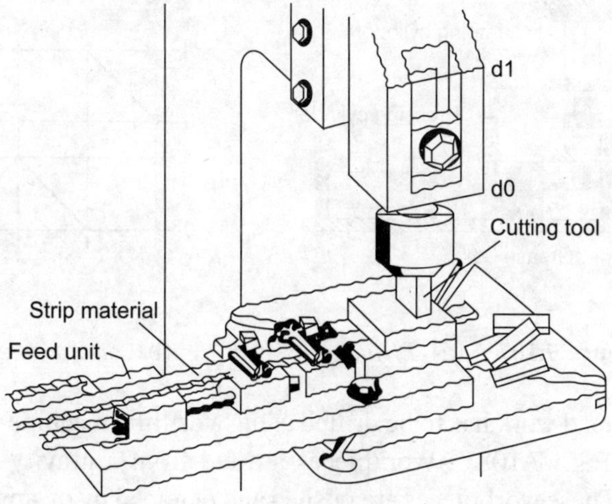

Figure 7.106

There are three cylinders used for this application.

Feeder gripper	Designated as cylinder A
Feeder cylinder	Designated as cylinder B
Clamping cylinder	Designated as cylinder C
Press tool	Designated as drive D

The feeder gripper cylinder and clamping cylinder are diaphragm cylinders. The feeder cylinder could be rodless with an adequate long stroke.

The sequential operation is: first the gripper cylinder [A+] advances and grips the sheet metal against a moving table. Then feeder cylinder [B+] advances and transfers the sheet metal to the work table or the die of the press tool. Subsequently the gripper cylinder is released [A−] and the clamping cylinder advances [C+] and clamps the sheet metal against the fixed die block. After the clamping operation, the press tool descends and the punching operation is carried out. At the end of this process a signal from a position dependent control such as a limit switch [D] conveys a signal to reset the press tool to the initial position. The clamping cylinder is released [C−], the feeder cylinder retracts [B−] and the cyclic process repeats.

2. The schematic diagram of an assembly fixture for greasing roller bearings is shown in Fig. 7.107(a). After assembling the individual roller bearing parts, roller bearings are to be pressed by a pneumatic cylinder A(1.0). Cylinder B(2.0) is required to operate a grease press twice for injecting grease into the roller bearing. The sequence diagram is shown in Fig. 7.107(b). Develop a pneumatic control circuit using the cascade method for implementing this control task.

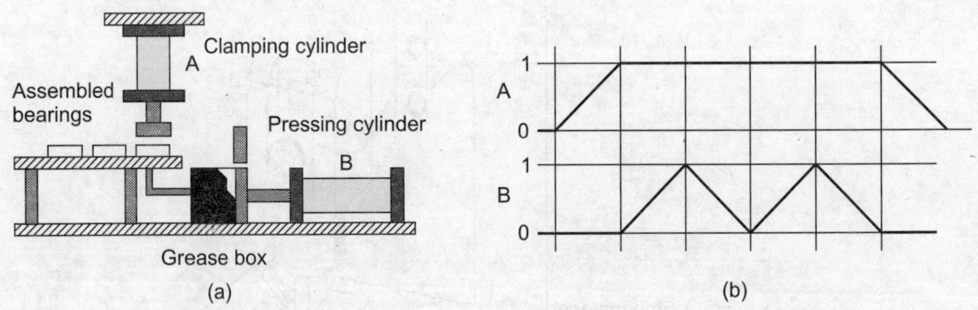

Figure 7.107 (a) *Positional sketch,* (b) *displacement–step diagram*

3. Square shaped work has to be drilled using a drilling machine which is pneumatically operated (Fig. 7.108). Workpieces are fed from a gravity magazine to a drilling machine. These workpieces are pushed and clamped by means of a clamping cylinder 1.0 (A); a hole is drilled by the drilling cylinder 2.0 (B), and the workpiece is ejected by ejecting cylinder 3.0 (C). The displacement–step diagram is shown in Fig. 7.109.

The sequence of operations has to be carried out either for one cycle or for continuous cycles with start and stop controls. Develop a pneumatic circuit to implement the given control task.

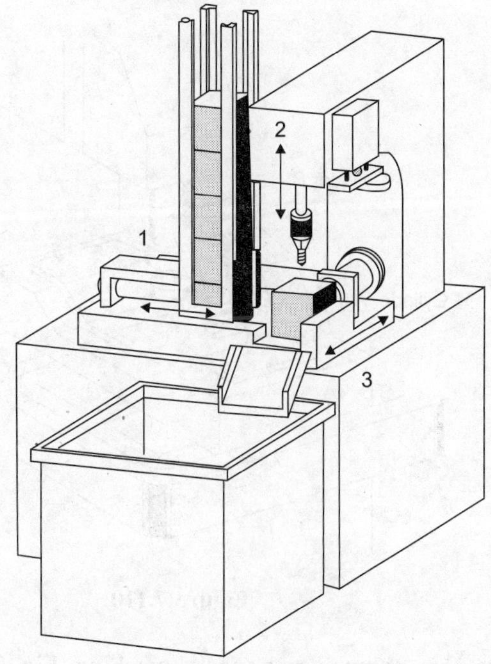

Figure 7.108

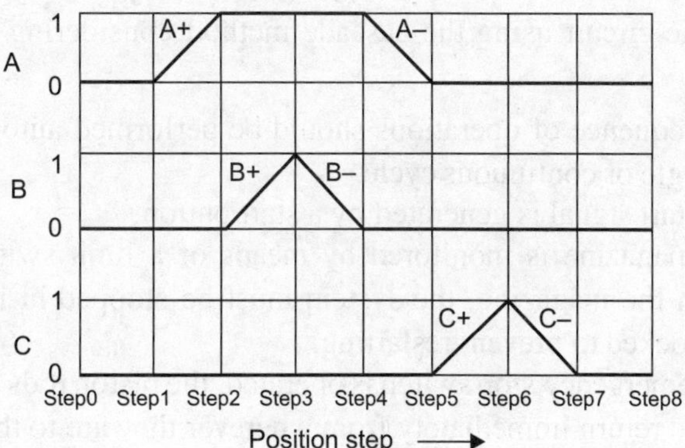

Figure 7.109

4. Two clips are to be riveted together on a semi-automatic press (Fig. 7.110). Components and rivets are positioned by hand, and after completion of the operation, removed by hand. The automated part of the working cycle consists of the hold-

ing and clamping of components (cylinder A) and also riveting (cylinder B), and the cycle to be performed ends at the starting position after a start button is operated.

Design the circuit using (a) classic method, (b) idle roller, (c) pressure sequence valve, and (d) cascade method.

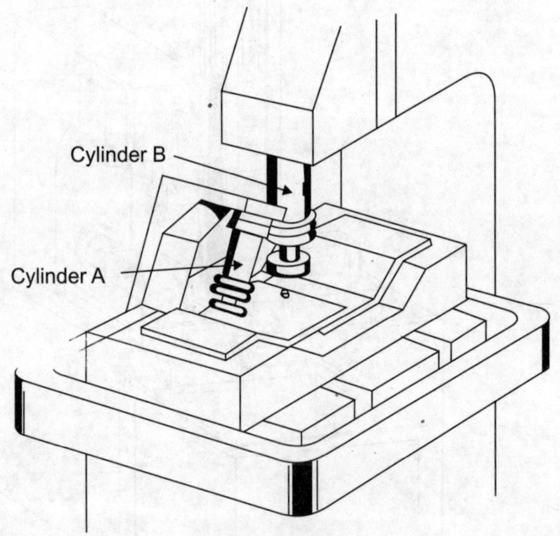

Figure 7.110

5. Rectangular parts are stamped on a special machine (Fig. 7.111). The parts are taken from a gravity feed magazine, pushed into the machine against a stop and clamped by cylinder A (1.0), stamped by cylinder B(2.0) and ejected by cylinder C(3.0). Design a pneumatic circuit using the cascade method considering the following auxiliary conditions:
    1. The sequence of operations should be performed automatically with a choice of single or continuous cycle.
    2. The start signal is generated by a start button.
    3. The magazine is monitored by means of a limit switch. If there is no part left in the magazine, the system must be stopped in its starting position and interlocked to prevent restarting.
    4. If an emergency stop switch is operated, the piston rods of all the three cylinders should return immediately from wherever they are to their starting position.

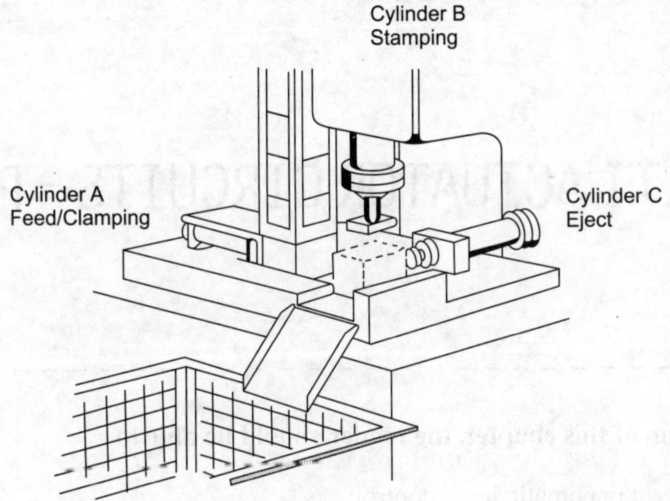

Figure 7.111

Answers

Fill in the blanks

1. more 2. step counter 3. programmable logic 4. step counter 5. limit switches

True or false

1. True 2. True 3. False 4. True 5. False

8 MULTI ACTUATOR CIRCUITS - PART II

Upon completion of this chapter, the reader should be able to

- Explain the term pneumatic logic control
- Understand the difference between pneumatic logic control and electrical logic control
- Explain the working of signal processing logic gates
- Simplify logic functions using the rules of Boolean algebra
- State DeMorgan's theorem and its use
- Explain NAND and NOR universality
- Build pneumatic circuits from a logic equation
- Draw a KV map for two, three and four variables
- Design circuits using the KV mapping method
- Design circuits using combinational logic method

8.1 Pneumatic Logic Control

Logic dates back to Aristotle, the Greek philosopher who was the first to examine in detail the form of reasoning called deductive logic. This is a logical analysis of a formal argument which consists of a major premise, a minor premise and a conclusion. The Irish mathematician and logician George Boole worked out a mathematical system in 1847 to solve problems in logic, probability and engineering. This system formulated logical statements symbolically so that they could be written and proven in a manner similar to that used in ordinary algebra. Logic was first applied to electrical relay logic and switching and is now also applied to pneumatics. The basic components of pneumatic logic control sensors and switching elements are called binary elements meaning that, at any given time there can only be in one of two positions: ON or OFF.

Pneumatic signal processing involves the use of pneumatic logic elements like air valves and other pneumatic devices for various control actions. These control actions are then applied to power systems to achieve the required actuation. A minimum amount of pneumatic force is required in the logic devices to send signals to power devices that operate high pressure control valves.

Compared to electrical logic devices, pneumatic logic devices offer certain advantages:

1. Simple design and construction
2. Safe for use in explosive environments
3. Longer life expectancy than electrical elements
4. High reliability
5. Faster switching response times (e.g. the response time of air logic systems is around 10−12 ms compared with 50-60 ms for most of industrial switches and relays)

Industrial automation often involves pneumatic controls across a wide range of applications. Pneumatic control systems use the same power medium for both control and work functions, thus reducing the complexity of the design. Use of pneumatic power is widespread in industries like welding, fabrication, automatic machine and press tools, assembling, material handling, packaging, food and beverages, printing and mining.

Logic circuits are divided into combinational and sequential circuits. In combinational circuits, each binary output is a function of the preset inputs only. In sequential circuits however, some or all of the outputs also depend on previous inputs, i.e., on the history of the system.

Sequential circuits can be synchronous or asynchronous. Synchronous circuits are time based, i.e., driven by a clock producing pulses at fixed intervals where each pulse triggers a new step. Asynchronous circuits are event based; a step occurs only after the previous step is successfully completed.

8.2 Signal Processing Logic Gates

The classic method is based on an understanding of pneumatic elements and logic. Each pneumatic element performs logic functions such as OR, AND, NAND, NOR, etc. Boolean algebra and DeMorgan's theorem are used in the analysis of control logics.

The seven basic logic gates are:

1. OR
2. AND
3. INVERTOR (NOT)
4. NAND
5. NOR
6. X-OR
7. X-NOR

A truth table shows all the possibilities for the inputs to a logic element, and their effects on the output in binary format (1 or 0). The number of possible input combinations is to 2^N , where N is number of inputs.

8.2.1 OR Gate

The output of an OR gate is logic 1 when any of its inputs are logic 1. The output of an OR gate is logic 0 when all of its inputs are logic 0. The Boolean equation for the OR operation is:

$$Y = A + B \text{ (2 inputs)}$$

$$Y = A + B + C \text{ (3 inputs)}$$

The symbols used for OR gate is shown in Fig. 8.1 and its truth table in Table 8.1.

Table 8.1 *Truth table for OR gate*

A	B	$Y = A + B$
0	0	0
1	0	1
0	1	1
1	1	1

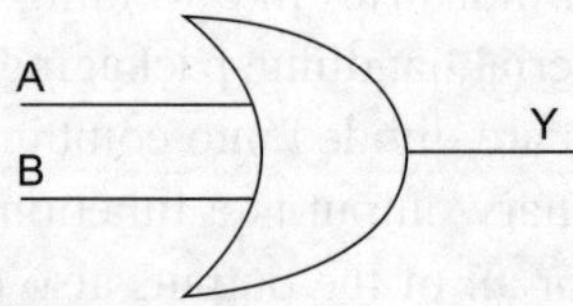

Figure 8.1 *Symbol for OR gate*

So, for the OR gate the output is high when any of its input is high. The OR function is represented by switches in parallel. The output lamps lights up when any of its input switches is ON. The OR gate is illustrated in Fig. 8.2.

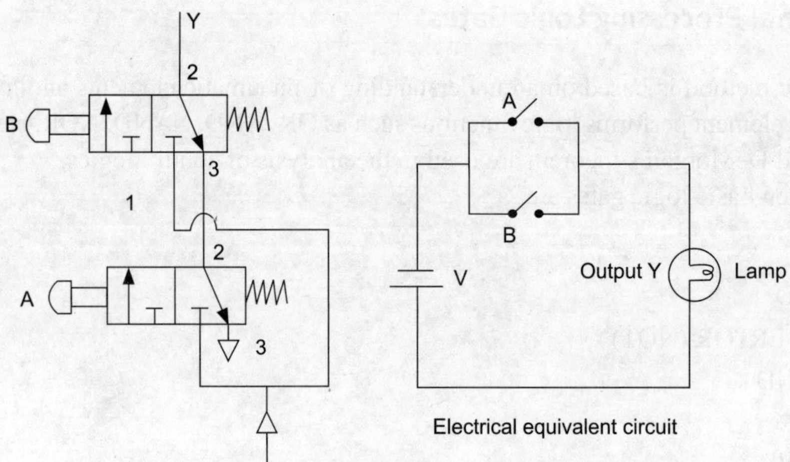

Figure 8.2 *Illustration of OR gate*

8.2.2 AND Gate

The output of an AND gate is logic 1 when all of its inputs are logic 1. The output of the AND gate is logic 0 when any of its inputs are logic 0. The Boolean equation for the AND operation is

$$Y = A \cdot B \text{ (2 inputs)}$$

$$Y = A \cdot B \cdot C \text{ (3 inputs)}$$

The symbol used for an AND gate is shown in Fig. 8.3 and its truth table is given in Table 8.2.

Table 8.2 *Truth table for an AND gate*

A	B	$Y = A \cdot B$
0	0	0
1	0	0
0	1	0
1	1	1

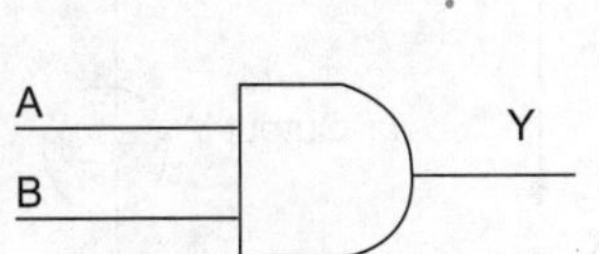

Figure 8.3 *Symbol for AND gate*

So for the AND gate, the output is high when all of its inputs are high. It is represented by switches in series. The output lamp lights up only when all the input switches are ON. An AND gate is illustrated in Fig. 8.4.

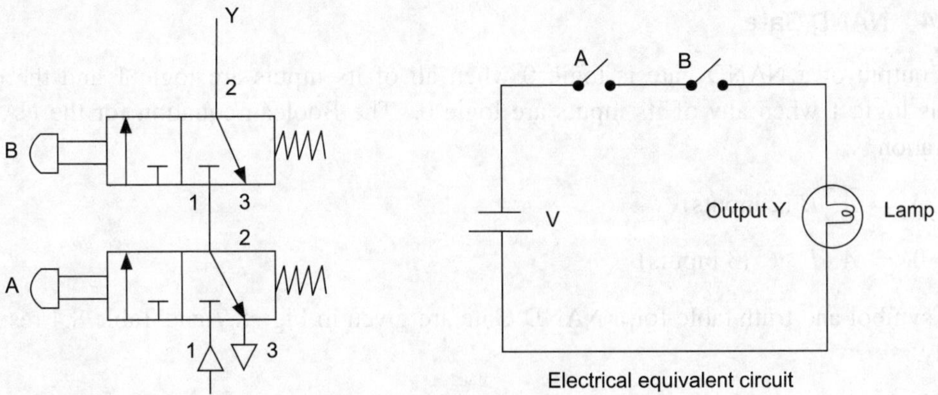

Figure 8.4 *Illustration of AND gate*

8.2.3 NOT (Inverter) Gate

The output of a NOT gate is the opposite logic level of the input. The Boolean equation for the NOT operation is:

$$Y = \overline{A}$$

The symbol used for a NOT gate is given in Fig. 8.5 and its truth table is given in Table 8.3.

Table 8.3 *Truth table for NOT gate*

A	$Y = \overline{A}$
0	1
1	0

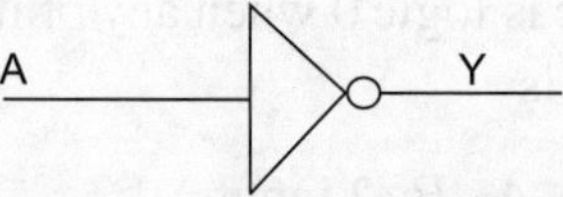

Figure 8.5 *Symbol for a NOT gate*

For a NOT gate the output is always the opposite of the input. It is represented by a switch in parallel. The output lamp lights up when the input power is OFF and is illustrated in Fig. 8.6.

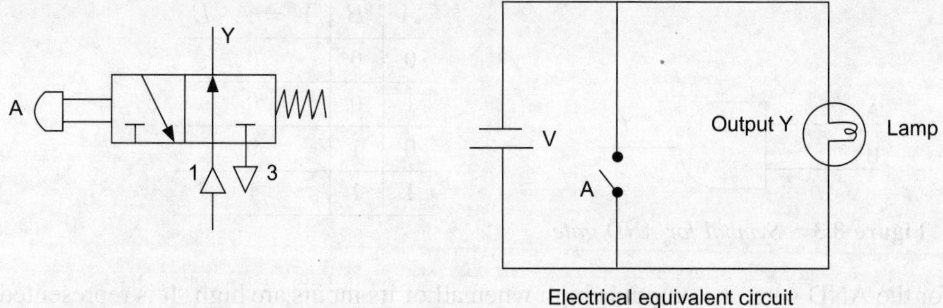

Figure 8.6 *Illustration of NOT gate*

8.2.4 NAND Gate

The output of a NAND gate is logic 0 when all of its inputs are logic 1 and the output is logic 1 when any of its inputs are logic 0. The Boolean equation for the NAND operation is:

$$Y = \overline{A \cdot B} \text{ (2 inputs)}$$

$$Y = \overline{A \cdot B \cdot C} \text{ (3 inputs)}$$

The symbol and truth table for a NAND Gate are given in Fig. 8.7 and Table 8.4 respectively.

Table 8.4 *Truth table for the NAND gate*

A	B	$Y = \overline{A \cdot B}$
0	0	1
1	0	1
0	1	1
1	1	0

Figure 8.7 *Symbol for NAND gate*

For a NAND gate, the output is high when any of its inputs are low and it is represented by switches in parallel as shown below. The output lamp lights up when any of the input switches are OFF. The NAND gate is illustrated in Fig. 8.8.

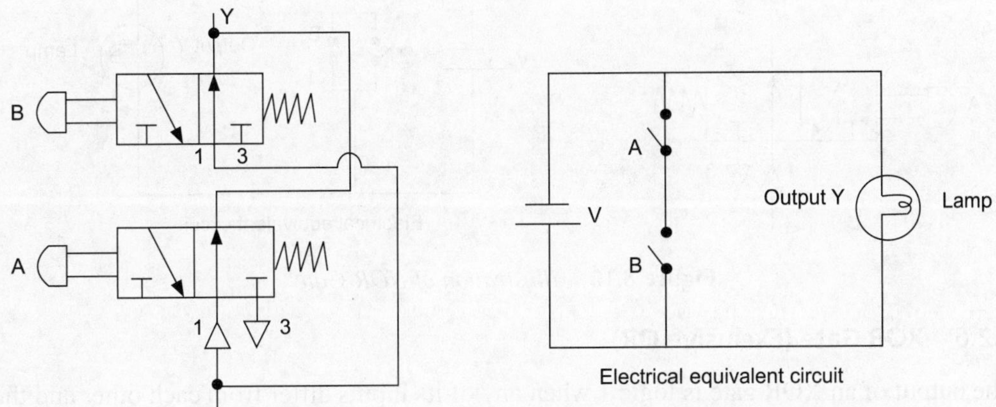

Figure 8.8 *Illustration of NAND Gate*

8.2.5 NOR Gate

The output of a NOR gate is logic 0 when any of its inputs are logic 1 and the output is logic 1 when all of its inputs are logic 0. The Boolean equation for the NOR operation is:

$$Y = \overline{A + B} \text{ (2 inputs)}$$

$$Y = \overline{A + B + C} \text{ (3 inputs)}$$

The symbol and the truth table for a NOR gate are shown in Fig. 8.9 and Table 8.5 respectively.

Table 8.5 *Truth table for NOR gate*

A	B	$Y = \overline{A + B}$
0	0	1
1	0	0
0	1	0
1	1	0

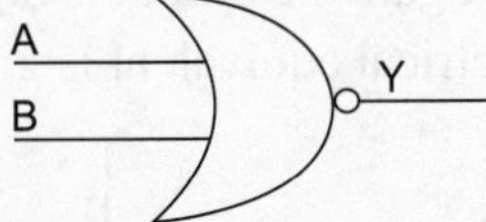

Figure 8.9 *Symbol for NOR gate*

For a NOR gate, the output is high when all of its inputs are low and it is represented by switches in parallel as shown in Fig. 8.9. The output lamp lights up when all input switches are OFF. The NOR gate is illustrated in Fig. 8.10.

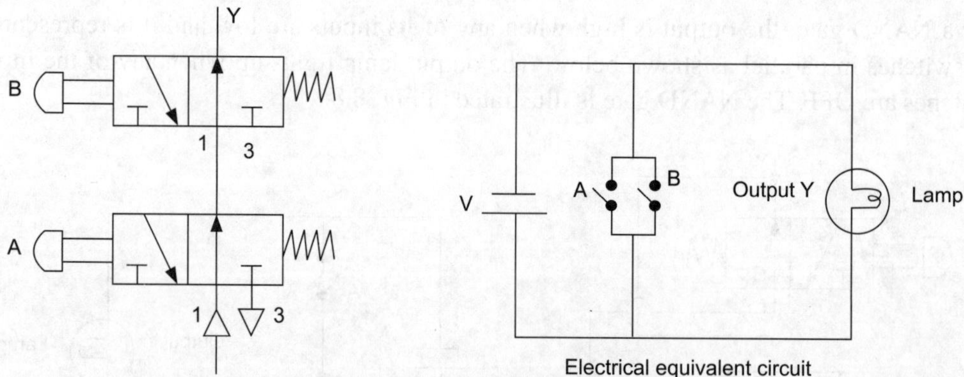

Figure 8.10 *Illustration of NOR Gate*

8.2.6 XOR Gate (Exclusive OR)

The output of an XOR gate is logic 1 when any of its inputs differ from each other and the output is logic 0 when inputs are the same. The Boolean equation for the XOR operation is:

$$Y = A \oplus B = \overline{A}B + A\overline{B} \text{ (2 inputs)}$$

The symbol and the truth table for an XOR gate are given in Fig. 8.11 and Table 8.6 respectively.

Table 8.6 *Truth table for XOR gate*

A	B	$A \oplus B$
0	0	0
1	0	1
0	1	1
1	1	0

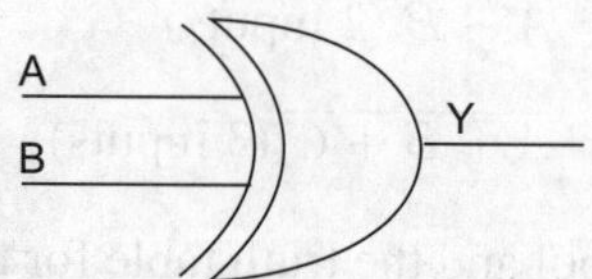

Figure 8.11 *Symbol for XOR gate*

For the XOR gate the output is high when its inputs are different and the output lamp lights up. The electrical equivalent is given in Fig. 8.12.

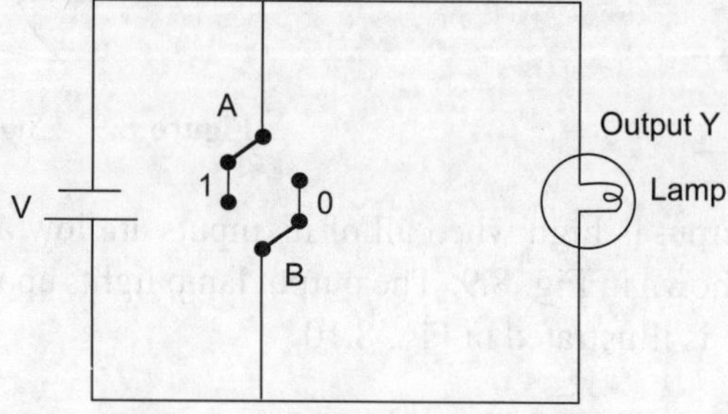

Figure 8.12 *Electrical equivalent circuits*

8.2.7 XNOR (Exclusive NOR) Gate

The output of an XNOR gate is logic 1 when the inputs are same and the output is logic 0 when the inputs are different. The Boolean equation for the XNOR operation is:

$$Y = \overline{A \oplus B} = AB + \overline{A}\,\overline{B} \text{ (2 inputs)}$$

The symbol and truth table for an XNOR gate is shown in Fig. 8.13 and Table 8.7 respectively.

Table 8.7 *Truth table of XNOR gate*

A	B	$\overline{A \oplus B}$
0	0	1
1	0	0
0	1	0
1	1	1

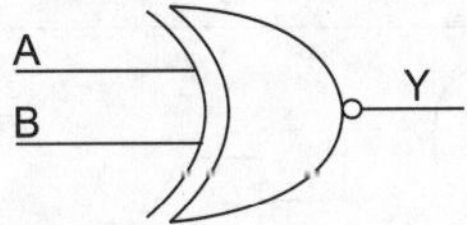

Figure 8.13 *Symbol for XNOR gate*

For the XNOR gate the output is high when its inputs are same and the output lamp lights up. The XNOR function is represented by switches in series as shown in the Fig. 8.14.

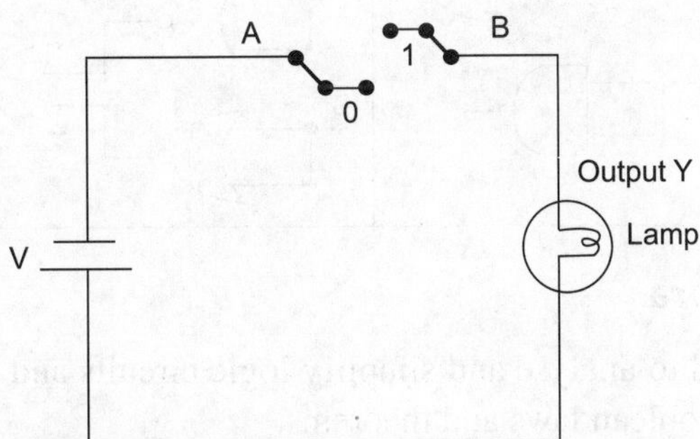

Figure 8.14 *Electrical equivalent circuits*

8.2.8 NAND and NOR Universality

When gates are used in combination with one another to form logic circuits, the technician must be able to analyse and troubleshoot them. This requires an understanding of some basic concepts employed in digital circuits, namely NAND/NOR university and alternate gate symbology. NAND and NOR gates exhibit properties that allows them to be used in combination with one another to create all the other basic logic functions. These techniques are used by designers for the optimal use of NAND or NOR gates on chips with minimal wastage or duplication. NAND and NOR universality is illustrated in Table 8.8.

Table 8.8 *NAND and NOR universality*

NAND universality	NOR universality
A, Y	A, Y
A, Y	A, B, Y, Y
A, B, Y	A, B, Y
A, B, Y	A, B, Y

8.3 Boolean Algebra

Boolean algebra is used to analyse and simplify logic circuits and express their operation mathematically using Boolean laws and theorems.

Boolean algebra is basically used to analyse a logic circuit, express its operation mathematically and simplify logic circuits by using certain Boolean laws and theorems.

a) Commutative Law

$$A + B = B + A$$
$$A \cdot B = B \cdot A$$

b) Associative Law

$$A + (B + C) = (A + B) + C$$
$$A \cdot (B \cdot C) = (A \cdot B) \cdot C$$

c) Distributive Law

$$A \cdot (B + C) = AB + AC$$
$$(A + B)(C + D) = AC + AD + BC + BD$$

8.3.1 Boolean algebra equalities or identities

Rule 1: $(A) \cdot 0 = 0$

Anything ANDed with zero is equal to zero.

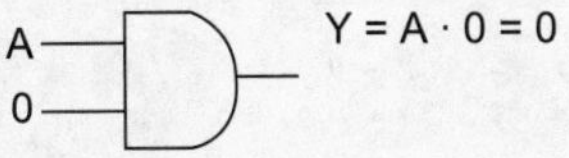

$Y = A \cdot 0 = 0$

Rule 2: $(A \cdot 1) = A$

$Y = A \cdot 1 = A$

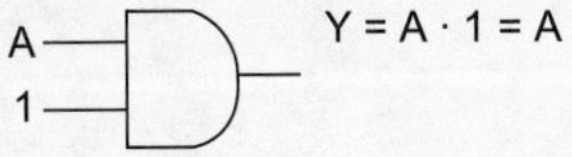

Rule 3: $(A \cdot A) = A$

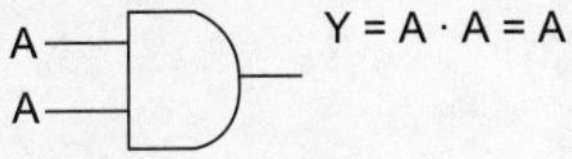

$Y = A \cdot A = A$

Rule 4: $(A \cdot \overline{A}) = 0$

$Y = \bar{A} \cdot A = 0$

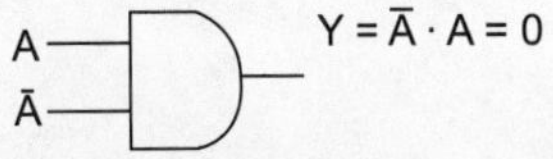

Rule 5: $(A + 0) = A$

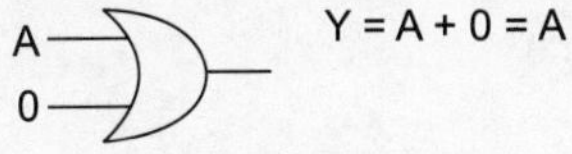

$Y = A + 0 = A$

Rule 6: $(A + 1) = 1$

$Y = A + 1 = 1$

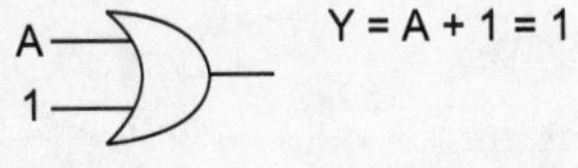

Rule 7: $(A + A) = A$

$Y = A + A = A$

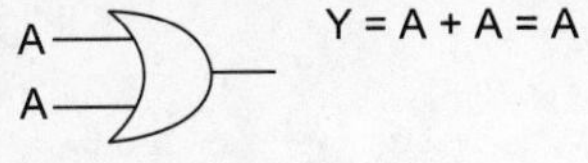

Rule 8: $A + \overline{A} = 1$

$Y = A + \bar{A} = 1$

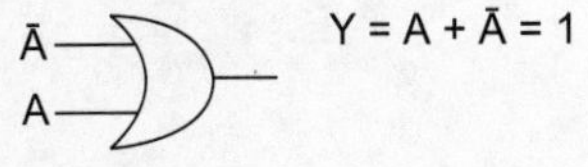

Rule 9: $\overline{\overline{A}} = A$

A — $\bar{A}$ — $\bar{\bar{A}}$ = A

Rule 10: $A + A \cdot B = A$

$$= A(1 + B)$$

$$= A(\because 1 + B = 1)$$

Rule 11: $A + \overline{A}B = A + B$

Using rule 10, we replace A

$$A + \overline{A}B = A + A \cdot B + \overline{A}B$$

Using rule 3, replace A

$A \cdot A + A \cdot B + \overline{A}B$, rewriting

$$(A \cdot A + A \cdot B + \overline{A}B) + 0$$

$$(A \cdot A + A \cdot B + \overline{A}B) + A\overline{A}$$

$$A \cdot A + A\overline{A} + A \cdot B + \overline{A}B$$

$$A(A + \overline{A} + B(A + \overline{A})$$

$$A + B$$

Rule 12: $(A + B)(A + C) = A + B \cdot C$

Expand using distributive law

$$(A \cdot A) + (A \cdot C) + (B \cdot A + (B \cdot C)$$

$$A + A \cdot C + B \cdot A + B \cdot C$$

$$A(1 + C) + B \cdot A + B \cdot C$$

$$A + B \cdot A + B \cdot C$$

$$A(1 + B) + B \cdot C$$

$$A + B \cdot C$$

8.3.2 DeMorgan's Theorem

DeMorgan's theorem is stated as follows:

Theorem 1: The complement of the product of variables is equal to sum of the complement of variables

$$\overline{A \cdot B} = \overline{A} + \overline{B}$$

Theorem 2: The complement of two ORed variables is equal to the AND of the individual complemented variables.

$$\overline{A + B} = \overline{A} \cdot \overline{B}$$

DeMorgan's theorem can be applied in three steps:

1. Complement the entire expression.
2. Change all OR functions to AND functions, and all AND functions to OR functions.
3. Complement each of the individual variables.

Example 8.1 Write the Boolean equation for the circuit in Fig. 8.15. Use DeMorgan's theorem and Boolean algebra rules to simplify the equation and draw the simplified circuit.

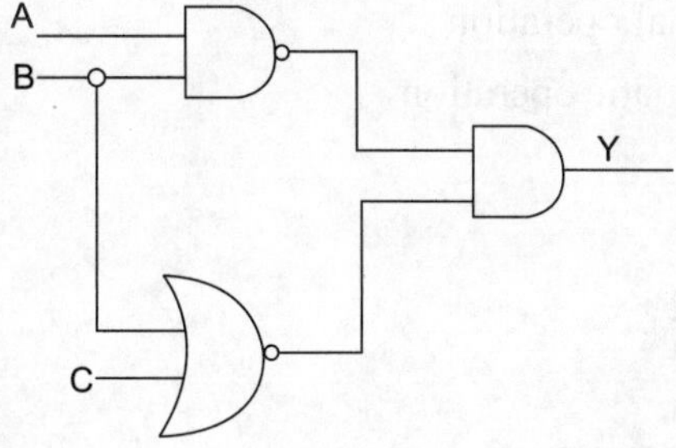

Figure 8.15

The Boolean equation is $X = \overline{AB} \cdot \overline{B + C}$

Applying DeMorgan's theorem

$$X = (\overline{A} + \overline{B}) \cdot \overline{B}\,\overline{C}$$

Notice the use of brackets to maintain proper grouping.

$$Y = \overline{A}\,\overline{B}\,\overline{C} + \overline{B}\,\overline{B}\,\overline{C}$$

$$Y = \overline{A}\,\overline{B}\,\overline{C} + \overline{B}\,\overline{C}$$

$$Y = \overline{B}\,\overline{C}\,(\overline{A} + 1)$$

$$Y = \overline{B}\,\overline{C}$$

The simplified diagram is given in Fig. 8.16.

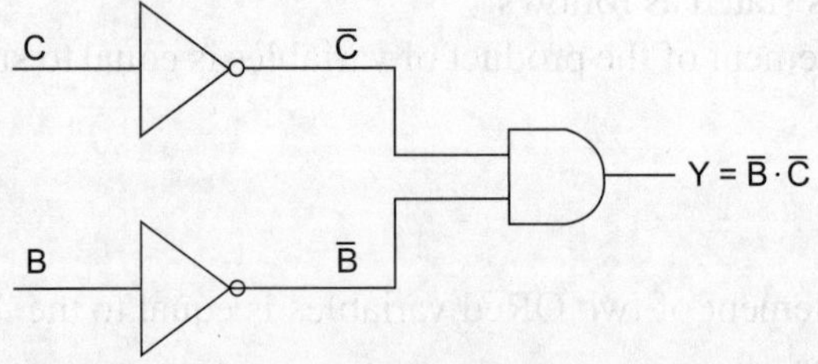

Figure 8.16

Example 8.2 The feed system of a press can be operated either manually or automatically with the following preconditions:

1. The press tool should be in position.
2. The work must be clamped.
3. No part of the operator's body should be in the working area.

Formulate the control equation.

S_1 = Switch for manual operation

S_2 = Switch for automatic operation

T = Tool

C = Clamp

H = Operator's hand

Y = Output signal

As per the pre-conditions:

$$Y = (S_1 + S_2) \cdot T \cdot C \cdot \overline{H}$$

That is, S_1 or S_2 and T and C and not H will give the output signal Y.

Inverting both sides,

$$\overline{Y} = (\overline{S1} \cdot \overline{S2}) + \overline{T} + \overline{C} + \overline{\overline{H}}$$

$$Y = \overline{(\overline{S1} \cdot \overline{S2})} \cdot T \cdot C \cdot \overline{H}$$

Using AND logic, the circuit diagram requires four input AND gates as shown in Fig. 8.17. The pneumatic diagram of this equation is shown in Fig. 8.18.

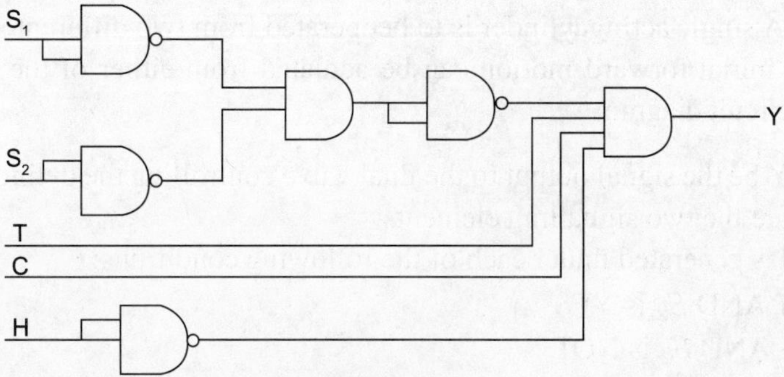

Figure 8.17

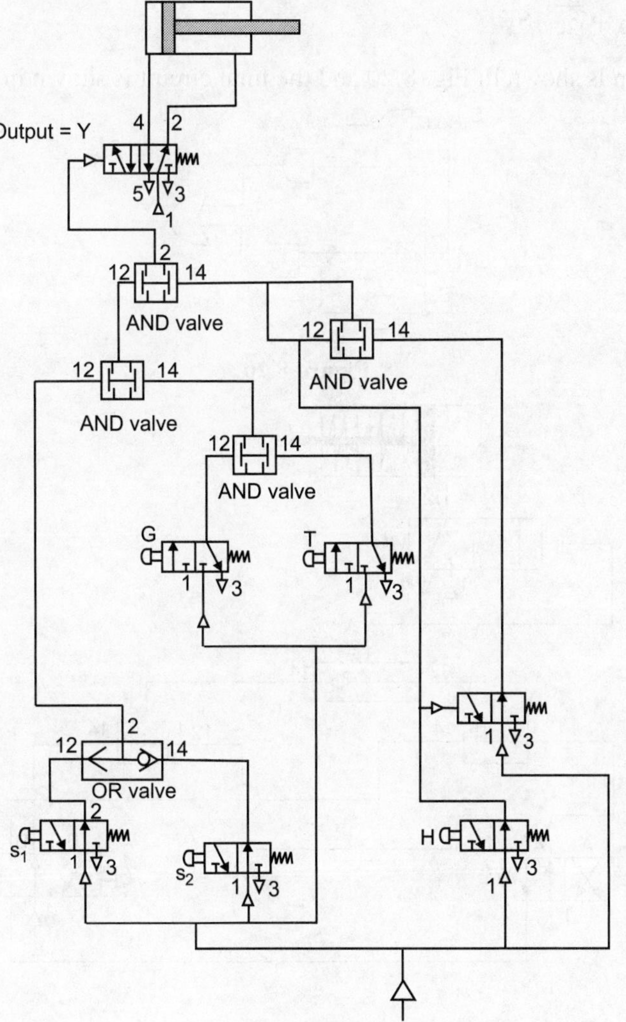

Figure 8.18

Example 8.3 A single actingcylinder is to beoperated from two different sources S_1 and S_2 such that its initial forward motion can be actuated from either of the two locations. Draw the logic circuit diagram.

Solution Let Y be the signal output to the final valve controlling the cylinder, and S_1 and S_2 (4/2 valves) are the two signalling elements.

Output signal is generated under each of the following conditions:

- S_1 is NOT AND S_2 is YES
- S_1 is YES AND S_2 is NOT.

The equation is

$$Y = \overline{S_1} \cdot S_2 + \overline{S_2} \cdot S_1$$

The logic diagram is shown in Fig. 8.20 and the final circuit is shown in Fig. 8.21.

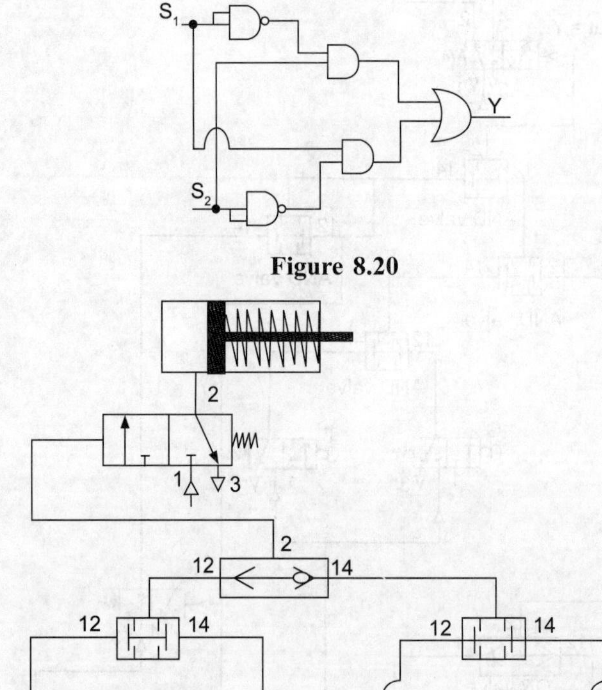

Figure 8.20

Figure 8.21

Example 8.4 For the circuit shown in Fig. 8.22, what combination of A, B, C and D results in an output of logic 1?

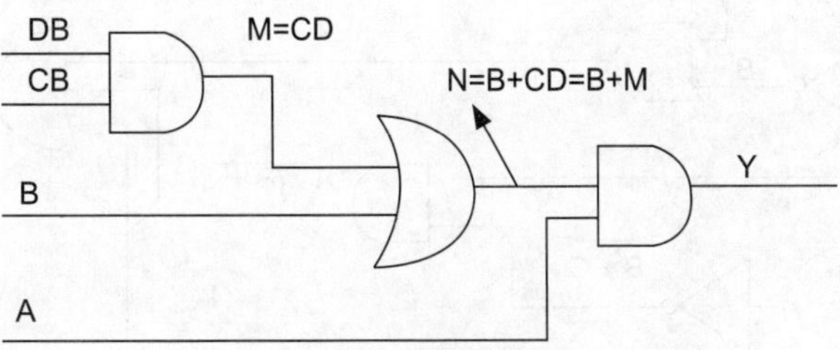

Figure 8.22

The output equation is $A \cdot (B + C \cdot D)$ and truth table is given in Table 8.9.

Table 8.9 *Truth table*

Serial No	D	C	B	A	M	N	Y
1	0	0	0	**0**	0	**0**	0
2	0	0	0	**1**	0	**0**	0
3	0	0	1	**0**	0	**1**	0
4	0	0	1	**1**	0	**1**	**1**
5	0	1	0	**0**	0	**0**	0
6	0	1	0	**1**	0	**0**	0
7	0	1	1	**0**	0	**1**	0
8	0	1	1	**1**	0	**1**	**1**
9	1	0	0	**0**	0	**0**	0
10	1	0	0	**1**	0	**0**	0
11	1	0	1	**0**	0	**1**	0
12	1	0	1	**1**	0	**1**	**1**
13	1	1	0	**0**	1	**1**	0
14	1	1	0	**1**	1	**1**	**1**
15	1	1	1	**0**	1	**1**	0
16	1	1	1	**1**	1	**1**	**1**

Note 1: CD = M is logic1 for 4 combinations
Note 2: N = B+M is logic 1 for 10 combinations
Note 3: Y = N+A is logic 1 for 5 combinations

Example 8.5 Simply the following Fig. 8.23 and draw the circuit.

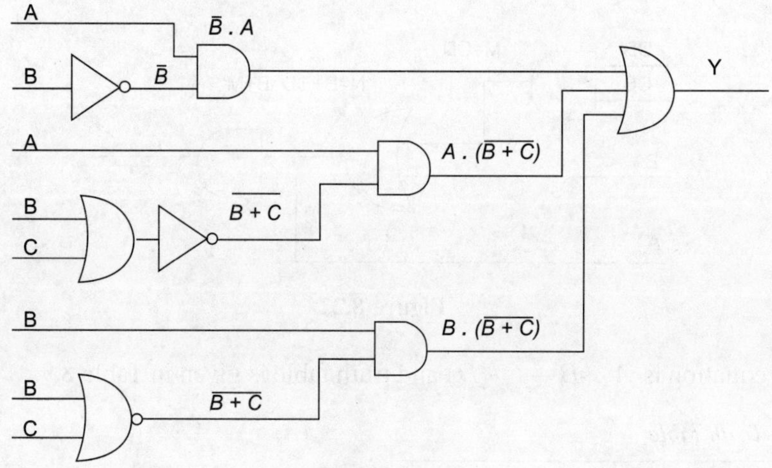

Figure 8.23

$$Y = A\overline{B} + A \cdot (\overline{B + C}) + B(\overline{B + C})$$

Using DeMorgan's theorem,

$$Y = A\overline{B} + A \cdot (\overline{B} \cdot \overline{C}) + B(\overline{B} \cdot \overline{C})$$

$$Y = A\overline{B} + A \cdot (\overline{B} \cdot \overline{C}) + 0 \cdot \overline{C} \quad [\because \; B\overline{B} = 0]$$

$$Y = A\overline{B}(1 + \overline{C}) + 0 \quad [\because \; (1 + \overline{C}) = 1]$$

$$Y = A\overline{B}$$

The simplified equation is given in Fig. 8.24.

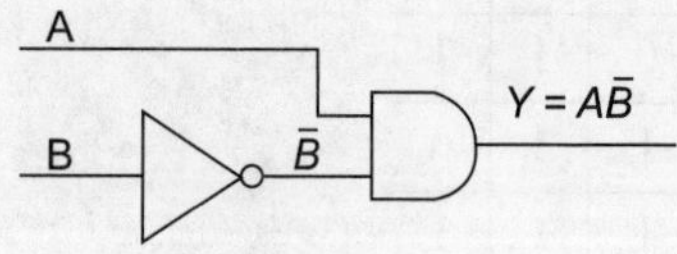

Figure 8.24

Example 8.6 Draw the truth table for $Y = \overline{A}B + A\overline{B}$

Solution Y is high when A and B are different which is equivalent to an exclusive OR. Truth table is given in Table 8.10

Table 8.10 *Truth table*

A	B	Y
0	0	0
0	1	1
1	0	1
1	1	0

Example 8.7 Draw the truth table for $Y = \overline{\overline{A}B + A\overline{B}}$

$$Y = \overline{(\overline{A}B)} \cdot \overline{(A\overline{B})}$$

$$Y = (\overline{A} + \overline{\overline{B}}) \cdot (\overline{\overline{A}} + \overline{B})$$

$$Y = (\overline{A} + \overline{\overline{B}}) \cdot (\overline{\overline{A}} + \overline{B})$$

$$Y = \overline{A}A + \overline{B}\,\overline{A} + BA + B\overline{B}$$

$$Y = 0 + \overline{B}\,\overline{A} + BA + 0$$

$$Y = \overline{B}\,\overline{A} + BA$$

This is equivalent to anexclusive NOR and the truth table is given in Table 8.11.

Table 8.11 *Truth table*

A	B	Y
0	0	1
0	1	0
1	0	0
1	1	1

8.3.3 Sum of Products (SOP)

Sum of products is an expression in which an addition operator is applied to two or more sub logic multiplication operator. For example: scalar addition and multiplication of two expressions like $AB + CD$. In general, SUM OF PRODUCTS = (PRODUCT) + (PRODUCT) + (PRODUCT)......

Example 8.8 Simplify the following using SOP.

$$Y = \overline{A}B\overline{C} + (\overline{A} + \overline{B}) \cdot (B + \overline{C} + \overline{AB})$$

$$Y = \overline{A}B\overline{C} + (\overline{A} + \overline{B}) \cdot (B + \overline{C} + \overline{A} + \overline{B})$$

$$Y = \overline{A}\,B\overline{C} + \overline{A}B + \overline{A}\,\overline{C} + \overline{A}\,\overline{A} + \overline{A}\,\overline{B} + \overline{B}B + \overline{B}\,\overline{C} + \overline{A}\,\overline{B} + \overline{B}\,\overline{B}$$

$$Y = \overline{A}B\overline{C} + \overline{A}B + \overline{A}\,\overline{C} + \overline{A} + \overline{A}\,\overline{B} + \overline{B}\,\overline{C} + \overline{A}\,\overline{B} + \overline{B} \quad [\because\ \overline{A}\,\overline{A} = \overline{A} \text{ and } \overline{B}B = 0]$$

$$Y = \overline{A}B(\overline{C} + 1) + \overline{A}(1 + \overline{C}) + \overline{A}\,\overline{B} + \overline{B}\,\overline{C} + \overline{B}(\overline{A} + 1) \quad [\because\ 1 + \overline{C} = 1]$$

$$Y = \overline{A}B + \overline{A} + \overline{A}\,\overline{B} + \overline{B}\,\overline{C} + \overline{B}$$

$$Y = \overline{A}B + \overline{A}(1 + \overline{B}) + \overline{B}(\overline{C} + 1)$$

$$Y = \overline{A}B + \overline{A} + \overline{B}$$

$$Y = \overline{A}(B + 1) + \overline{B}$$

$$Y = \overline{A} + \overline{B}$$

$$Y = \overline{A \cdot B}$$

Example 8.9 Simplify the following:

$$Y = A\overline{B}C + \overline{A}\,\overline{B} + AB\overline{C}D$$

Note that in first term there is no D and in second term there are no C and D.

Rewriting the first term:

$$A\overline{B}\,C = A\overline{B}\,C \cdot 1 = A\overline{B}\,C \cdot (D + \overline{D}) = A\overline{B}\,CD + A\overline{B}\,C\overline{D}$$

Rewriting the second term:

$$\overline{A}\,\overline{B} = \overline{A}\,\overline{B} \cdot 1.1 = \overline{A}\,\overline{B}(C + \overline{C}) \cdot (D + \overline{D})$$

$$\overline{A}\,\overline{B} = \overline{A}\,\overline{B}(C + \overline{C}) \cdot (D + \overline{D})$$

$$= \overline{A}\,\overline{B}\,CD + \overline{A}\,\overline{B}\,C\overline{D} + \overline{A}\,\overline{B}\,\overline{C}D + \overline{A}\,\overline{B}\,\overline{C}\,\overline{D}$$

Rewriting all terms:

$$Y = A\overline{B}\,CD + A\overline{B}\,C\overline{D} + \overline{A}\,\overline{B}\,CD + \overline{A}\,\overline{B}\,C\overline{D} + \overline{A}\,\overline{B}\,\overline{C}\,D + \overline{A}\,\overline{B}\,\overline{C}\,\overline{D} + AB\,\overline{C}D$$

Each product is equal to logic 1 for only one possible combination of input valves. This combination is listed in Table 8.12.

Table 8.12

Terms	Combination of ABC and D which gives "1"
$A\overline{B}\,CD$	1011
$A\overline{B}\,C\overline{D}$	1010
$\overline{A}\,\overline{B}\,CD$	0011
$\overline{A}\,\overline{B}\,C\overline{D}$	0010
$\overline{A}\,\overline{B}\,\overline{C}D$	0001
$\overline{A}\,\overline{B}\,\overline{C}\,\overline{D}$	0000
$AB\,\overline{C}D$	1101

$$Y = A\overline{B}\,CD + A\overline{B}\,C\overline{D} + \overline{A}\,\overline{B}CD + \overline{A}\,\overline{B}C\overline{D} + \overline{A}\,\overline{B}\,\overline{C}D + \overline{A}\,\overline{B}\,\overline{C}\,\overline{D} + AB\,\overline{C}D$$

$$Y = 1011 + 1010 + 0011 + 0010 + 0001 + 0000 + 1101$$

8.3.4 Product of Sums (POS)

Products of sums is an expression in which a multiplication operator is applied to two or more sub logic addition operator. For example: scalar addition and multiplication of two expressions like (A+B).(C+D). In general

PRODUCTS OF SUMS = (SUMS) × (SUMS) × (SUMS)......

Note: All the variables should be in the standard form.

Example 8.10 Simplify the following using POS

$$Y = (A + (\overline{B}) + C) \cdot (A + B + \overline{D}) \cdot (A + \overline{B} + \overline{C} + D)$$

Note that in first term there is no D and in second term there is no C.

Rewriting the first term:

$$Y = (A + \overline{B} + C) = (A + \overline{B} + C + 0) = (A + \overline{B} + C + D \cdot \overline{D})$$

$$Y = (A + \overline{B} + C + D \cdot \overline{D})$$

Using the property of Boolean Algebra,

$$M + NP = (M + N)(M + P)$$

Where $M = A + \overline{B} + C$

$$N = D$$

$$P = \overline{D}$$

Therefore, $Y = A + \overline{B} + C + D \cdot \overline{D} = (A + \overline{B} + C + D) \cdot (A + \overline{B} + C + \overline{D})$

We shall rewrite second term as

$$Y = (A + B + \overline{D}) = A + B + \overline{D} + 0$$

$$Y = (A + B + \overline{D}) = A + B + \overline{D} + C \cdot \overline{C}$$

Using the property of Boolean Algebra,

$$M + NP = (M + N)(M + P)$$

Where $M = A + B + \overline{D}$

$$N = C$$

$$P = \overline{C}$$

Therefore, $Y = A + B + \overline{D} + C \cdot \overline{C} = (A + B + \overline{D} + C) \cdot (A + B + \overline{D} + \overline{C})$

Rewriting all terms,

$$Y = (A + (\overline{B}) + C) \cdot (A + B + \overline{D}) \cdot (A + \overline{B} + \overline{C} + D)$$

$$Y = (A{+}\overline{B}{+}C{+}D)\cdot(A{+}\overline{B}{+}C{+}\overline{D})\cdot(A{+}B{+}\overline{D}{+}C)\cdot(A{+}B{+}\overline{D}{+}\overline{C})\cdot(A{+}\overline{B}{+}\overline{C}{+}D)$$

We can find all possible combination of binary values, whereby that particular sum is equal to zero. Basically, each sum expression is only equal to zero for one combination of values of binary variables. Simplifying,

$$Y = (0100) \cdot (0101) \cdot (0010) \cdot (0011) \cdot (0110)$$

8.3.5 Analysis Using SOP and POS and a Truth Table

Example 8.11 For the following truth table (Table 8.13), write the SOP equation.

Table 8.13

A	B	C	Y
0	0	0	0
0	0	1	0
0	**1**	**0**	**1**
0	1	1	0
1	0	0	0
1	**0**	**1**	**1**
1	1	0	0
1	1	1	0

Sum of products is given by:

$$Y = \overline{A}B\overline{C} + A\overline{B}C' = 010 + 101 = \text{LOGIC } 1$$

Example 8.12 For the following truth table (Table 8.14), write the SOP equation and draw the circuit diagram.

Table 8.14

A	B	C	Y
0	0	0	0
0	0	1	0
0	1	0	0
0	**1**	**1**	**1**
1	**0**	**0**	**1**
1	0	1	0
1	**1**	**0**	**1**
1	**1**	**1**	**1**

Sum of product expression is

$$Y = \overline{A}BC + A\overline{B}\,\overline{C} + AB\overline{C} + ABC = 011 + 100 + 110 + 111 = \text{LOGIC 1}$$

The circuit diagram is shown in Fig. 8.25.

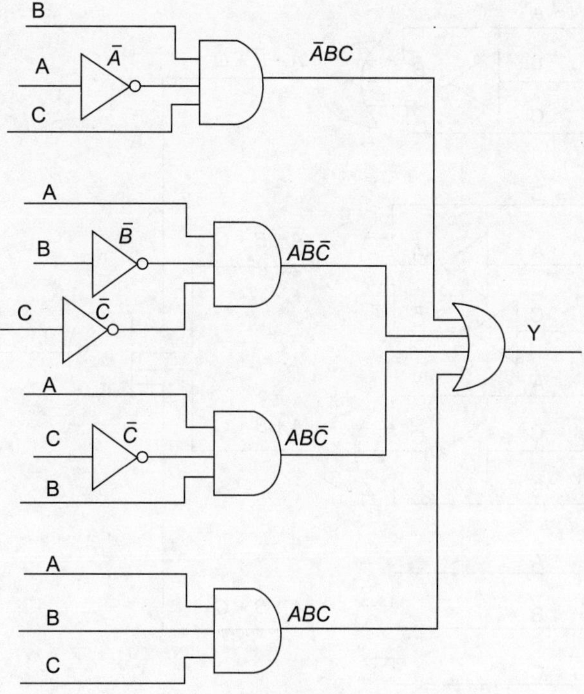

Figure 8.25

Example 8.13 For the following truth table (Table 8.15), write the POS equation and draw the circuit diagram.

Table 8.15

A	B	C	Y
0	**0**	**0**	**0**
0	**0**	**1**	**0**
0	**1**	**0**	**0**
0	1	1	1
1	0	0	1
1	**0**	**1**	**0**
1	1	0	1
1	1	1	1

$$Y = (000) \cdot (001) \cdot (010) \cdot (101) = \text{LOGIC } 0.$$

$$Y = (A + B + C) \cdot (A + B + \overline{C}) \cdot (A + \overline{B} + C) \cdot (\overline{A} + B + \overline{C})$$

The circuit diagram is shown in Fig. 8.26.

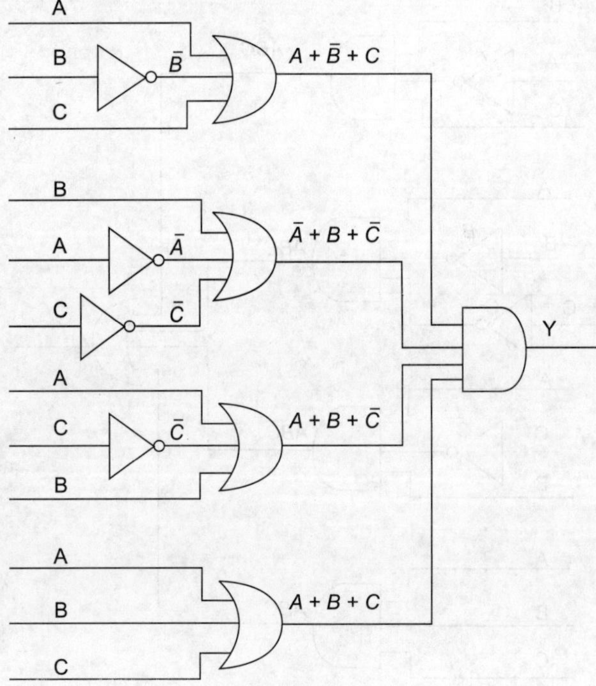

Figure 8.26

8.4 General Analysis of Karnaugh–Veitch Maps

In this method, the Boolean expressions are transferred from a truth table form. Each cell position in truth table represents combination of input conditions, while each cell value represents the corresponding output value. These terms are expression in either SOP or POS to get a minimal Boolean expression which gives the required logic.

A sum-of-products expression can always be implemented using AND gates feeding an OR gate, and a product-of-sums expression leads to OR gates feeding an AND gate.

Examples of 1- variable KV map, 2-variable KV map and 3-variable KV are given below.

2-variable KV map

A↓ B→	**0**	**1**
0	00	01
1	10	11

3-variable KV map

AB↓ C→	**0**	**1**
00	000	001
01	010	011
11	110	111
10	100	101

4-variable KV map

AB↓ CD→	**00**	**01**	**11**	**10**
00	0000	0001	0011	0010
01	0100	0101	0111	0110
11	1100	1101	1111	1110
10	1000	1001	1011	1010

Each cell value is the intersecting values of the rows and columns.

8.4.1 Condition of Adjacency

Consider the four-variable KV map shown in Table 8.16.

Table 8.16 *Four variable KV*

AB↓ CD→	**00**	**01**	**11**	**10**
00				
01		0101		
11	1100	**1101**	1111	
10		1001		

As seen in the table, the adjacent elements for 1101 are 0101 (north), 1111 (east), 1100 (west) and1001 (south). Note that for each adjacent element of 1101, there is only one variable that changes in the pattern. For example, with the north element **0**101 the only change is in the first digit. Similarly, the rest of the adjacent elements only differ in one digits (if different positions).

In the example in Table 8.17, the central element 0001 has adjacent elements of 1001(north), 0101(south), 0000 (west) and 0011(east). Notice that there is only one digit that changes for each of the adjacent elements when compared to 0001.

Table 8.17

AB↓ CD→	**00**	**01**	**11**	**10**
00	0000	**0001**	0011	
01		0101		
11				
10		1001		

Similarly in Table 8.18, element 0110 has adjacent elements of 0010, 1110, 0111 and 0100, each only differing by one digit.

Table 8.18

AB↓ CD→	**00**	**01**	**11**	**10**
00				0010
01	0100		0111	**0110**
11				1110
10				

Example 8.14 Draw the KV map for the following equation:

$$Y = \overline{A}BC + A\overline{B}C + A\overline{B}\,\overline{C}.$$

Step 1: Check for standard form; it is standard SOP form because each product contains A, B and C.

Step 2: Make the truth table for the three-variable KV map.

$$Y = \overline{A}BC + A\overline{B}C + A\overline{B}\,\overline{C}$$

$$Y = \overline{A}BC + A\overline{B}C + A\overline{B}\,\overline{C} = 011 + 101 + 100$$

This is illustrated in Table 8.19.

Table 8.19

AB↓ C→	0	1
00	000	001
01	010	(011)
11	110	111
10	(100)	(101)

Example 8.15 Draw the KV map for the following equation:

$$Y = \overline{A}BC\overline{D} + ABC\overline{D} + AB\overline{C}\,\overline{D} + ABCD,$$

Step 1 Check whether equation is the standard form of SOP. It is, since all variables are present in each product.

Step 2 Make the truth table for a four-variable KV map.

$$Y = 0110 + 1110 + 1100 + 1111$$

This is illustrated in Table 8.20 and Table 8.21.

Table 8.20

AB↓ CD→	00	01	11	10
00				
01				(0110)
11	(1100)		(1111)	(1110)
10				

Table 8.21

AB↓ CD→	00	01	11	10
00				
01				(1)
11	(1)		(1)	(1)
10				

Example 8.16 Draw the KV map for the following equation:

$$Y = BC + \overline{A}\,\overline{C}$$

Step 1: Check whether equation is the standard form of SOP. The first product term is missing variable A,and the second product term is missing variable B. Therefore this equation is not the standard form of SOP.

Step 2: Convert the equation to standard SOP form using Boolean algebra rules.

$$Y = BC + \overline{A}\,\overline{C}$$

$$Y = 1.BC + 1.\overline{A}\,\overline{C}$$

$$Y = (A + \overline{A}) \cdot BC + (B + \overline{B}) \cdot \overline{A}\,\overline{C}$$

$$Y = (ABC + \overline{A}BC) \cdot +(B\overline{A}\,\overline{C} + \overline{A}\,\overline{B}\,\overline{C})$$

$$Y = ABC + \overline{A}BC + \overline{A}B\overline{C} + \overline{A}\,\overline{B}\,\overline{C}$$

$$Y = (111) + (011) + (010) + (000)$$

$$Y = (111) + (011) + (010) + (000)$$

Step 3: Make the truth table for a 4-variable KV map. See tables below.

Table 8.22

AB↓ C→	0	1
00	(000)	001
01	(010)	(011)
11	110	(111)
10	100	101

Table 8.23

AB↓ C→	0	1
00	(1)	
01	(1)	(1)
11		(1)
10		

Example 8.17 Draw the KV map for the following equation:

$$Y = A + \overline{C}D + AC\overline{D} + \overline{A}BC\overline{D}$$

Step 1: Check whether the equation is the standard form of SOP. The first product term is missing variables B, C and D. The second product term is missing variables A and B. The third product term is missing variable B. Therefore this equation is not the standard form of SOP.

Step 2 Convert the given equation to the standard form of SOP.

$$Y = A + \overline{C}D + AC\overline{D} + \overline{A}BC\overline{D}$$

The first term contains only A and therefore not in standard form of SOP. The possible combinations of the first term are as follows. A is 1 and the possible combinations are listed in Table 8.24.

Table 8.24

1BCD
1000
1001
1010
1011
1100
1101
1110
1111

Table 8.25

$AB\overline{C}D$
1001
0001
1101
0101

Table 8.26

$ABC\overline{D}$
1010
1110

The second term $(\overline{C}D)$ again is not in standard form of SOP and contains only $\overline{C}$ and D. The Values of $\overline{C}$ and D are 0 and 1 (fixed) The possible combinations of A are 0 and 1. The possible values of B are 0 and 1. Truth table for second term is given in Table 8.25.

The possible combinations of the third term are as follows. Values of A, $C\overline{D}$ are 1, 1 and 0 (fixed). The possible values of B are 0 and 1 (Table 8.26).

The final KV map is given in Table 8.27.

Table 8.27a

AB↓ CD→	00	01	11	10
00		**0001**		
01		**0101**		**0110**
11	**1100**	**1101**	**1111**	**1110**
10	**1000**	**1001**	**1011**	**1010**

Table 8.27b

AB↓ CD→	00	01	11	10
00		1		
01		1		1
11	1	1	1	1
10	1	1	1	1

Grouping of 1s:

Rule 1: Maximise the size of the group and minimisethe number of groups. Size must an exponent of 2 (e.g. 2, 4,8,16 and 32). For a 3-variable KV map,it means a maximum of 8 cells and for a 4-variable KV map it is a maximum of 16 cells.

Rule 2: Each cell must be adjacent to one or more cells, but not all cells should be adjacent to each other.

Rule 3: Always use the largest possible group.

Rule 4: Each one must be group and can overlap. An overlapping group includes non-common ones. Correct logic statement (see Table 8.29 for instance)

Example 8.18 Draw the KV map and groups for the following expression:

$$Y = BC + \overline{A}\,\overline{C}$$

Step 1: Check whether equation is the standard form of SOP.
Step 2: Make the truth table for a 4-variable KV map.

$$Y = 1.BC + 1.\overline{A}\,\overline{C}$$

$$Y = (A + \overline{A}) \cdot BC + (B + \overline{B}) \cdot \overline{A}\,\overline{C}$$

$$Y = (ABC + \overline{A}BC) \cdot +(B\overline{A}\,\overline{C} + \overline{A}\,\overline{B}\,\overline{C})$$

$$Y = ABC + \overline{A}BC + \overline{A}B\overline{C} + \overline{A}\,\overline{B}\,\overline{C}$$

$$Y = (111) + (011) + (010) + (000)$$

$$Y = (111) + (011) + (010) + (000)$$

The final KV map is given in Table 8.28 and Table 8.29.

Table 8.28

AB↓ C→	0	1
00	000	001
01	010	011
11	110	111
10	100	101

Table 8.29

AB↓ C→	0	1
00	1	
01	1	1
11		1
10		

Example 8.19 Group the following KV map (Table 8.30).

Table 8.30

AB↓ C→	0	1
00	1	1
01	1	
11		1
10	1	1

Example 8.20 Group the following KV map (Table 8.31).

Table 8.31

AB↓ CD→	00	01	11	10
00	1			1
01	1	1		1
11	1	1		1
10	1		1	1

Grouping of 4

Grouping of 2

8.4.2 Logic Circuit Design with Karnaugh–Veitch Mapping Method

While the classic algebraic method of simplifying binary functions is useful for simple cases, it becomes time consuming for complex functions. So a graphical design method using Karnaugh–Veitch maps is used to simplify intricate control problems and produce the simplest possible result. This method is generally much faster than the algebraic one.

The KV map can contain a large number of control variables in their algebraic and binary logic modes. The diagram can be used to form a number of corresponding Boolean equations which are mapped together and minimised by applying the rules of set algebra. This results in the final control equation in its simplest form.

Steps involved in the design procedure for logic sequential circuits:

1. Make a statement of the sequence.
2. Draw a transverse-time diagram including all the cylinders in the sequence.
3. Prepare a signal flow diagram.
4. Plot a signal flow path.
5. Prepare a KV map for every memory (DCV) and for every auxiliary memory appearing in the signal flow diagram.
6. Extend the presence of memory designation by in-filling and form loops.
7. Extract minimised memory set and reset the equation.
8. Construct the circuit using pneumatic, hydraulic or logic symbols.

Control problem 1 Design a pneumatic circuit for sequence A+B+B−A− using the KV mapping method.

Step 1 Make a statement of sequence This problem involves an automatic drilling machine with cylinders A and B. Cylinder A extends to clamp the workpiece, then cylinder B extends to drill a hole and retracts. Cylinder A then unclamps the workpiece and retracts.

Step 2 Draw a traverse–time diagram Label the diagram according to the following rules. For each time unit, list only those memory designations which are required in order for the particular cylinder to move as desired.

Step 3 Prepare a signal flow diagram A KV map is required to accommodate the signal flow path. Horizontal lines on the map represent all the possible position sequence valve combinations. Hence their number is determined by the definition (lines = 2^n).

where n is the number of cylinders. The vertical columns represent auxiliary memory outlets and their number is defined by the signal flow path.

The KV map has 4 horizontal lines for the two cylinders. Through a step-by-step transfer, all the position sequence valve combinationsare transferred into the prepared KV map to get the signal flow diagram. Each square in the KV map is termed as a field. This field represents a memory designation;(A1 B1 A0 and B0) for example field a0b0 is responsible for ordering the cylinder A to move forward to the memory designation A1.

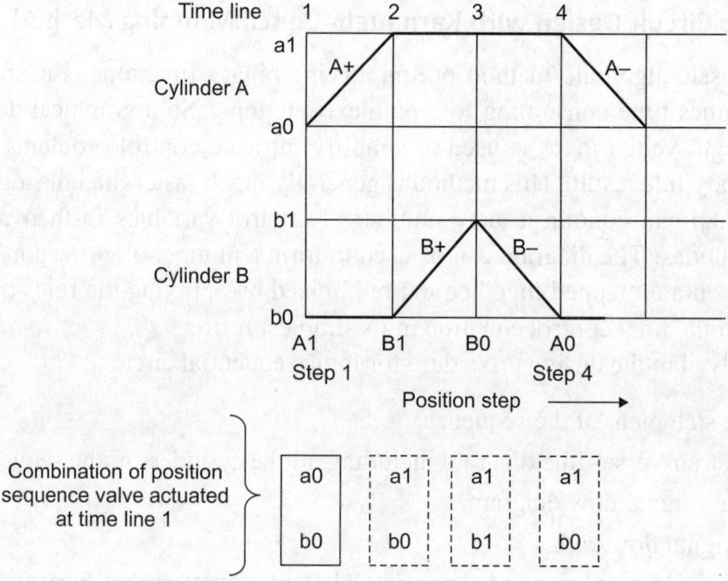

Figure 8.27 *Transverse-time diagram*

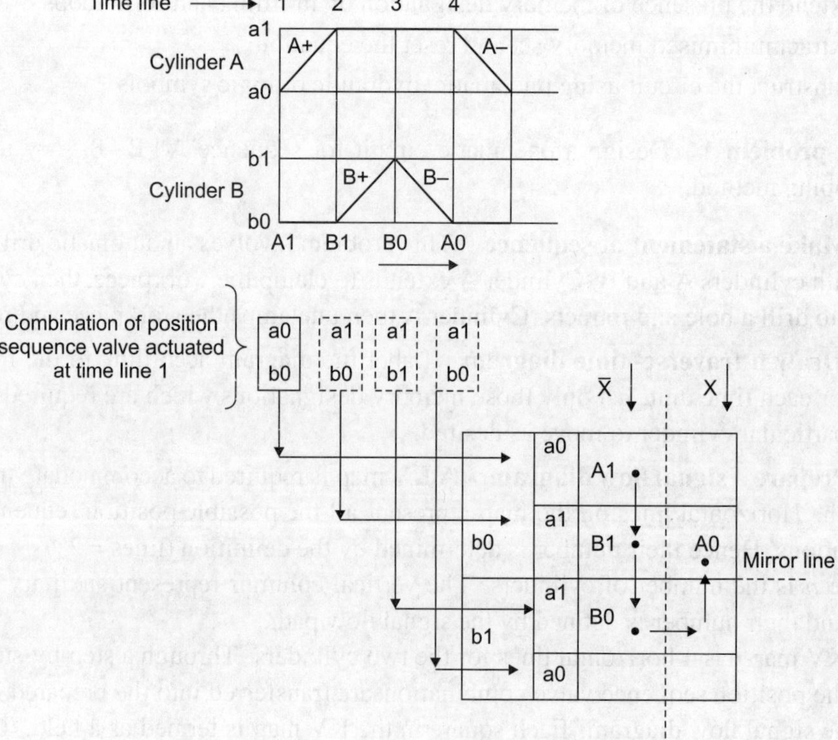

Figure 8.28

Step 4 Plot a signal flow diagram

Rule 1: A field in the signal flow diagram is occupied onlyonce. If a memory designation (A1, A0, B1, etc.) falls into an already occupied field, the map must be enlarged to have twice the number of current columns. The new columns represent the auxiliary memory which is needed to differentiate between the identical position sequence valve combinations. The variables X (logic 1) and $\overline{X}$ (logic 0) are called secondary variables.

Rule 2: Memory designations which fall into the already occupied field (A0) are moved into the new column along with the preceding memory designation (B0). The now vacant field is used to set the required auxiliary memory X with its set designation X1.

Rule 3: The set designations of auxiliary memories are plotted in a bar column and conversely the reset designations in a non-bar column.

Rule 4: Prior to the simultaneous movement of two or more cylinders, an auxiliary memory must be set or reset. This avoids premature cancellations of primary variable combinations, which could be associated with the moving cylinders.

Rule 5: Where a memory designation is immediately succeeded by its complement, an auxiliary memory must be set or reset to hold the opposing signal commands at least one step apart.

Rule 6: Auxiliary memories can be used for the purpose of discrimination as well as to avoid premature signal cancellation, and once an auxiliary memory has served its purpose, it may be reset to serve another function.

Logic equations now can be extracted directly from the signal flow diagram. Signal B1 for example, orders cylinder B to extend and is located in the field $b0\ a1\ \overline{X}$. Hence $B1 = b0\ a1\ \overline{X}$ is an equation which can properly control the extension of cylinder B. Therefore, the equation for problem will read:

$$A1 = b0\ a0\ \overline{X}$$

$$A0 = b0\ a1\ X$$

$$B1 = b0\ a1\ \overline{X}$$

$$B0 = b1\ a1\ X$$

$$X1 = b1\ a1\ \overline{X}$$

$$X0 = b0\ a0\ X$$

If these logic equations seem unduly complex for the relatively simple task performed, it is because they are indeed more complex than necessary. The KV mapping method lends itself to minimising these equations by using the looping method, which is illustrated in the following section (Fig. 8.29 and Fig. 8.30)

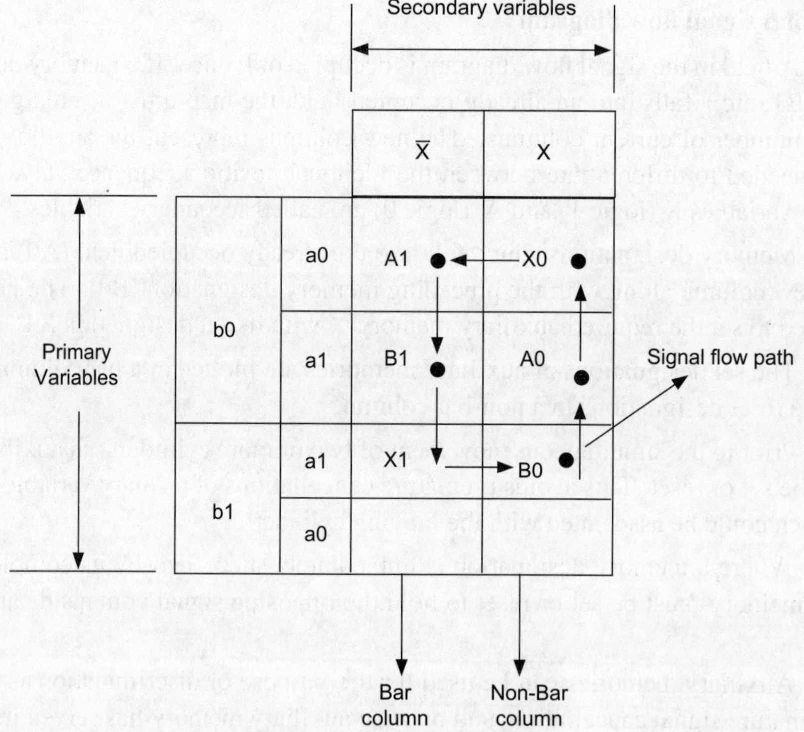

Figure 8.29 *Signal flow path*

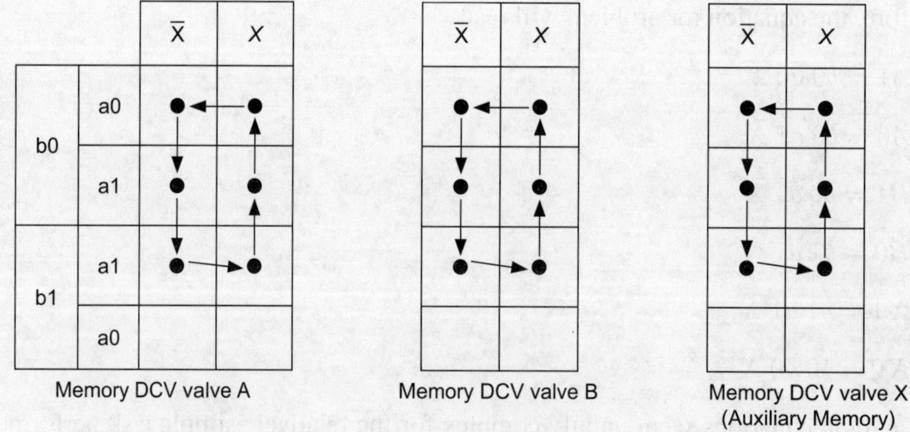

Figure 8.30 *Signal flow path for each memory valve*

8.4.2.1 *Minimisation of logic equation*

Draw a KV map for each memory component appearing in the signal flow diagram. Insert the signal flow path from the figure into every map (no designations are required at this stage).

Start with the map for memory A. Label the signals $A1$ and $A0$ (for some circuits these signals could appear several times if the cylinder has repetitive movements within the same cycles).

Extend the presence of the signal as far as possible along the signal flow path until it disappears, to permit counter signaling by $A0$. This has to occur in the direction of the arrows. Shade all the circles where $A1$ is present, and proceed with contrast shading made it clear in the following figure for AO (Fig. 8.31).

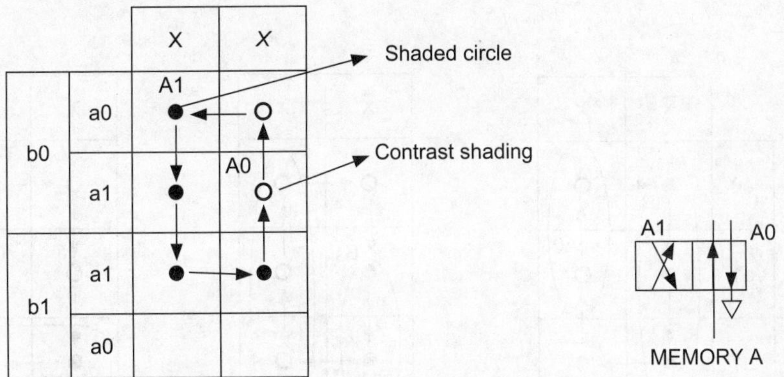

Figure 8.31 *Signal flow path for memory A*

Continue in a similar manner with the maps for memory B and the auxiliary memory X. Draw a loop around the circles to indicate the presence of signal $A1$.

8.4.2.2 *Rules for looping*

1. Unoccupied fields such as $b1\ a0\ \overline{X}$ or b1a0X may be included in the loop, if they are relevant to implementing the rules explained in this section.
2. All circles of the same colour or shade need not be included in the same loop. However, the circle labeled $A1$ must be included.
3. Circles of same shade not incorporated in the loop are meaningless and do not contribute to equation.
4. The number of fields encompassed by the loop must beaexponent of 2 (i.e. 2, 4, 8, 16, 32).
5. Draw the loop as large as possible to include the maximum number of fields. This produces shorter logic equations which means fewer AND/OR elements.
6. The loop must be symmetrical about the mirror lines, if it extends beyond them.

7. Loops with respect to mirror line must be in symmetrical shapes (i.e. no U shape, sentence added to make it clear L shape or other unsymmetrical shapes).
8. Separate loops with the same designation which cover each other when the map is folded along a mirror line may be treated as one loop. 9. Individual self-contained loops dealing with identical memory designations are linked in the final equation with an OR sign(+).

8.4.2.3 *Looping of memory designation after simultaneous actuation*

Where two or more cylinders are actuated simultaneously, the memory designation in the flow-path immediately thereafter must contain all the primary variables actuated by all the cylinders in its final equation. This also applies to cases where this designation appears several times along the flow path and can be encompassed into one common loop (Fig. 8.32).

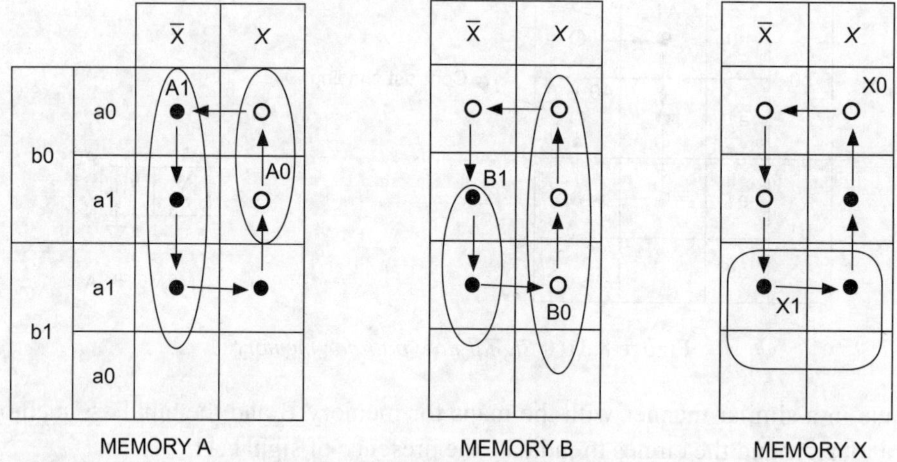

Figure 8.32 *Looping each memory valve*

Next, loop all other memory designation. If the circuits are very complex it is advisable to draw a separate map for each memory designation. Variables that change their binary value (0 to 1) within the loop render each other valueless and drop out of the minimised equation. Therefore the derived equation reads $A1 = \overline{X}$. If single cycling is selected for this circuit, then a start signal to prevent continuous cycling must be included. The final equation now reads

$A1 = \overline{X}$. start (Fig. 8.33, Fig. 8.34 and Fig. 8.35)

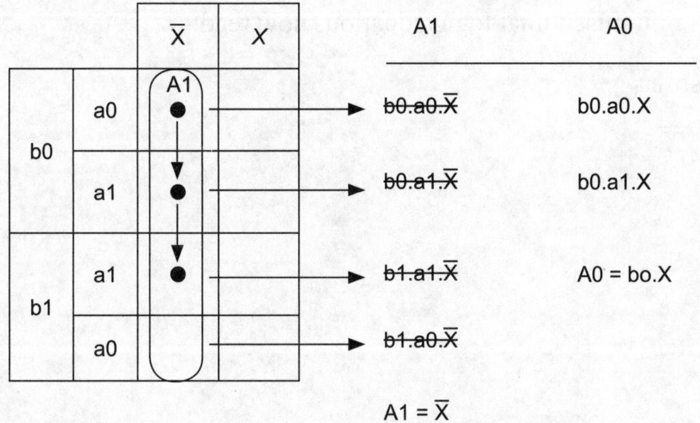

Figure 8.33 *Minimisation of equation for valve A*

$\overline{X}$ | X

B1 | B0

~~b0.a0.X~~

~~b0.a0.$\overline{X}$~~

~~b0.a1.X~~

~~b0.a1.$\overline{X}$~~

~~b1.a1.X~~

B1 = a1.$\overline{X}$

~~b1.a0.X~~

B0

B0 = X

Figure 8.34 *Minimisation of equation for valve B*

$\overline{X}$ | X

X1 | X0

~~b0.a0.X~~

X0

b1.a1.$\overline{X}$

~~b0.a0.X~~

b1.a1.X

~~b1.a0.$\overline{X}$~~

b1.a0.$\overline{X}$

~~b1.a0.X~~

X1

b1.a0.$\overline{X}$

X0 = a0

X1 = b1

Figure 8.35 *Minimisation of equation for valve X*

The derived and minimised final logic equations now read:

$$A1 = \overline{X}.start$$

$$B1 = a1 \cdot \overline{X}$$

$$X1 = b1$$

$$A0 = bo \cdot X$$

$$Bo = X$$

$$X0 = ao$$

The final circuit diagram is shown in Fig. 8.36.

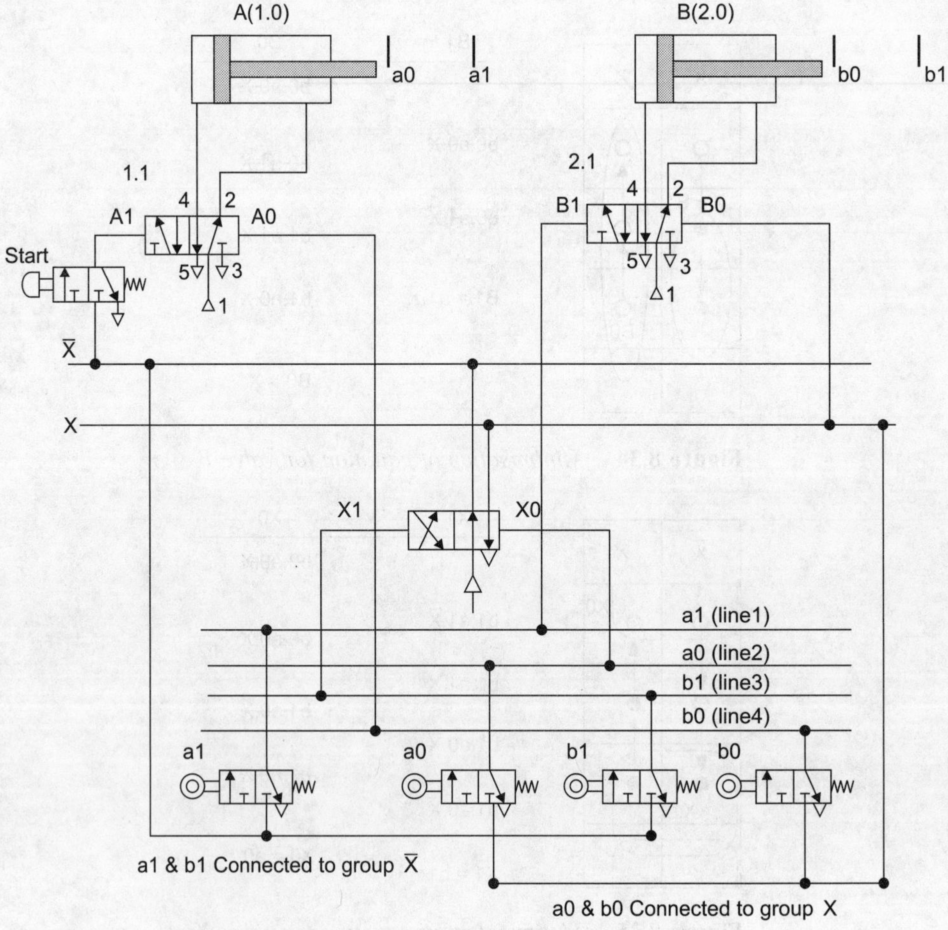

Figure 8.36 *Pneumatic circuit for* $A + B + B - A-$

Control problem 2 Design a pneumatic circuit for the sequence A+A−B+B− using the KV mapping method.

Step 1 Make a statement of sequence This problem considers an automatic drilling machine with two cylinders A and B. Cylinder A extends first and then retracts. Cylinder B then extends to do perform work operations and retracts.

Step 2 Draw a traverse–time diagram (Figure 8.37) Label the diagram according to the following rules. Under each time unit, list only those memory designations required to move the particular cylinder as desired.

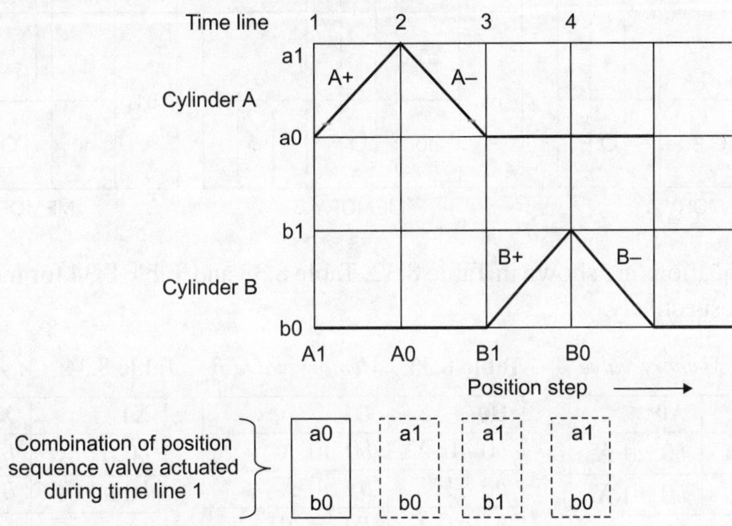

Figure 8.37 *Displacement–time diagram*

Step 3 Prepare a signal flow diagram (Figure 8.38)

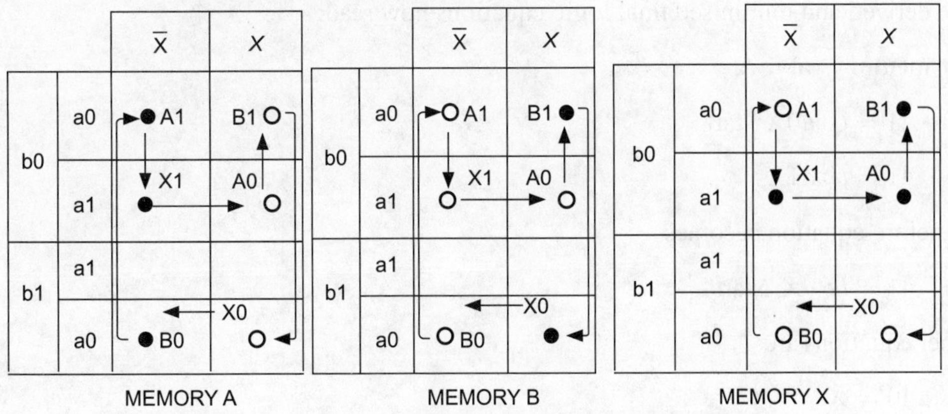

Figure 8.38 *Signal flow diagrams*

Step 4 Equation for the minimisation of logic For each memory appearing in the signal flow diagram, an individual minimisation map as shown below is required.

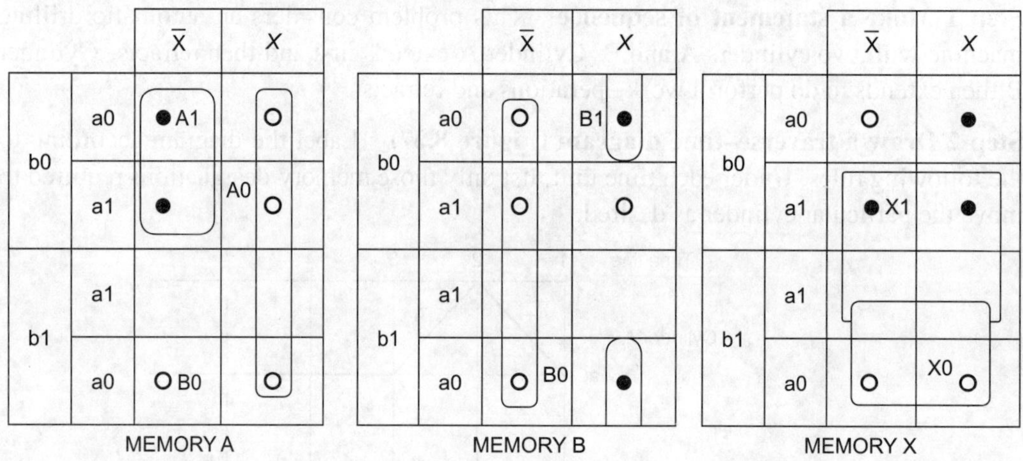

Minimised equations are shown in Table 8.32, Table 8.33 and Table 8.34 for memory valves A, B and X respectively.

Table 8.32 *Memory valve A*

A1	**A0**
$b0\ a0\ \overline{X}$	$b0\ a0\ X$
$b0\ a1\ \overline{X}$	$b0\ a0\ X$
$A1 = b0\ \overline{X}$	$b1\ a1\ X$
	$b1\ a0\ X$
	$\mathbf{A0 = X}$

Table 8.33 *Memory valve B*

B0	**B1**
$b0\ a0\ X$	$b0\ a0\ X$
$b0\ a1\ X$	$b1\ a0\ X$
$b1\ a1\ X$	$\mathbf{B1 = a0 \cdot X1}$
$b1\ a0\ X$	
$\mathbf{B0 = \overline{X}}$	

Table 8.34 *Memory valve X*

X1	**X0**
$b0\ a1\ \overline{X}$	$b1\ a1\ \overline{X}$
$b0\ a1\ X$	$b1\ a1\ X$
$b1\ a1\ \overline{X}$	$b1\ a0\ \overline{X}$
$b1\ a1\ X$	$b1\ a0\ X$
$\mathbf{X1 = a1}$	$\mathbf{X0 = b1}$

The derived and minimised final logic equations now read:

For memory valve A

$$A1 = b0\ a0\overline{X}.\text{start}$$

$$A1 = bo\ a1\overline{X}.$$

Therefore equation becomes

$$A1 = bo \cdot \overline{X}\ .\text{start}$$

Other equations are

$$A0 = X$$

$$Bo = \overline{X}$$

$B1 = ao.X$

$X1 = a1$

$X0 = b1$

Final circuit diagram is shown in Fig. 8.39.

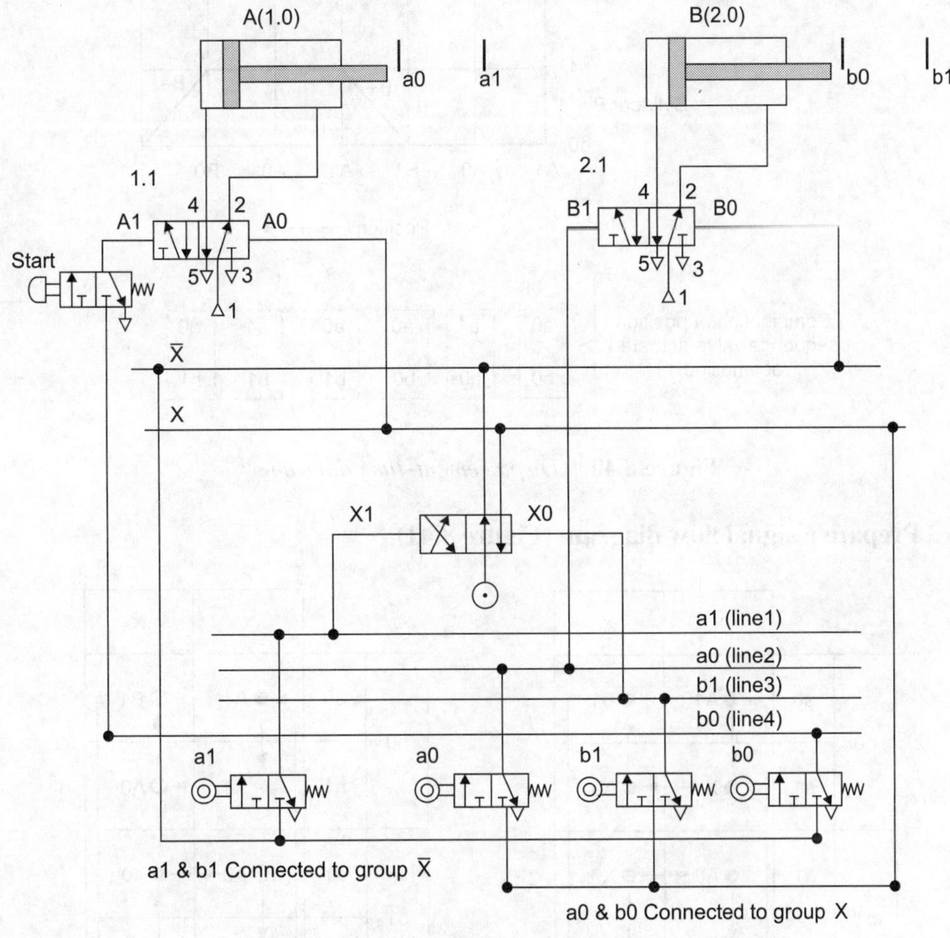

Figure 8.39 *Pneumatic circuit for* $A + A - B + B-$

Control problem 3 Design a pneumatic circuit for sequence A+A-B+A+A-B- using the KV mapping method

Step 1 Make a statement of sequence A control circuit is to be designed for the given transverse time diagram. A start valve must be included to prevent continuous cycling. The designation S for the start is inserted into the field where the commencing step is plotted.

Step 2 Draw a traverse–time diagram (Figure 8.40) Label the diagram according to the following rules. Under each time unit, list only those memory designations required to move the particular cylinder as desired.

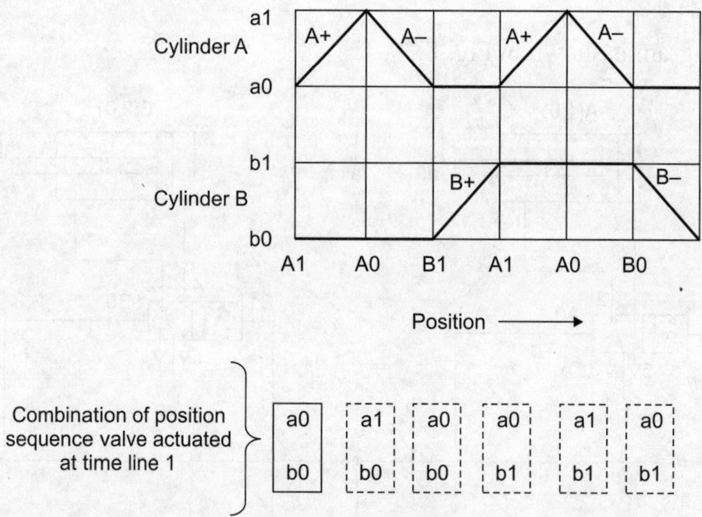

Figure 8.40 *Displacement–time diagram*

Step 3 Prepare a signal flow diagram (Figure 8.41).

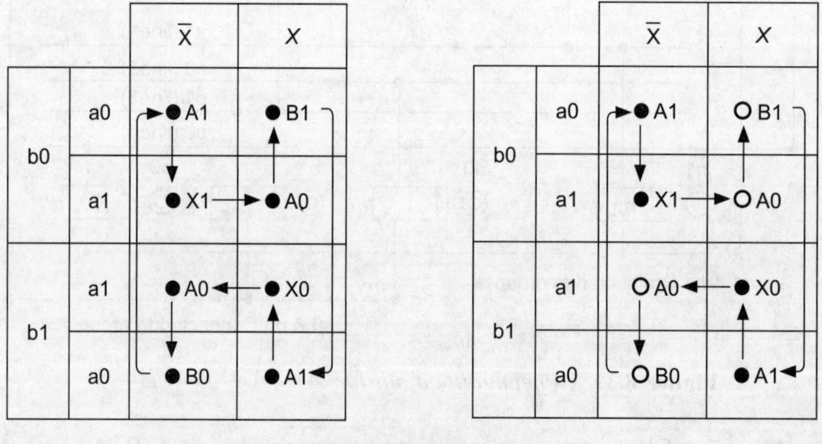

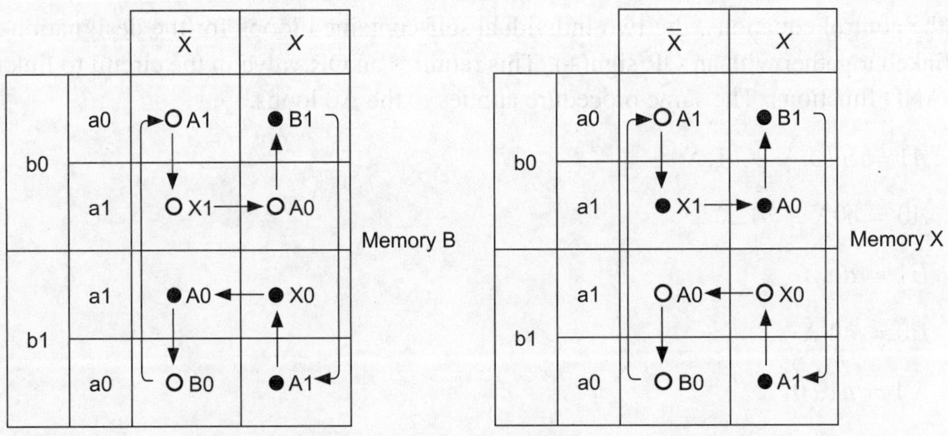

Figure 8.41 *Signal flow diagram for each valve*

Step 4 Equation for minimisation of logic For each memory appearing in the signal flow diagram an individual minimisation map is required.

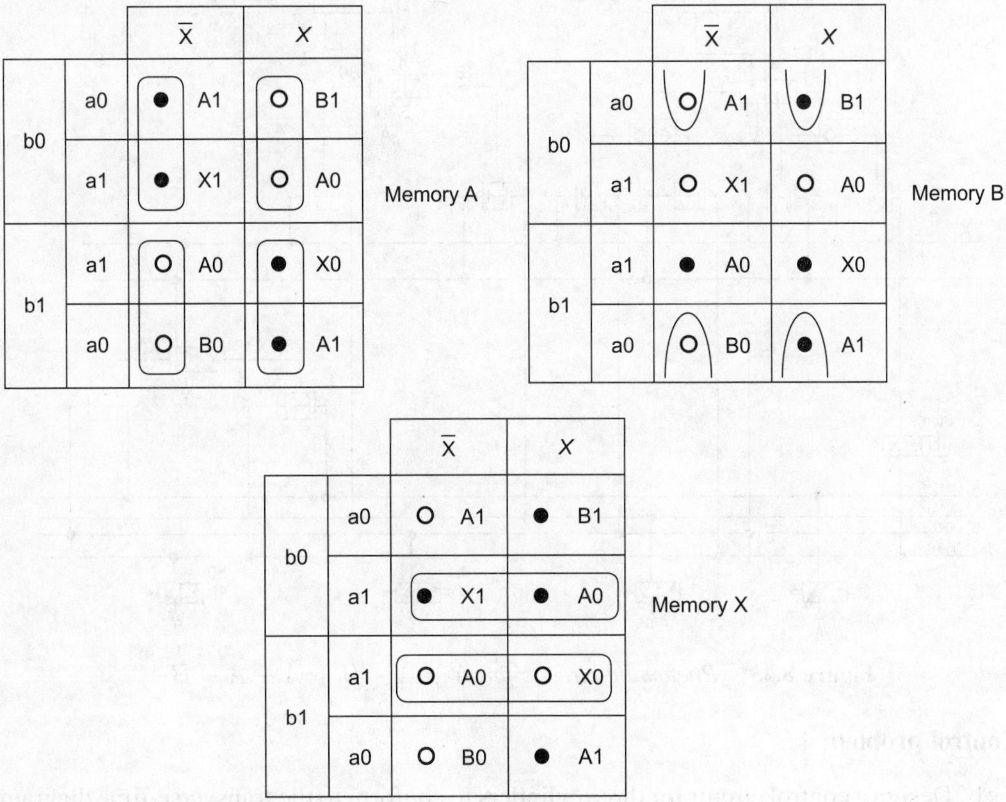

Memory A

		$\overline{X}$	X
b0	a0	● A1	○ B1
	a1	● X1	○ A0
b1	a1	○ A0	● X0
	a0	○ B0	● A1

Memory B

		$\overline{X}$	X
b0	a0	○ A1	● B1
	a1	○ X1	○ A0
b1	a1	● A0	● X0
	a0	○ B0	● A1

Memory X

		$\overline{X}$	X
b0	a0	○ A1	● B1
	a1	● X1	● A0
b1	a1	○ A0	○ X0
	a0	○ B0	● A1

Figure 8.42 *Minimisation of logic*

For the control equations, the two individual self-contained loops for the designation A1 are linked together with an OR sign(+). This requires an OR valve in the circuit to link the two AND functions. The same procedure applies to the A0 loops.

$$A1 = S.bo.\overline{X} + b1.X$$

$$A0 = b0X + b1.\overline{X}$$

$$B1 = ao.X$$

$$B0 = a0.\overline{X}$$

$$X1 = a1.b0$$

$$X0 = a1.b1$$

The final circuit diagram is shown in Fig. 8.43.

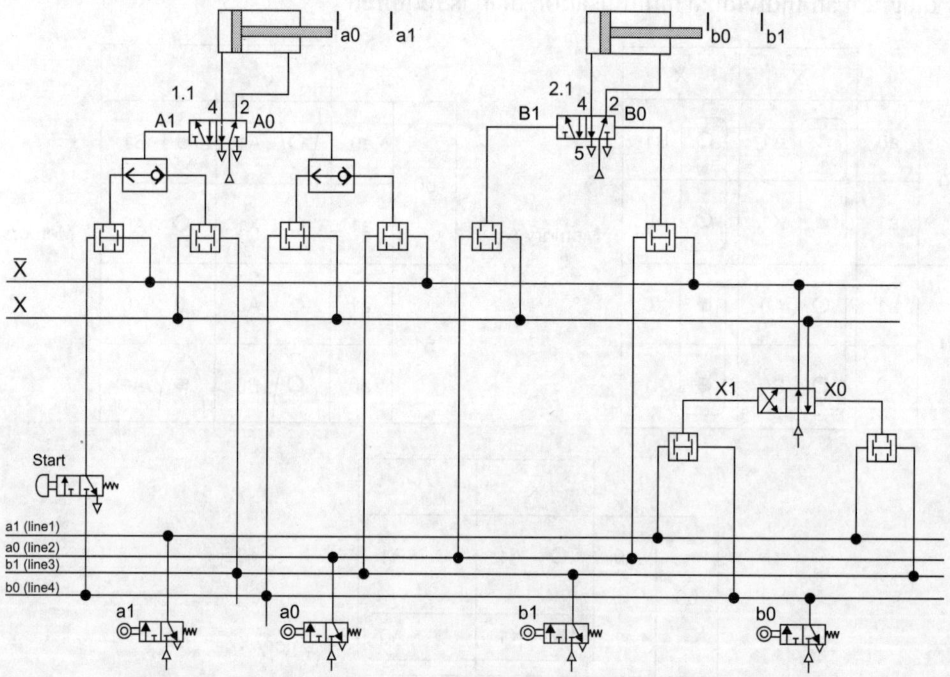

Figure 8.43 *Pneumatic circuits for* $A + A - B + A + A - B-$

Control problem 4

1. Design a control circuit for three cylinders to conform to the transverse-time diagram in Fig. 8.44.
2. Produce the fully labeled transverse-time diagram.

3. Draw minimisation maps for all the memories and loop designations. Also produce loops for the optical solutions.
4. Formulate switching equations for all the memories including auxiliary memories and for the optional loops.
5. Compare all the equations and select the most economical group for the final control circuit.
6. Construct the control circuit.

The procedure to solve the problem is to first produce the fully labeled transverse-time diagram. Then establish the minimisation map for all the memories and loop designations. Loops should produce optimal solutions and lead to the switching equations for all the memories including auxiliary memories and for the optional loops. These equations are compared to each other and the most economical group for the final control circuit is established. Using these groups of equations the final control circuit is drawn.

The following section details the solution of the control problem.

Step 1 Make a statement of sequence Given transverse time diagram (Fig. 8.44) clearly gives the sequence of operation.

Step 2 Draw a traverse–time diagram Label the diagram according to the following rules. Under each time unit, list only those memory designations which are required to control the movement of a particular cylinder.

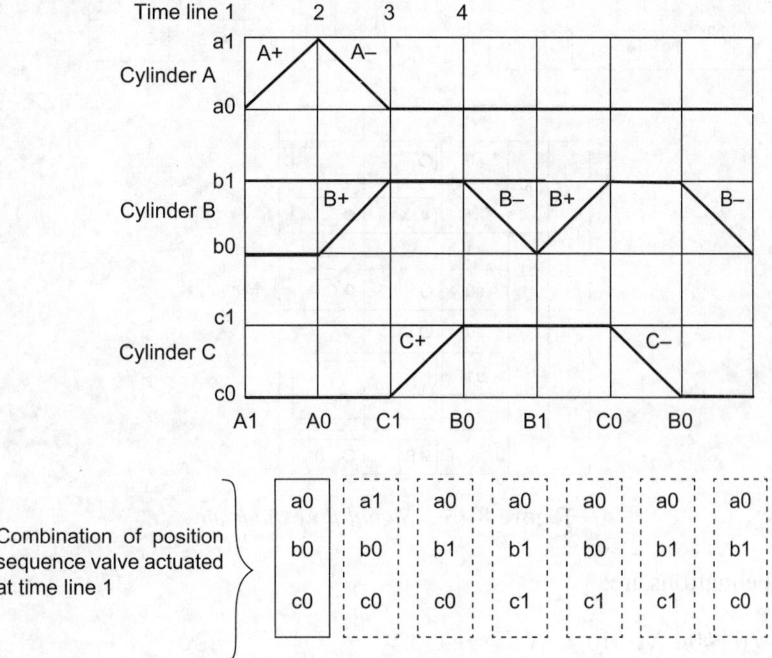

Figure 8.44 *Traverse time diagram*

Step 3 Prepare a signal flow diagram (Figure 8.45)

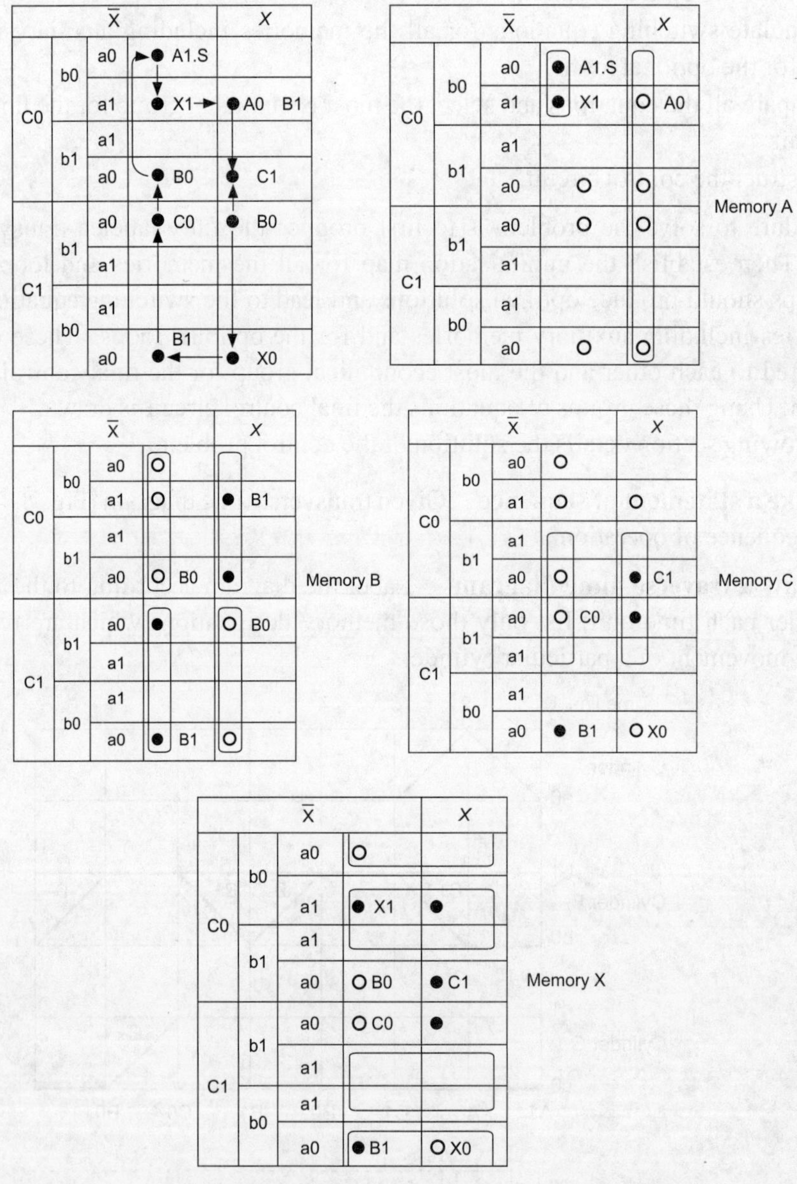

Figure 8.45 *Signal flow diagram*

The control equations are:

$$A1 = c0 \cdot b0 \cdot \overline{X};\ A0 = X$$

$$B1 = c0 \cdot X + c1 \cdot;\ B0 = c0 \cdot \overline{X} + c1 \cdot \overline{X}$$

$$C1 = b1 \cdot X \cdot a0;\; C0 = b1 \cdot X$$

$$X1 = a1;\; X0 = a0 \cdot b0$$

Final circuit diagram is shown in Fig. 8.46.

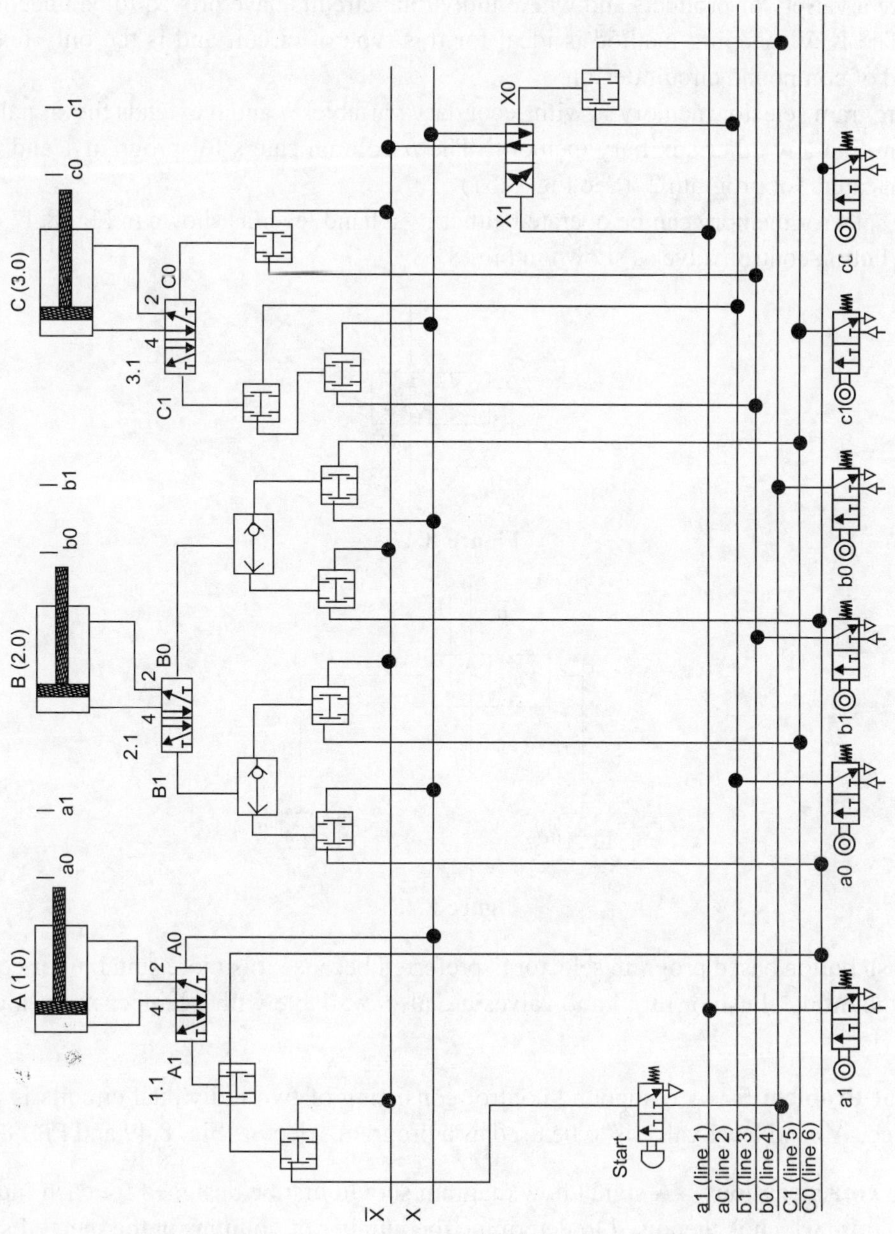

Figure 8.46 *Pneumatic circuits for* $A + (A - B+)C + B - B + C - B-$

8.5 Compound Circuit Design

A compound circuit is sequential control circuits consisting of several individual sequential programs merged into one common circuit, with a selection provision for each individual program. This type of circuit design is often used in multipurpose machines, each of which produce a variety of products and where individual circuits have proved to be uneconomical. The K V mapping method is ideal for this type of circuit and is the only feasible method of compound circuit design.

A program selector memory H with secondary variables $\overline{h}$ and h extends the signal flow diagram like any other auxiliary memory. The $\overline{h}$ column caters for program 1 and the h column caters for program 2. (See Fig. 8.51)

The selector memory can be operated either by a hand lever as shown in Fig. 8.47,or by a push button control valve as shown in Fig. 8.48

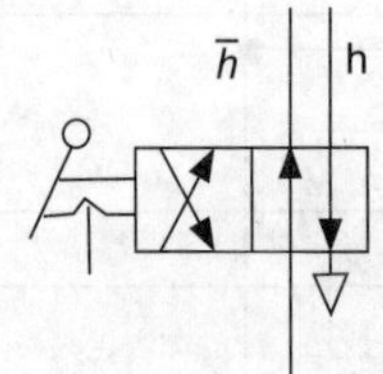

Figure 8.47

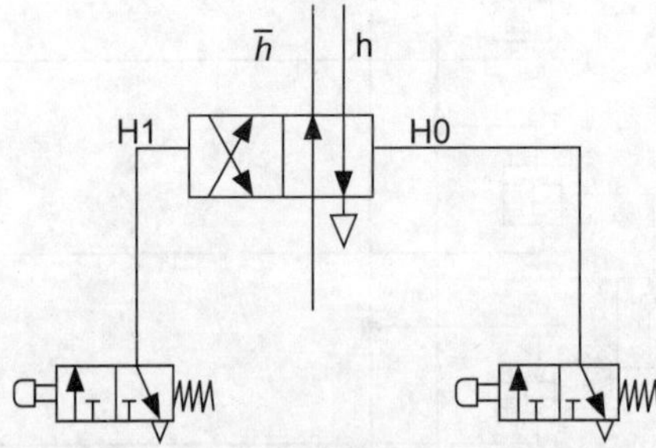

Figure 8.48

The push button based program selector is preferred because it is more suitable for control panel mounting. Push or turn knob valves are also available with a tamper-proof locking key device.

Control Problem 5 A compound control consisting of two individual circuits is to be designed. A hand lever valve is to be used as a program selector (Fig. 8.49 and Fig. 8.50).

Design considerations A signal flow diagram should first be designed for each individual program,which is then used to determine the number of columns in the merged signal flow diagram. Program 1 requires an auxiliary memory and so needs two columns in its individual KV map. Therefore, the merged signal flow diagram has four columns. This is because the selector memory H extends the maps to twice their size. The merged map for the circuit that has the most columns should be dealt with first, since this circuit determines the total number of columns in the merged signal flow diagram.

Program 2 needs no auxiliary memory and therefore it starts in the column where the auxiliary memory is reset. This consideration is important since the circuit would not start if it began in the X column.

Signal flow and looping is shown in Fig. 8.51.

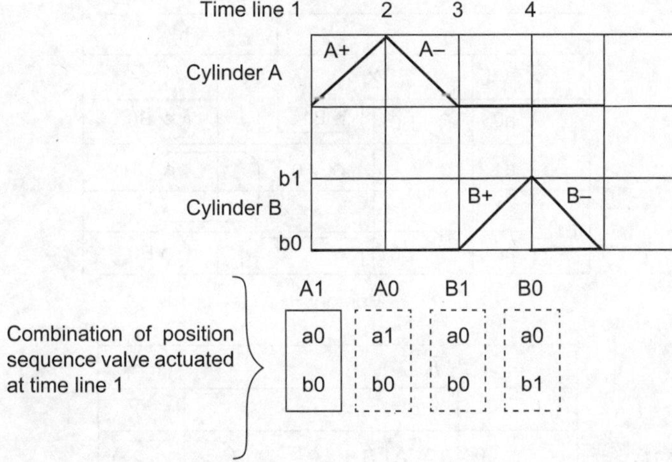

Figure 8.49 *Traverse time diagram for program 1*

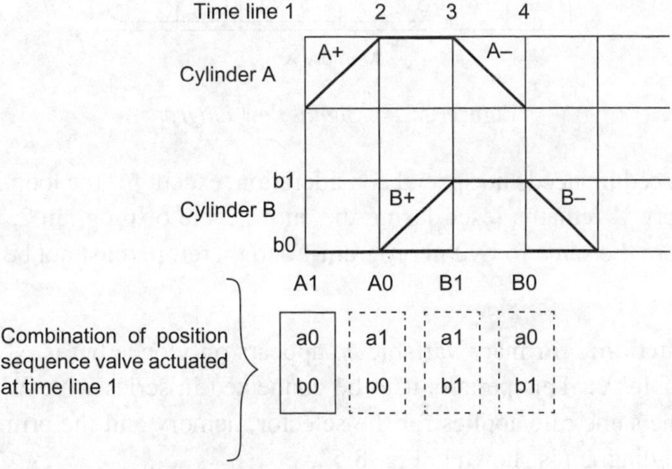

Figure 8.50 *Traverse time diagram for program 2*

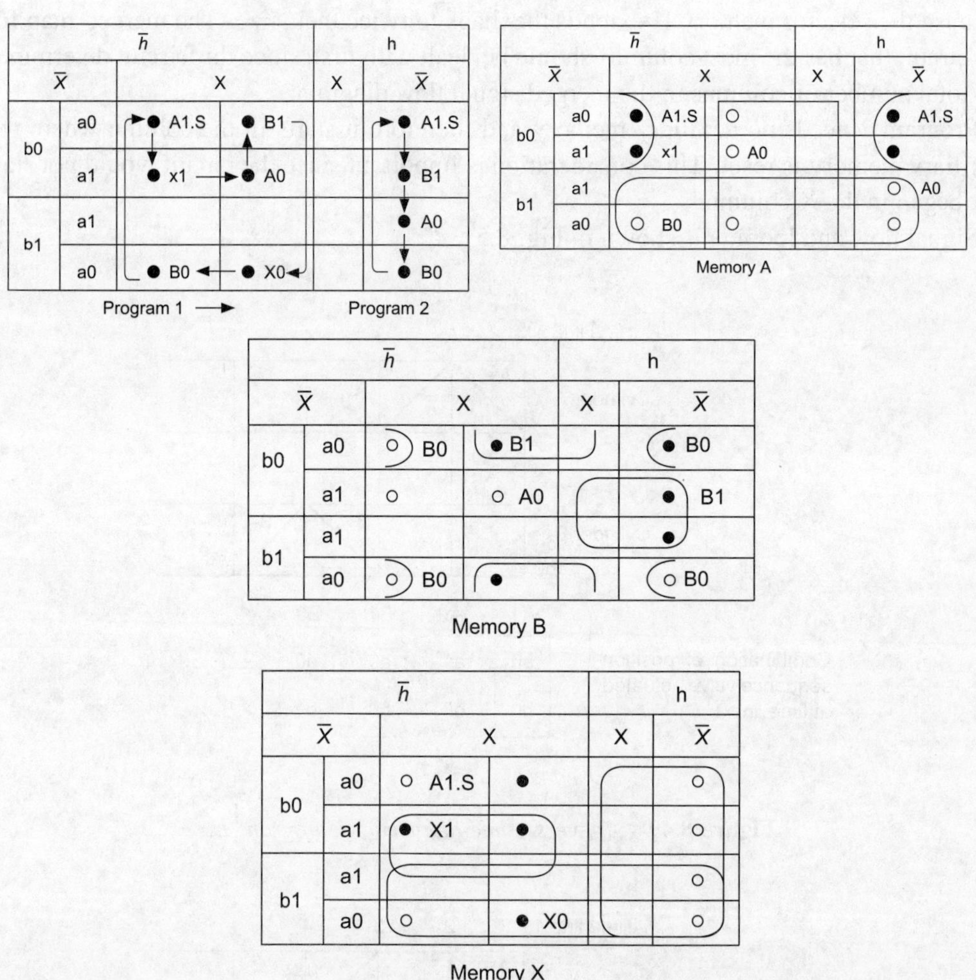

Figure 8.51 *Signal flow diagram*

The looping procedure needs no special consideration, except for the loop X1. Because the auxiliary memory X remains reset during the entire cycle of program 2, all its switching condition dots are the same for X0 in program 1 and therefore must not be included for the loop X1.

Circuit construction: Primary variable *bo* appears only once but is AND connected to secondary variable $\overline{X}$. This permits it to be connected in series with $\overline{X}$, which saves an AND valve. The same rule applies for the selector memory and the primary variable $a1$. The final circuit diagram is shown in Fig. 8.52.

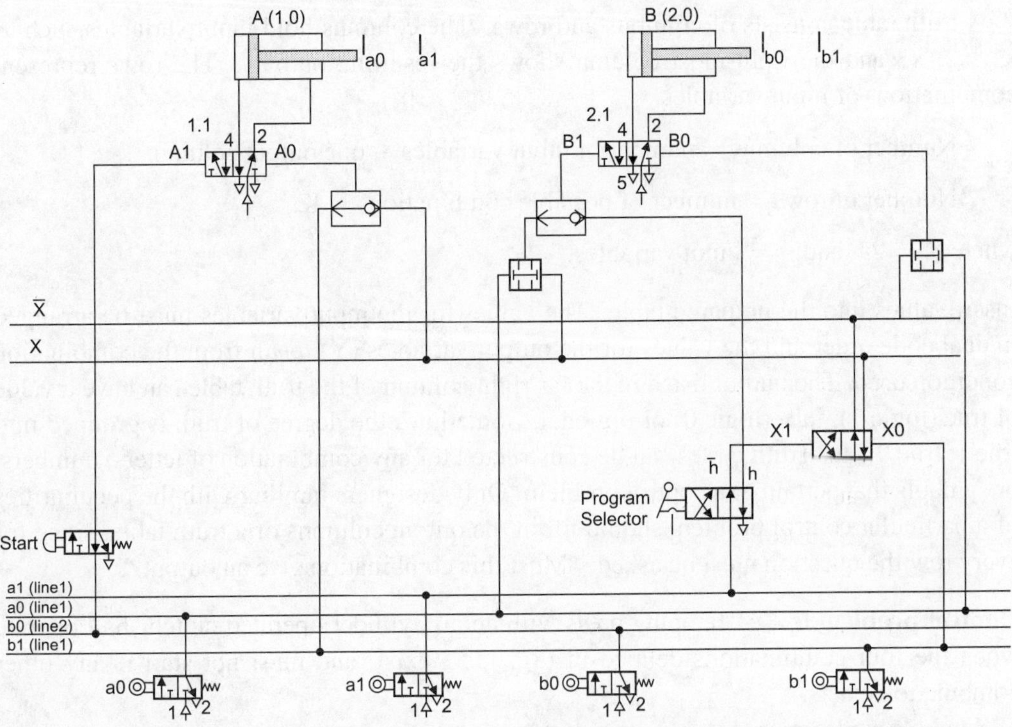

Figure 8.52 *Pneumatic circuit diagram*

8.6 Combinational Circuit Design

The output signal of the combinational circuit depends solely on the momentary state of its input signals or input variables; time delay and signal memorisation are not taken into consideration. This fact makes combinational circuits relatively simple to design and the KV map is once again used to minimise the equation.

Design procedure

1. Analyse all the input combinations for the condition which must provide an output signal, because all the remaining combinations must inhibit the output signal.
2. Draw a truth table to ensure that all possible combinations have been considered and form the un-minimised equation which provides an output signal.
3. Construct a KV map for all the participating input variables.
4. Minimise the equation and construct the circuit.

Truth table construction To achieve a better understanding of the problem of combinational input relationships, links of the variables are tabulated. The table is called a value table, or more commonly a truth table.

A truth table consists of columns and rows. The columns hold input variables such as x1, x2, x3 and the right-most column shows the resulting outputs. The rows represent combinations of input variables.

Number of columns = number of input variables + one output column.

Number of rows = number of possible combinations = K,

where K = 2^n, and n = input variables.

Insert values into the prepared table. The values for the input variables must be arranged in dual code order and the values for the output variables (Y) result from the combination condition used. The output listed in the far right column of the truth table can have a value of true (logic 1), false(logic 0) or optional. Sometimes the degree of truth is grouped into true 1, true 2, etc. Truth tables can be constructed for any combination of letters, numbers, or symbols that pertain to a control problem. Only designers familiar with the peculiarities of a particular control problem should fill in the output columns of a truth table, since for every row the question must be asked: "Must this combination give an output?"

Control problem 6 A stamping press with an air cylinder operated clutch shall operate when the four combinations defined in Fig. 8.53 exist, and must not start if any other combination exists.

Symbols explanation

G = Signal from the valve operated by the guard

T = Signal from the valve operated by the operator

W = Signal from the valve operated by a workpiece in the machine

R = Signal from the valve operated by a remote start

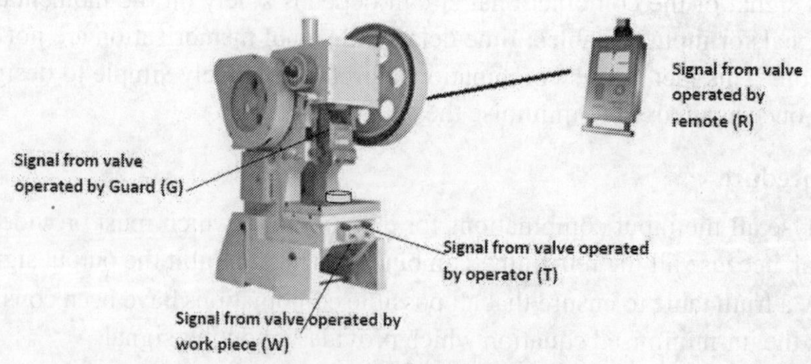

Figure 8.53 *Positional layout*

Truth Table evaluation The combination that produces logic 1 in the output cclumn from the un-minimised Boolean switching function:

$$\text{Start} = \overline{G}\,\overline{R}\,\overline{W}T + GR\overline{W}\,\overline{T} + G\overline{R}WT + GRW\overline{T}$$

To minimise the start equation, a KV map is needed. The map could be drawn as a single column map, a double column map with eight fields in each column (Table 8.35), or as a matrix type map as shown in Table 8.36. The form or labelling order of the map does not influence the minimisation process.

Table 8.35

G	R	W	T	ST	
0	0	0	0	0	
0	0	0	1	**1**	START
0	**0**	1	**0**	**0**	
0	0	1	1	1	START
0	1	0	0	0	
0	1	0	1	0	
0	1	1	0	0	
0	1	1	1	0	
1	0	0	0	0	
1	0	0	1	0	OPTIONAL
1	0	1	0	0	
1	0	1	1	1	START
1	1	0	0	0	
1	1	0	1	0	OPTIONAL
1	1	1	0	1	START
1	1	1	1	0	OPTIONAL

Plot the four un-minimised combinations extracted from the truth table into the fields of the prepared truth table and form loops. For combinational circuit design, all the plotted designations must be taken into consideration and all the plotting dots contribute towards the final minimised equation. Plotting dots which cannot be included in loops remain and must appear in the minimised equation.

Individual loops and circles are now linked with OR functions and grouped into the final minimised equation.

Start = ▯ + ▭ + ○

$$\text{Start} = \overline{G}\,\overline{R}T + W\overline{R}T + G\overline{T}WR$$

Table 8.36

		$\overline{R}$		R	
$\overline{G}$		G		G	G
$\overline{T}$	$\overline{W}$				
	W			S	
T	W	S	S		
	$\overline{W}$	S			

Final circuit is shown in Fig. 8.54

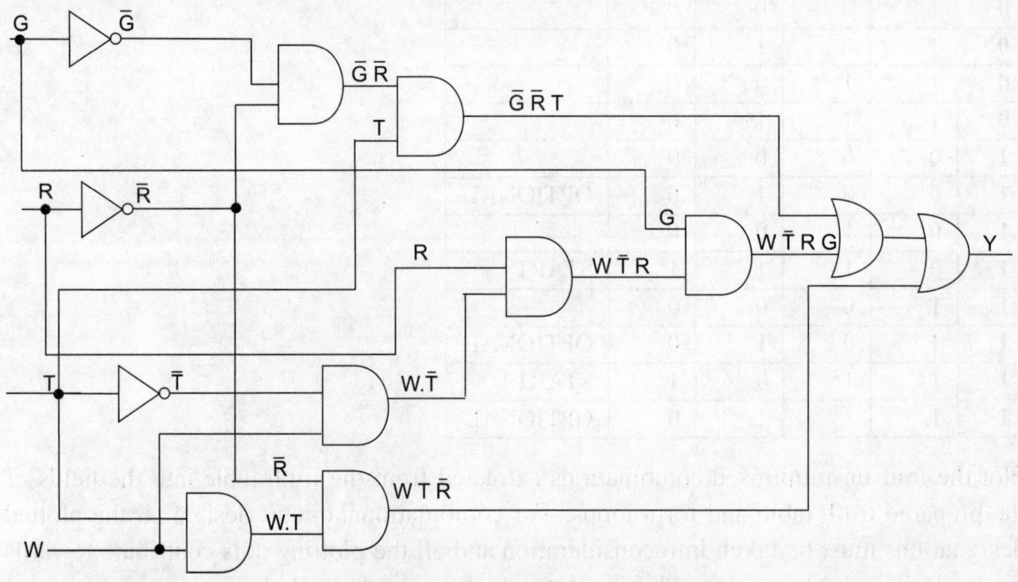

Figure 8.54

Summary

1. Pneumatic signal processing involves the use of pneumatic logic elements. Pneumatic logic uses air valves and other pneumatic devices for various control actions. These control actions are then applied to power systems to obtain the required actuation.
2. Logic circuits are divided into sequential and combinational circuits. In combinational circuits, each binary output is a function of the preset inputs only. In sequential circuits, on the other hand, some or all outputs also depend on previous inputs, i.e., on the past history of the system.

3. The classic method is based on the knowledge of pneumatic logic elements and the application of logic. In this circuit, each pneumatic element is a logic element performing various logic functions like OR, AND, NAND, NOR etc. The principle of Boolean algebra and DeMorgan's theorem are employed to solve problems in the analysis of control logics.
4. NAND and NOR gates exhibit properties that allows these gates to be used in combination with one another to create all the other basic logic functions. This techniqueuses left over NAND or NOR gates on a chipinstead of installing a new chip just for one gate.
5. The graphical design method using Karnaugh–Veitch maps is the basis for sequential logic circuit design. It is used to simplify Boolean equations. It is an array of cells in which each cell represents a binary value of input variables.
6. A compound circuit is sequential control circuits consisting of several individual sequential programs merged into one common circuit, with a selection provision for each individual program. This type of circuit design is often used on multipurpose machines which each produce a variety of products and where individual circuits have proved to be uneconomical.
7. Combinational circuit output signals depend solely on the momentary state of their input signals or input variables and time delay and signal memorisation are not taken into consideration. This fact makes combinational circuits relatively simple to design and the KV map is once again used to minimise equation.

EXERCISES

Fill in the blanks

1. In the KV mapping method, the vertical columns represent ——— outlets and their number is defined by the ——— signal flow path.
2. When hardware costs are not a factor, the step counter method is used. When minimum hardware is required then ——— method is used.
3. For design of compound circuits and ————— circuits, the logic design method is used.
4. The experience level required of the designer is ——— for logic circuit design than for step counter.
5. In combinational circuits, each binary output is a function of only the preset inputs. In sequential circuits, on the other hand, ———— also depend on previous inputs, i.e., on the past history of the system.

State whether true or false

1. KV mapping method is used to simplify the Boolean equation.
2. KV method is used for any number of variables.

3. In KV map, number of horizontal lines is equal to number of cylinders.
4. Time delay and signal memorisation are taken into consideration in combinational circuits.
5. Logic design method ensures that minimum hardware is required.

Review questions

1. Explain the difference between combinational circuits and sequence circuits.
2. Briefly explain the advantages of pneumatic logic control.
3. Compare electric logic control with pneumatic logic control.
4. List the four important characteristics of the logic design method.
5. Compare step counter method with the logic design method
6. Compare the cascade method with logic design method.
7. Explain the principle of KV mapping with a suitable sequence example.
8. Briefly explain the principles of looping in KV mapping.
9. Draw a KV map for 2, 3 and 4 variables.
10. Explain with the help of neat sketch the process of minimisation of logic equations using the KV method.
11. Compare compound circuit design with combinational circuit design.

Problems

Note: Readers are advised to solve Problem 22 to Problem 30 after studying the chapter 10 on PLC control in this book.

1. Simplify the following using Boolean algebra rules:

$$Y = \overline{A \cdot B + C}$$

2. Write the Boolean equation for the circuit shown in Fig. 8.55. Use DeMorgan's theorem and then Boolean algebra rules to simplify the equation. Draw the simplified circuit.

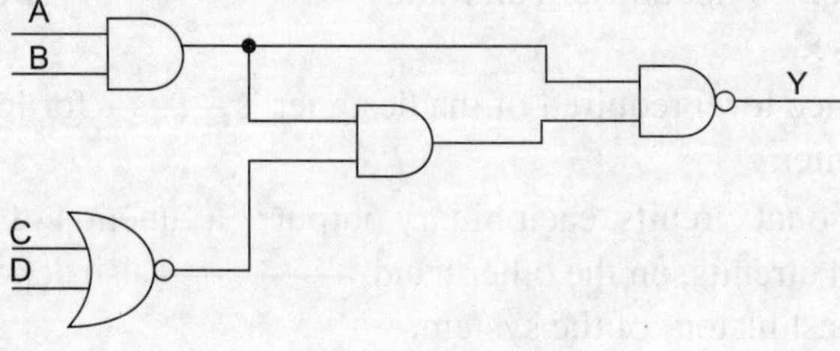

Figure 8.55

3. Simplify the following:

$$Y = \overline{AB} + \overline{AC} + \overline{\overline{A}BC}$$

4. Which unique binary combination of following SOP gives a logic output "1"

$$Y = \overline{X}YZ + X\overline{Y}Z + XY\overline{Z} + \overline{X}Y\overline{Z} + XYZ$$

5. For the following Truth Table (Table 8.37), write the POS equation

Table 8.37

A	B	C	X
0	0	0	1
0	**0**	**1**	**0**
0	**1**	**0**	**0**
0	1	1	1
1	0	0	1
1	0	1	1
1	1	0	1
1	**1**	**1**	**0**

6. Group the KV map given in Table 8.38.

Table 8.38

A↓ B→	00	01	11	10
00		1		
01		1		1
11	1	1	1	1
10	1	1	1	1

7. Simplify using KV map:

$$Z = f(A, B, C) = \overline{A}\,\overline{B}\,\overline{C} + \overline{A}B + AB\overline{C} + AC$$

8. Simplify using KV map:

$$Z = f(A, B, C) = \overline{A}\,\overline{B} + B\overline{C} + BC + A\overline{B}\,\overline{C}$$

9. Simplify using KV map:

$$Z = f(A, B) = \overline{A}\,\overline{B} + A\overline{B} + \overline{A}B$$

10. Write the simplified equation for the given truth table (Table 8.39) using the KV method.

Table 8.39

CD \ AB	00	01	11	10
00		1		
01		1	1	1
11	1	1	1	
10			1	

11. Simplify using a KV map:

$$\text{Out} = \overline{A}\,\overline{B}CD + \overline{A}BCD + ABCD + A\overline{B}CD + AB\overline{C}\,\overline{D} + AB\overline{C}D + ABC\overline{D}$$

12. Simplify using a KV map:

$$\text{Out} = \overline{A}\,\overline{B}\,\overline{C}\,\overline{D} + \overline{A}\,\overline{B}C\overline{D} + A\overline{B}\,\overline{C}\,\overline{D} + A\overline{B}C\overline{D}$$

13. Simplify using a KV map:

$$\text{Out} = \overline{A}\,\overline{B}\,\overline{C}\,\overline{D} + \overline{A}\,\overline{B}\,\overline{C}D + \overline{A}\,\overline{B}CD + \overline{A}\,\overline{B}C\overline{D} + A\overline{B}\,\overline{C}\,\overline{D} + A\overline{B}\,\overline{C}D + A\overline{B}C\overline{D}$$

14. Simplify using a KV map:

$$\text{Out} = \overline{A}\,\overline{B}\,\overline{C}\,\overline{D} + \overline{A}\,\overline{B}\,\overline{C}D + \overline{A}\,\overline{B}C\overline{D} + B\overline{C}\,\overline{D} + BC\overline{D} + A\overline{B}\,\overline{C}D + A\overline{B}D + A\overline{B}C\overline{D}$$

15. Simplify using a KV map:

$$\text{Out} = \overline{A}\,\overline{B}\,\overline{C}\,\overline{D} + \overline{A}\,\overline{B}C\overline{D} + A\overline{B}\,\overline{C}\,\overline{D} + ABCD$$

16. Simplify using a KV map:

$$\text{Out} = \overline{A}\,\overline{B}\,\overline{C}\,\overline{D} + \overline{A}\,\overline{B}\,\overline{C}D + \overline{A}B\overline{C}D + \overline{A}BCD + ABCD + ABC\overline{D} + A\overline{B}\,\overline{C}D + A\overline{B}C\overline{D}$$

17. Set up the KV map for the truth table given in Table 8.40.
18. Use a KV map to simplify the truth table given in Table 8.40 and implement it in ladder logic.

Table 8.40

A	B	C	D	Result
0	0	0	0	0
0	0	0	1	0
0	0	1	0	0
0	0	1	1	1
0	1	0	0	1
0	1	0	1	1
0	1	1	0	1
0	1	1	1	1
1	0	0	0	0
1	0	0	1	0
1	0	1	0	1
1	0	1	1	1
1	1	0	0	0
1	1	0	1	0
1	1	1	0	1
1	1	1	1	1

Table 8.41

A	B	C	D	Result
0	0	0	0	0
0	0	0	1	0
0	0	1	0	0
0	0	1	1	0
0	1	0	0	0
0	1	0	1	0
0	1	1	0	1
0	1	1	1	1
1	0	0	0	0
1	0	0	1	0
1	0	1	0	0
1	0	1	1	0
1	1	0	0	0
1	1	0	1	0
1	1	1	0	1
1	1	1	1	1

19. Write the simplest Boolean equation for the KV map given in Table 8.42.

Table 8.42

	CD	$C\overline{D}$	$\overline{C}\,\overline{D}$	$\overline{C}D$
AB	1	0	0	1
$A\overline{B}$	0	0	0	0
$\overline{A}\,\overline{B}$	0	0	0	0
$\overline{A}B$	0	1	1	0

20. Given the truth table (Table 8.43), find the most efficient ladder logic to implement it. Use a structured technique such as Boolean algebra or Karnaugh maps.

Table 8.43

A	B	C	D	X	Y
0	0	0	0	0	0
0	0	0	1	0	0
0	0	1	0	0	0
0	0	1	1	0	0
0	1	0	0	0	0
0	1	0	1	0	0
0	1	1	0	0	1
0	1	1	1	0	1
1	0	0	0	1	0
1	0	0	1	1	0
1	0	1	0	0	0
1	0	1	1	0	0
1	1	0	0	1	0
1	1	0	1	1	0
1	1	1	0	0	1
1	1	1	1	0	1

Answers

Fill in the blanks

1. auxiliary memory 2. logic 3. combinational 4. more 5. some or all outputs

True or false

1. True 2. False 3. False 4. False 5. True

9 ELECTRO-PNEUMATIC CONTROL

Upon completion of this chapter, the reader should be able to

- Explain the various steps involved in electro-pneumatics
- List seven electrical devices used in electro-pneumatics
- Describe the constructional details of solenoid valves
- Explain the operations of control devices like limit switches, sensors, timers, counters and pressure switches
- Differentiate between capacitive and inductive proximity sensors
- Differentiate between dominant on and off latching circuits
- Design single actuator electro-pneumatic circuits
- Design a sequence circuits using two or more cylinders

9.1 Introduction

Electro-pneumatics is now commonly used in many areas of industrial low-cost automation. It is also used extensively in production, assembly, pharmaceutical, chemical and packaging systems. There has been a significant change in control systems; relays have increasingly been replaced by programmable logic controllers (PLC) in order to meet the growing demand for more flexible automation.

Electro-pneumatic control consists of electrical control systems operating pneumatic power systems. Solenoid valves are used as the interface between electrical and pneumatic systems. Devices like limit switches and proximity sensors are used as feedback elements.

Electro-pneumatic control is more widely used for large applications. The signal medium in these systems is the electrical signal which is either AC or DC. The working medium is compressed air. Operating voltages from around 12 V to 220 V are common and the final control valve is activated by solenoid actuation. The resetting of the valve is either by spring valve (single solenoid) or using a double solenoid valve. More commonly, the valve actuation and reset is achieved by a pilot assisted solenoid actuation to reduce the size and cost of the valve.

Control of electro-pneumatic systems is achieved using either using a combination of relays and contactors, or with the help of Programmable Logic Controllers (PLC). A relay is often is used to convert signal input from sensors and switches to a number of output signals (either normally closed or normally open). Signal processing can be easily achieved using relay and contactor combinations.

A programmable logic controller provides a convenient way to get output based on specific requirements. The output signals are supplied to the solenoids activating the final control valves which control the movement of the cylinders. The greatest advantage of electro-pneumatics is the integration of various types of proximity sensors (electrical) and PLC for very effective control. As the signal speed with electrical signal is much higher, cycle time is reduced and signal can be conveyed over long distances.

Electro-pneumatic controls involve three important steps:

- **Signal generation using input devices** Signal is generated using devices such as switches, contactors, and sensors.
- **Processing of Signals** Signals are processed using contactors and relays, or Programmable Logic Controllers (PLC).
- **Signal Outputs** Output from signal processing units are used to activate output devices like solenoids, indicators or audible alarms.

9.2 Control Electrical Devices

Seven electrical devices are commonly used in the control of fluid power systems:

1. Manually actuated push button switches
2. Limit switches
3. Pressure switches
4. Solenoids
5. Relays
6. Timers
7. Temperature switches

Other devices also used in electro-pneumatics are:

- Proximity sensors
- Electric counters

9.2.1 Push button switches

A push button (PB) is a switch used to open or close an electric control circuit. It is primarily used for starting and stopping operation of machinery and also to provide a manual override during emergencies. Push button switches are actuated by pushing the actuator into the housing causing a set of contacts to open or close.

Push buttons are of two types:

(i) Momentary push button
(ii) Maintained contact or detent push button

Momentary push buttons return to their unactuated position when they are released. Maintained push buttons have a latching mechanism to hold them in the selected position.

Push buttons are also distinguished according to contact behaviour:

Normally open (NO) The contacts are open in the normal position, disconnecting energy flow through them. In the actuated position the contacts are closed, permitting energy flow through them.
Normally closed (NC) The contacts are closed in the normal position, permitting energy flow through them. The contacts are open in the actuated position, disconnecting energy flow through them.
Change-over (CO). A change-over contact is a combination of NO and NC contacts.

Cross-section views of push buttons in the normal and actuated positions and their symbols are shown in Fig. 9.1.

Type of devices	***Terminal Numbers***	
	Normally closed contacts	**Normally open contacts**
Push buttons and Relays	1 and 2	3 and 4
Timers and Counters	5 and 6	7 and 8

Designation of the push buttons

Type of contact	Designation
Momentary contact type PB station (2NO +2 NC) 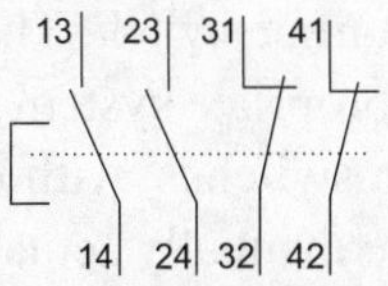 13 23 31 41 14 24 32 42	First digit indicates the function of contact. Second digit represents a serial ordering. 3 and 4 represents NO contacts and 1 is the serial no.  3 4
Maintained Contact type PB station (2NO+2 NC) 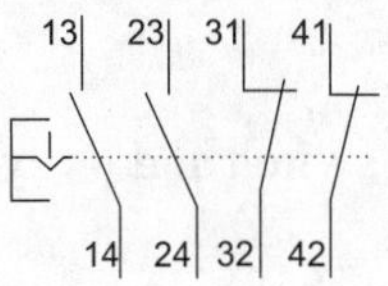 13 23 31 41 14 24 32 42	First digit indicates the function of contact. Second digit represents a serial ordering. 3 and 4 represents NO contacts and 1 is the serial no.  3 4

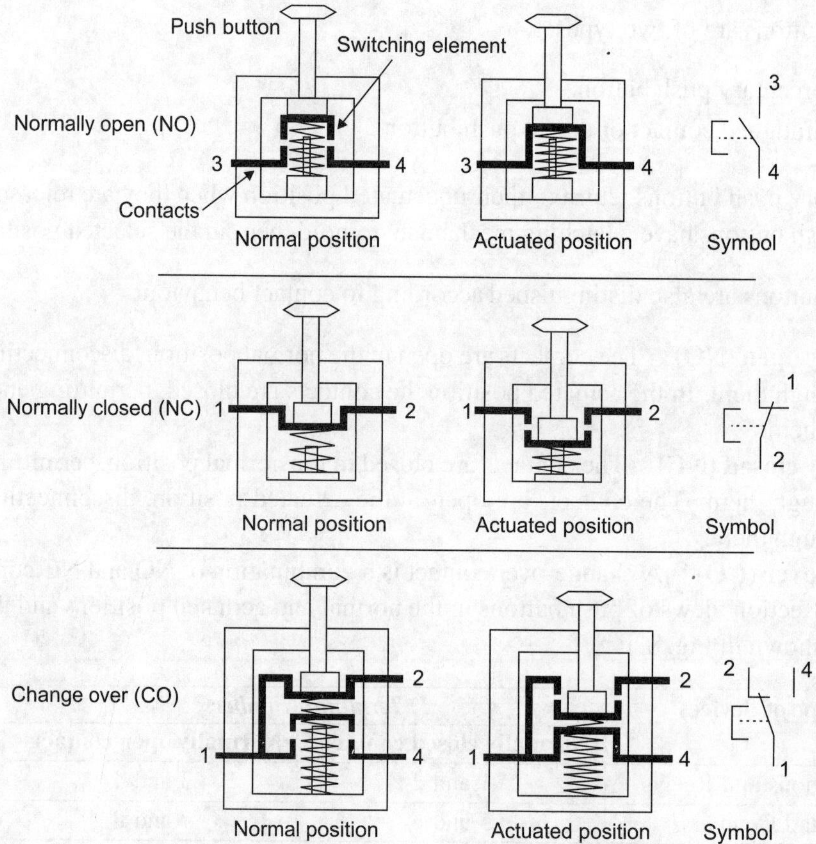

Figure 9.1 *Push buttons and their symbols*

9.2.2 Limit switches

Any switch that is actuated by the position of a fluid power component (usually a piston rod, a hydraulic motor shaft, or the position of a load) is termed a limit switch. The actuation of a limit switch provides an electrical signal that causes an appropriate system response.

Limit switches perform the same function as push button switches, with the difference that they are mechanically actuated whereas push buttons are manually actuated.

Limit switches are classified by the method of actuation of the contacts:

- Lever actuated: where contacts are operated slowly
- Spring loaded: where contacts are operated rapidly

Figure 9.2 shows a simplified cross-sectional view of a limit switch and its symbol.

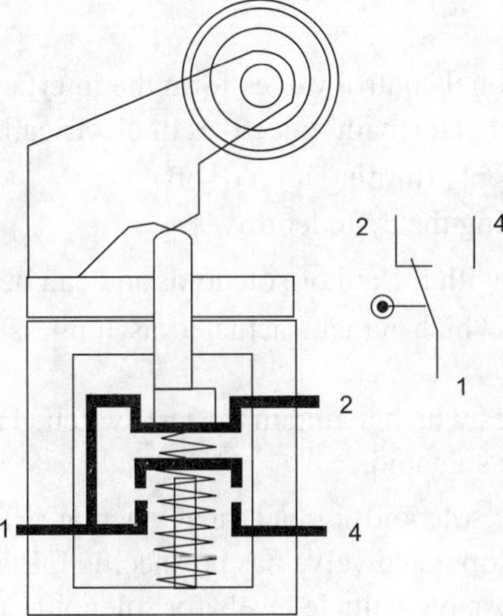

Figure 9.2 *Cross-sectional view of a limit switch*

9.2.3 Pressure switches

A pressure switch is an electro pneumatic signal converter. It is used to sense a change in pressure, and opens or closes an electrical switch when a predetermined pressure is reached. Bellows or diaphragms are used to sense the change of pressure and expand or contract in response. Figure 9.3 shows a diaphragm type of pressure switch. Pressure is applied at the inlet and when the preset pressure level is reached, the diaphragm expands and pushes the spring loaded plunger to make or break the contact.

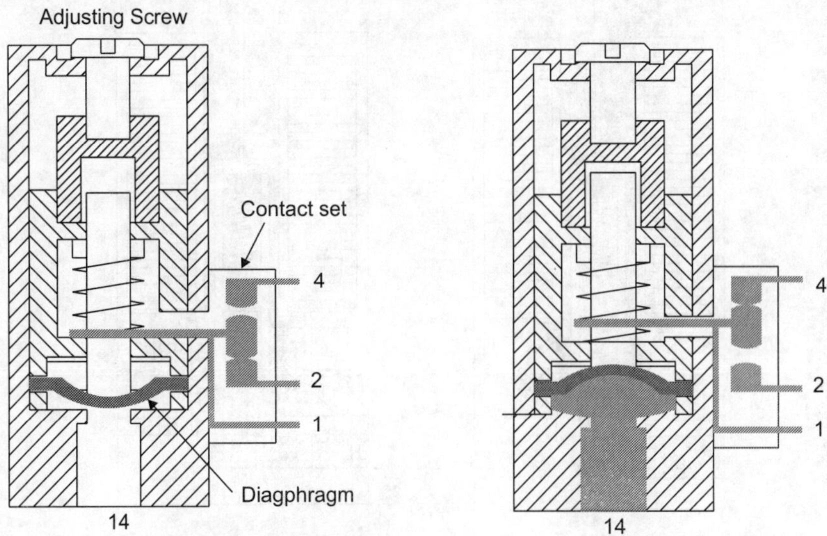

Figure 9.3 *Cross-sectional view of a pressure switch*

9.2.4 Solenoids

Electrically actuated directional control valves form the interface between the two parts of an electro-pneumatic control. The main objectives of electrically actuated DCVs include:

- Controlling the air supply (turning it on or off)
- Extending and retracting the cylinder drives

These DCVs are controlled with the aid of solenoids and can be divided into two groups:

- Spring return valves which remain actuated as long as current flows through the solenoid
- Double solenoid valves which maintain the last switched position even when no current flows through the solenoid.

In the initial position, all the solenoids of electrically actuated DCVs are de-energised and inactive. A double solenoid operated valve has no specific initial position since it does not have a return spring. The possible voltage levels for solenoids are: 12 V DC, 12 V AC, 12 V 50/60 Hz, 24 V 50/60 Hz, 110/120 V 50/60 Hz and 220/230 V 50/60 Hz.

(a) 3/2 single solenoid valve, spring return The cross-sectional view of a 3/2 single solenoid valve in the normal and actuated positions are shown in Fig. 9.4. In the normal position, port 1 is blocked and port 2 is connected to port 3. When the rated voltage is applied to the coil, the armature is lifted away from the valve seat and pulled towards the centre of the coil. The compressed air now flows from port 1 to port 2, and port 3 is blocked. When the voltage supply to the coil is stopped, the valve returns to the normal position. Figure 9.5 shows a 2/2 solenoid operated valve.

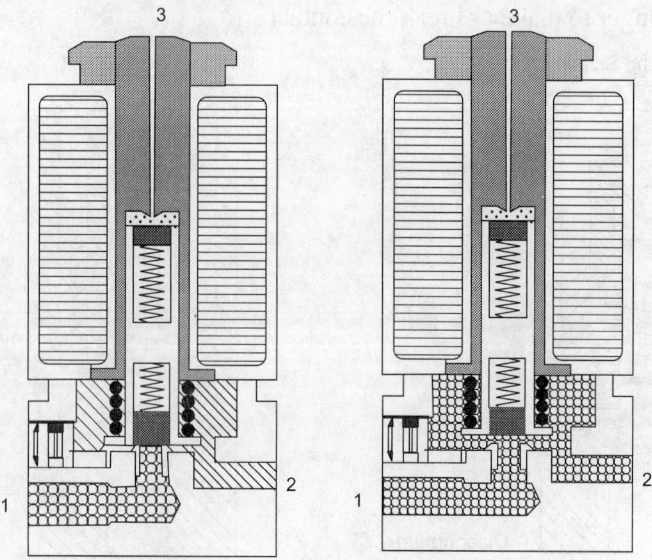

Figure 9.4 *Cross-sectional view of a 3/2 single solenoid valve*

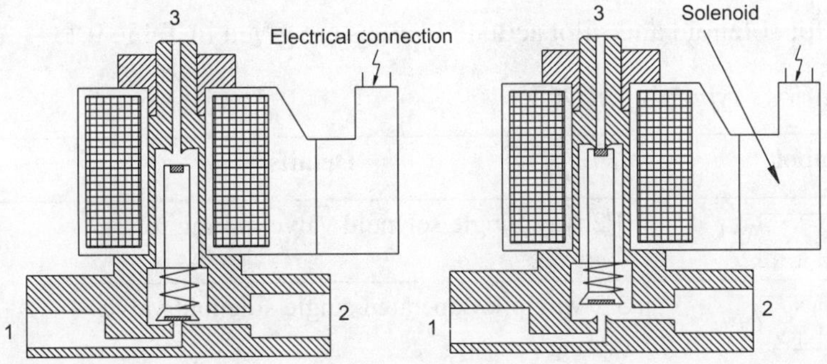

Figure 9.5 *Cross-sectional view of a 2/2 way solenoid operated valve*

(b) 5/2 single solenoid valve, spring return The cross-sectional view of a 5/2 single solenoid in the normal and actuated positions is shown in Fig. 9.6. In the normal position, port 1 is connected to port 2, port 4 to port 5, and port 3 is blocked. When the rated voltage is applied to coil 14, the valve is actuated through an internal pilot valve. In the actuated position, port 1 is connected to port 4, port 2 is connected to port 3, and port 5 is blocked. The valve returns to the normal position when the voltage to the armature coil is removed. This type of valve is normally used as the final valve to control double acting cylinders.

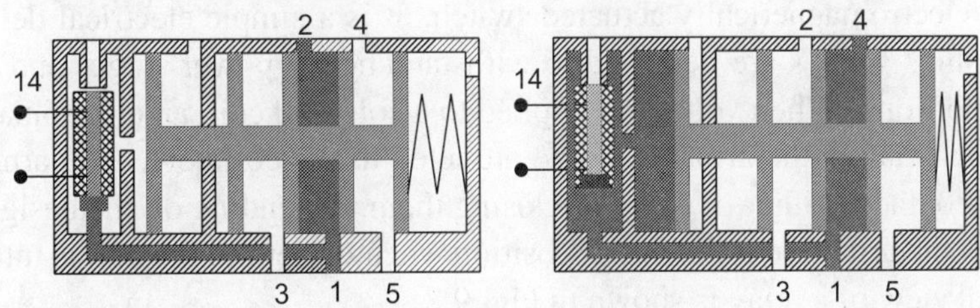

Figure 9.6 *Cross-sectional view of a 5/2 single solenoid operated valve*

(c) 5/2 double solenoid valve The cross-sectional view of a 5/2 way double solenoid in the normal and actuated positions is shown in Fig. 9.7. When the rated voltage is applied to coil 14, the valve is actuated to one switching position with port 1 connected to port 4, port 2 connected to port 3, and port 5 blocked. When the rated voltage is applied to coil 12, the valve is actuated to the other switching position with port 1 connected to port 2, port 4 connected to port 5, and port 3 blocked.

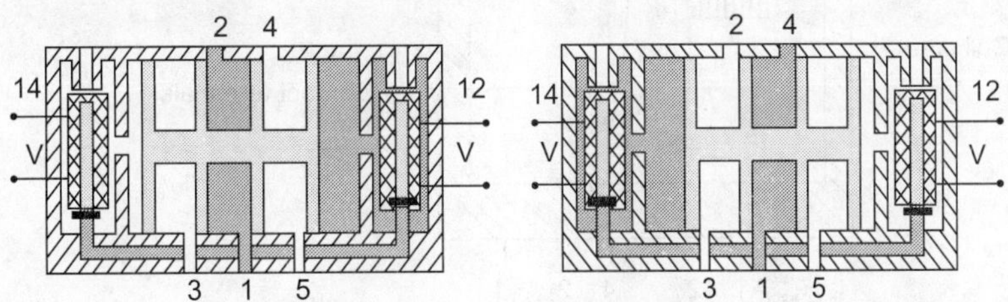

Figure 9.7 *Cross-sectional view of a 5/2 double solenoid operated valve*

The symbols for solenoid and pilot actuated valves are given in Table 9.1.

Table 9.1 *Various symbols for DCVs*

Symbol	Details
	3/2 way Single solenoid valve (spring return)
	3/2 way pilot operated single solenoid valve (spring return)
	5/2 way single solenoid valve (spring return)
	5/2 way double solenoid valve.
	5/2 way piloted operated double solenoid valve.

9.2.5 Relays

A relay is an electromagnetically actuated switch. It is a simple electrical device used in signal processing. Relays are designed to withstand heavy power surges and harsh environmental conditions. When voltage is applied to a solenoid coil, an electromagnetic field is created which causes the armature to be attracted to the coil core. The armature actuates the relay contacts, either opening or closing them, depending on the design. A return spring restores the armature to its initial position when current to the coil is interrupted. A cross-sectional view of a relay is shown in Fig. 9.8.

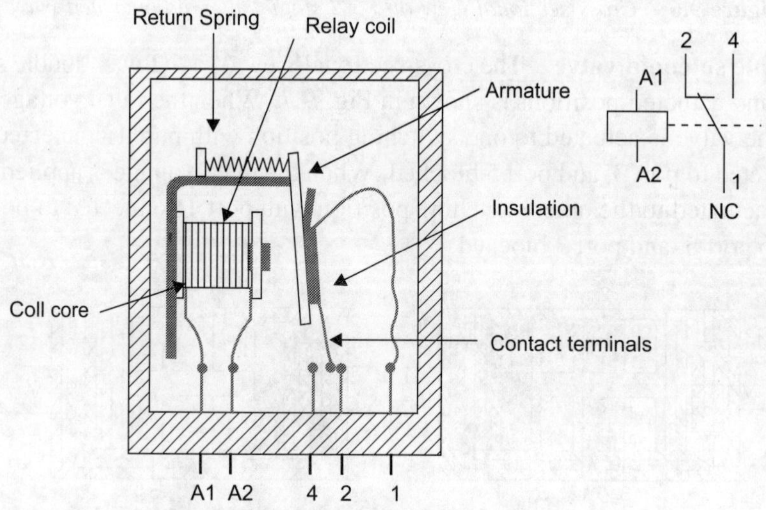

Figure 9.8 *Cross-sectional view of a relay*

A large number of control contacts can be incorporated in to relays, in contrast to push button stations. Relays possess interlocking capability—an important safety feature in control circuits, and are commonly designated as K1, K2, K3 etc. Interlocking avoids the simultaneous switching of certain combination of coils.

9.2.6 Time delay relays (Timers)

Timers are required in control systems to introduce time delays between work operations via the associated control element. Electronic timers are most commonly used. There are two types of time delay relays:

- Pull-in delay (on-delay timer)
- Drop-out delay (off-delay timer)

In the on-delay timer shown below, when a push button is pressed (ON), the capacitor is charged through potentiometer R1 as diode D is reverse-biased. The time taken to charge the capacitor depends on the resistance of the potentiometer (R1) and the capacitance (C) of the capacitor. Time delay can be controlled by adjusting the resistance of the potentiometer. When the capacitor is sufficiently charged, coil K is energised and its contacts are operated after the preset time delay. When the push button is released (OFF), the capacitor discharges quickly through a small resistance (R2) as the diode bypasses resistor R1, and the contacts of the relay (K) return to their normal position without any delay.

In an off-delay timer, the contacts are operated without any delay when the push button is pressed (ON). The contacts return to the normal position after the preset delay when the push button is released (OFF).

The construction and symbols of the on-delay and off-delay timers are given in Fig. 9.9. The timing diagram is shown in Table 9.2.

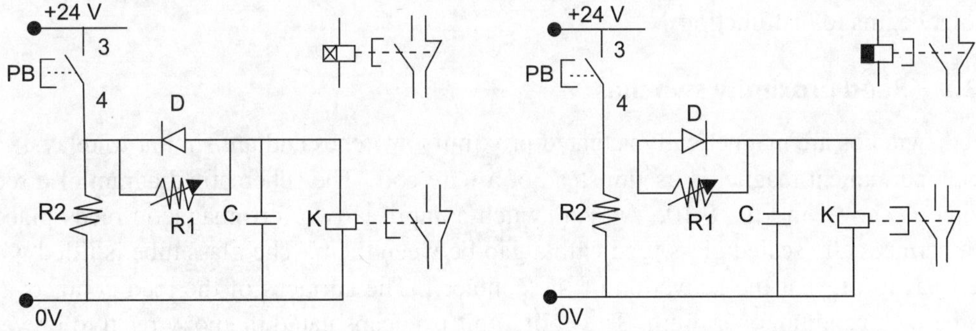

Figure 9.9 *Construction features of a timer and its symbols*

Table 9.2 *Timing diagram for on-delay and off-delay timer*

Timing diagram	Symbol
Voltage at relay coil Open Contact Closed t = on-delay 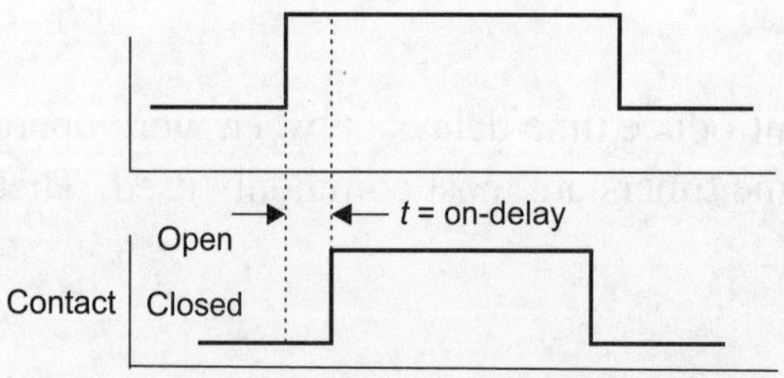	On-delay timer circuit 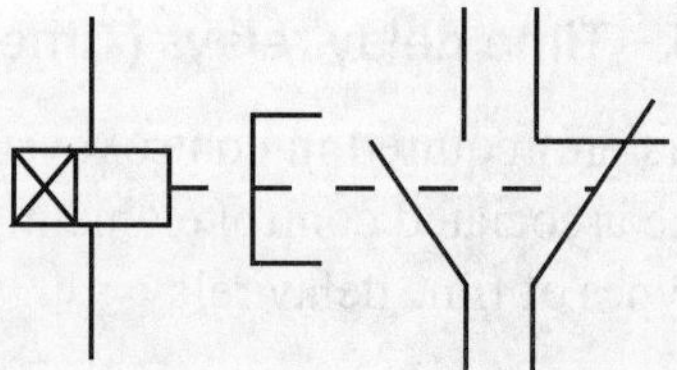
Voltage at relay coil Off-delay Closed Contact Open t 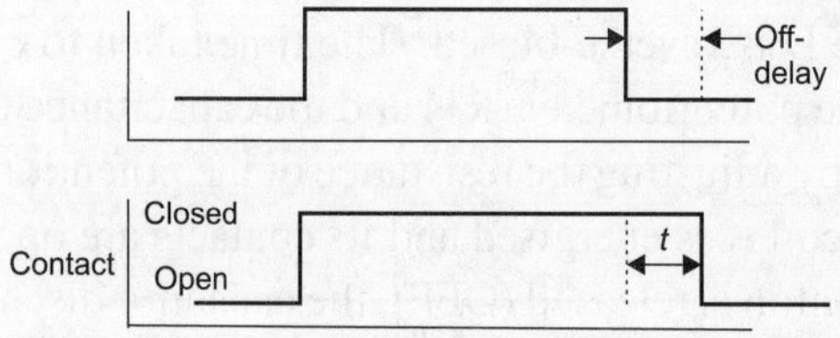	Off-delay timer circuit 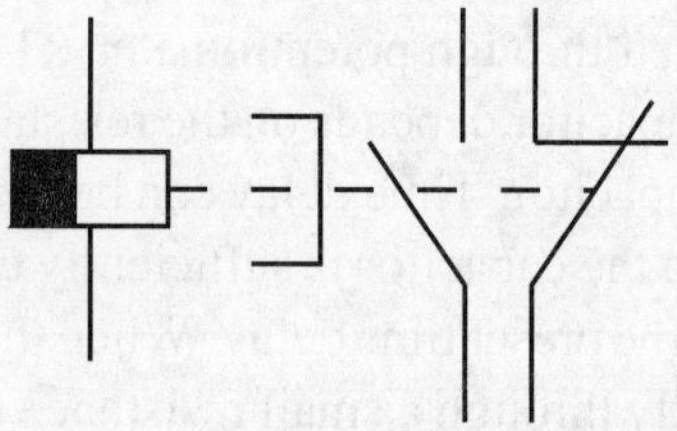

9.2.7 Temperature switch

A temperature switch automatically senses a change in the temperature and opens or closes an electrical switch when a predetermined temperature is reached. It can be wired to be either normally open or normally closed. Temperature switches can be used to protect a fluid power system from serious damage when a component such as a pump, strainer or cooler begins to malfunction.

9.2.8 Reed proximity switches

Reed switches are magnetically actuated proximity switches and are similar to relays, except a permanent magnet is used instead of a wire coil. The schematic diagram of a reed switch is shown in Fig. 9.10. A reed switch comprises two ferromagnetic reeds placed in a hermetically sealed glass tube with a gap between them. The glass tube is filled with inert gas to prevent the activation of the contacts. The surfaces of the reed contacts are plated with rhodium or iridium. The entire unit is encapsulated in epoxy resin to prevent mechanical damage to the switch, and an LED indicator displays the switching status.

When the magnet is at the farthest, the switch is open and when the magnet is brought near, the switch is closed. The reed switch is operated by the magnetic field of an energised coil, or a permanent magnet which induces north (N) and south (S) poles on the reeds.

The reed contacts are closed by this magnetic attractive force. When the magnetic field is removed, the reed elasticity causes the contacts to open the circuit. The transfer type reed switch is normally ON due to mechanical bias of the common lead (COM), which is between the normally closed (NC) reed contact and the normally open (NO) reed contact. When an external magnetic field is induced, the NC blade is not affected because it is non-magnetic but the COM lead is attracted by the NO lead and moves. When the magnetic field is removed, the COM lead again moves to the NC lead by mechanical force.

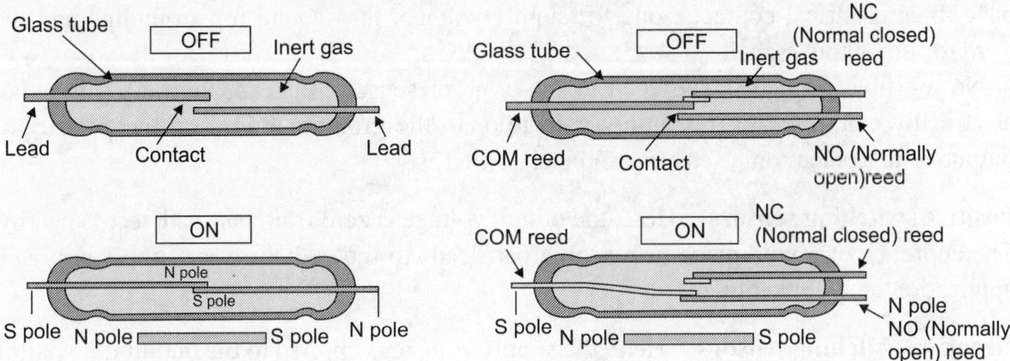

Figure 9.10 *Construction features of a reed switch*

The two-wire reed switch consists of two reeds. One reed is connected to the positive terminal of the electrical supply and other is connected to the signal output. The three-wire reed switch consists of three reed contacts. The first is connected to the positive terminal of the electric supply, the second is connected to the negative terminal of the electric supply and the third is connected to the signal output. The symbol of the three-wire and two-wire reed switch is shown in Fig. 9.11.

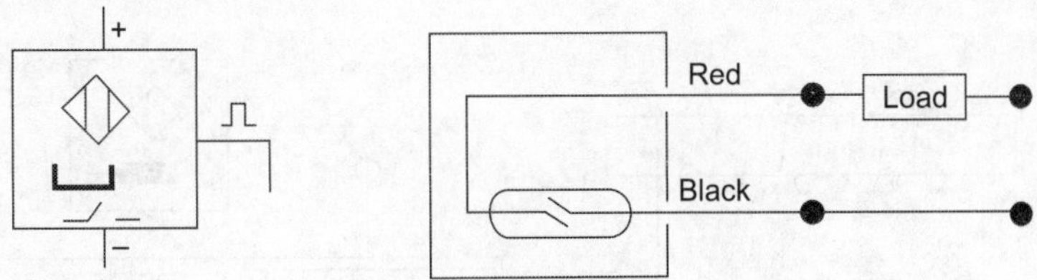

Figure 9.11 *Symbols of 3 wire and 2 wire reed switches*

Advantages of reed switches

1. Reed switches are cheap.
2. They have a long service life.
3. They have shorter switching times (0.2 to 0.3 milliseconds).
4. They are compact and maintenance-free.

Disadvantages of reed switches

1. Cannot be used in environments with magnetic fields (like resistance welding machines)
2. Closing of contacts in a reed switch is not bounce-free

9.2.9 Electronic sensors

Inductive, optical and capacitive proximity switches are electronic sensors. They normally have three electrical contacts, one for supply voltage, the second for grounding and the third for the output signal.

No movable contact is switched in this type of sensor. Instead, the output is either electrically connected to the supply voltage or to the ground. Based on the polarity of output voltage, electronic sensors can be classified as:

Positive switching sensors Here, the output voltage is zero if no part is detected nearby. The approach of a workpiece or machine part leads to the switch-over of the output and application of supply voltage.

Negative switching sensors Here, the supply voltage is applied to the output if no part is detected nearby. The approach of a workpiece or machine part leads to switch-over of the output, switching the output voltage to zero.

(a) Inductive sensors An inductive sensor uses current induced by magnetic fields to detect nearby metal objects. It uses a coil or inductor to generate a high frequency magnetic field as shown in Fig. 9.12. If there is a metal object near the changing magnetic field, current flows into the object. This current flow sets up a new magnetic field that opposes the original field. The net effect is a change in the inductance of the coil in the inductive sensor. By measuring this change, the sensor detects the presence of metal nearby.

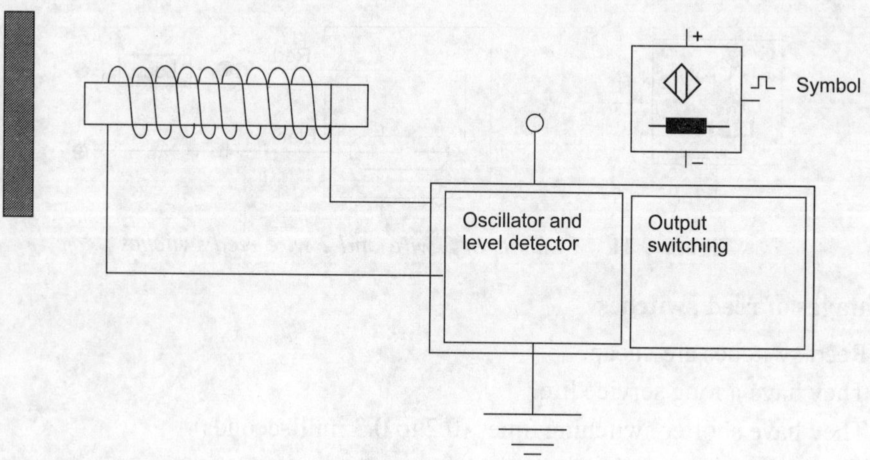

Figure 9.12 *Inductive sensor*

These sensors can detect all metals; therefore multiple sensors are required to sense multiple metal objects or parts. In addition to metals, graphite can also be sensed. It is important to note that these sensors work by setting up a high frequency field. When a target nears the field, it induces eddy currents. These currents consume power because of resistance resulting in a loss of energy in the field, and as a result the signal amplitude decreases. The detector examines field magnitude to determine when it has decreased enough to switch.

The sensors can detect objects a few centimetres away from the end regardless of the direction as shown in Fig. 9.13. The magnetic field of the uncovered sensors extends over a large volume around the head of the coil. Adding a shield (a metal jacket around the sides of the coil) results in a smaller but more directed magnetic field. Inductive sensors use shields to improve their directionality and accuracy.

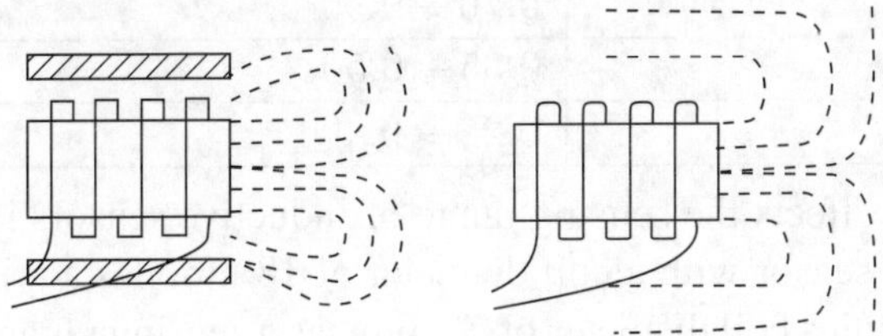

Figure 9.13 *Shielded versus unshielded inductive sensors*

Advantages of proximity sensors:

1. They are self-contained, rugged and extremely reliable
2. They have a long service life
3. They have shorter switching times
4. They are compact and maintenance-free

Disadvantages of proximity sensors:

- They cannot be used in environments with magnetic fields (like resistance welding machine)

Applications of proximity sensors
Proximity sensors are used in the following applications:

1. Sensing the end position of linear actuators (cylinders, semi-rotary actuators)
2. To detect metallic pieces (or their absence) on conveyors
3. To detect the end position in a press
4. To monitor breakages during drilling
5. Used to provide feedback in speed measuring devices

Factors that influence the sensing distance The switching distance of inductive sensors depends on the conductivity and permeability of the metal part whose presence or absence is to be detected. This distance varies with the material composition of the target object; using mild steel as the reference material, the reduction factor is deduced (Table 9.3). The reduction factor is the factor by which the sensing range of the inductive sensor is reduced, based on the material composition of the object to be sensed when compared to steel [FE 360].

Table 9.3 *Reduction factors for various materials*

Material	Reduction factor
Stainless steel	0.80 – 0.85
Nickel steel	0.70 – 0.90
Aluminium and brass	0.35 – 0.50
Copper	0.25 – 0.40

Another factor which affects the sensing range of inductive sensors is the diameter of the sensing coil. A small sensor with a coil diameter of 18 mm has a typical range of 1 mm, while a large sensor with a coil diameter of 75 mm has a sensing range up to 50mm or even more.

(b) Capacitive sensors Capacitive sensors are able to detect most materials at distances up to a few centimetres.

Using the following equation:

$$\text{Capacitance} = \frac{\text{Area of plates} \times \text{Dielectric constant}}{\text{Distance between plates (electrodes)}}$$

The area of the plates and distance between them is fixed in this type of sensor. But the dielectric constant of the space around them varies as the different material is brought close to the sensor. An illustration of a capacitive sensor is shown in Fig. 9.14 where an oscillating field is used to determine the capacitance of the plates. When this changes beyond the selected sensitivity the sensor output is activated.

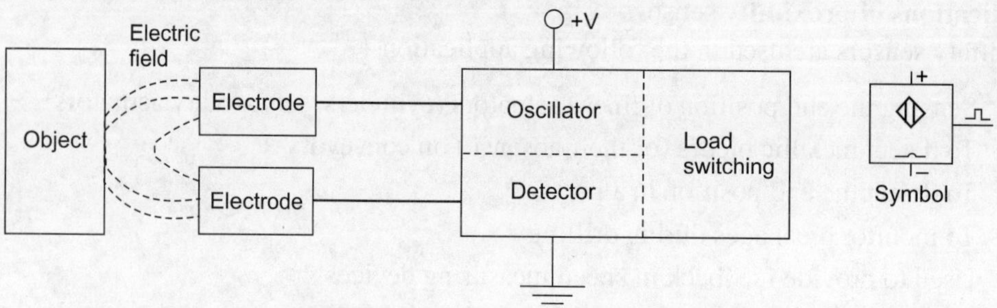

Figure 9.14 *Capacitive sensors*

The proximity of any material near the electrodes increases the capacitance in a capacitive sensor. The increased capacitance varies the magnitude of the oscillating signal and the detector decides when it is great enough to determine proximity.

These sensors work well for insulators (such as plastics) that tend to have high dielectric coefficients, thus increasing the capacitance. But they also work well for metals because the conductive materials in the target appear as larger electrodes, thus increasing the capacitance as shown in Fig. 9.15. In total, the capacitance changes are normally in the order of pF.

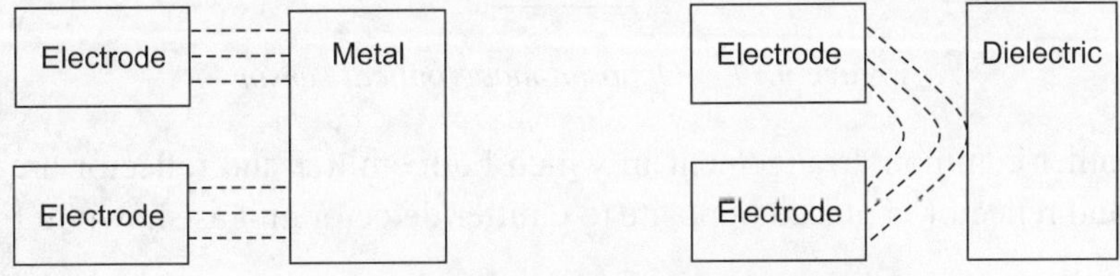

Figure 9.15 *Capacitive sensors for metals and dielectrics*

Advantages of proximity sensors

1. They are widely used because of their ability to react with a wide range of materials.
2. They are suitable for detecting non-metallic objects.
3. They can be used to sense and monitor levels in storage containers.

Disadvantages of proximity sensors

1. They are sensitive especially in humid environments.
2. Without the compensator ring, the sensor is very sensitive to dirt, oil and other contaminants that might stick to the sensor.

(c) Optical proximity sensors Optical sensors require emitter and detector. Emitters emit light using LED or laser diodes. Photodiodes or phototransistors are commonly used detectors. A basic optical sensor is shown in Fig. 9.16

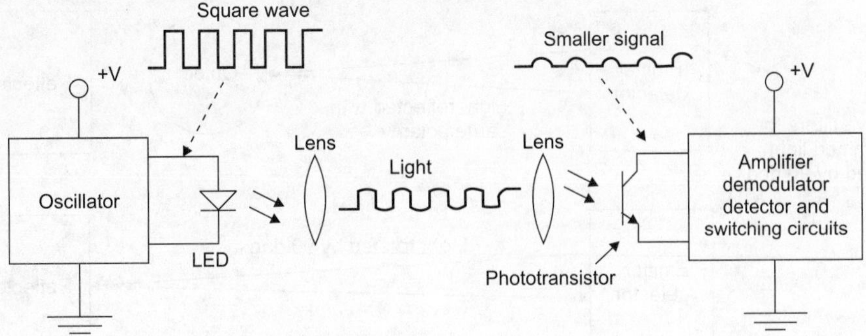

Figure 9.16 *A basic optical sensor*

A beam is transmitted from a emitter and it is received by a receiver as shown in Fig. 9.16. If the object is not present between emitter and detector, the beam is not broken. If the beam then is broken the detector will indicate that an object is present.

If the emitter and detectors are placed as shown in Fig. 9.17, then the mode of operation is called opposed mode. This is because emitter and detector are opposite to each other. Opposed modes are used for opaque and reflective objects.

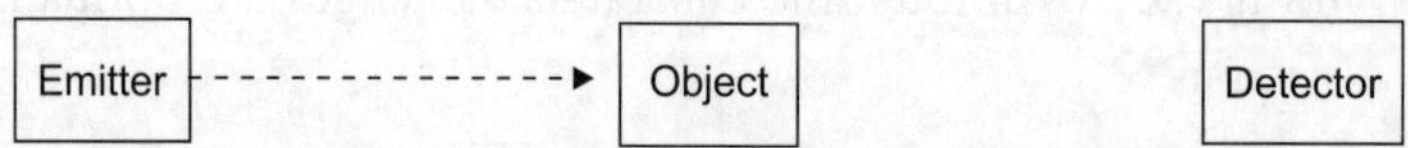

Figure 9.17 *Opposed mode optical sensor*

There is another common arrangement in which both emitter and reflector are housed in single unit and reflector is placed opposite to emitter/detector unit as shown in Fig. 9.18.

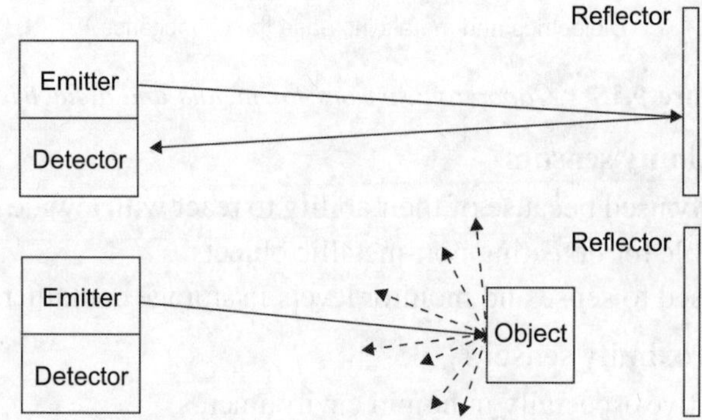

Figure 9.18 *Emitter and detector in one unit*

When reflective object is used in such arrangement (Fig. 9.18), a beam is returned. This problem can be overcome by using polarised light as shown in Fig. 9.19. However, a reflector is still needed in retro-reflective type as shown in Fig. 9.19.

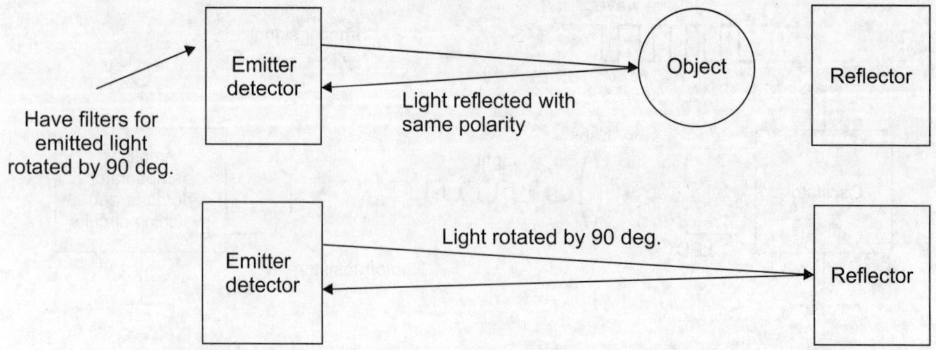

Figure 9.19 *Polarised light in retro-reflective sensors*

A diffuse sensor is a single unit that does not use a reflector, but uses focused light as shown in Fig. 9.20

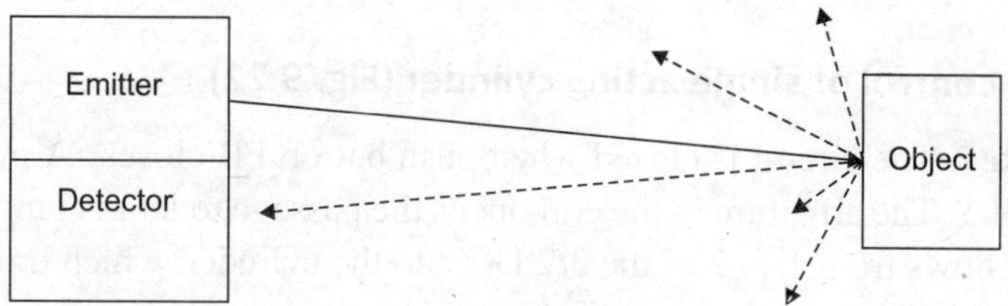

Figure 9.20 *Diffuse sensor*

In diffuse sensors, reflected light is scattered, therefore less amount of light is returned to the detector. Since less amount of light is returned, an amplication unit is necessary to operate it effectively.

9.2.10 Electric counters

An electric counter consists of a coil, associated circuits and contacts, a reset coil, manual reset, a release button and a display window. To preset the counter value, the release button of the counter is pressed and the desired value is entered. The counter valve is displayed in the window. There are two types of counters: up and down.

Up counter An up counter counts electrical signals upwards from zero. For each electrical pulse input to the coil, the counter value is incremented by 1. When the predetermined valve has been reached, the relay picks up and the contact set is actuated.

Down counter A down counter counts down electrical signals from a preset valve. When the count reaches zero the relay picks up and the contact set is actuated. The counter can be reset manually by pressing the reset button or electrically by applying a reset pulse to the reset coil. The preset value is maintained when the counter is reset. The symbols of up and down counters are shown in Fig. 9.21(a) and Fig. 9.21(b).

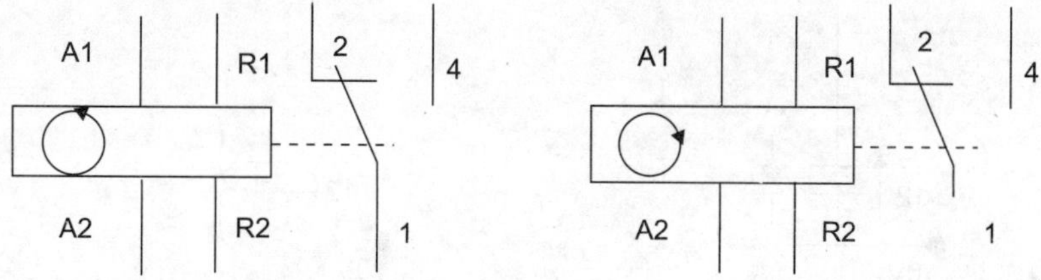

Figure 9.21 *(a) Up counter, (b) Down counter*

9.3 Electro-pneumatic Circuits for Single Actuators

Control of single cylinders using electro-pneumatics

9.3.1 Direct control of single acting cylinder (Fig. 9.22)

Forward stroke The circuit is closed when push button PB closes. A magnetic field is produced in coil Y. The armature in the coil opens the passage to let in compressed air. The compressed air flows from 1 to 2 of the 3/2 DCV to the cylinder, which travels to the final forward position.

Return stroke When push button PB is released, the circuit is interrupted. The magnetic field at coil Y collapses, the 3/2 valve switches back to its original position as shown in Fig. 9.22. The compressed air in the cylinder then exhausts through port 3 of the DCV and the cylinder travels to the final rear position.

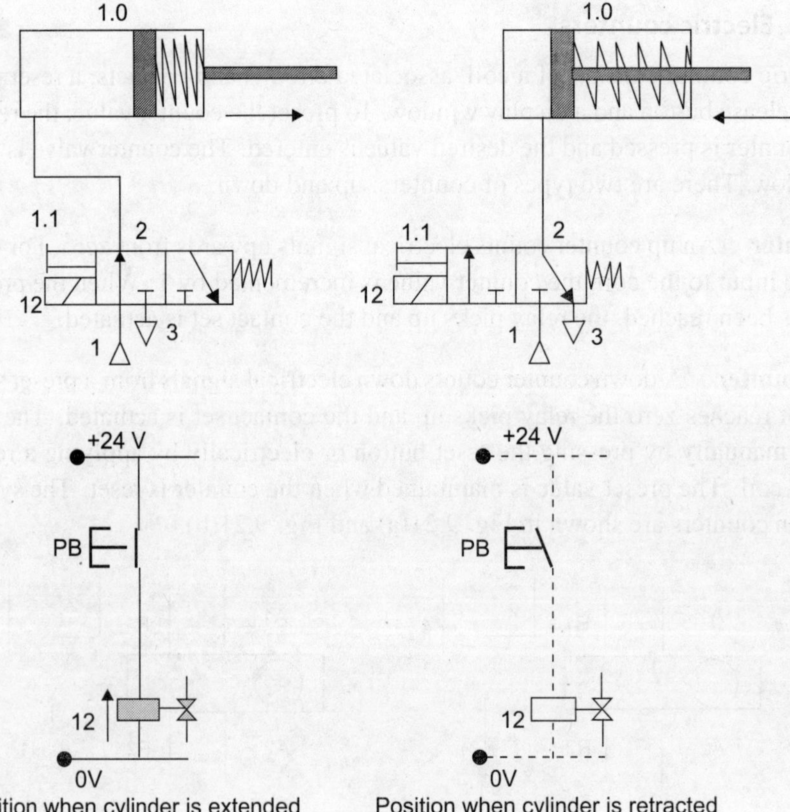

Figure 9.22 *Direct control of single acting cylinder*

9.3.2 Indirect control of a single acting cylinder (Fig. 9.23)

Forward stroke The circuit is closed when push button PB closes which energises NO relay K1 which in turn indirectly energises coil Y. The magnetic field produced in the armature of coil Y opens the passage to let in compressed air. The compressed air flows from 1 to 2 of the 3/2 DCV to the cylinder, which travels to the final forward position.

Return stroke When push button PB is released, the circuit is interrupted which de-energises relay K1 causing the magnetic field at coil Y to collapse. The 3/2 valve switches back to its original position as shown in Fig. 9.23. The compressed air in the cylinder then exhausts through port 3 of the DCV and the cylinder travels to its final rear position.

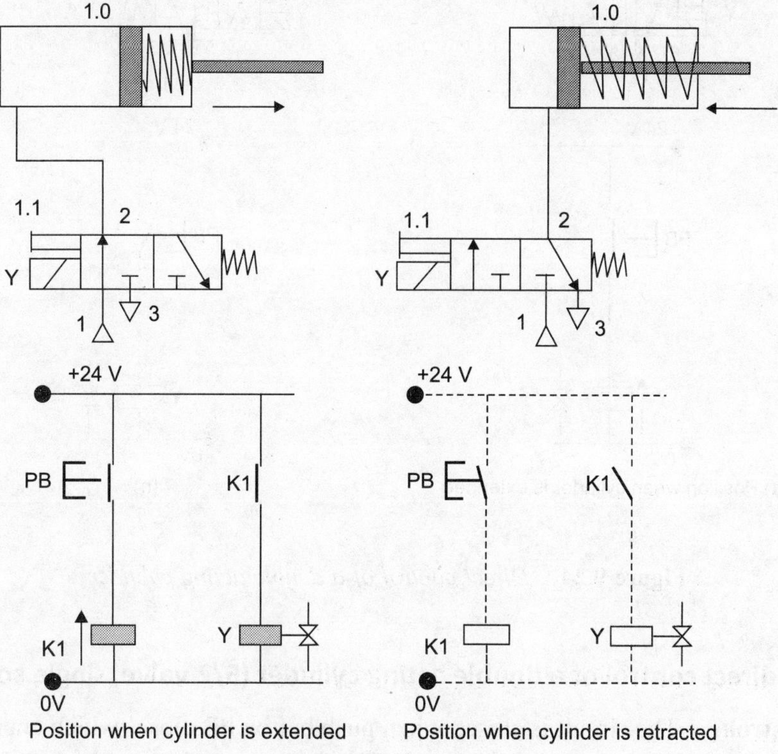

Figure 9.23 *Indirect control of a single acting cylinder*

9.3.3 Direct control of a double acting cylinder (Fig. 9.24)

Forward stroke The double acting cylinder is controlled by a 5/2 valve. When push button PB is pressed, coil Y is energised and the DCV is activated. The piston travels to the final forward position.

Return stroke When push button PB is released, the circuit is interrupted and the magnetic field at coil Y collapses. The return spring of 5/2 becomes active and it switches back to its original position as shown in Fig. 9.24. The compressed air in the cylinder then exhausts through port 5 of the 5/2 DCV and the cylinder travels to the final rear position.

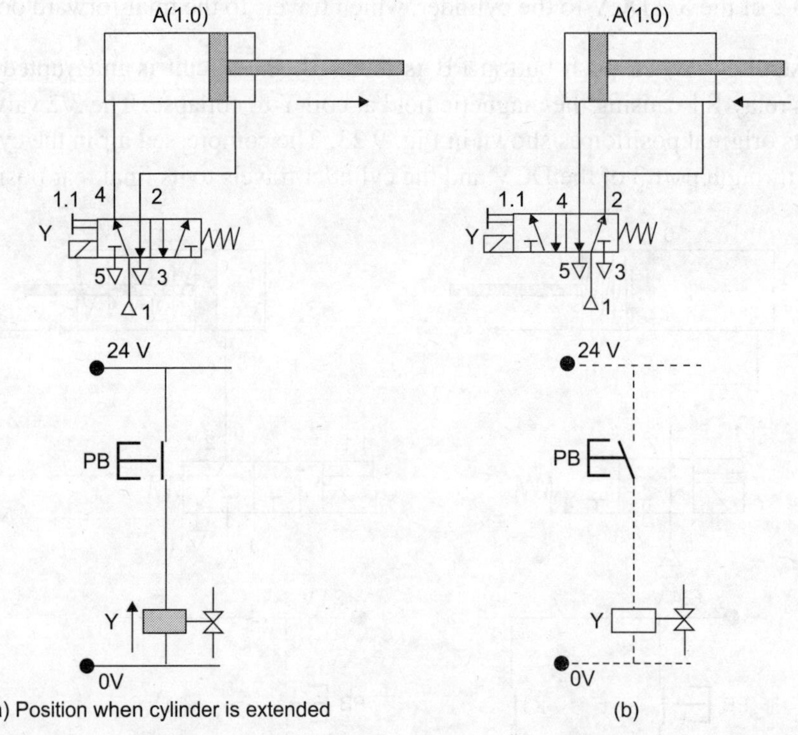

Figure 9.24 *Direct control of a double acting cylinder*

9.3.4 Indirect control of a double acting cylinder (5/2 valve, single solenoid)

Forward stroke The circuit is closed when push button PB closes which energises relay K1 which in turn indirectly energises coil Y. The magnetic field produced in the armature of the coil Y opens the passage to let in compressed air through the internal pilot. The compressed air flows from 1 to 4 of the 5/2 DCV to the cylinder, which travels to the final forward position.

Return stroke When push button PB is released, the circuit is interrupted which de-energises relay K1 causing the magnetic field at coil Y to collapse. The 5/2 valve switches back to its original position as shown in Fig. 9.25. The compressed air in the cylinder then exhausts through port 5 of the DCV and the cylinder travels to the final rear position.

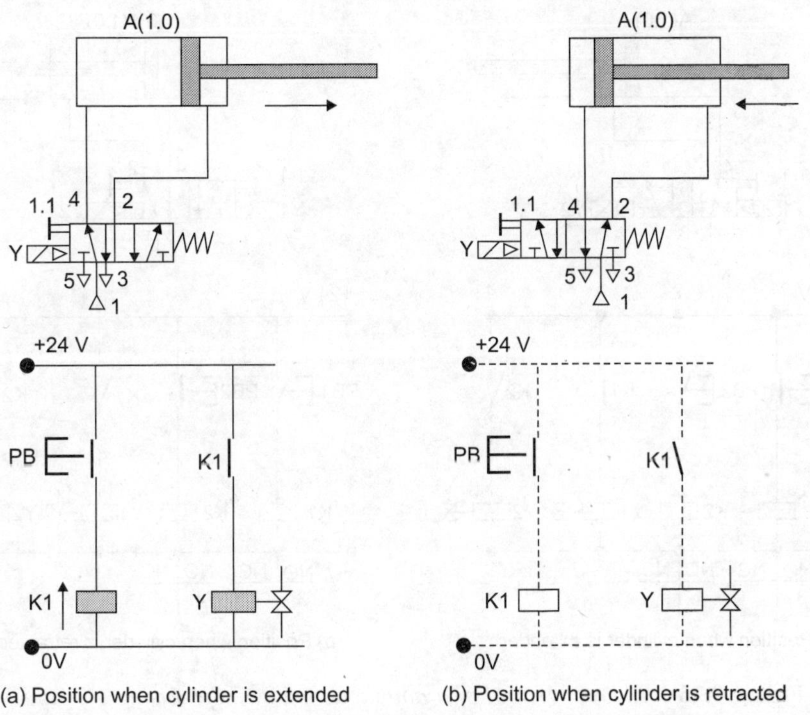

Figure 9.25 *Indirect control of a double acting cylinder*

9.3.5 Indirect control of a double acting cylinder (5/2 valve, double solenoid)

Forward stroke When push button PB1 is pressed, coil Y1 is energised and the 5/2 DCV changes over. The piston travels out and remains in the final forward position until a signal is applied to coil Y2. The 5/2 DCV remains in the last attained position since it is a double solenoid valve with no return spring.

Return stroke When push button PB1 is released and PB2 is pressed, PB1 de-energises relay K1 and the open contact causes the magnetic field at coil Y1 to collapse. The closing of PB2 energises Y2 and the piston returns to its original position as a result of the changeover of the 5/2 valve. The 5/2 DCV does not switch over if there is an active opposing signal. For example, if Y1 is switched on and a signal is given to Y2, there will be no reaction as there would be an opposing signal.

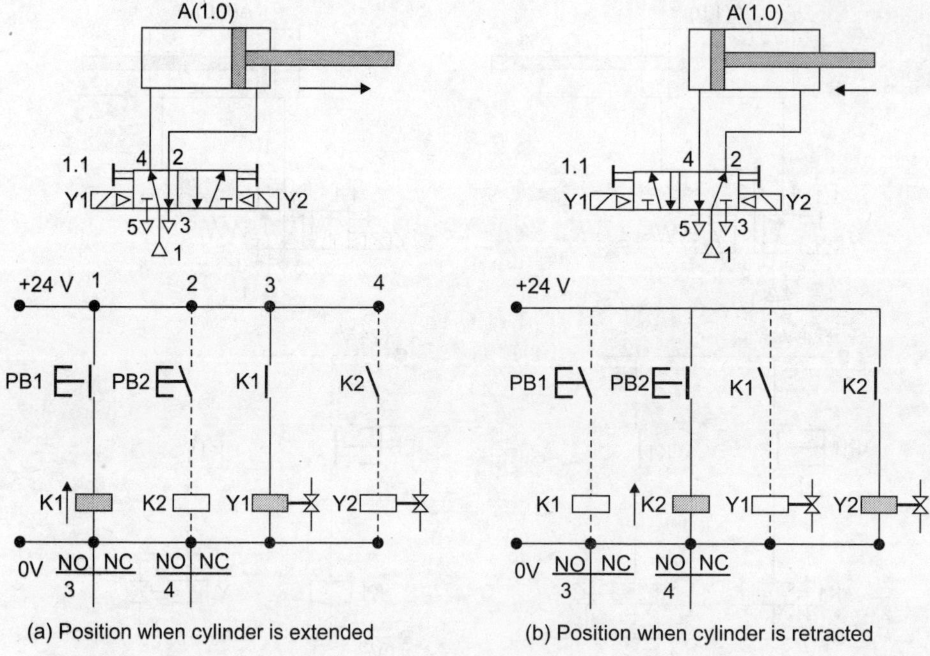

Figure 9.26 *Indirect control of a double acting cylinder (using 5/2 valve)*

9.3.6 Control of a double acting cylinder, OR logic (Parallel circuit)

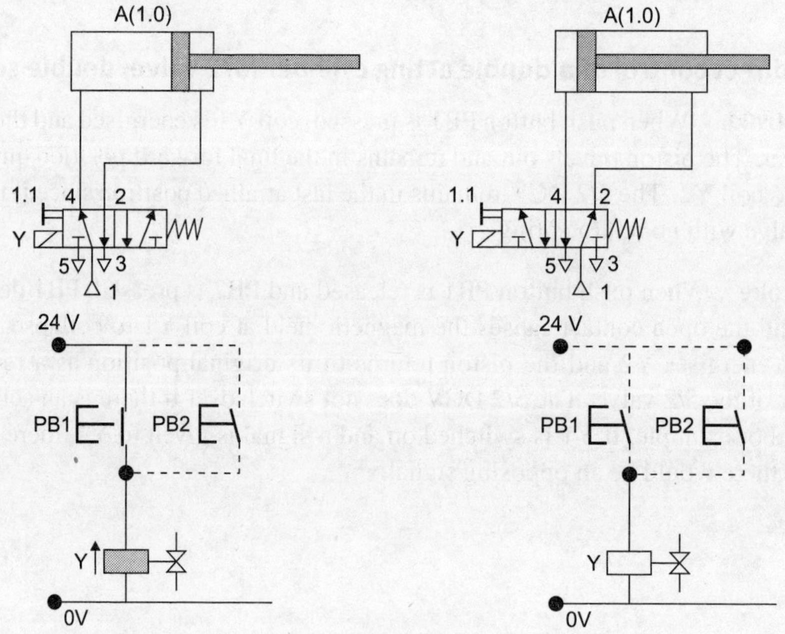

Figure 9.27 *OR logic*

The piston of a double acting cylinder travels out when either one of the two push buttons is pressed. It returns when both are released. When push button PB1 or PB2 is pressed coil Y1 is energised. The DCV switches over and the piston travels to the final forward position. When both push buttons are released, the signal is removed from Y1 and the cylinder travels back to its original position.

9.3.7 Control of a double acting cylinder, AND logic (Fig. 9.28)

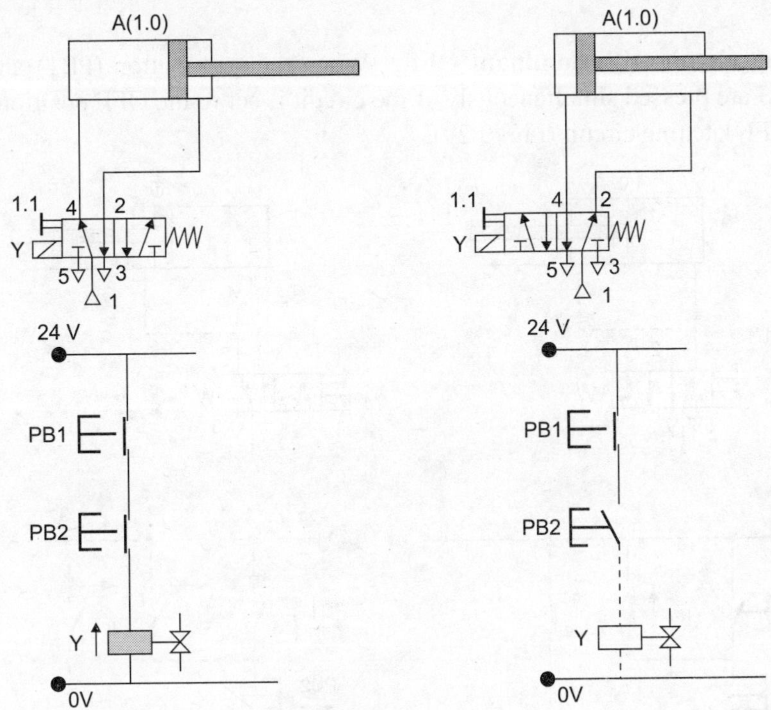

Figure 9.28 *AND logic*

9.3.8 Latching circuits

Example 9.1 A double acting cylinder is to be controlled using a 5/2 DCV, a single solenoid and a spring return. When push button PB1 is pressed, the cylinder should extend and remain in that position when PB1 is released. The cylinder is to retract completely when PB2 is pressed. In addition, the cylinder is to remain in the retracted position even when PB2 is released. Develop an electro-pneumatic control circuit with an electrical latching with (a) dominant OFF (b) dominant ON

Solution In the given pneumatic circuit, a double acting cylinder is controlled by 5/2 valve. When Y1 is energised, the cylinder moves forward. When Y1 is de-energised, the cylinder retracts to its initial position.

The latching circuit can be constructed using the following electrical components:

1. NO push button for start button control (ON)
2. NC pushbutton for stop button control (OFF)
3. Relay

The latching circuit can be dominant ON or dominant OFF. The dominant position refers to the state of the relay coil (circuit) when both the start and stop signals are applied simultaneously.

a) Latching circuit with dominant OFF When the start button (PB1) and the stop button (PB2) are pressed simultaneously, if the circuit is set to the OFF position, then its a dominant OFF latching circuit (Fig. 9.29).

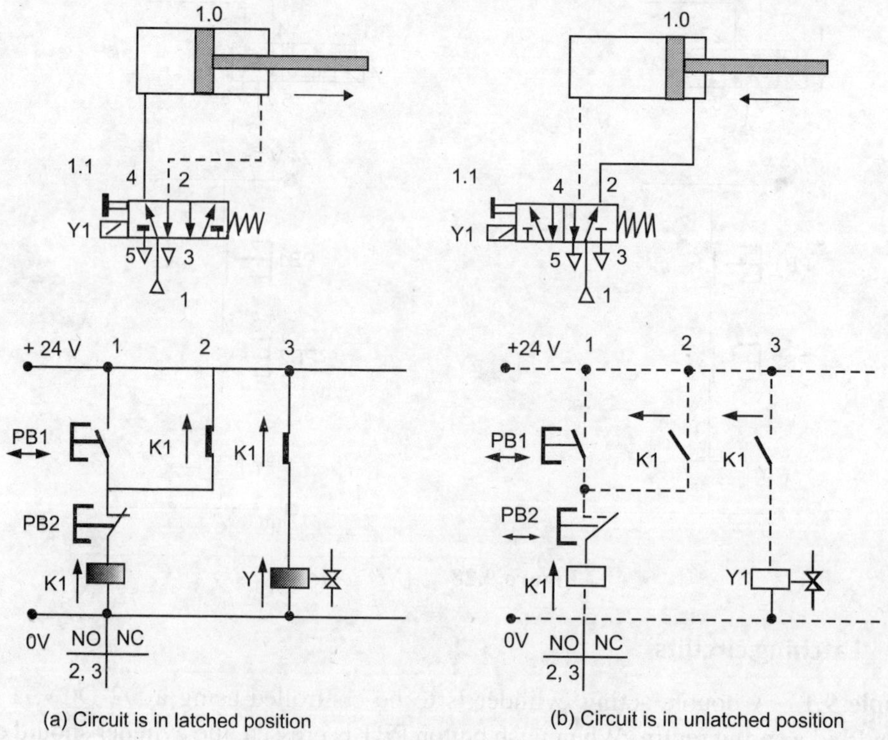

Figure 9.29 *Dominant OFF circuit*

a) When the start pushbutton PB1 is pressed and released, the following operations occurs:

   1. Relay coil K1 in branch 1 (vertical) is energised. All contacts of K1 close.
   2. Note that there is a NO contact of K1 in branch 2, which is connected in parallel to PB1. This NO contact of K1 latches the start push button, so even if the PB1 is released, NO contact of K1 in branch 2 keeps coil K1 energised.

3. There is another NO contact in branch 3, which is connected to Y1. When push button PB1 is pressed, this remains closed. As a result, cylinder moves forward and remains there until stop button PB2 is pressed.

b) When stop is pressed, push button PB2 is pressed momentarily and released and the following operations occur:

1. Relay coil K1 in branch 1 (vertical) is de-energised. All contacts in K1 are open.
2. NO contact of K1 in branch 2, which is connected in parallel to PB1 is now open. This NO contact of K1 stops latching the start push button.
3. NO contact in branch 3 is also open now and de-energised. As a result, the cylinder moves back to its home position and remains there until the start button PB1 is pressed again.

When start button (PB1) and stop button (PB2) are pressed simultaneously, K1 contact opens and the circuit goes to the OFF position, and is therefore termed as dominant OFF latching circuit.

b) Latching circuit with dominant ON When the start button (PB1) and the stop button (PB2) are pressed simultaneously, if the circuit is set to the ON position, then it is termed a dominant ON latching circuit (Fig. 9.30).

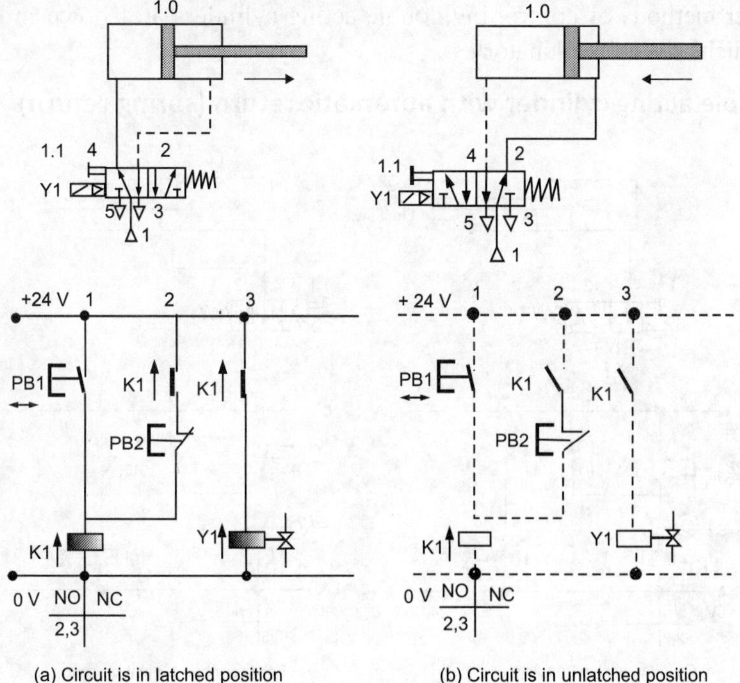

Figure 9.30 *Dominant ON circuit*

a) When pushbutton PB1 is pressed and released and the following operations occur:
   1. Relay coil K1 in branch 1 (vertical) is energised. All contacts in K1 close.
   2. Note that there is a NO contact of K1 in branch 2, which is connected in parallel to PB1 and in series with PB2. This NO contact of K1 latches the start push button, so that even if PB1 is released, the NO contact of K1 in branch 2 keeps coil Y1 energised.
   3. There is another NO contact in branch 3 which is connected to Y1. When push button PB1 is pressed, this also remain closed, and as a result the cylinder moves forward and remains there until the stop button PB2 is pressed.

b) When stop is pressed, push button PB2 is pressed momentarily and released. Then the following operations occur:
   1. Relay coil K1 in branch 1 (vertical) is de-energised. All contacts of K1 open.
   2. NO contact of K1 in branch 2, which is connected in parallel to PB1 is now open and stops latching the start pushbutton.
   3. NO contact in branch 3 is also open now and de-energised. As a result, the cylinder moves back to its home position and remains there until the start button PB1 is pressed again.

When start button PB1 and stop button PB2 are pressed simultaneously, K1 contacts open and the circuit is set to the OFF position, and therefore is termed as a dominant OFF latching circuit. Other methods of controlling double acting cylinders are shown in Fig. 9.31 to Fig. 9.36, which are self explanatory.

9.3.9 Double acting cylinder with automatic return (spring return)

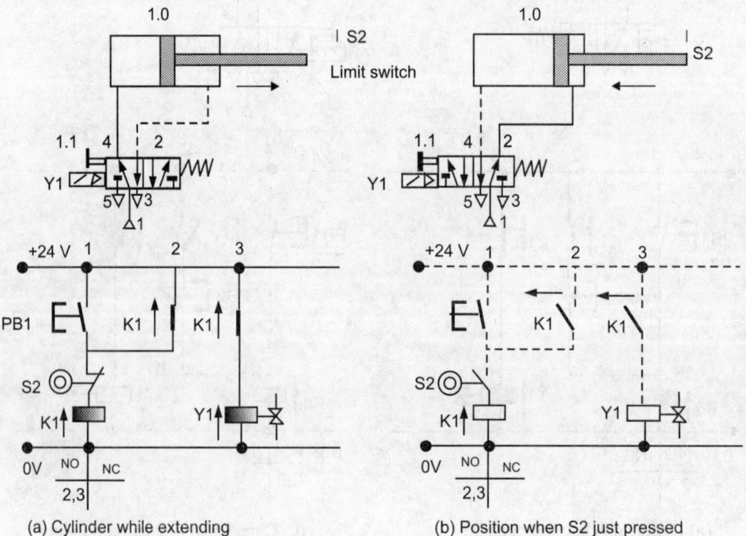

Figure 9.31 *Double acting cylinder with automatic return using single solenoid*

9.3.10 Direct control of a double acting cylinder with automatic return (double solenoid)

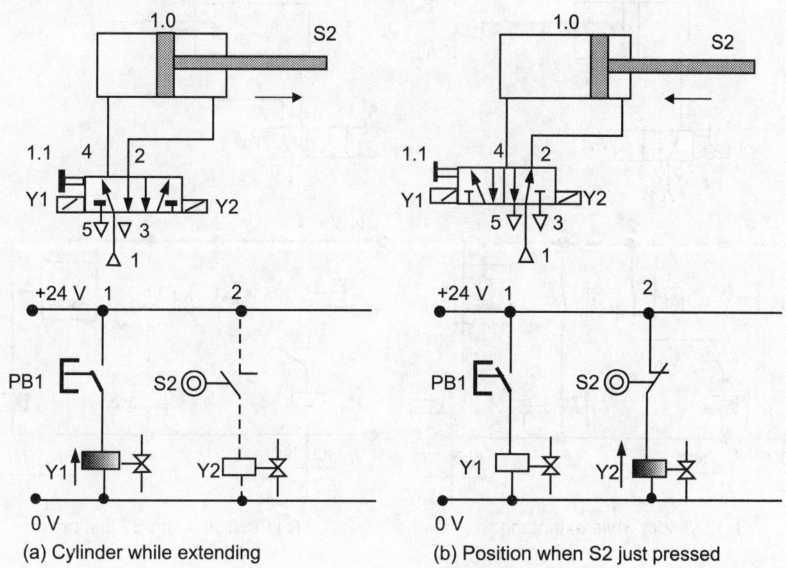

Figure 9.32 *Direct control of a double acting cylinder with automatic return*

9.3.11 Indirect control of a double acting cylinder with automatic return (double solenoid)

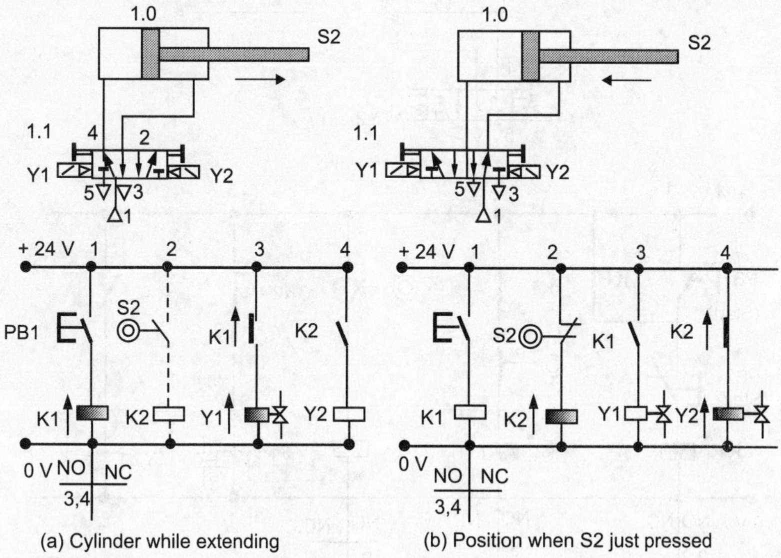

Figure 9.33 *Indirect control of a double acting cylinder with automatic return (double solenoid)*

9.3.12 Automatic return of a double acting cylinder (using proximity switch)

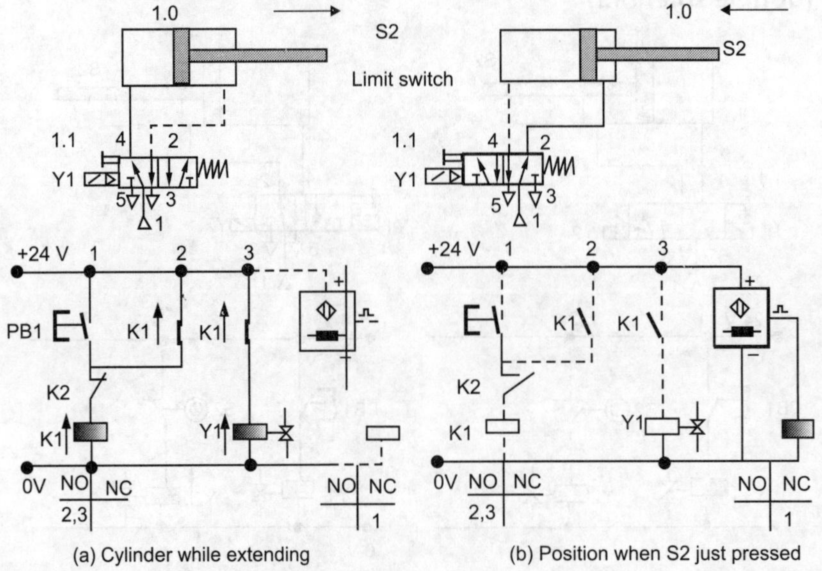

Figure 9.34 *Automatic return of a double acting cylinder using a proximity switch*

9.3.13 Oscillating motion of a double acting cylinder (forward)

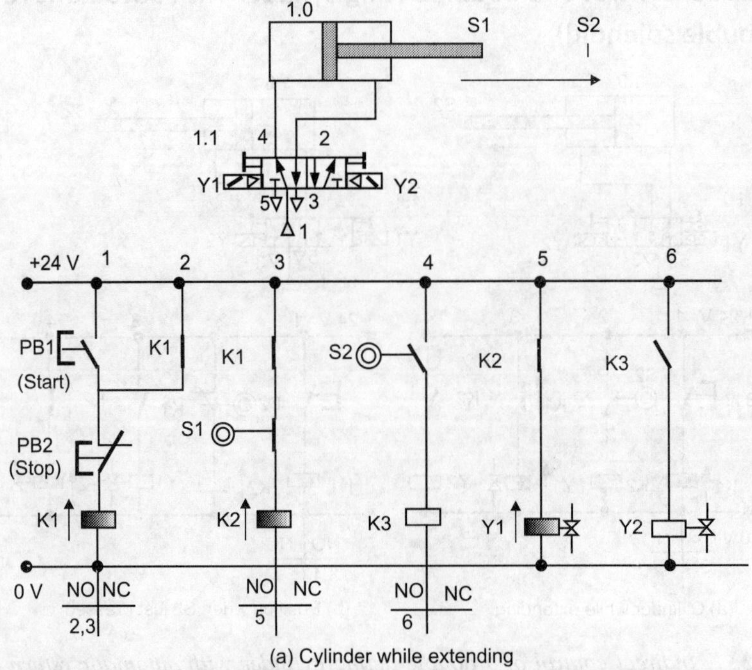

Figure 9.35 *Oscillating motion of double acting cylinder (forward motion)*

9.3.14 Oscillating motion of a double acting cylinder (Return)

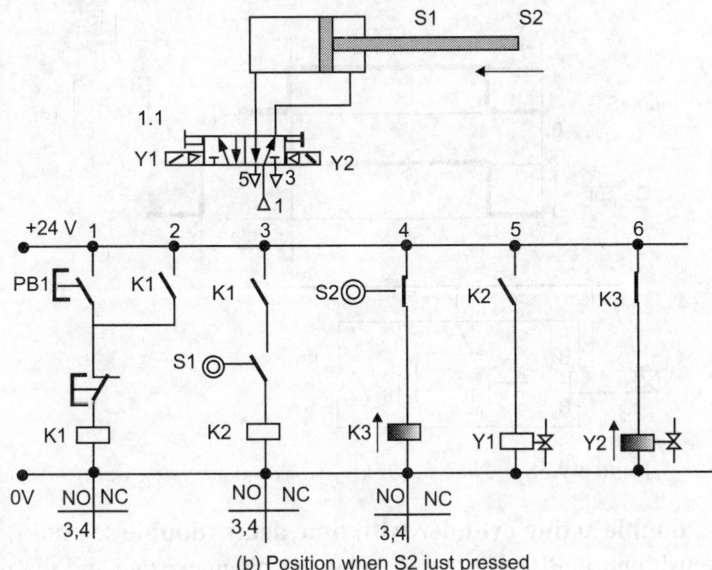

Figure 9.36 *Oscillating motion of double acting cylinder (return motion)*

9.3.15 Control systems with timed response

Control systems which are assigned a particular timing sequence must be equipped with electrical time-lag relays. There are control systems which are purely affected by time or a combination of path scanning and time.

Time-lag relays, which are electronic nowadays, have two basic types of timed responses: with energising delay and with de-energising delay are shown in Figs 9.37 and 9.38 respectively.

a) Time-lag relay with energising delay

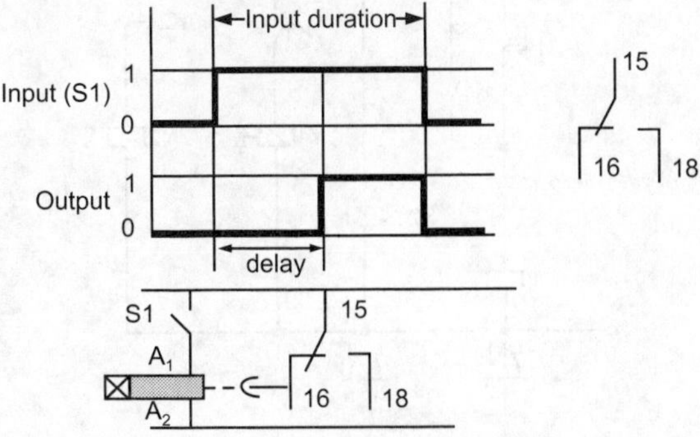

Figure 9.37 *Time-lag relay with energising delay*

b) Time-lag relay with de-energising delay

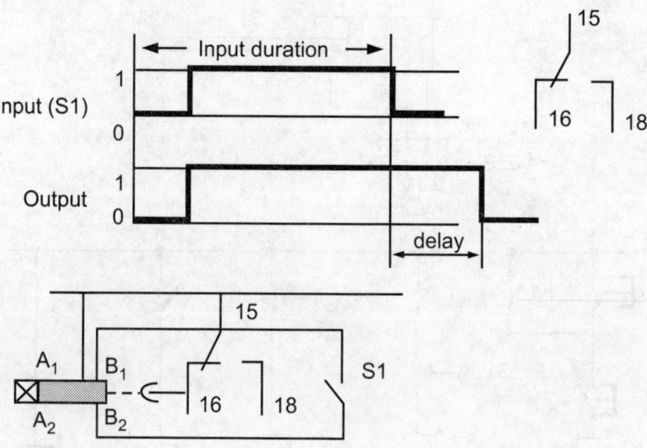

Figure 9.38 *Time lag relay with de-energising delay*

c) Control of a double acting cylinder with time delay (double solenoid)

When manual push button PB1 is pressed, relay K1 changes state and the normally open (NO) contact K1 of the relay is connected to solenoid coil Y1. When the NO contact closes, the solenoid valve changes state and the cylinder travels to its final forward position where it actuates the limit switch S2. This limit switch starts the time-lag relay K2 (with energising delay).

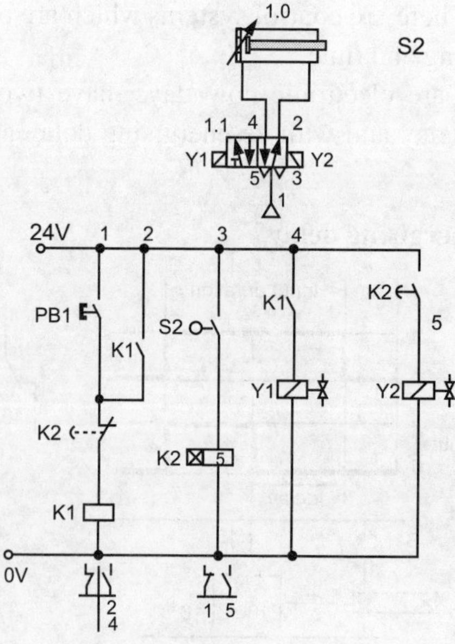

Figure 9.39 *Control of a double acting cylinder with time delay*

After 5 seconds the NO contact of the time-lag relay energises solenoid coil Y2 of the directional control valve. The valve changes over and causes the piston to travel to its final rear position.

d) Control of a double acting cylinder using timer (single solenoid)

A latching circuit is used to implement the necessary memory function. The position of the circuit when push button PB1 is pressed and then released is given in Fig. 9.40(a). The cylinder extends to its forward end position and automatically actuates the limit switch S2. As the return motion is to be delayed, the on-delay timer is used to introduce the time delay. Limit switch S2 controls the timer coil T. After the preset delay, the timer contact interrupts the latching circuit causing the return motion of the cylinder as shown in Fig. 9.40(b).

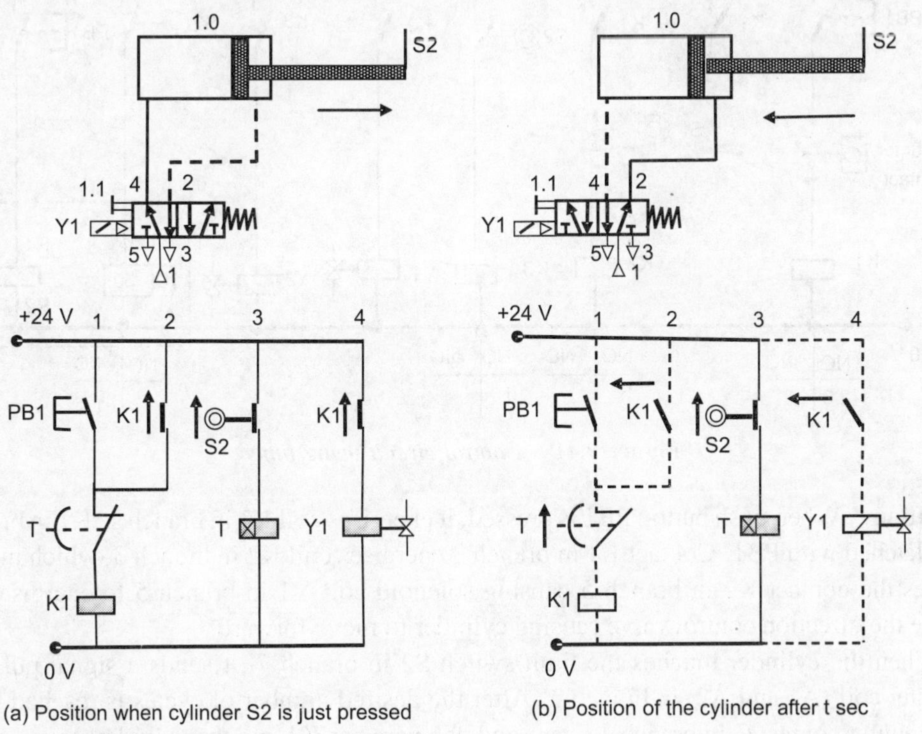

Figure 9.40 *Control of a double acting cylinder with time delay*

9.3.16 Control of a double acting cylinder using an electric counter with two end sensors

Example 9.2 Design an electro-pneumatic circuit for a double acting cylinder performing continuous to and fro motion. The cylinder has to stop automatically after 50 back and forth cycles.

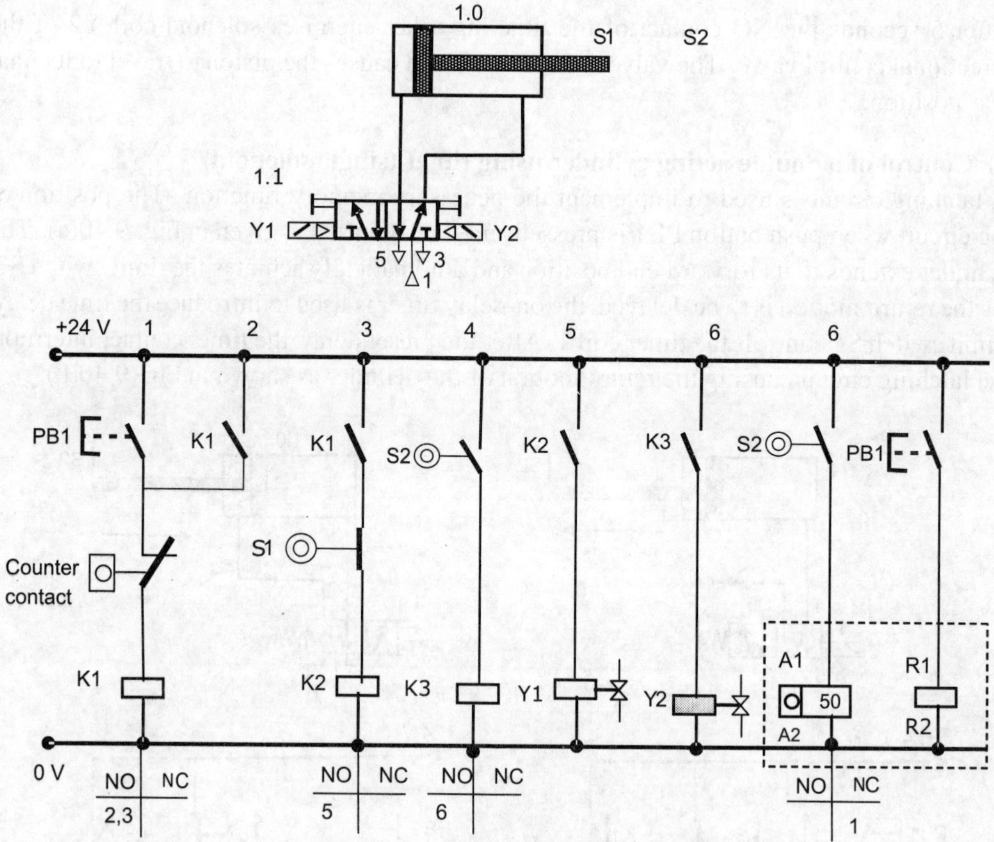

Figure 9.41 *Control circuit using timer*

Solution When push button PB1 is pressed, it energises coil K1 in branch 1. K1 in branch 2 is latched with PB1. Contact K1 in branch 3 energises coil K2 in branch 3, which in turn closes the contact K2 in branch 5 causing solenoid coil Y1 in branch 5 to energise and move the direction control valve causing cylinder to move forward.

When the cylinder touches the limit switch S2 in branch 7, it sends a signal pulse to counter coil (A1 and A2) in branch 7. After the desired number of cycles is reached (50), the counter contact C in branch 1 opens and de-energises K1 and the cylinder stops.

9.3.17 Oscillation of a double acting cylinder using proximity switches

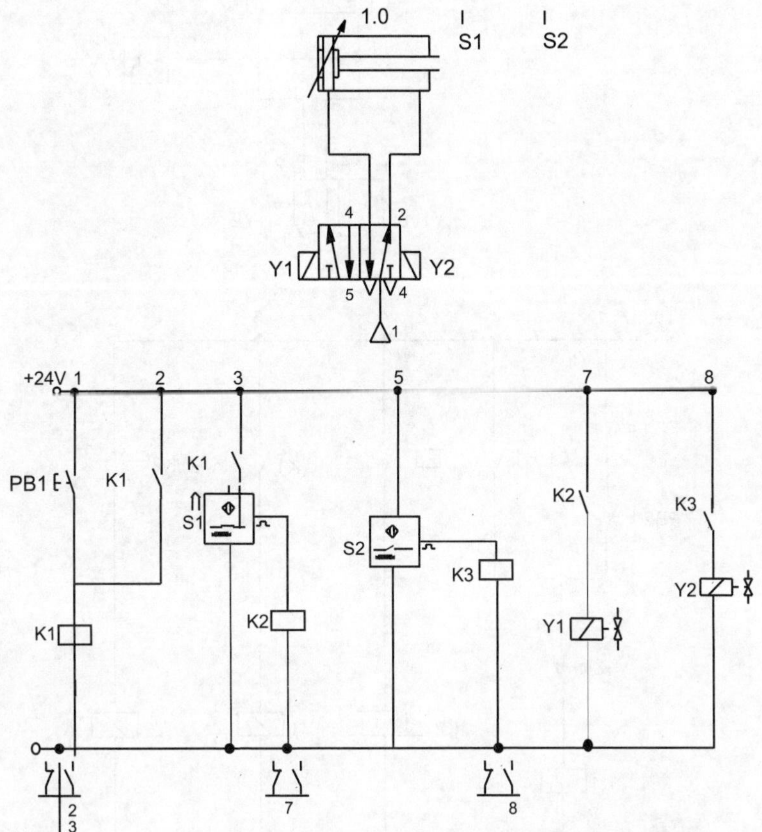

Figure 9.42 *Oscillation of a double acting cylinder using proximity switches*

9.3.18 Control of a double acting cylinder using a pressure switch

Example 9.3 Components are to be stamped using a stamping machine. When two push buttons are simultaneously pressed, a double acting cylinder pushes the attached die down to a fixture. The die is to return to the initial position upon reaching sufficient stamping pressure, as sensed by a pressure switch, followed by a one second delay. Develop an electro-pneumatic control circuit to implement the control task for the stamping operation.

Solution is shown in Fig. 9.43, which is self-explanatory.

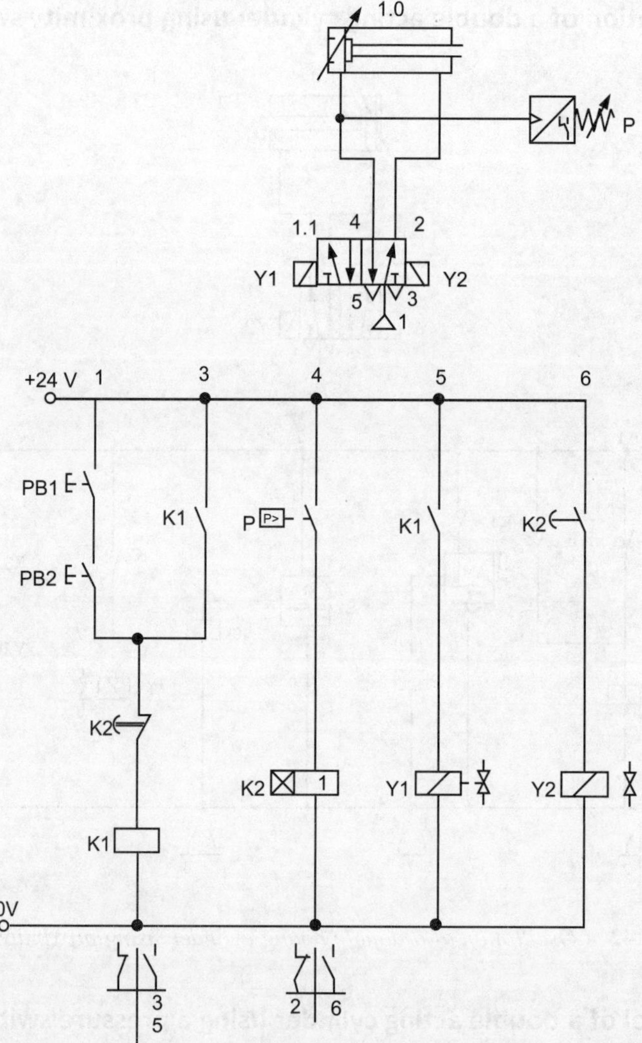

Figure 9.43 *Control of a double acting cylinder using a pressure switch*

9.3.19 Control of a double acting cylinder using a delay on and off timer and counter

Example 9.4 Components are to be stamped using a stamping machine. A double acting cylinder is used to push the attached die down to a fixture one second after the push button is pressed. The die is to return to the initial position after a stamping time of two seconds. This cycle should stop automatically after 5 cycles. The start button should reset the counter. Develop an electro-pneumatic control circuit to implement the control task for the stamping operation.

The solution is shown in Fig. 9.44, which is self-explanatory.

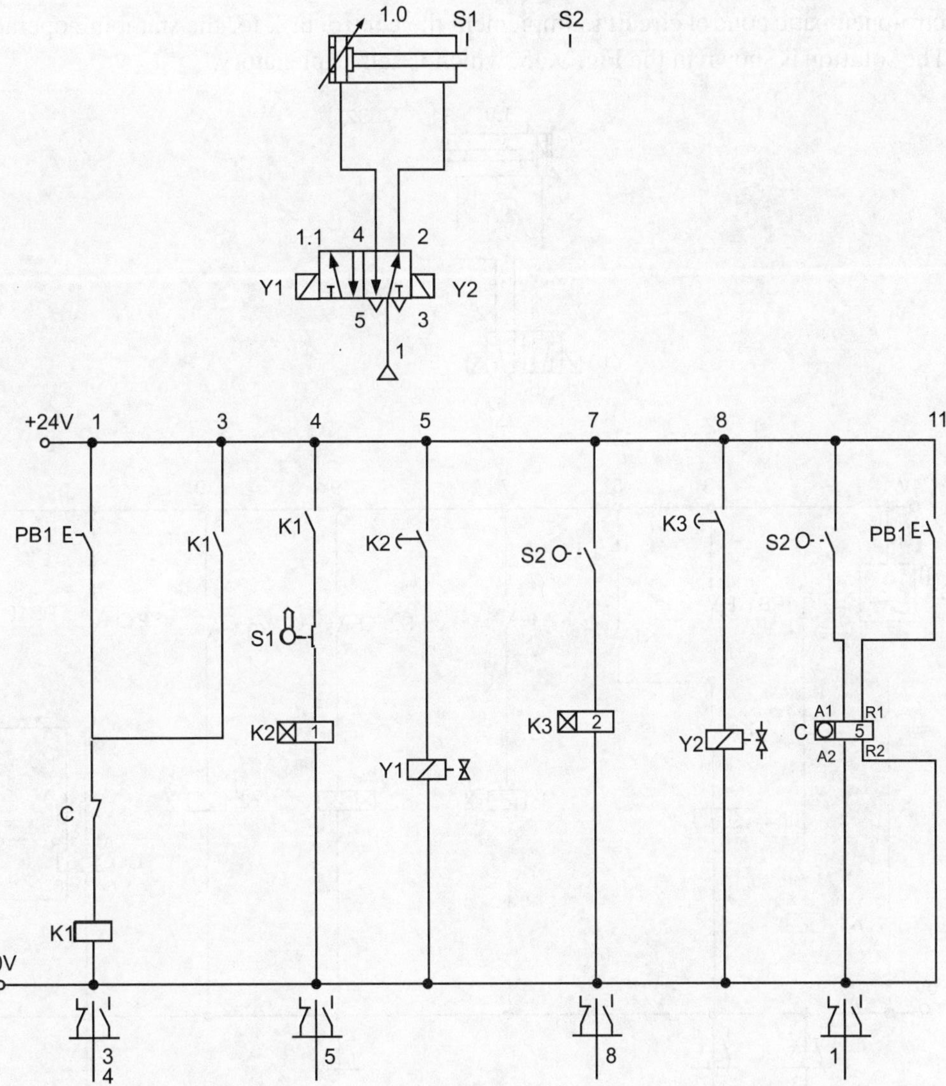

Figure 9.44 *Control of a double acting cylinder using counter and delay*

9.3.20 Control of a double acting cylinder using a reed switch for initial sensing

Example 9.5 Components are to be stamped using a stamping machine. A double acting cylinder is used to push the attached die down to a fixture one second after the push button is pressed. The die is to return to the initial position after a sufficient stamping time of two seconds. This cycle should stop automatically after 5 cycles. The start button should reset the counter. The initial position sensing is done by a magnetic reed switch. Develop an electro-pneumatic control circuit to implement the control task for the stamping operation.

The solution is shown in the Fig. 9.45, which is self-explanatory.

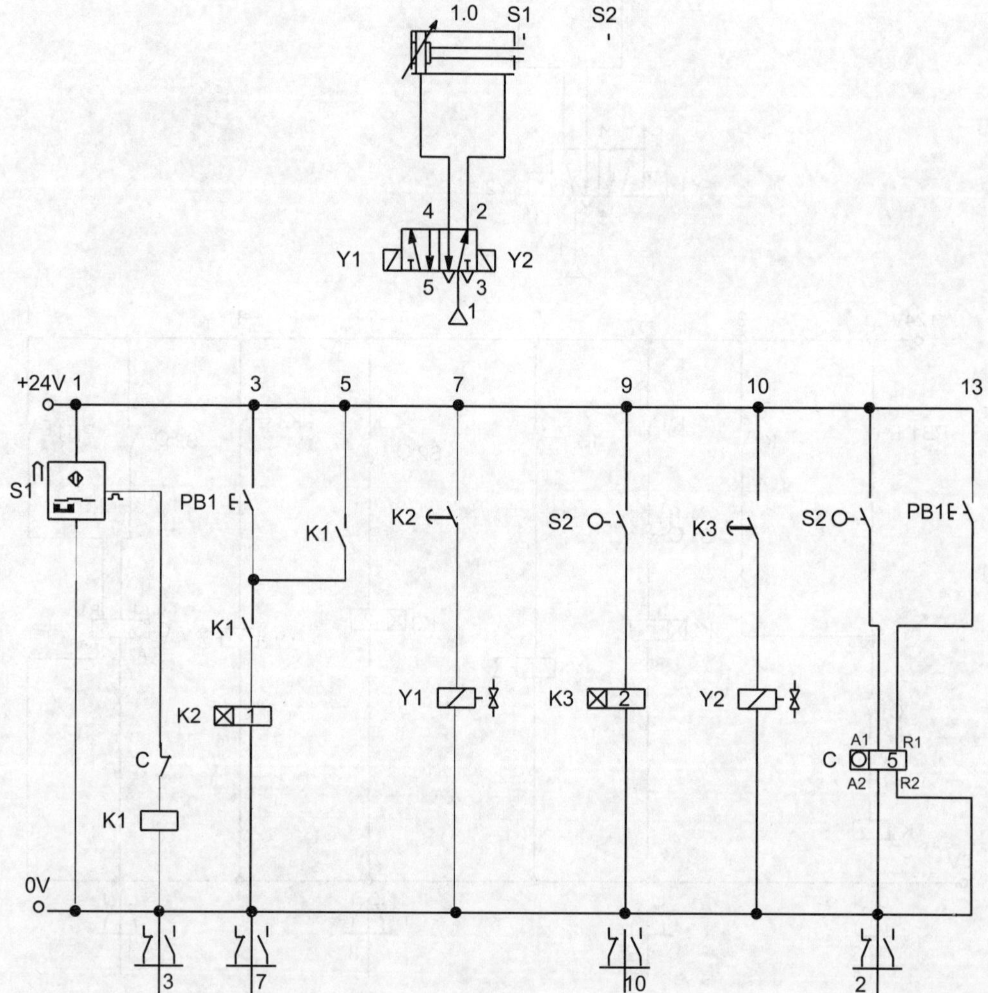

Figure 9.45 *Control of a double acting cylinder using a reed switch for initial sensing*

Example 9.6 A pneumatically controlled press with a stamping die shown in Fig. 9.46 is used for producing badges from a thin metal sheet. A double acting cylinder is to extend when push buttons PB1 and PB2 are pressed simultaneously. The cylinder is to retract automatically after reaching the forward end position and attaining a preset pressure to get consistent quality. The cylinder should retract immediately if emergency push button PB3 is pressed. Develop a electro pneumatic control circuit to implement the control task.

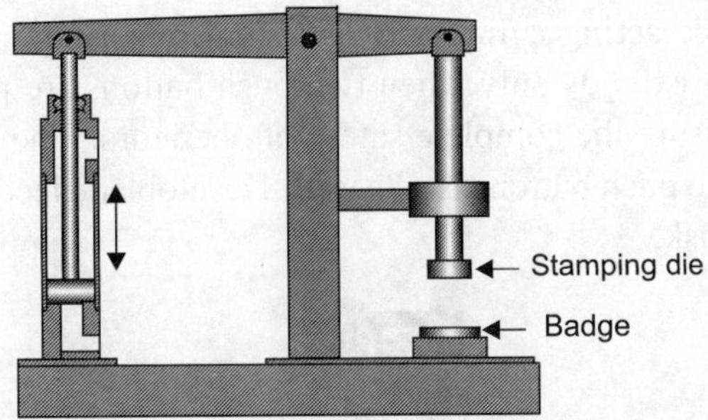

Figure 9.46 *Stamping machine for making badges*

The solution is shown in Fig. 9.47, which is self-explanatory.

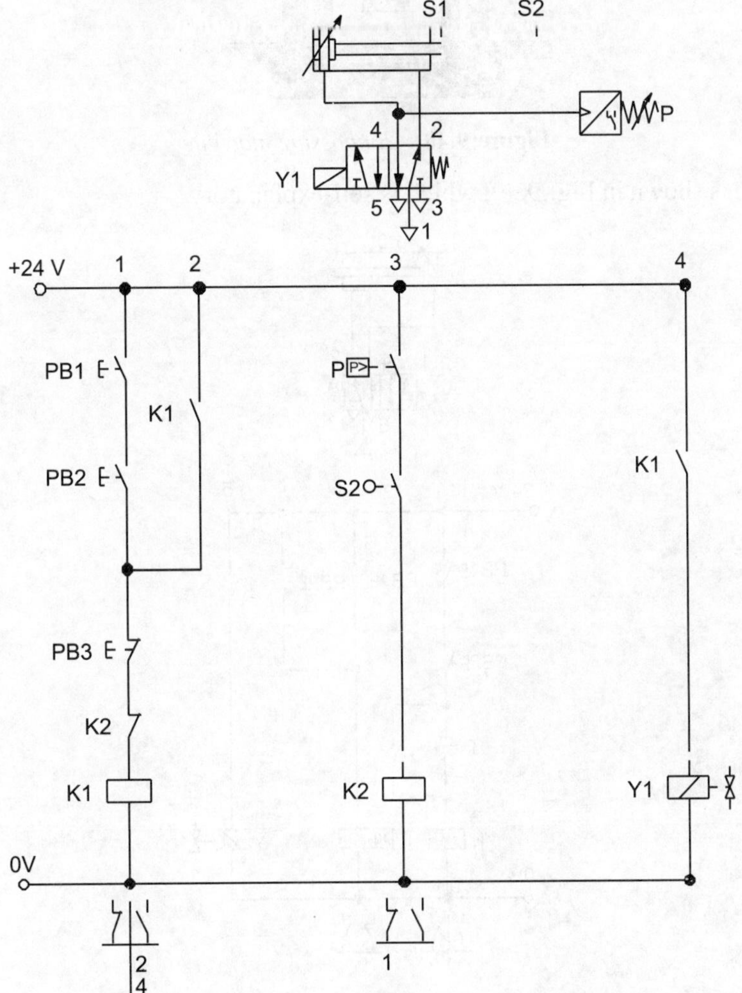

Figure 9.47 *Stamping machine with pressure sensing*

Example 9.7 A double acting cylinder is used for embossing slide rules as shown in Fig. 9.48. The cylinder extends only when two push buttons are pressed simultaneously within one second to ensure the complete safety of operators. The cylinder should retract immediately if any of the push buttons is released. Develop an electro-pneumatic circuit to implement the control task.

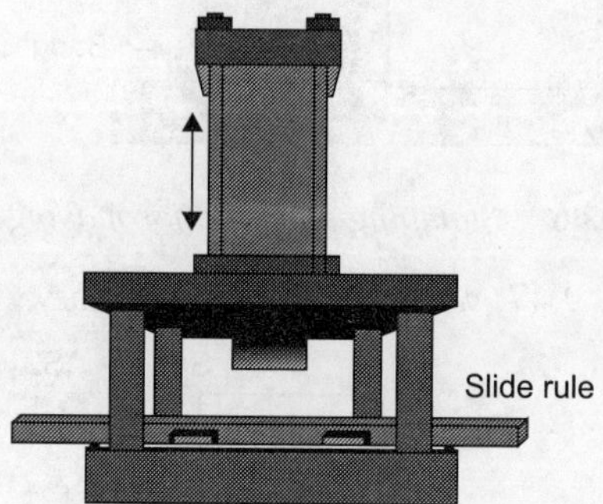

Figure 9.48 *Embossing machine*

The solution is shown in Fig. 9.49, which is self-explanatory.

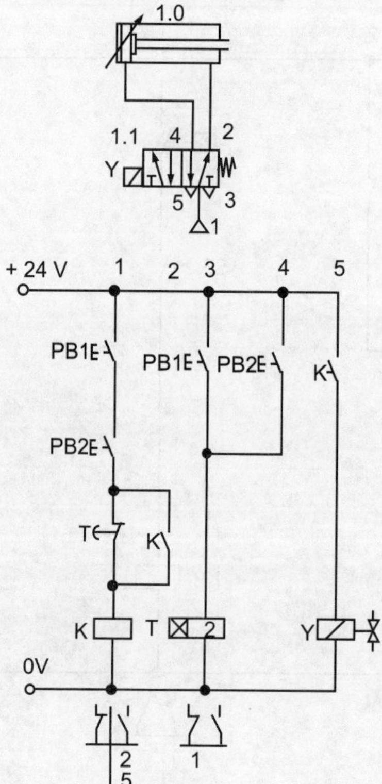

Figure 9.49 *Control circuit for an embossing machine*

Example 9.8 Figure 9.50 shows the arrangement for quenching heat treated valves for automobile engines. A double acting cylinder is used to move the container filled with valves up and down in a bath for a few cycles. The operator provides the start signal manually. The heat treatment of the valves is turned off automatically after a preset time. Develop a electro pneumatic control circuit to implement the control task.

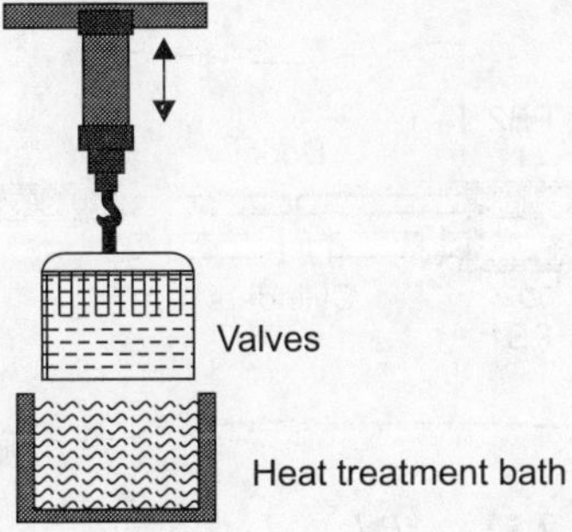

Figure 9.50 *Quenching device*

The solution is shown in Fig. 9.51, which is self-explanatory.

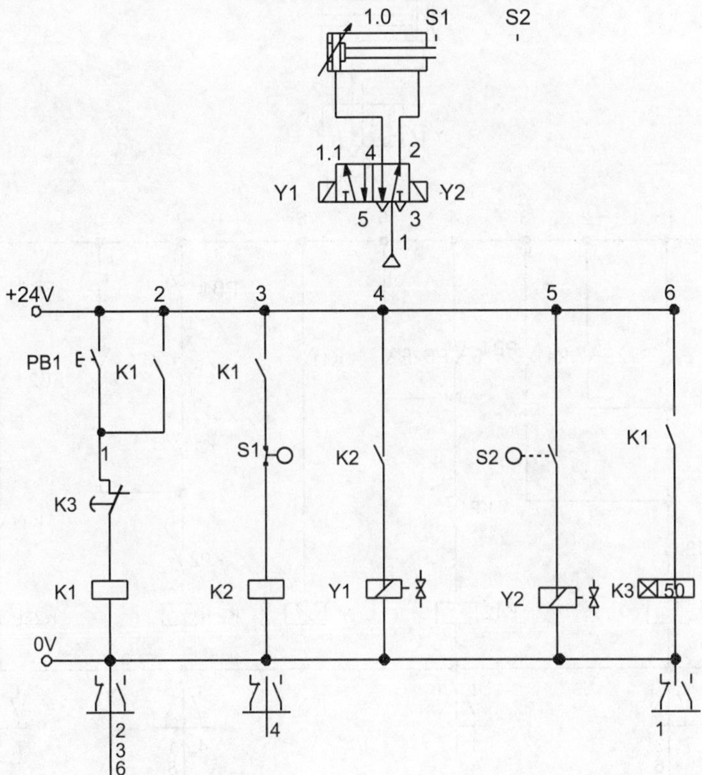

Figure 9.51 *Control circuit for quenching device*

Example 9.9 A pneumatically controlled double sliding door is used in the metrology lab (Fig. 9.52). The door is to be controlled either from the inside by push button PB1 or from the outside by pressing push button PB2. The opening and closing operations are initiated by the same push button in each case. The speeds of the double sliding door movements are to be adjustable. Develop a electro pneumatic control circuit to implement this given task.

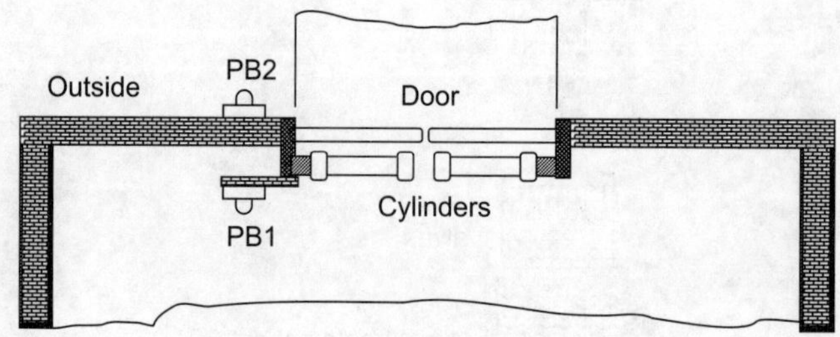

Figure 9.52 *Sliding door in metrology lab*

The solution is shown in Fig. 9.53, which is self-explanatory.

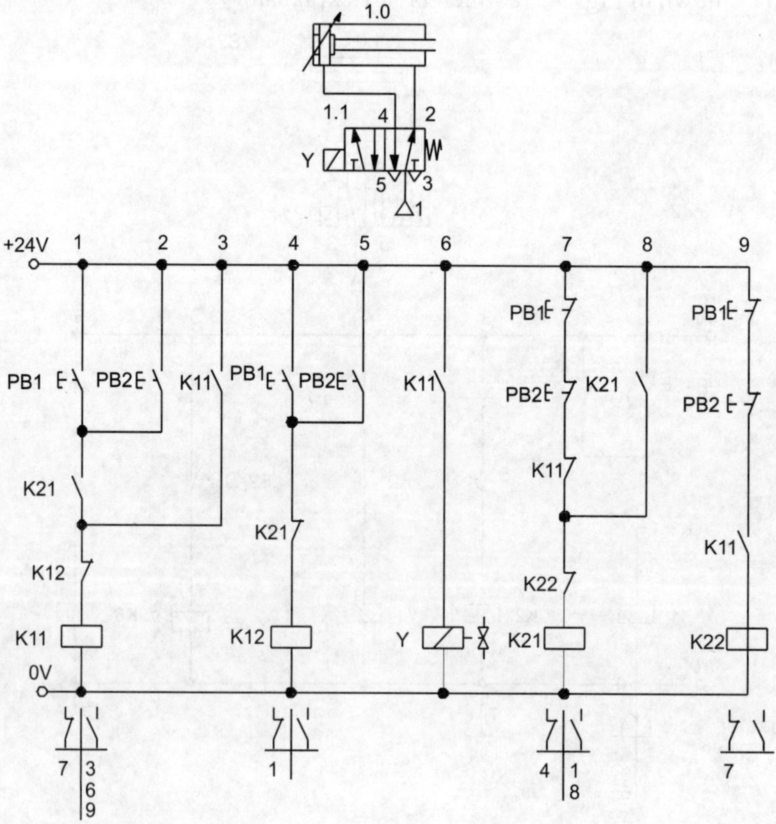

Figure 9.53 *Control circuit for sliding door in metrology lab*

9.4 Electro-pneumatic Circuits for Multi Actuators

Understanding basic circuits is essential before moving forward to designing sequencing circuits. Sequencing circuits usually have two or more cylinders that extend or retract in a predefined sequence. As in pure pneumatic circuits, signal conflict has to be eliminated in electro-pneumatic circuits as well. A simple circuit for a two-group cascade using only relays is discussed first followed by a discussion of three-group cascading circuits.

9.4.1 Two group electro-pneumatic circuits

Case 1: Design of a sequencing circuit using **pneumatic memory** for the sequence A+B+A−B−

Step 1: Draw the cylinders A and B with 5/2 double solenoid directional control valve. Connect the compressed power supply. Indicate the position of the limit switch (Fig. 9.54).

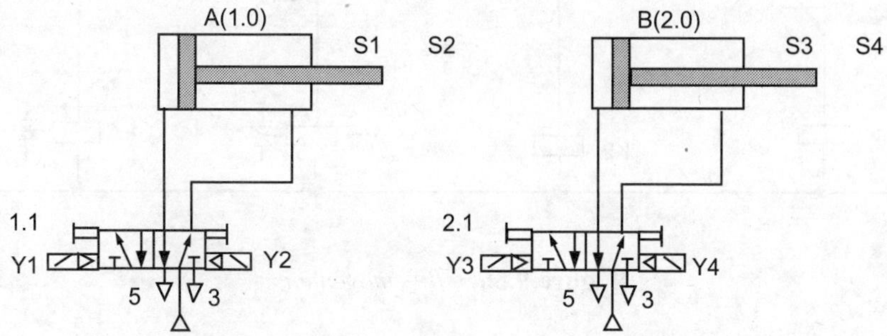

Figure 9.54 *Cylinders with solenoid valves*

Step 2: A+ movement (Fig. 9.55)

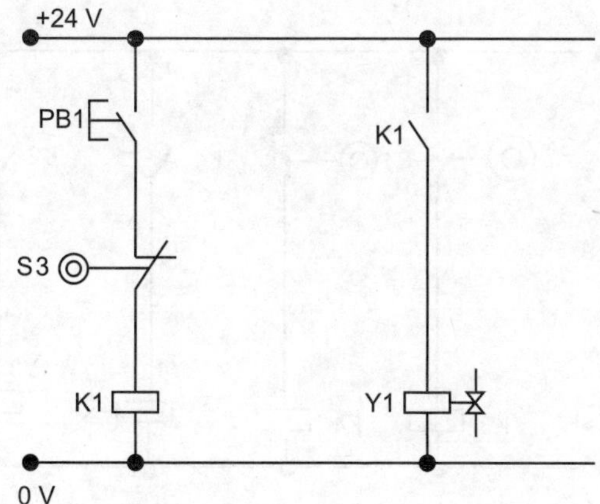

Figure 9.55 *A+ movement*

In the control circuit in Fig. 9.55, relay K1 is energised via the ON push button PB1 and limit switch S3. In the main circuit, a normally open contact of K1 closes the circuit. Solenoid Y1 is energised and causes cylinder A to move forward.

Step 3: B+ movement (Fig. 9.56)
Limit switch S2 is actuated at the final position of cylinder A. This causes relay K2 to energise. A normally open contact of K2 energises solenoid Y3. The valve 2.1 (Fig. 9.54) changes state and cylinder B extends.

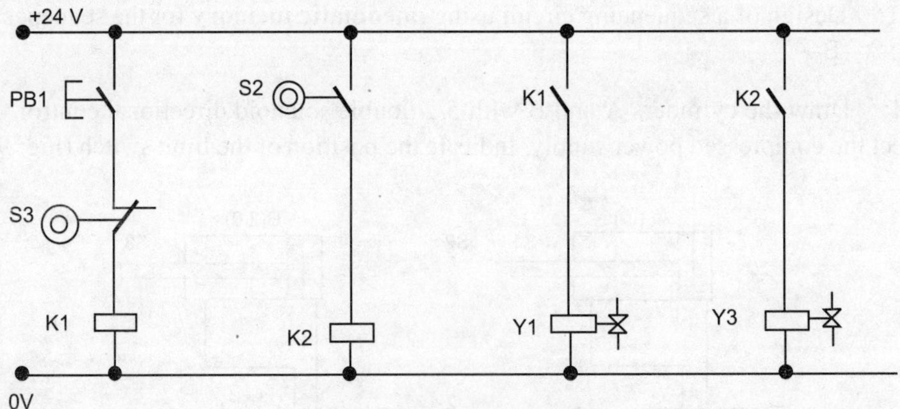

Figure 9.56 *B+ movement*

Step 4: A− movement and B− movement (Fig. 9.57)
Draw the next rung of the electrical circuit and the main circuit. It is similar to Step 3 with different limit switches.

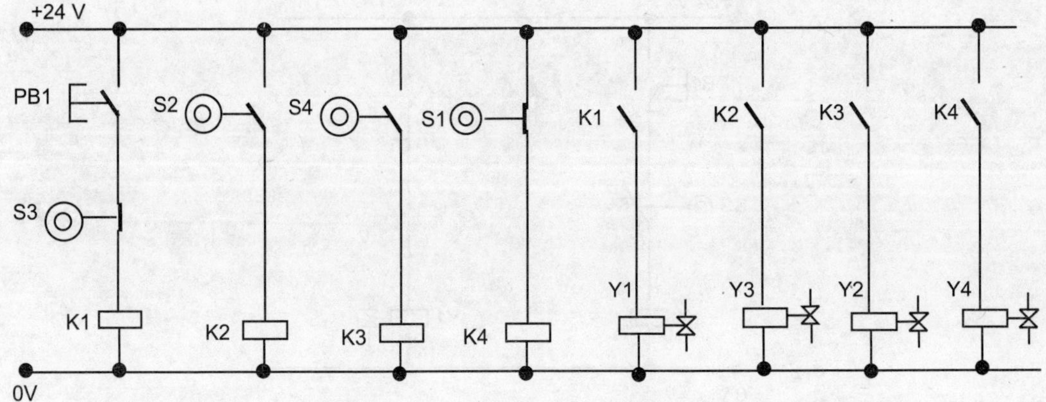

Figure 9.57 *A− and B− movements*

Cylinder B extends and at its forward position it actuates limit switch S4 which energises relay K3, and a normally open contact in K3 energises solenoid Y2 and cylinder A retracts.

As cylinder A retracts, it activates limit switch S1 and energises relay K4 which in turn activates solenoid Y4. Cylinder B then retracts and is ready for a new cycle. A new cycle starts with the activation of S5.

The limit switches can be replaced with proximity switches for this method. The solution is shown in Fig. 9.58 which is self-explanatory.

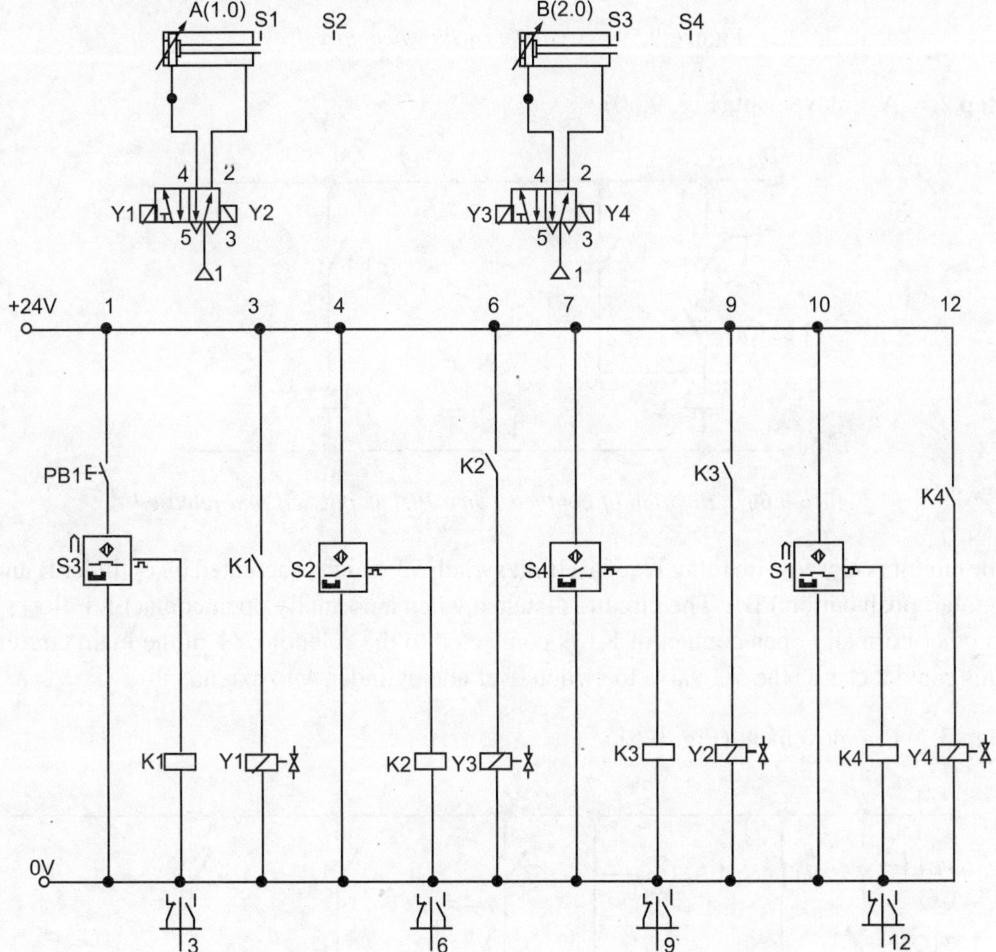

Figure 9.58 *A+B+A−B− with reed switch control*

Case 2: Design of a sequencing circuit using **electric memory** for the sequence A+B+A−B−

Step 1: Draw cylinders A and B with a 5/2 double solenoid directional control valve. Connect the compressed power supply. Indicate the position of the limit switch.

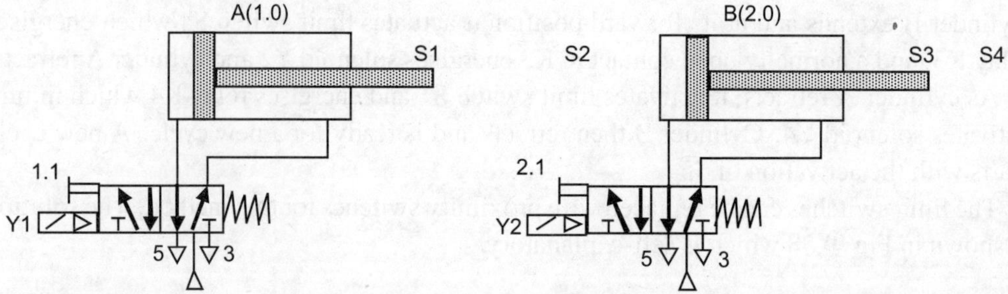

Figure 9.59 *Cylinders with solenoid valves*

Step 2: A+ movement (Fig. 9.60)

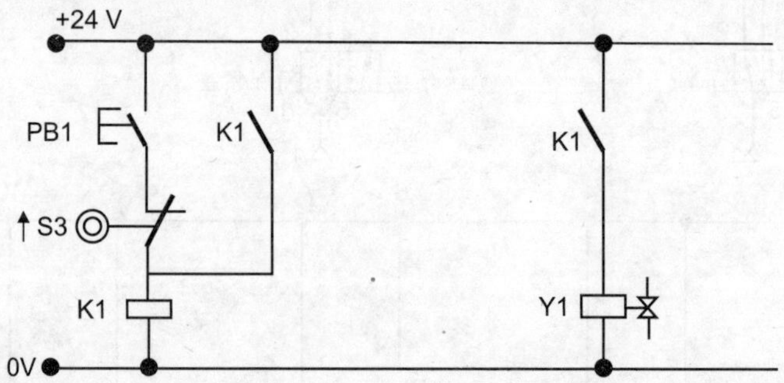

Figure 9.60 *Position of contacts when PB1 is pressed and released*

The circuit is closed with relay K1 via a limit switch S3 which is actuated by cylinder B and the start push button PB1. The circuit is latched when a normally open contact K1 closes . Another normally open contact of K1 is connected to the solenoid Y1 in the main circuit. This contact causes the 5/2 valve to switch over and cylinder A to extend.

Step 3: B+ movement (Fig. 9.61)

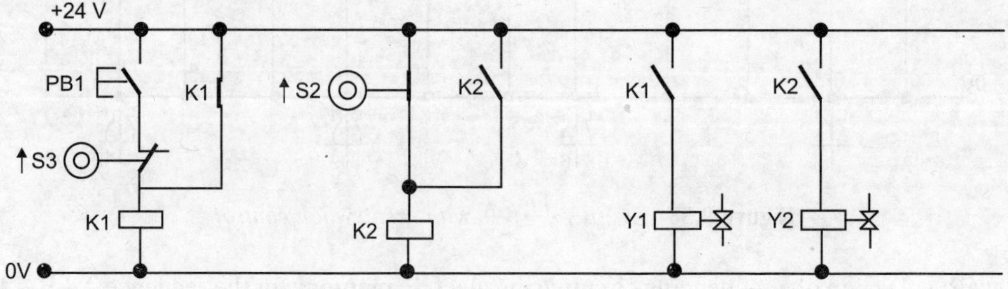

Figure 9.61 *Position of contacts when S2 is pressed, causing B+ motion*

Draw relay K2 and solenoid coil Y2.

Cylinder A travels to its forward position and makes contact with limit switch S2, which energises K2 in the control circuit. Connected in parallel with this is the latching circuit for relay K2 via a normally open contact of K2.

When the normally open contact of K2 is closed in the main circuit, solenoid coil Y2 is energised and the 5/2 directional control valve switches over and cylinder B travels out.

Step 4: A– movement (Fig. 9.62)
Draw in limit switch S4 and switch-off relay K1.

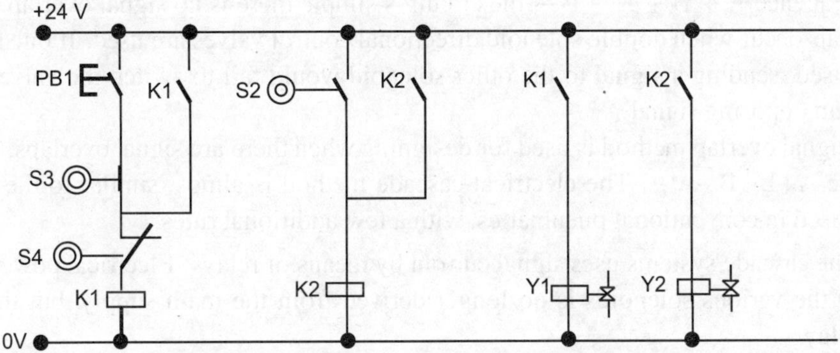

Figure 9.62 *Position of contacts when S4 is pressed, de-energising Y1 and causing A–*

Limit switch S4 is arranged ahead of relay K1. When S4 (normally closed) is actuated by cylinder B, the latching circuit is interrupted and relay K1 is de-energised. The normally open contact of K1 in the main circuit returns to its normal position, coil Y1 is de-energised, the 5/2 directional control valve switches back and cylinder A retracts.

Step 5: B– movement (Fig. 9.63)
Draw in limit switch S1 and switch off relay K2.

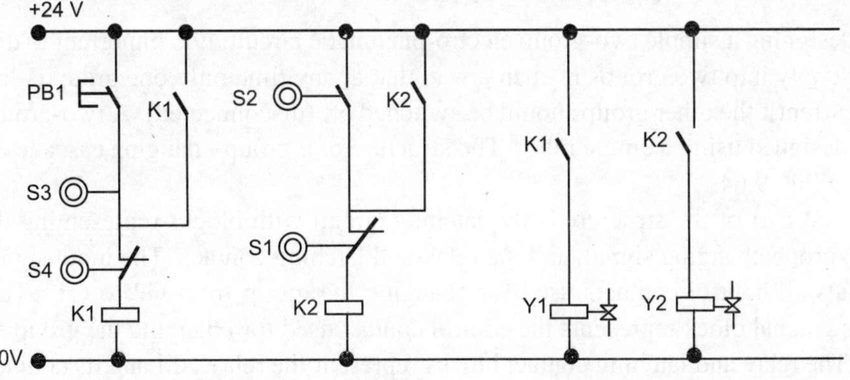

Figure 9.63 *A+B+A–B– circuit diagram*

Limit switch S1 is arranged ahead of relay K2. When S1 (normally closed) is actuated by cylinder B, the latching circuit is interrupted and relay K2 is de-energised. The normally open contact K2 in the main circuit returns to its normal position, coil Y2 is de-energised, the 5/2 directional control valve switches back and cylinder B retracts.

Limit switch S3 is actuated again and a new cycle can be started with start push button PB1.

Case 3: Design of a sequencing circuit using electric memory for the sequence A+B+ B−A−, using single relay coil

In the sequence A + B + A − B−, the circuit is simple there is no signal overlap. Opposing signals occur when double solenoid directional control valves are used. If one solenoid is energised, sending a signal to the other solenoid would fail to switch the valve over as there is an opposing signal.

The signal overlap method is used for designing when there are signal overlaps, as in the sequence A+B+B−A−. The electrical cascade method is almost similar to the cascade system used in conventional pneumatics, with a few additional rules:

1. The cascade systems uses signal cut-out by means of relays. Electrical power supply to the various solenoids is no longer derived from the main supply, but through a relay.
2. In the design of cascades, the working principle is that only one supply line is energised at a time. The procedure for constructing an electrical cascade follows.
3. Define the sequence of motion in abrreviated notation A+B+B−A−
4. Divide the sequence into groups. Here there are two groups: A+B+ together in group G1 and B−A− together in group G2.
5. Divide the groups such that a cylinder operation occurs only once in a group, and keeping the number of groups to a minimum. So A+ and A− cannot be in same group. One supply line per group is required.
6. Number of relays = Number of groups −1

When designing a simple two-group electro-pneumatic circuit, it is important to divide the power supply into two groups in such a way that at any time only one group is live (with active current); the other group should be switched off (disconnected). A two-group circuit can be designed using a single relay. The structure of a group-changing cascade circuit is shown in Fig. 9.64.

The first part of the structure is the latching circuit with blocks representing the input signals, group changing signal, and the relay and latching contact. The input signal block represents all control contacts used for changing the group from G2 to G1. The group changing signal block represents the control contact used for changing the group from G1 to G2. The relay and latching contact blocks represent the relay coil and its contact which implement the memory function of the cascade operation. The relay coil also controls the group changing function of the cascade circuit through its NO and NC contacts.

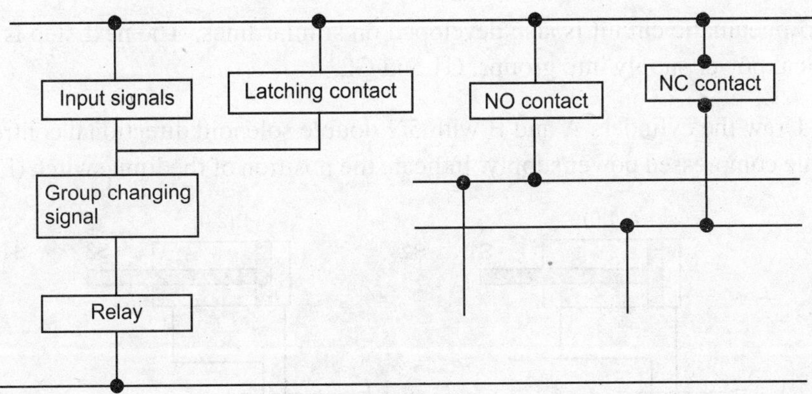

Figure 9.64 *Methodology to design a sequence circuit using a single relay*

The second part of the structure consists of two blocks, one each of NO contacts and NC contacts of the relay used to switch groups G1 and G2 respectively. Initially group G2 is live through the NC contact of the relay. The group changing concept is with relays is illustrated schematically in Fig. 9.65.

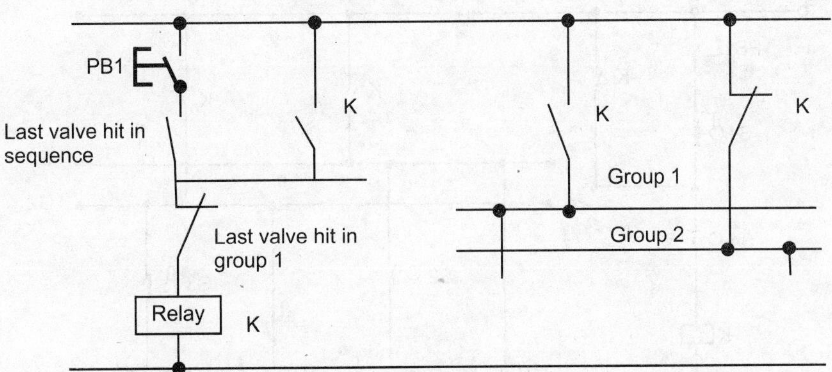

Figure 9.65 *Group changing concept using single relay*

There is a signal conflict in this case because B + B– appear together, and need to be separated into two groups as shown in Fig. 9.66.

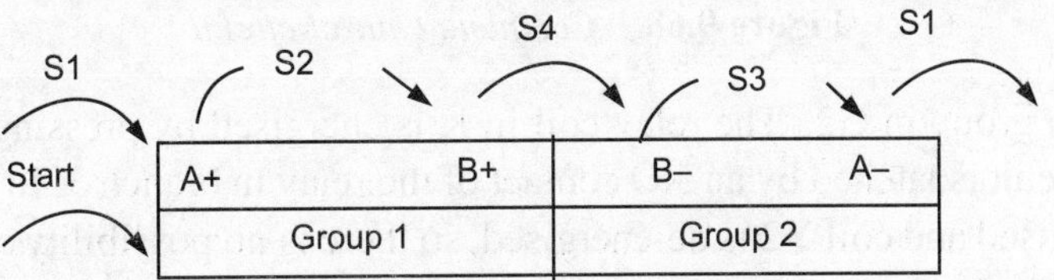

Figure 9.66 *Grouping*

An electro-pneumatic circuit is also developed on similar lines. The next step is to divide the electrical power supply into groups, G1 and G2.

Step 1: Draw the cylinders A and B with 5/2 double solenoid directional control valves. Connect the compressed power supply. Indicate the position of the limit switch (Fig. 9.67).

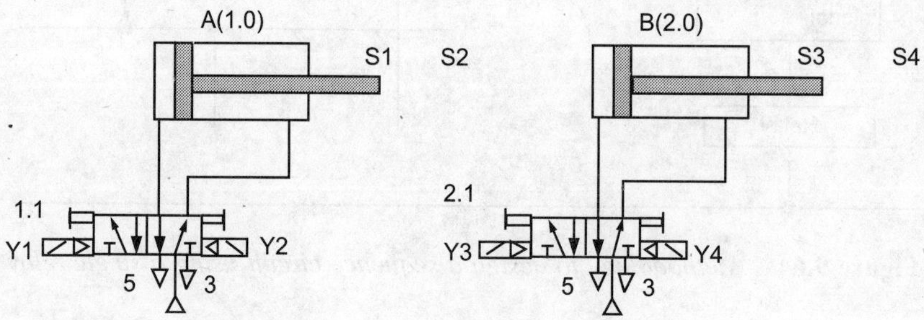

Figure 9.67 *Cylinders with solenoid valves*

Step 2: Complete circuit is shown in the Fig. 9.68

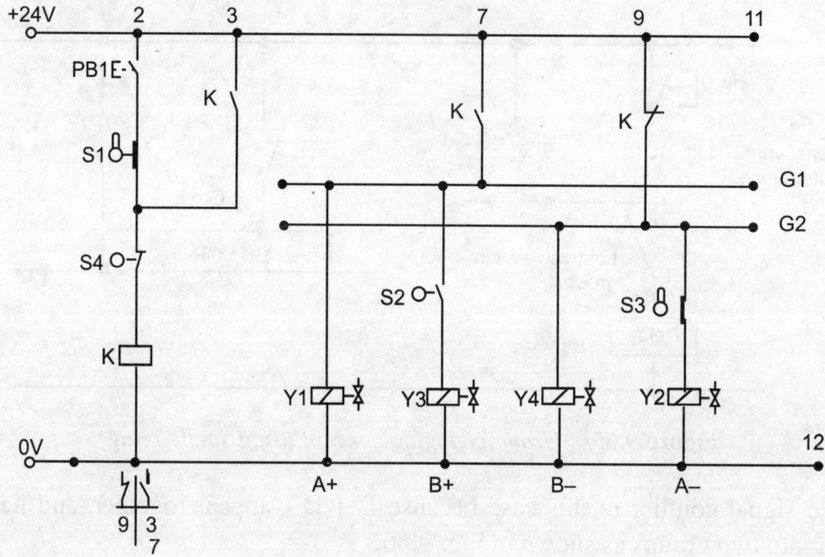

Figure 9.68 *Complete control circuit*

Supply starts in group in G2. The relay coil in K is energised by pressing the start button PB1 and the circuit is latched by an NO contact of the relay in branch 3. In the main circuit, coil Y1 is energised and coil Y2 is de-energised, so there is no possibility of signal conflict. Valve 1.1 (Fig. 9.67) changes its state and cylinder A travels out (A+). Limit switch S2 is actuated thereby energising coil Y3. As a result, valve 2.1 (Fig. 9.67) switches over and cylinder B travels out (B+). Limit switch S4 is now actuated. The signal from sensor S4 must perform a signal shut-off and therefore its NC contact is included in branch 2 of the circuit. The latched circuit relay coil K is cleared by the actuated limit switch S4. In the main circuit, group G1 is again disconnected through the release of the NO contact of relay K in branch 7. Group G2 is now supplied with current again through the NC contact of relay K in branch 9.

Coil Y4 is energised and coil Y3 is de-energised, with no possibility of signal conflict here. Valve 2.1 changes its state and cylinder B returns to the rear position (B−). Limit switch S3 is actuated thereby energising coil Y2. Valve 1.1 switches over and cylinder A returns to the rear position (A−). The cycle of cylinder actions can now be repeated.

Example 9.10 Design a sequencing circuit using electric memory for the sequence A + B + A − B− using a single relay coil

The complete circuit is shown in Fig. 9.69.

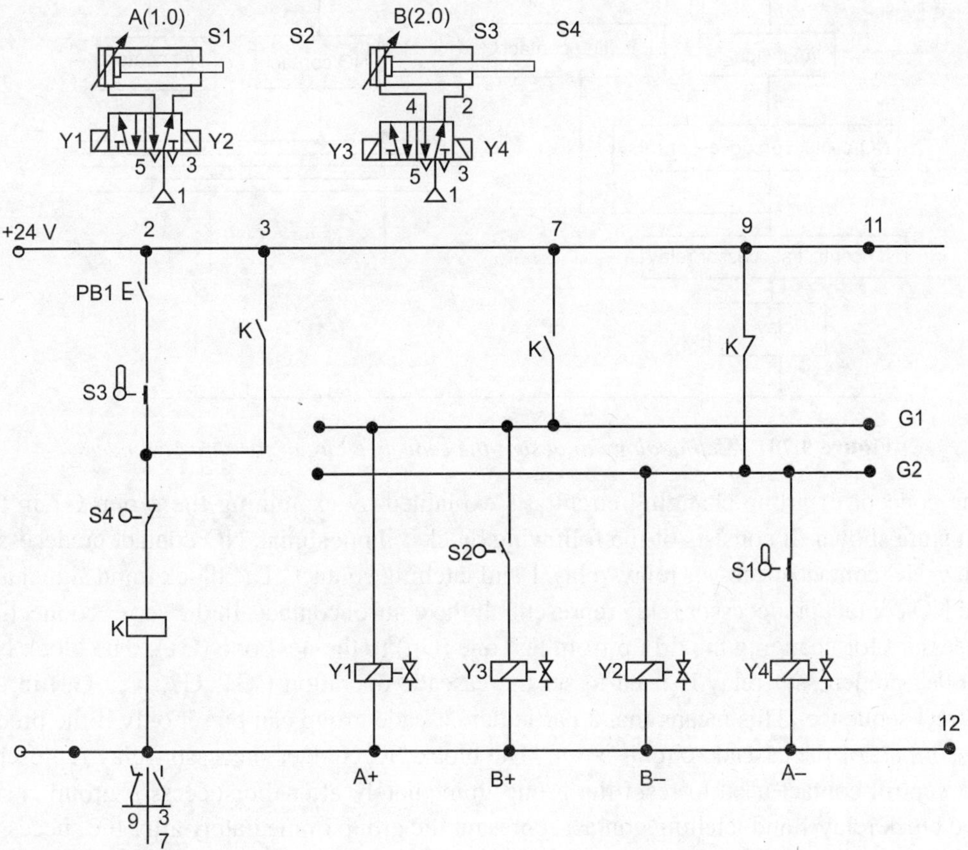

Figure 9.69 *Complete control circuit*

9.4.2 Design of three-group electro-pneumatic circuits

When designing electro-pneumatic circuits with three or more cascade groups, the power supply should be divided into three or more groups in such a way that at any point of time only one group is live, that is, the other groups should be switched off. A three group circuit is designed using three relays. The illustration of a group changing cascade circuit for n groups using as many relays is shown in Fig. 9.70

The structure shown in the figure is divided into a number of identical groups, G1 G2 ...Gn. The NO contacts of relays 1 to n are used to switch power supply from G1 to Gn respectively. The G1 portion of the structure controls the G1 supply through relay 1, the block G2 controls the group G2 supply through relay 2 and so on.

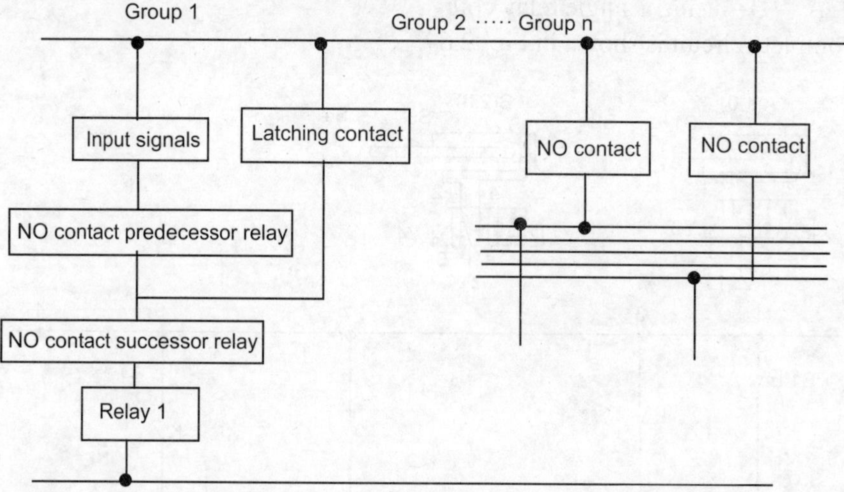

Figure 9.70 *Methodology to design the sequence circuit for three cylinders*

The concept of group changing circuits is explained by examining the group G1 in the structure shown. It consists of the following blocks: input signal, NO contact predecessor relay, NC contact successor relay, relay 1 and latching contact. The block input signal and the NO contact predecessor relay represent all the control contacts in the series connection necessary for changing the group from last one (Gn) to the next one (G1). The block NO contact predecessor relay is used to set the cascade operation (G1, G2,......Gn) in the correct sequence. This means that a particular cascade group can be set only if the predecessor part of the cascade circuit is set. The block NC contact successor relay represents the control contact used to reset the group immediately after the successor group is set. The block relay 1 and latching contact represent the group immediately after the successor group is set. The blocks relay 1 and latching contact represent the relay and its contact that implement the required memory function of the cascade operation.

The final part of the structure consists of blocks, NO contact relay 1, NO contact relay 2, etc. These blocks represent NO contacts of the relays 1, 2, etc., respectively. These contacts are used to switch the respective groups G1, G2 etc. As usual, in pneumatic systems the last group should be set initially. The cascade circuit then changes the group in the sequence G1, G2...Gn in response to appropriate input signals. This concept can be applied at any number of cylinders.

Another method of arranging groups for three and four cylinders is shown in Fig. 9.71 and Fig. 9.72 respectively.

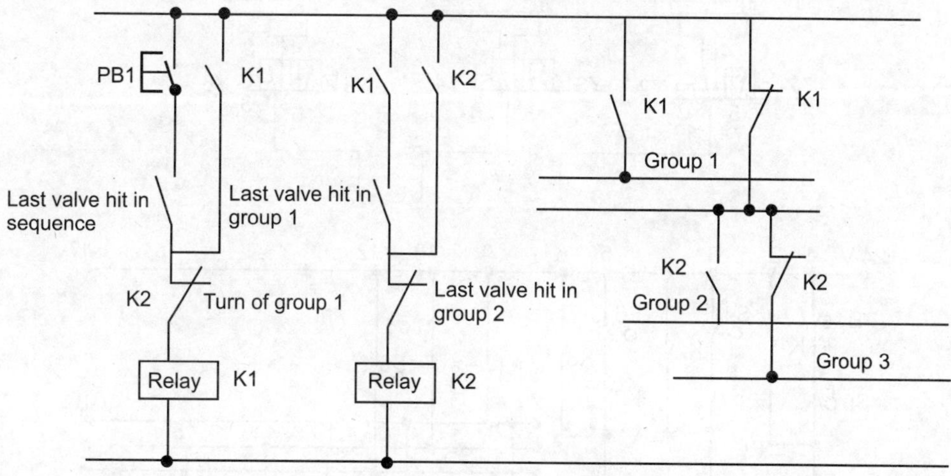

Figure 9.71 *Methodology to design the sequence circuit for three groups*

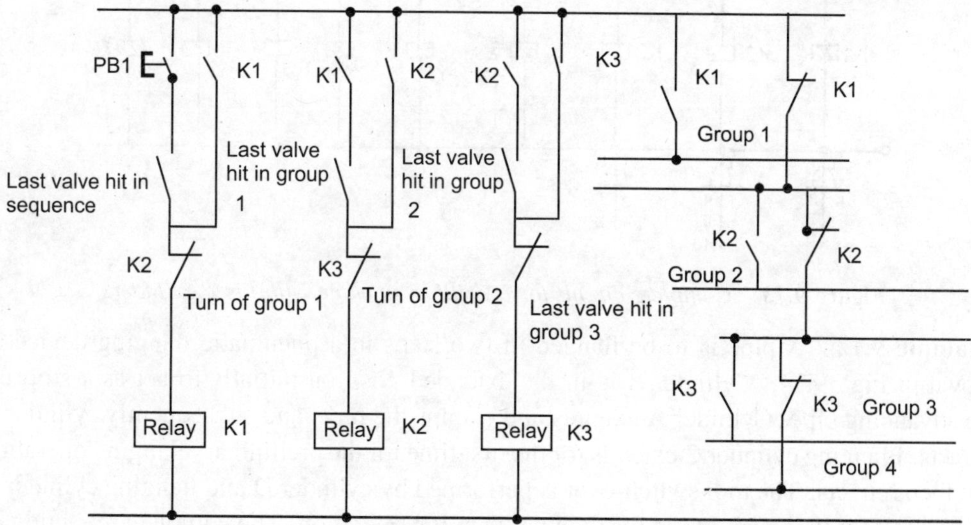

Figure 9.72 *Methodology to design a sequence circuit for four groups*

Example 9.11 Workpieces are to be drilled using a pneumatically controlled drilling machine. The workpieces are arranged in a gravity feed magazine. These parts are to be pushed and clamped by clamping cylinder A and drilled by drilling cylinder B and ejected by ejecting cylinder C. Develop an electro-pneumatic circuit to implement the given control task.

The complete solution is shown in Fig. 9.73 which is self-explanatory.

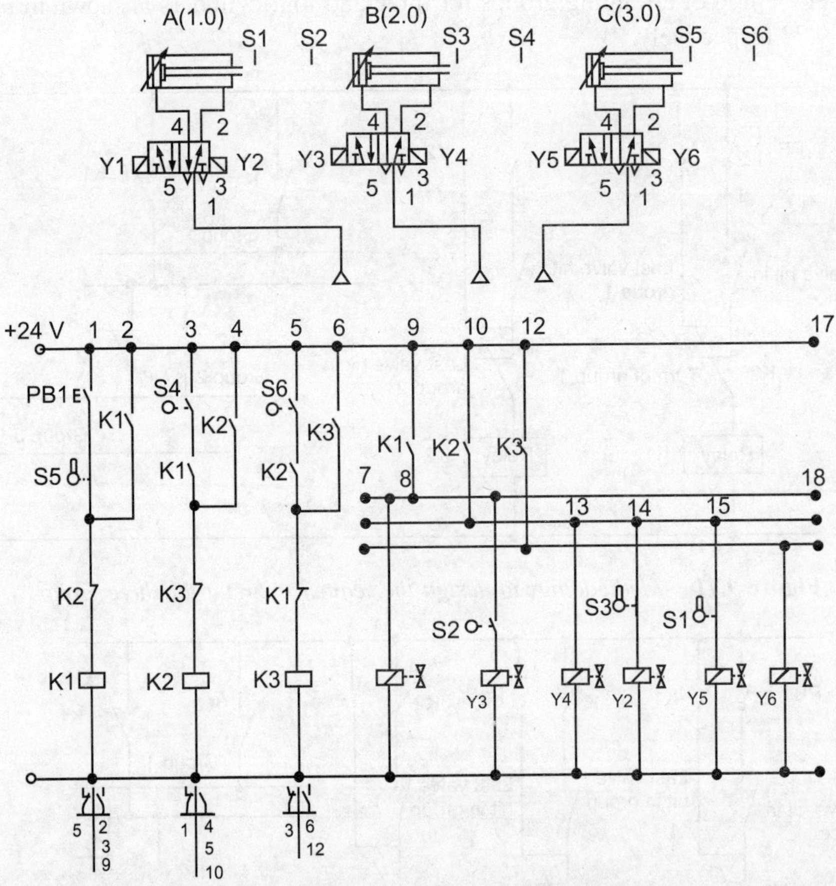

Figure 9.73 *Complete circuit for a drilling machine with three cylinders*

Example 9.12 A pipe is to be flanged in two steps in a pneumatic flanging device as shown in Fig. 9.74. Cylinder B is in the extended position initially to act as a stop for the advancing pipe. Cylinder A extends and clamps the pipe, and subsequently cylinder B retracts. Flanging cylinder C extends for the first time for the preliminary flanging operation and then retracts. The tool switch-over is performed by cylinder D and flanging cylinder C is in action for the second time for the final flanging operation. The complete retraction of cylinder C triggers the return motion of clamping cylinder A and tool switching cylinder D, and the forward motion of positioning cylinder B. The sequence of operations is shown in the sequence diagram. Develop a electro pneumatic control circuit to implement the control task.

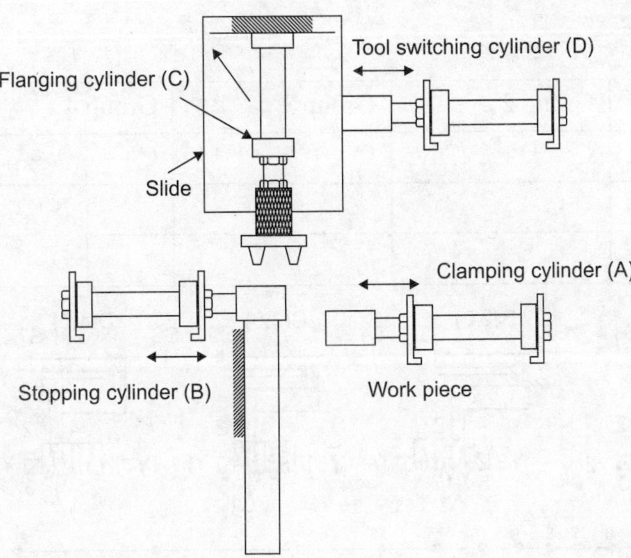

Figure 9.74 *Positional layout*

Step 1: Represent the control task using notational form
Cylinder A advancing is designated as A+
Cylinder A retracting is designated as A−
Cylinder B advancing is designated as B+
Cylinder B retracting is designated as B−
Cylinder C advancing is designated as C+
Cylinder C retracting is designated as C−
Cylinder D advancing is designated as D+
Cylinder D retracting is designated as D−

Step 2: Draw the displacement step diagram (Fig. 9.75)

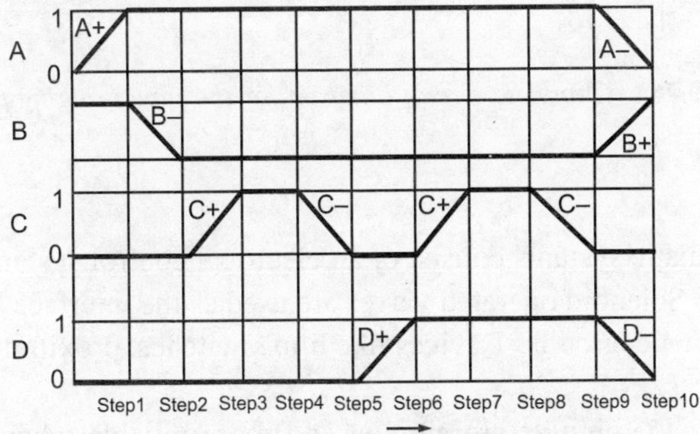

Figure 9.75 *Displacement–step diagram*

Grouping

Group 1	Group 2	Group 3	Group 4	
A + B − C + (t)	C − D+	C + (t)	C−	A−
				B+
				D−

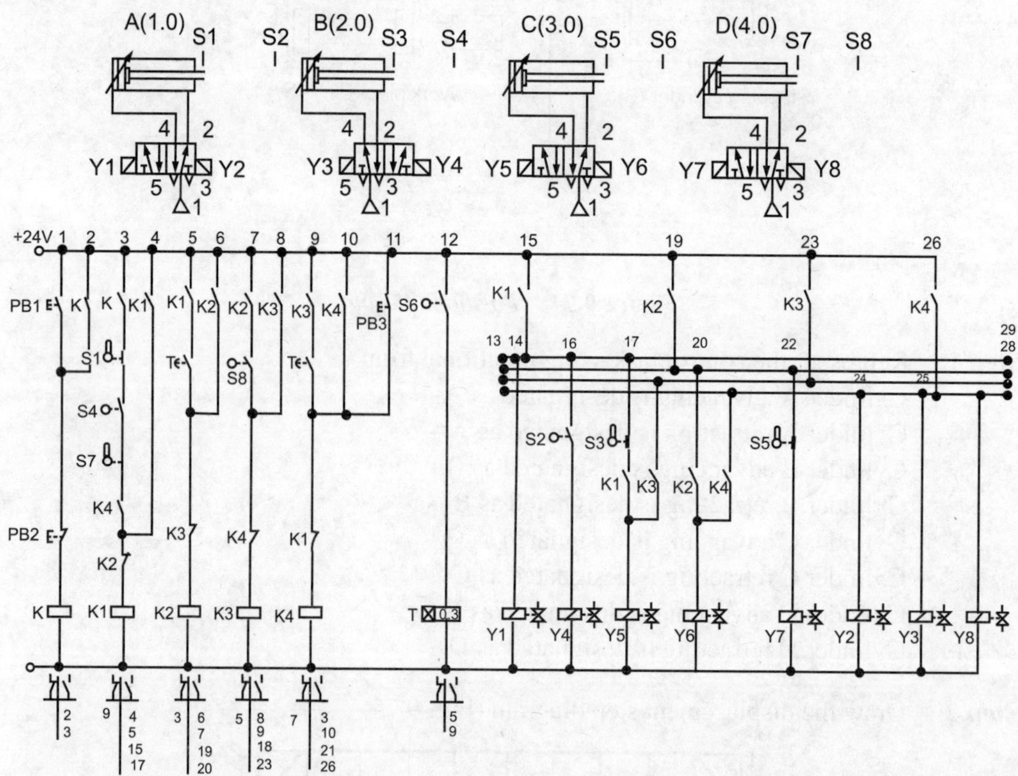

Figure 9.76 *Complete circuit diagram for the pipe flanging problem*

Summary

1. Electro-pneumatic systems consist of an electrical control system operating a pneumatic system. Solenoid operated valves are used as the interface between the electrical and pneumatic circuits. Devices like limit switches, proximity switches are used as input devices.
2. Electro-pneumatic circuits are simpler and more reliable when compared to pure pneumatic circuits. Electro-pneumatic circuits involve less planning and commissioning effort and also are easier to install.

3. A push button is a switch used to open and close electric control circuits. It consists of a set of mixed and movable contacts, and a retaining spring. A relay can be considered as an electro-magnetically operated switch operated by another electrical switch. It is mainly used for signal processing.
4. It is important to remember that double piloted directional control valves are susceptible to problems of signal conflict or signal overlaps.
5. The main difference between dominant OFF and dominant ON circuits is that when both start and stop push buttons are pressed simultaneously, circuits are set to off in dominant OFF mode, and circuits are set to on in dominant ON mode.
6. The main difference between a limit switch and a push button is that former is operated automatically while the latter is operated manually.
7. A reed switch is also known as a magnetically actuated proximity switch. Inductive proximity switches are used for metals. They are used where it is necessary to record or count moving objects or workpieces on machines or conveyors. Owing to their ability to handle a wide range of materials, capacitive proximity sensors are used more universally and are suitable for detecting non-metallic objects, monitoring filling levels of storage containers etc.
8. An optical sensor uses optical and electronic means to sense objects. It consists of an emitter and a receiver. In diffuse sensors, the emitter and receiver are fitted in the same housing.
9. Timers are required in control systems to introduce a time delay between work operations. This is possible by delaying the operations of the associated control element through a timer.
10. An electric counter is used for counting operations. There are two types of counters: up counters and down counters. A down counter counts electrical signal downwards from a preset value. An up counter counts electrical signal upwards from zero to a preset value.

EXERCISES

Fill in the blanks

1. A relay is considered an electro ———— actuated switch.
2. When both start and stop buttons are pressed simultaneously, circuits goes to OFF condition in dominant —— circuit.
3. Inductive sensors cannot be used to detect ———————- materials.
4. Push buttons are operated manually, whereas limit switches are operated ———— ————
5. ————— switches are magnetically actuated proximity switches.

State whether true or false

1. Electro-pneumatic circuits are less reliable than pure pneumatic circuits.
2. A pressure switch is an electric-pneumatic signal convertor.
3. A relay is an electro-pneumatically operated switch that operates under the control of additional electrical circuits.
4. An electrical latching relay circuit is an example of memory function.
5. Limit switches are usually operated manually.

Review questions

1. Explain briefly the working principle of an electromagnetic relay.
2. Draw the symbol for an electromagnetic relay with 2 NOs and 1 NC.
3. What are the different ways to implement memory functions in electro-pneumatic circuits?
4. Explain the step displacement diagram for the sequence A+B+B-A-.
5. List the disadvantages of using electro-pneumatic circuits.
6. Explain the principle of cascade method using electro-pneumatics with a suitable sequence example.
7. What are the functions of sensors?
8. Explain the working principle of a limit switch.
9. Explain the working principle of a reed switch.
10. Draw the symbols for the following:
    (a) limit switch (b) inductive proximity switch
    (c) capacitive proximity switch (d) optically operated sensor
11. Write the advantages of diffuse sensors.
12. Compare inductive, capacitive and optical types of sensors.
13. Explain the functions of on-delay timers with a suitable circuit.
14. Explain the functions of off-delay timers with a suitable circuit.
15. Differentiate between the behaviour of on-delay timer and off-delay timers with the help of a timing diagram.
16. Explain the function of a pressure switch.
17. Explain the function of up counters and down counters.
18. Draw a group changing cascade circuit for two groups, three groups, and four groups.
19. Explain the construction and working of a pressure sequence valve with the help of neat sketch.
20. What are the different ways to sense the end position and movement of cylinders?

Problems

1. A double acting cylinder is used to press parts for 20 seconds (Fig. 9.77). A push button PB1 is used to control the forward stroke of the cylinder. After pressing the components for 20 seconds, the cylinder is to retract automatically. The return stroke must occur even if the start button is still actuated. A new start signal may only be effective after the initial position of the piston is reached and the push button is released. Develop a pneumatic circuit to implement the control task.

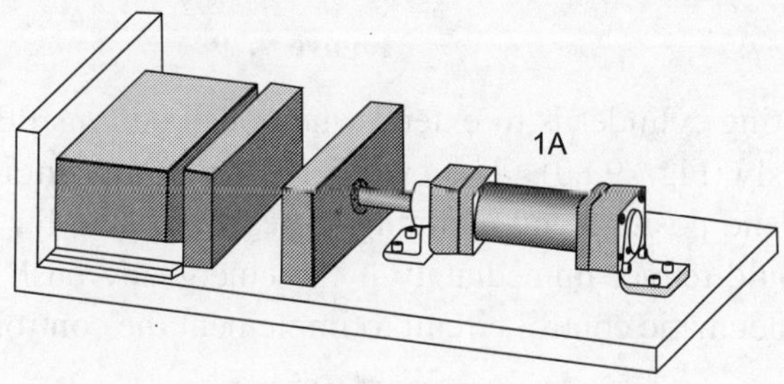

Figure 9.77

2. Figure 9.78 shows the arrangement for cleaning washers for injection pumps in a cleaning bath. A double acting cylinder is used to move the container filled with washers up and down in the bath several times. The operator provides a start signal manually. The washing operation is turned off automatically after a preset time. Develop a pneumatic control circuit to implement this control task.

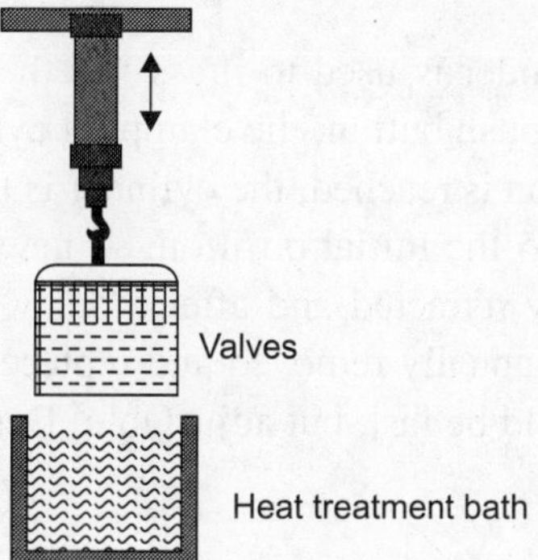

Figure 9.78

3. A double acting cylinder is used for embossing slide rules as shown in Fig. 9.79. The cylinder extends only when two push buttons are pressed simultaneously within one second to ensure the complete safety of the operators. The cylinder should retract immediately if any of the push buttons is released. Develop a pneumatic circuit to implement the control tasks.

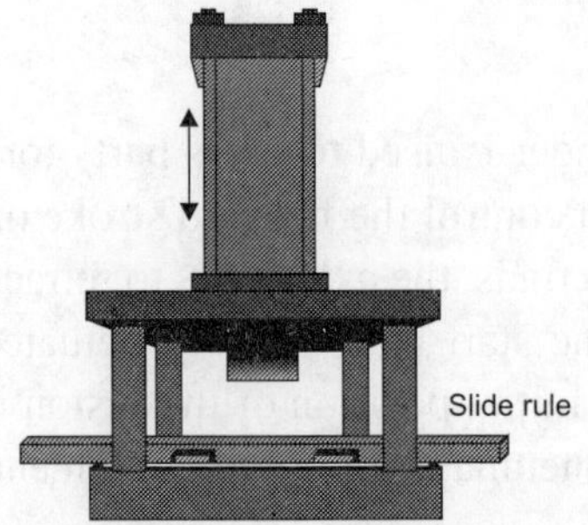

Figure 9.79

4. A double acting cylinder is to extend when push buttons PB1 and PB2 are pressed simultaneously (Fig. 9.80). The cylinder is to retract automatically after reaching the forward end position and attaining a preset pressure for consistent quality. The cylinder should retract immediately if the emergency push button PB3 is pressed. Develop a pneumatic control circuit to implement the control task.

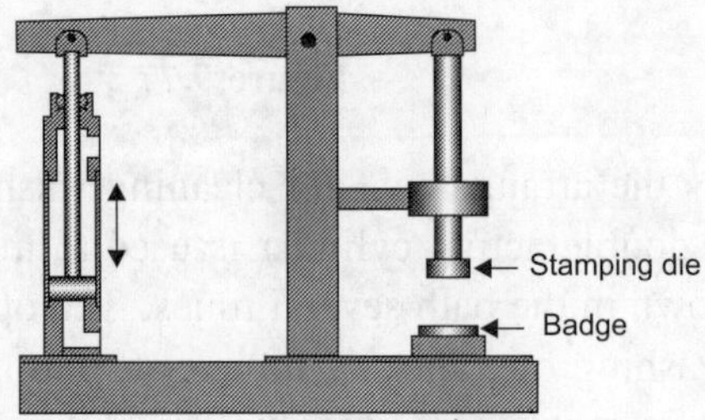

Figure 9.80

5. A double acting cylinder is used to press together glued components (Fig. 9.81). Upon operation of a push button, the clamping cylinder slowly advances. Once the fully extended position is reached, the cylinder is to remain for 10 seconds and then immediately retract to the initial position. A new start cycle is only possible after the cylinder has fully retracted and after a delay of 5 seconds. During this delay the finished part is manually removed and replaced with new parts for gluing. The retracting speed should be fast, but adjustable. Design and draw the circuit diagram for the problem.

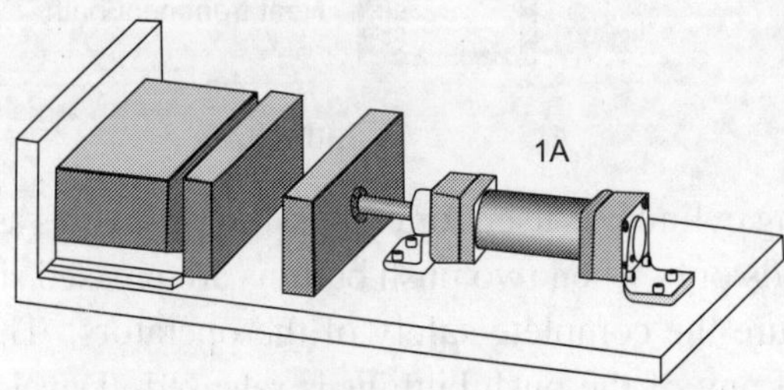

Figure 9.81

6. Plastic components are embossed using a die powered by a double acting cylinder (Fig. 9.82). The return of the die is to be effected when the cylinder rod has fully extended to the embossing position and the preset pressure is reached. A roller lever valve is to be used to confirm full extension. The signal for retracting must only be generated when the piston rod has reached the embossing position. The pressure in the piston chamber is indicated by a pressure gauge. Design and draw the circuit diagram for the problem.

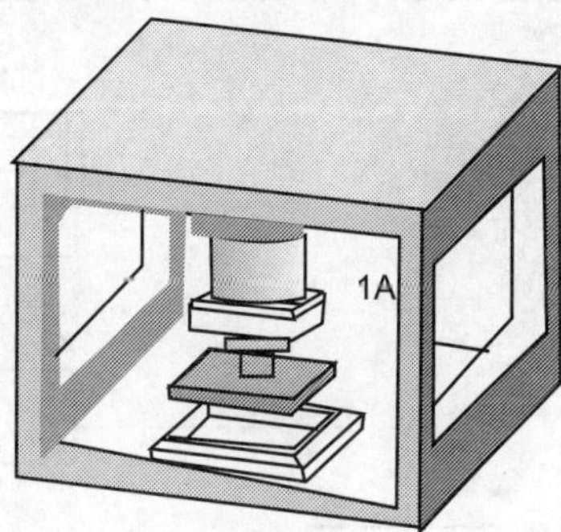

Figure 9.82

7. In Fig. 9.83, the piston rod of the cylinder 1.0(A) is to advance only if a workpiece is inserted in the workpiece retainer, a guard has been lowered and the operator presses the push button valve. Upon the release of the push button or if the guard is no longer in its lower position; the cylinder 1.0(A) is to retract to the initial position. Design and draw the circuit diagram for the problem.

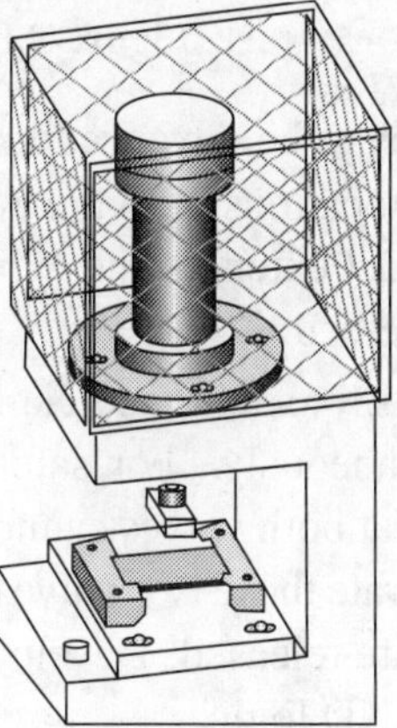

Figure 9.83

8. A double acting cylinder is to extend when a push button is operated. Upon release of the push button the cylinder is to retract (Fig. 9.84). The cylinder is of small bore (25 mm diameter) requiring a small flow rate to operate at the correct speed.

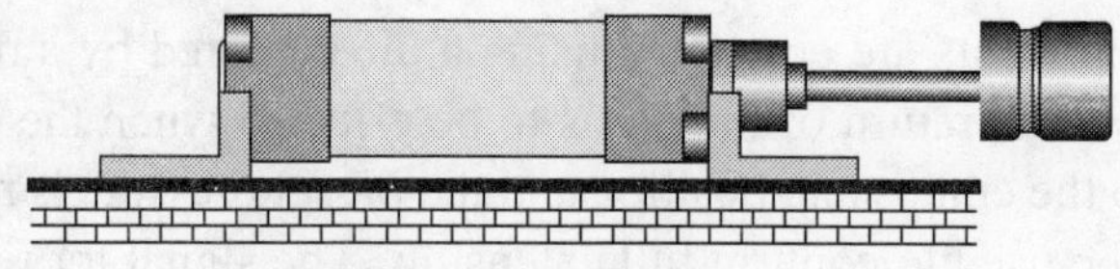

Figure 9.84

9. Draw the electro-pneumatic circuit for the following circuit (Fig. 9.85).

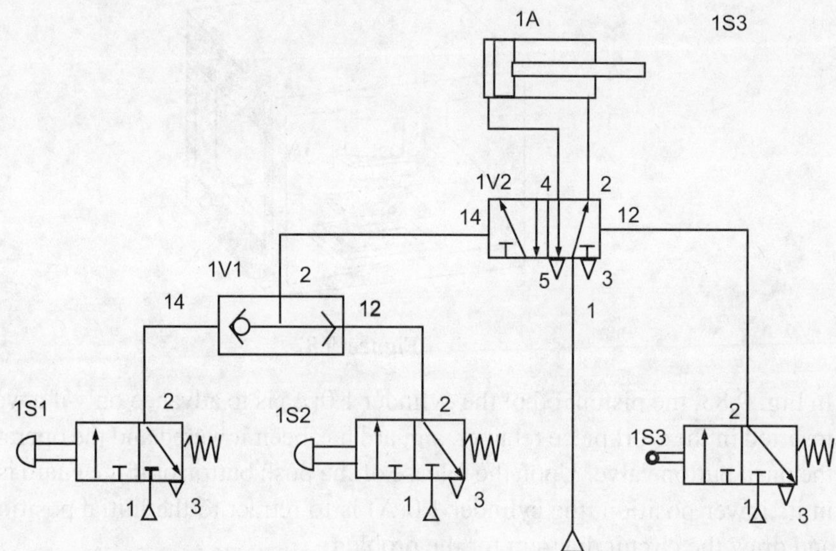

Figure 9.85

10. A double acting cylinder is to be advanced using three hand operated valves which are positioned at different locations from the cylinder. If any of the three push buttons is pressed, the cylinder must extend. The cylinder must be in retracted condition if all push buttons are not pressed.
11. A double acting cylinder is to extend and clamp a workpiece when two push button valves are pressed simultaneously. For safety reasons, these push buttons valves are installed such a way that both valves cannot be operated with one hand, i.e., both hands must be used to operate these two valves. The cylinder is to retract when either one or both push buttons are released. Develop a pneumatic circuit to implement the given control task using AND logic.
12. A currency coin is embossed using a die driven by a double acting cylinder. The die is to advance and emboss the coin when a push button is operated (Fig. 9.86). The return of the die is to be effected when a preset pressure is reached. The embossing pressure is to be adjustable so that different coins can be embossed in the same machine.

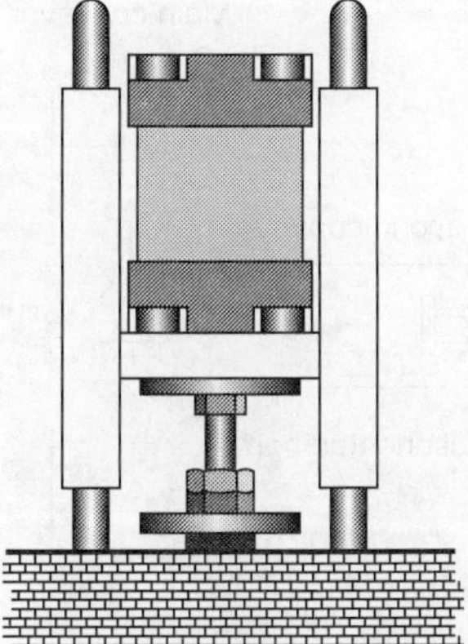

Figure 9.86

13. A plastic component is embossed using a die driven by a double acting cylinder (Fig. 9.90). The die is to advance and emboss the plastic when a push button is operated. The return of the die is to be effected when a preset pressure is reached. The embossing pressure is to be adjustable.
14. A double acting cylinder is used to press together glued components (Fig. 9.91). Upon operation of a push button, the clamping cylinder extends. Once the fully extended position is reached, the cylinder is to remain for 6 seconds and then immediately retract to the initial position. The cylinder retraction is to be adjustable. A new start cycle is only possible after the cylinder has fully retracted.

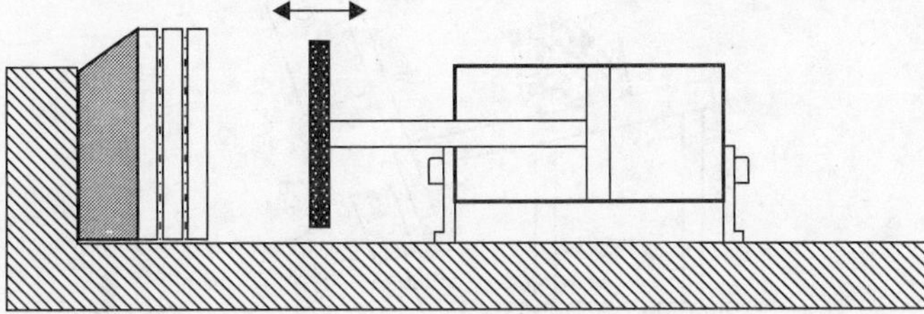

Figure 9.87

15. A single acting cylinder is used to sort disoriented metal stampings (Fig. 9.88). If the metal stampings are disoriented they are pushed to a second conveyor. If the push button is pressed, the single acting cylinder must actuate to push stamping to the second conveyor which is inclined downwards; the cylinder has to retract if the push button is released. The forward motion of the cylinder takes 0.5 seconds. A pressure gauge is to be fitted before and after the one-way flow control valve. Design a pneumatic circuit.

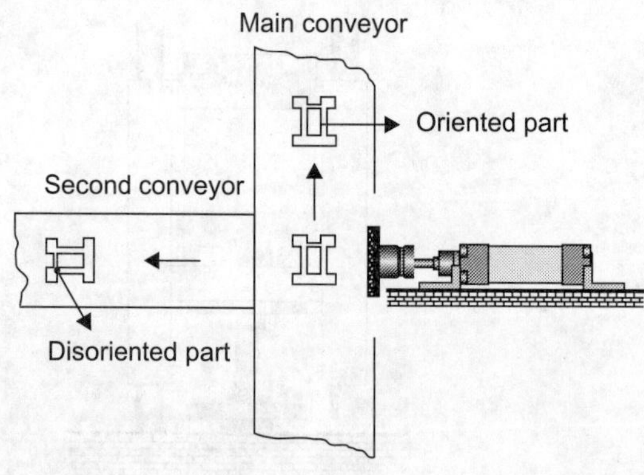

Figure 9.88

16. In the parcel delivery room of a central post office, a parcel separating device feeds post from a sloping conveyor to an X-ray checking unit before delivery (Fig. 9.89). Operating the push button causes very rapid retraction of the single acting actuator. Forward motion time is 10 seconds. A pressure gauge has to be fitted before and after the flow control valve.

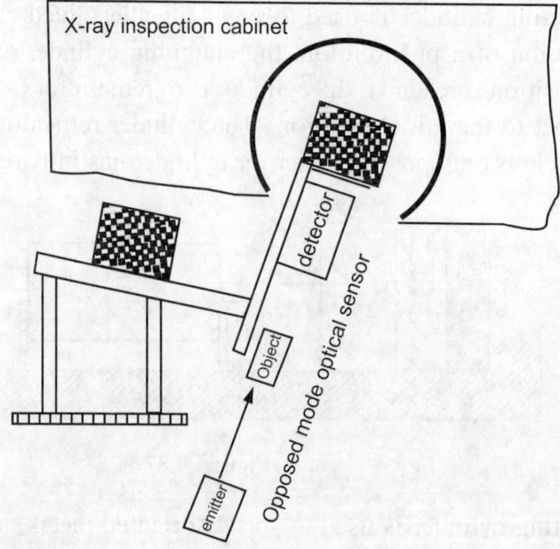

Figure 9.89

17. A swiveling slide is used to supply soft drinks packets either to the upper or the lower conveyor, according to the selection (Fig. 9.90). Design a pneumatic circuit. The designation of the swiveling slide is decided by a valve with a selector switch. The upward motion of the double acting cylinder takes 10 seconds and the downward motion takes 5 seconds. Pressure on both sides of the piston is indicated. In the initial position, the cylinder is in the retracted position.

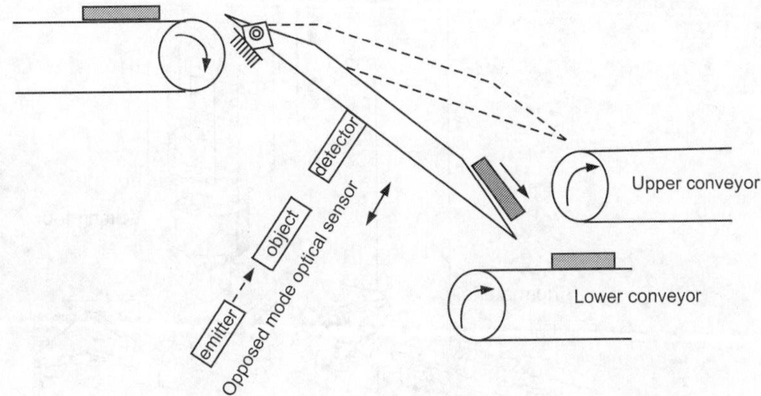

Figure 9.90

18. A double acting cylinder guides cylinder pins towards a measuring device. The pins are separated by means of a continuous to and fro movement. The oscillating motion can be started by means of a valve with a selector switch (Fig. 9.91). The duration of the forward stroke of the cylinder is to be 6 seconds. The return stroke is 4 seconds. The cylinder is to remain in the forward end position for 10 seconds, resulting in a cycle time of 20 seconds.

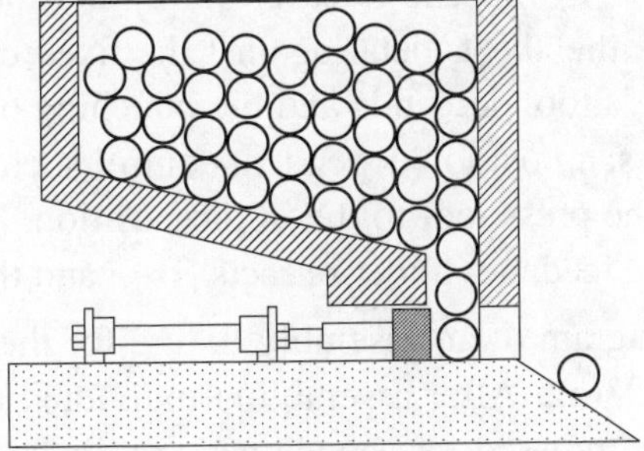

Figure 9.91

19. In a press shop, sheet metal strips are continuously fed to a press tool work table and clamped before press tool operation. Figure 9.92 shows the working of the linear indexing for feeding sheet metal strip to the press tool. Four cylinders are used for this application.

Feeder gripper	Designated as cylinder A
Feeder cylinder	Designated as cylinder B
Clamping cylinder	Designated as cylinder C
Press tool	Designated as drive D

The feeder gripper cylinder and the clamping cylinder are both diaphragm cylinders for either gripping or clamping. The feeder cylinder could be a rodless cylinder with an adequately long stroke.

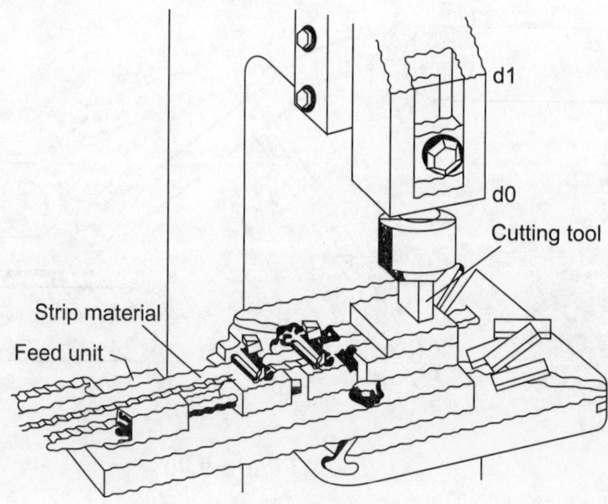

Figure 9.92

The required sequential operation is first the gripper cylinder [**A**+] advances and grips the sheet metal against a moving table. Then the feeder cylinder [**B**+] advances and transfers the sheet metal to the work table or the die of the press tool. Subsequently the gripper cylinder is released [**A**−] and the clamping cylinder advances [**C**+] and clamps the sheet metal against the fixed die block. After the clamping operation, the press tool descends and the punching operation is carried out. At the end of this process, a position dependent control such as a limit switch [**D**] conveys a signal to reset the press tool to the initial position. Then the clamping cylinder is released [**C**−], the feeder cylinder retracts [**B**−] and the cyclic process repeats.

20. The schematic diagram of an assembly fixture for the greasing of roller bearings is shown in Fig. 9.93(a). After assembling individual roller bearing parts, the roller bearings are to be pressed by a pneumatic cylinder A (1.0). Cylinder B (2.0) is required to operate a grease press twice for injecting grease into the roller bearing. The required sequence diagram is shown in Fig. 9.93(b). Develop a pneumatic control circuit using the cascade method for implementing this control task.

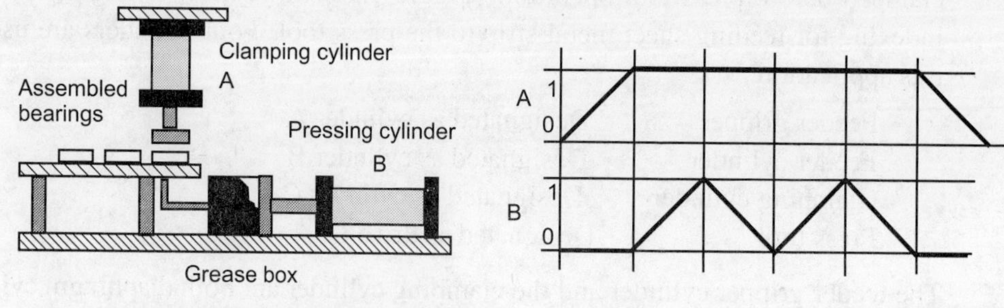

Figure 9.93 *(a) Positional sketch* *(b) Displacement step diagram*

21. Square shaped work has to be drilled using a drilling machine which is pneumatically operated (Fig. 9.94). Workpieces are fed from a gravity magazine to a drilling machine. These workpieces are pushed and clamped by means of clamping cylinder 1.0(A). A hole is drilled by the drilling cylinder 2.0(B) and the workpiece is ejected by ejecting cylinder 3.0(C). The displacement step diagram is shown in Fig. 9.95. The sequence of operations has to be carried out either for one cycle or for continuous cycles with start and stop controls. Develop a pneumatic circuit to implement the given control task.

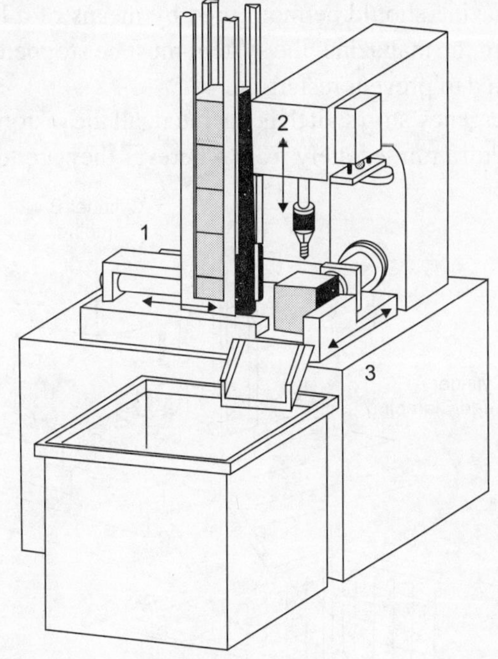

Figure 9.94

A 1 0 A+ A–
B 1 0 B+ B–
C C+ C–
Step 0 Step 1 Step 2 Step 3 Step 4 Step 5 Step 6 Step 7 Step 8

Figure 9.95

22. Rectangular parts are stamped on a special machine (Fig. 9.96). The parts are taken from a gravity feed magazine, pushed into the machine against a stop and clamped by cylinder A (1.0), stamped by the second cylinder B (2.0) and ejected by the ejector cylinder C (3.0). Design a pneumatic circuit using the cascade method, considering the auxiliary conditions:
    1. The sequence of operations should be performed automatically, with choice being provided for either a single cycle or continuous cycles.
    2. The start signal is input by means of a start button.
    3. The magazine should be monitored by means of a limit switch. If there is no part left in the magazine, the system must be stopped in its starting position and interlocked to prevent restarting.
    4. If an emergency stop switch is operated, all the piston rods of the three cylinders should return immediately from wherever they are to their starting position.

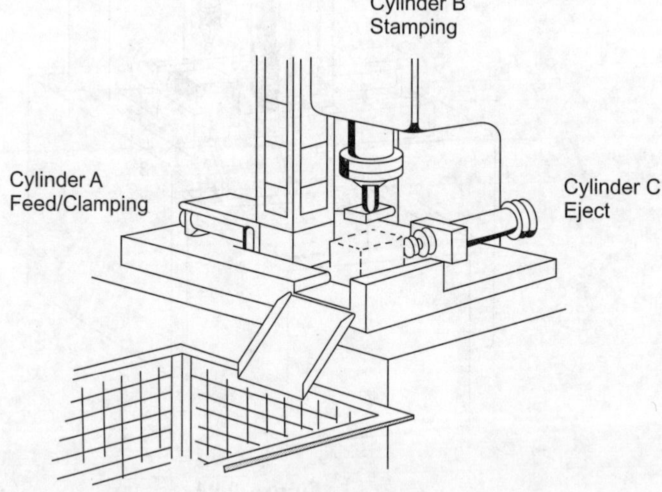

Figure 9.96

23. An arrangement for feeding, clamping, processing and ejecting of workpieces is shown in Fig. 9.97. Workpieces are taken from the gravity feed magazine and placed into the device by cylinder 1.0. Next, cylinder 2.0 extends and clamps the workpiece. Cylinder 3.0 is used to carry out processing. The processing time can be adjusted by a timer. Cylinder 4.0 is used to eject the workpiece. The sequence of operations is shown in the sequence diagram (Fig. 9.98). Develop a pneumatic control circuit to implement the control task.

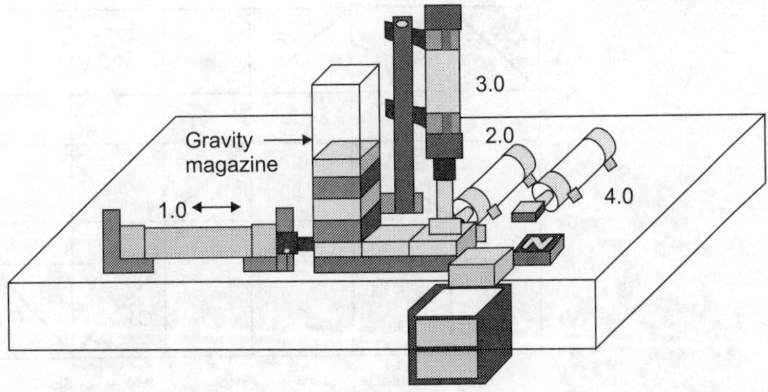

Figure 9.97

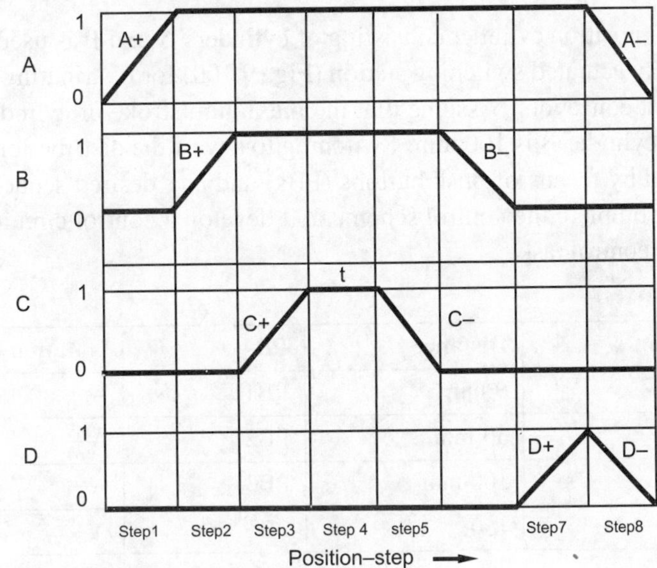

Figure 9.98

24. Automatic forming machine is used in a press tool shop (Fig. 9.99). There are three cylinders: a parts-transfer cylinder A (1.0), a parts-forming cylinder B (2.0) and a parts-ejection cylinder C (3.0). The sequence of operation is expressed in notational form as A + B + A − B − C − C+. To control the cylinders, 5/2 double piloted DC valves are used. Roller valves are mounted at both ends of each cylinder to provide a feedback signal with the exception of the forward end position of cylinder B. A pressure sequence valve is used to confirm the extended position of Cylinder B. A push button (PB) starts the cycle. Develop a pneumatic control circuit to implement the given control task.

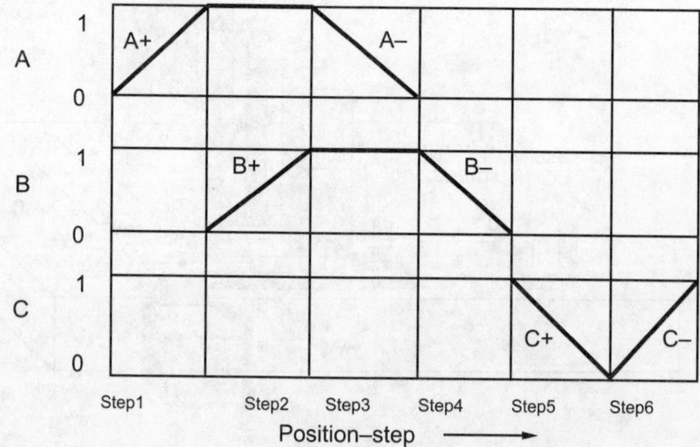

Figure 9.99

25. A multi position cylinder consisting of cylinders A and B is used to operate a pneumatically actuated switching station (Fig. 9.100) for distributing workpieces to four different conveyors. Assume that the maximum stroke of cylinder A is 50 mm and that of cylinder B is 100 mm. Switching to any of the distribution conveyors is to be initiated by means of push buttons (PBs) and any desired sequence should be possible. Complete the control scheme and develop a control circuit to accomplish the desired control task.

Solution

Position	Distance	PB	Notation	
1	0 mm	PB1	A–	A+
2	50 mm	PB2	A+	B–
3	100 mm	PB3	A–	B+
4	150	PB$	A+	B+

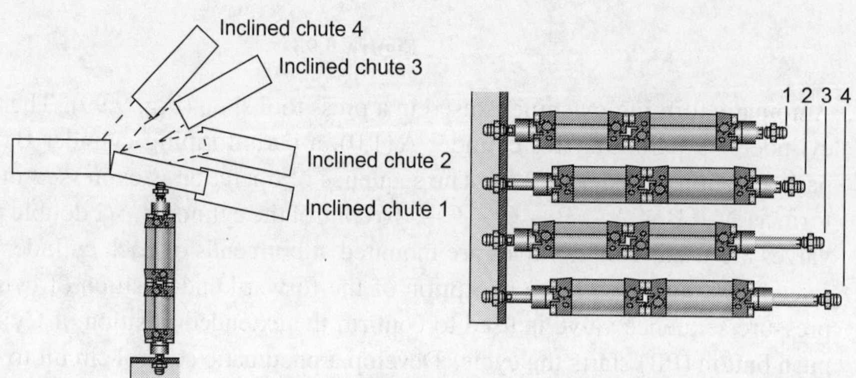

Figure 9.100

26. The arrangement for a semi-automatically operated pneumatic lathe is shown in Fig. 9.101. Cylinder 1.0 shifts the entire feed magazine to the transfer station. Cylinder 2.0 pushes the workpiece into the lathe chuck. Cylinders 1.0 and 2.0 then retract. Cylinder 3.0 moves the slide rest forward and back. Cylinder 4.0 then ejects the workpiece following the release of the lathe chuck. The sequence of operation of these cylinders is shown in the displacement step diagram (Fig. 9.102). Develop a pneumatic control circuit to implement the given control task for: (i) one cycle and (ii) continuous sequence of operation.

Solution

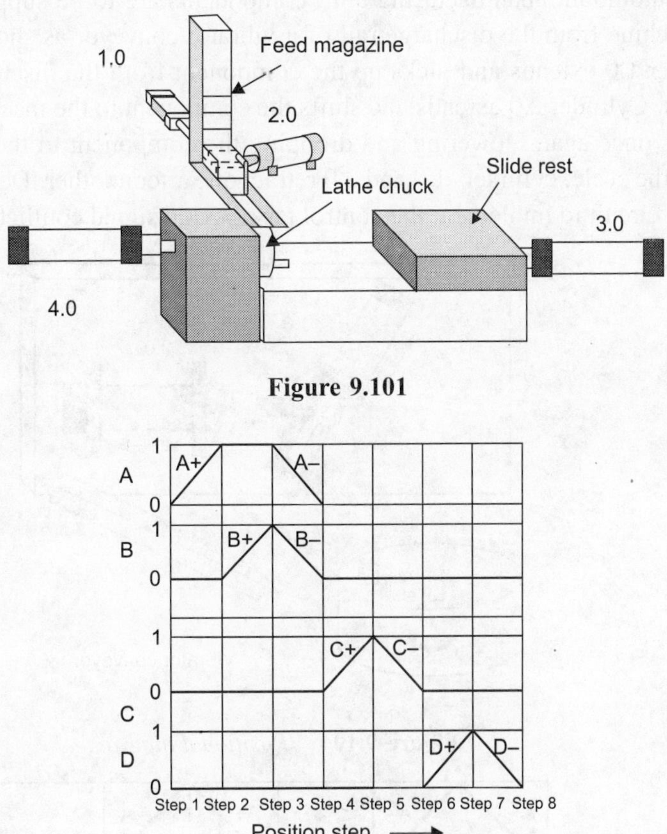

Figure 9.101

Figure 9.102

27. A pneumatic system is to be designed for reaming of drill holes as shown in Fig. 9.103. Cylinder 1.0 clamps the workpiece. Feed cylinder comes fully forward and backward. The total number of feed cycles is 3. The required sequence diagram is shown. Develop a pneumatic control circuit to implement the control task.

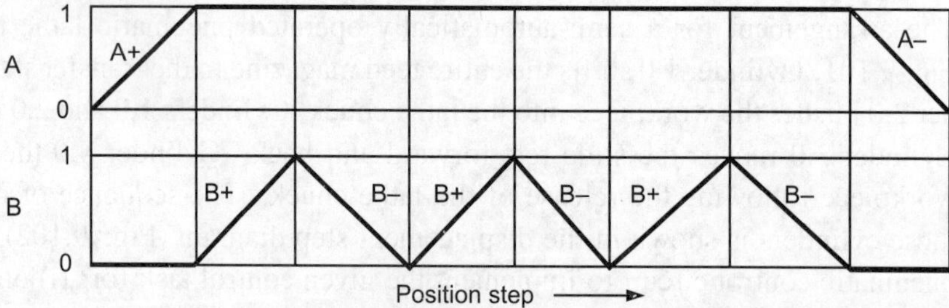

Figure 9.103

28. In an automobile manufacturing unit, components are to be supplied to a processing machine from the discharge rail of a vibrator conveyor as shown in Fig. 9.104. Cylinder 1.0 extends and picks up the component from the discharge rail and then retracts. Cylinder 2.0 extends and shifts the component to the machine. Cylinder 1.0 extends once again, lowering and dropping the component to the machine. At the end of the cycle, cylinders 1.0 and 2.0 retract one after another. Develop a pneumatic control circuit to implement the control task. Avoid signal conflicts in the circuit.

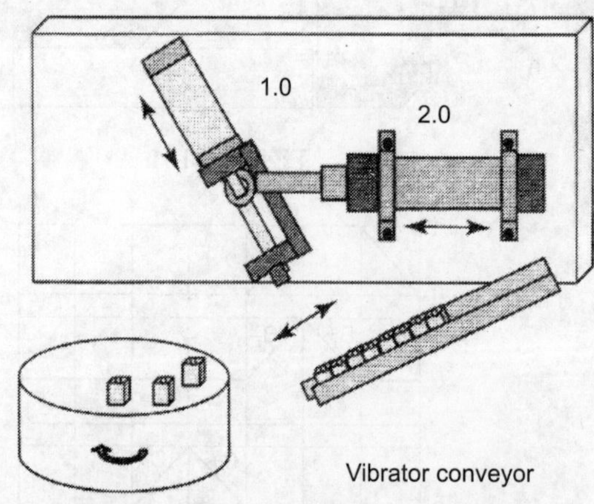

Figure 9.104 *Positional diagram*

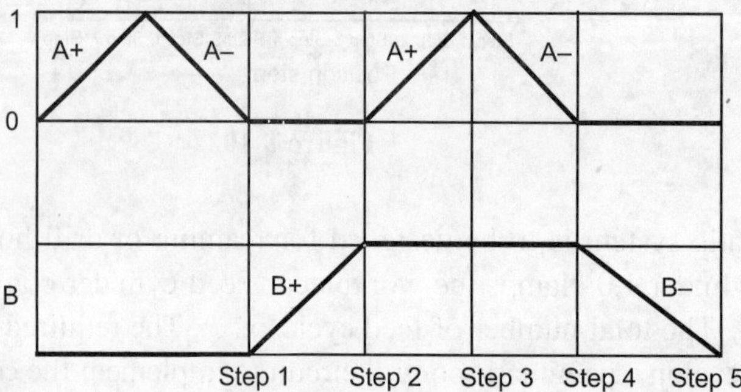

29. In a production process, workpieces are to be fed into a jig from a gravity magazine using a double acting cylinder (Fig. 9.105). Feeding occurs when a push button is pressed and the cylinder returns automatically to its start position after the workpiece has reached the jig position.

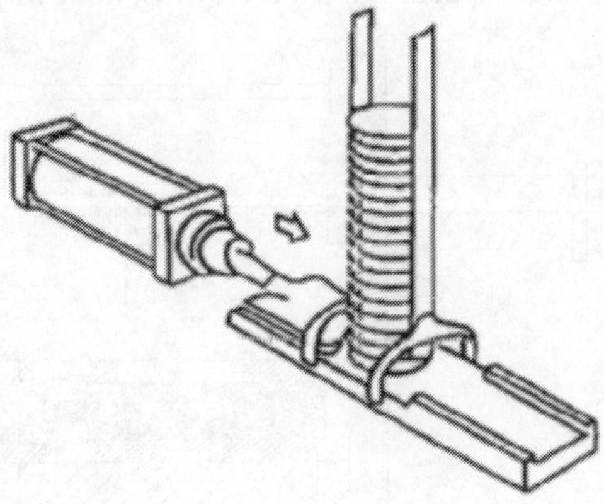

Figure 9.105

30. The tested parts are distributed to three conveyors (Good/Reject/Rework) as seen in Fig. 9.106. Shifting to the required position must be possible with any sequence by push buttons.

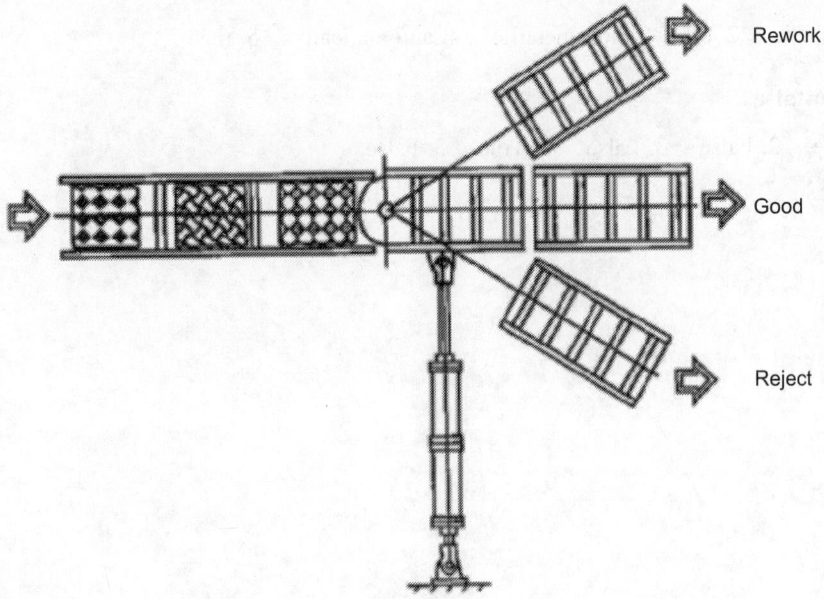

Figure 9.106

31. Parts are to be punched using a double acting cylinder via mechanical linkage. Punching is only possible when the workpiece is in position (mandrel), the safety cover is closed and either a push button or a foot pedal is actuated. After the punching cylinder has reached the rear end position, it is automatically returned to its extended position. The cylinder speeds are to be very fast both ways.

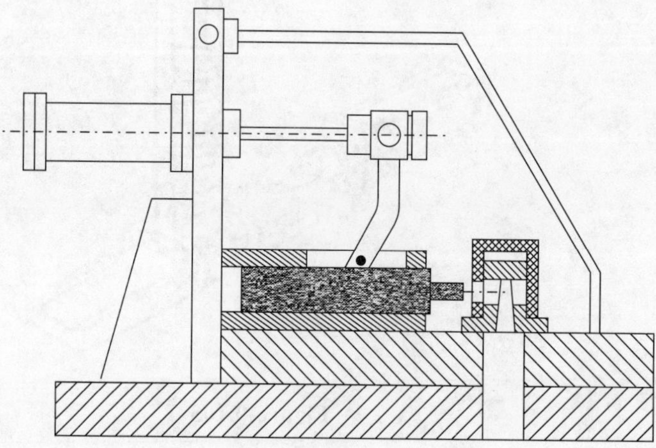

Figure 9.107

Answers

Fill in the blanks

1. magnetic 2. off 3. non-metallic 4. automatically 5. reed

True or false

1. False 2. False 3. False 4. True 5. False

10 PNEUMATIC CIRCUIT DESIGN USING PLCs

Upon completion of this chapter, the reader should be able to

- Define PLC and its function
- Explain the difference between hardwired control and PLC control
- Explain the different sections of PLC
- List the advantages of PLC over electromagnetic relays
- Explain the functions of major components of PLC
- Explain various programming approaches used in PLC
- Describe the working of memory functions, timers and counters
- Convert logic functions into ladder diagrams
- Design PLC circuits for single and multi actuators

10.1 Introduction

A programmable logic controller (PLC) is essentially a user friendly microprocessor-based microcomputer with hardware and software designed to control the operation of industrial equipment and processes. An important advantage of a PLC is that it can be easily programmed and reprogrammed. It has tremendous impact on industrial control and instrumentation due to its high reliability and flexibility during the design and implementation stages. The decreasing cost of microprocessors coupled with an expanding feature set has widened the scope and applications of PLCs. In recent years, PLCs are being used in place of electromechanical relays and cam operated logic controllers to control fluid power systems. Modern day PLCs are incorporated into sophisticated, highly versatile control system components capable of performing complex mathematical functions and operating at high microprocessor speeds. Some leading PLC manufacturers are ABB, Allen Bradley, Honeywell, Siemens, GE Fanuc, Mitsubishi, Modicon and Omron.

10.2 Features of Programmable Logic Controllers

PLC can be defined as a digital electronic device that uses a programmable memory to store instructions and implement functions such as logic, sequencing, counting, timing and arithmetic in order to control machines, processes and instrumentation. In other words, it is a user-friendly digital computer used for making logic decisions and providing output. It consists of solid state digital elements and is a replacement for hardwired electromechanical relays to control pneumatic systems.

The term 'programmable logic controller' is defined as follows by IEC 1131 (PLC standard) part 1:

"A digitally operating electronic system, designed for use in an industrial environment, which uses a programmable memory for the internal storage of user-oriented instructions for implementing specific functions such as logic, sequencing, timing, counting and arithmetic, to control through digital or analog inputs and outputs, various types of machines or processes. Both the PC and its associated peripherals are designed in such a way that they can be easily integrated into an industrial control system and easily used with all their intended functions."

PLCs are quite similar to digital computers. They also have certain features which are specific to logic controllers such as

1. They are rugged and designed to withstand vibrations, temperature variations, humidity and noise.
2. The interface for the input and output is part of the controller (PLC).
3. They are easily programmable and primarily use logic and switching functions.

10.3 Hardwired Control System

Hardwired control systems use relays. For example, in electrical control the wiring of control elements such as sensors, solenoids, counters etc. are through relay controls. Such relay controlled systems are termed hardwired control systems because any modification in the control program involves rewiring the circuit. Therefore, hardwired controls are cumbersome and difficult to modify when production requirements change regularly. Hardwired control systems are difficult to maintain because any small problem in design could be a major problem in terms of tracing and rewiring.

Hardwired control systems have three sections.

1. **Input section** Consists of pushbuttons, switches and sensors which transfer signals to the processing section.

2. **Processing section** Consists of relay coils and contacts which determine the relationship between the inputs received and the outputs required.
3. **Output section** Consists of solenoids, lamps, contactor coils, etc. The processed signals are transferred to this section.

10.4 PLC Systems

PLC systems offer numerous advantages over hardwired electromechanical relay control systems. Unlike electromechanical relays, PLCs are not hardwired to perform specific functions. So when system operating requirements change, a software program is easily modified instead of physically rewiring relays. In addition, PLCs are more reliable, faster in operation, smaller in size, and can be readily expanded.

PLC systems consist of three sections

1. **Input section** Consists of push buttons, switches and sensors which are connected to specific input addresses in the program. They transfer address information to the processing section.
2. **Processing section** The microprocessor receives signals from the input sections and executes the information (called instructions) using software programs and sends the processed signals to the output section.
3. **Output section** Takes the signal from the processing section and modifies it to operate output devices connected to specific output addresses.

Advantages of PLCs over electromechanical relays

a) PLCs are faster and more reliable.
b) They are compact and can be expanded easily.
c) Power consumption is lower.
d) They are less expensive when compared to hardwired systems with similar control functions.
e) Hardwired electromechanical relays lack flexibility and require rewiring when system requirements change.
f) PLCs have very few hardware failures when compared to electromechanical relays.
g) Special functions such as time-delay actions and counters can be easily implemented using PLCs.

A comparison between relays and PLCs is presented in Table 10.1.

Table 10.1 *Comparison between relays and PLCs*

Features	Electromechanical relay	PLC
Size of the controller	Bulky	Compact
Programming	Time consuming	Easy
Flexibility	Rewiring required	Reprogramming required
Cost	Expensive	Less expensive
Maintenance	Poor	Minimum
Fault finding and troubleshooting	Difficult	Easy to identify fault and repair

10.5 Major Components of PLC

As discussed earlier, a PLC is essentially a microcomputer consisting of hardware and software. The major components are:

1. Power supply module
2. Input module
3. Central processing unit
4. Output modules
5. Software

1. Power supply module The input/output modules usually require 24V DC and the processor requires 5V DC supply. The power supply is an integral part of PLC. The power supply unit converts 120/230V AC line voltage to a standard supply of 24V DC or 5V DC using standard rectifier circuits.

2. Input module Input devices include push buttons, sensors, potentiometers, pressure switches, etc. The function of the input module is to convert high voltages from input devices to low level logic voltages that the CPU uses internally for processing. The input module can process both analog and digital inputs. Digital inputs are preferred in the industry.

Analog input modules are used to convert analog signals from analog devices like temperature sensors, pressure sensors etc., to digital signals using an ADC (analog-to-digital convertor). Analog signals vary in voltage between 0-12 V or in current between 5-20 mA. These values of current or voltage are converted into integer values (e.g., a 16 bit word).

Input modules are used to convert digital input signals to 5V digital signals that the CPU uses internally to execute a user program.

3. Central processing unit The central processing unit (CPU) controls and processes all operations within the PLC and is called the brain of the PLC. It can perform a range of arithmetic and data manipulation functions with local and remotely located input/output sections. Further, the processor can perform many communication functions needed to interface with a personal computer, remote input/output, other PLCs and peripheral devices.

Functions of a CPU:

1. It receives input from various sensing devices and switches
2. It executes user programs
3. It makes various decisions to control the operation of the equipment or the process
4. It performs various arithmetic and data manipulation functions
5. It delivers corresponding output signals to various load control devices such as relay coils and solenoids

4. Output module Output devices include contactor coils, solenoid coils, lamps, etc. The output module amplifies low level logic signals generated by the CPU and passes these modified signals to the final control elements to operate the output devices.

5. Software PLC software consists of two parts: the operating system (OS) and the user programs. The PLC operating system provides effective support ranging from the creation of project structure to the creation of user programs. The OS is accessed through a graphical user interface window(also known as the main window). The main window contains all the functions needed to set up a project, configure the hardware, and write and test programs. User programs can be written in any standard PLC programming language like ladder diagram or instruction list (also known as statement list).

While processing a PLC program, the CPU scans and executes the main program cyclically. A program scan cycle consists of sequential operations that include input scan, program scan, and output scan. During the input scan, the CPU updates the process image input table. During the output scan, the CPU updates the process image output table. After the completion of each scan cycle, the CPU returns to the start of the next cycle and repeats it. The time taken to scan one program is called scan cycle time.

10.6 Programming of PLC

There are various approaches for programming a PLC:

1. Ladder diagram
2. Low level programming based on Boolean expressions
3. Functional blocks
4. High level languages

Most of the programming methods used today for PLCs are based on the ladder logic diagram which is explained in the following sections.

PLC programming using the ladder diagram involves writing a program in a manner similar to drawing a switching circuit. The ladder logic diagram is converted into a PLC ladder diagram using the conventions of ladder diagram construction. This method only requires a basic keyboard and a CRT with minimum graphics capability to display component symbols and their relationships in a ladder logic diagram. The components are of two types, contacts and coils. Contacts are used to represent input switches, relay contacts and similar elements. Coils are used to represent load such as solenoids, relays, timers, counters, etc. The programmer inputs the ladder diagram rung by rung into the PLC memory with the CRT displaying the results for visual verification.

The ladder diagram has two vertical sides (also called rungs) as shown in Fig. 10.1. The line on the left represents positive voltage and the line on the right represents zero voltage. Between the two vertical sides are the horizontal rungs representing the power flow. The symbols representing the various program elements are placed along horizontal rungs to implement the control task.

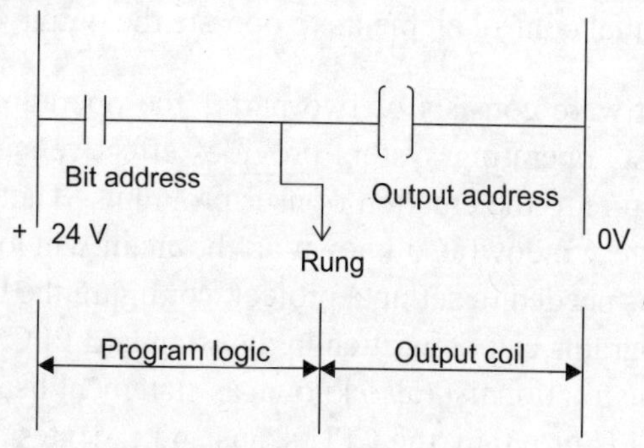

Figure 10.1 *Ladder diagram*

There are five program elements or operations commonly used in a PLC ladder diagram:

1. PLC bit logic operations
2. Timer operations
3. Counter operations
4. Comparison operations
5. Arithmetic operations

1. PLC bit logic operations:
   The important programming elements for bit logic operations are:

   1. NO contact
   2. NC contact
   3. Coil

Each of these elements can be selected from the program window. NO and NC elements should not be confused with hardware NO and NC contacts of switching devices.

NO Contact The PLC representation of an NO contact is given in Fig. 10.2. This contact scans for the signal state ON (1) at the specified bit address and if this condition is satisfied, power flows through the NO contact. This contact is used for scanning the signal state of input or output devices and other internal program elements.

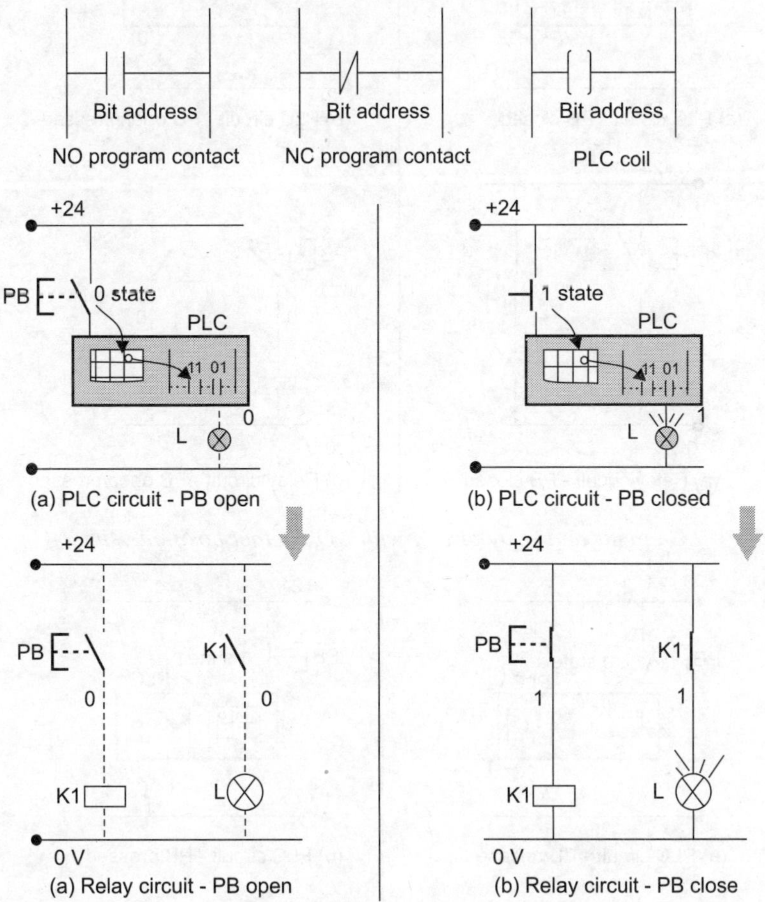

Figure 10.2 *PLC circuit and relay circuit with NO contact position using NO push button*

NC Contact Figure 10.3 shows NO contact positions thing NC push button. This contact scans for the signal state OFF (0) at the specified bit address and if this condition is satisfied, power flows through NC contact. This contact is used for scanning the signal state of input and output devices and other internal program elements. Figure 10.4 shows a PLC circuit with an NC contact position using an NO push button. Figure 10.5 shows a PLC circuit with an NC contact position using an NC push button.

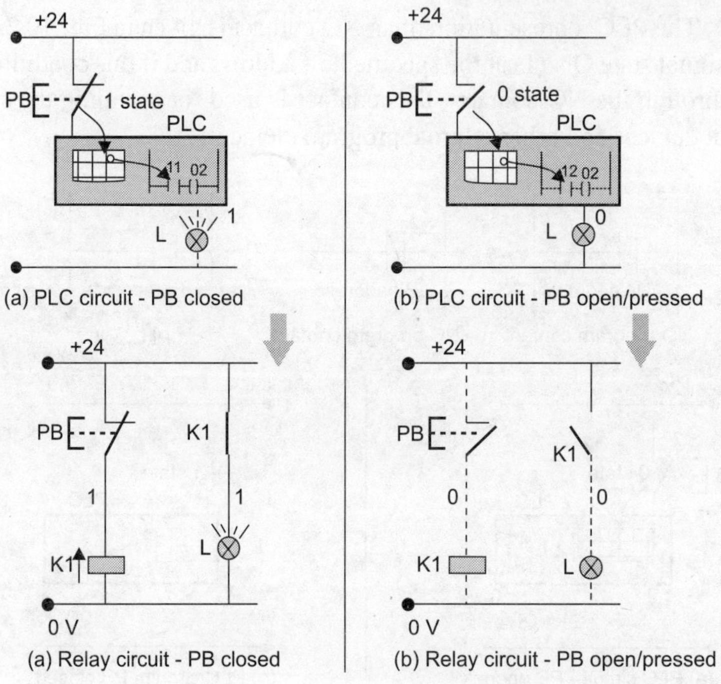

Figure 10.3 *PLC circuit and relay circuit with NO contact position using NC push button*

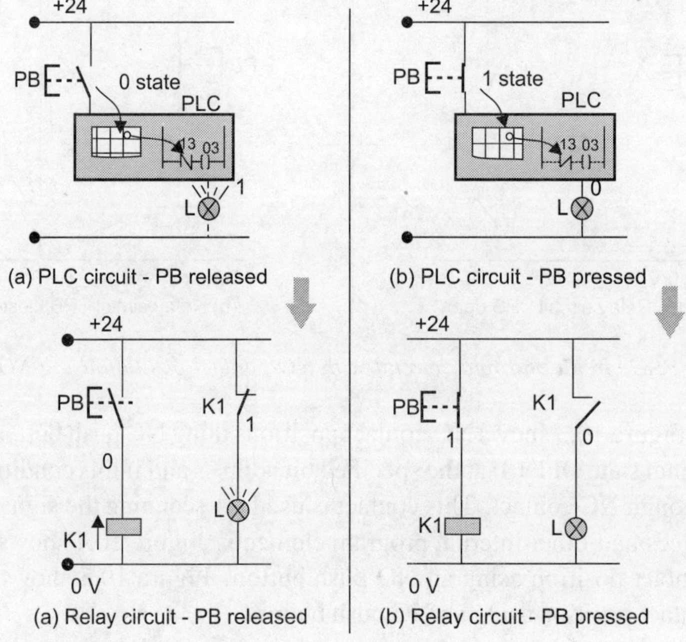

Figure 10.4 *PLC circuit and relay circuit with NC contact position using NO push button*

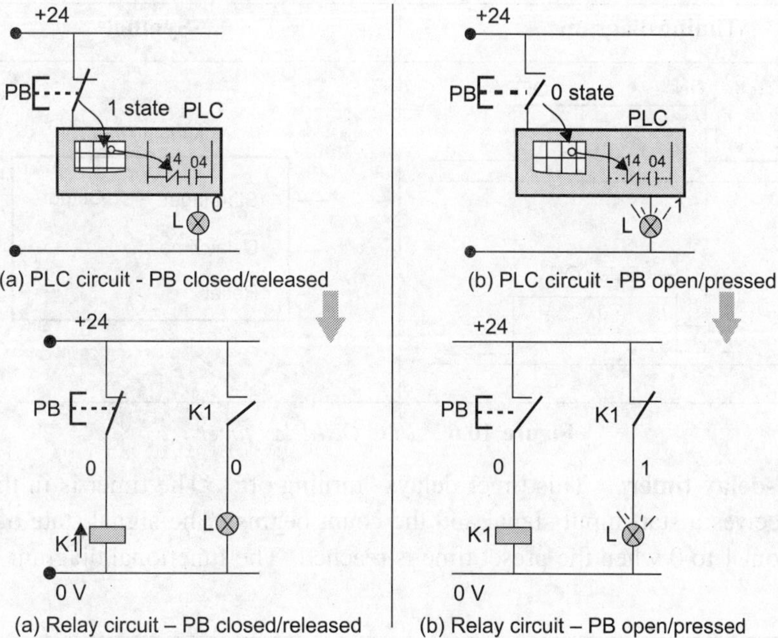

Figure 10.5 *PLC circuit and relay circuit with NC contact position using NC push button*

10.7 PLC Timers

Many control tasks involve programming for time related logic. For example, cylinder 2 is to extend if cylinder 1 is retracted but only after a specified delay. The timers of a PLC are realised in the form of software modules and are based on digital timing. Memory space is allocated in system memory to store the values of the time delay. The representation of the timer address varies from manufacturer to manufacturer. For simplicity, timer addresses are referred to here as T1, T2, etc. Commercial PLCs include timers in blocks of 64, 128, 256, 512, etc. To explicitly reset a timer, an RLO (Result of Logic operation) of 1 is applied at the reset port.

There are two types of PLC timers:

PLC ON-delay timer: This timer delays the "turning on". It is in the OFF state when it receives a start input signal, then starts counting up the preset time delay and when it is reached, the signal state of the output changes from 0 to 1.

The timer starts when input signal state changes from 0 to 1. After the specified time elapses, output signal changes from 0 to 1 if input signal is still 1. The functional diagram is shown in Fig. 10.6.

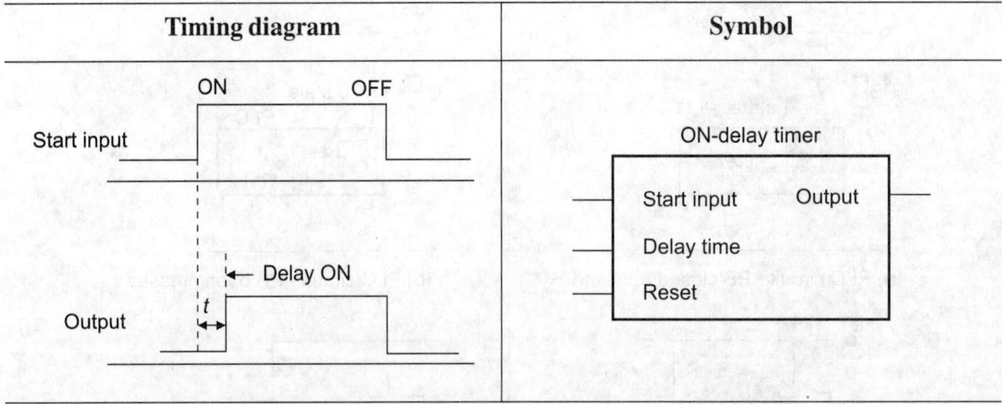

Figure 10.6 *PLC ON-delay timer*

PLC OFF-delay timer: This timer delays "turning off". The timer is in the ON state when it receives a start input signal and the count begins. The signal state of the output changes from 1 to 0 when the preset time is reached. The functional diagram is shown in Fig. 10.7.

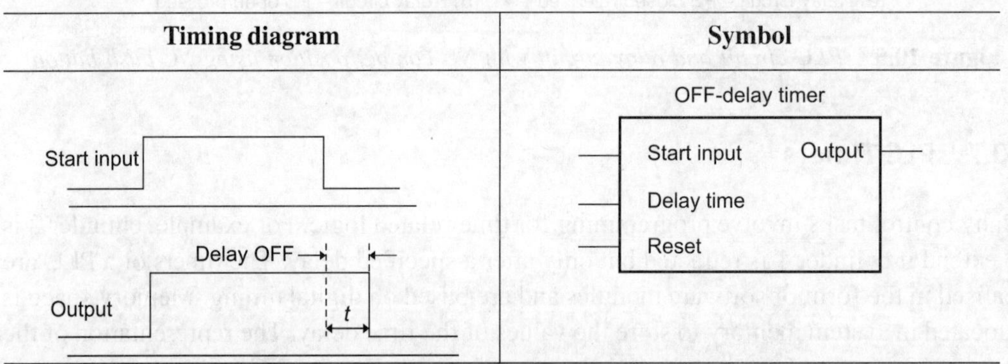

Figure 10.7 *PLC OFF-delay timer*

10.8 PLC Counters

Counters are used to detect workpieces, numbers and events. In practice, controllers are frequently operated with counters; e.g., a counter is required in a circuit if exactly 20 identical components are to be conveyed to a conveyor belt via a sorting device.

There are two basic types of counters: (a) count-up and (b) count-down.

When the input to a count-up counter is true, the accumulator value increases by 1 (no matter how long the input is true). If the accumulator value reaches the preset value the counter bit is set. A count-down counter decreases the accumulator value until the preset value is reached. The symbols are shown in Fig. 10.8. Count-down counters are most commonly used in industries.

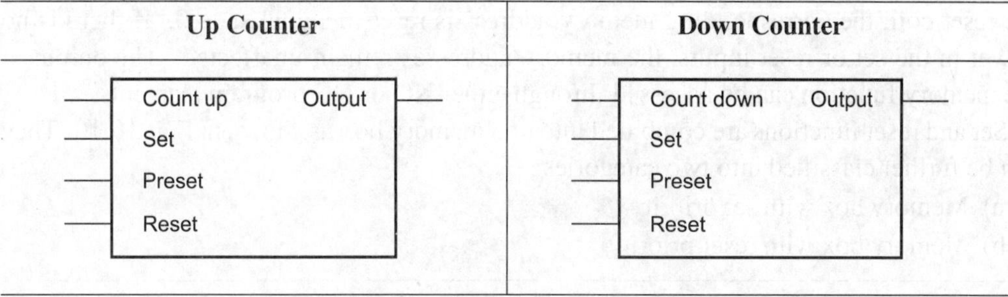

Figure 10.8 *Up and down counters*

10.9 PLC Memory Elements

Memory elements are used to store intermediate values. The memory function is achieved using flags (bit memory locations) and system memory. Specified memory bits can be set or reset using a set coil. In ladder logic, a latch uses one instruction to latch and a second instruction to unlatch, as shown in Fig. 10.9. The output S turns ON output D when input A turns true. D stays ON even if A turns OFF. Output D turns OFF if input B becomes true and output R becomes true.

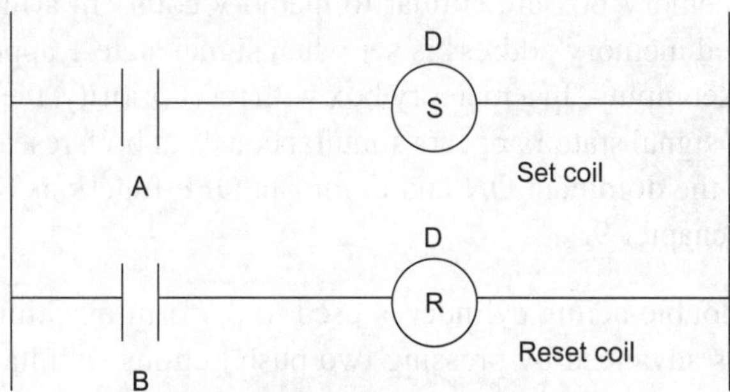

Figure 10.9 *Ladder logic latch*

Both the set and reset coils can be combined in one box as shown in Fig. 10.10. The following instructions apply when writing a program for memory functions:

- Set up a memory location
- Reset a memory location

The memory address locations vary from manufacturer to manufacturer. For the sake of simplicity, memory locations are represented as M1, M2, . . . here. Commercial PLCs offer bit memories in blocks of 1024, 2048, 8192, etc.

If power flows, either momentarily or continuously to the set coil, the corresponding memory address is set to signal state 1. If power flows momentarily or continuously to the reset coil, the corresponding memory address is reset to signal state 0. If there is no power in the set or reset inputs, the memory addresses remain unaffected. The output of the memory function can be accessed through either NO or NC program elements.

Set and reset functions are combined into one memory box as shown in Fig. 10.10. They can be further classified into two categories:

a) Memory box with set priority
b) Memory box with reset priority

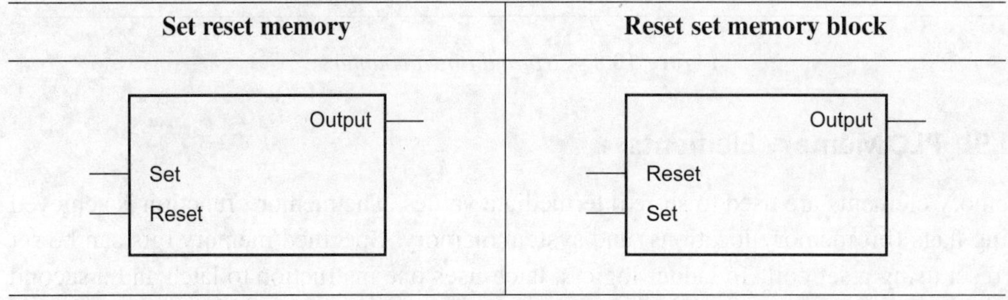

Figure 10.10 *Memory box representation*

The functions of a memory box are similar to memory coils. In a memory box with set priority, the associated memory address is set when signal state 1 appears simultaneously at both the set and reset inputs. In a memory box with reset priority, the associated memory address is reset when signal state 1 appears simultaneously at both reset and set inputs. This concept is similar to the dominant ON and dominant OFF functions of electrical latching circuits discussed in chapter 9.

Example 10.1 A double acting cylinder is used to perform machining operations. The pneumatic cylinder is advanced by pressing two push buttons simultaneously. If any one of the push buttons is released, the cylinder comes back to the start position. Draw the pneumatic circuit, PLC wiring diagram and ladder diagram to implement this task.

The solution is shown in Fig. 10.11.

As shown in the PLC wiring diagram, the push buttons PB1 and PB2 are connected to memory addresses I1 and I2 respectively. I1 and I2 are connected in series in the ladder diagram to implement the AND logic function.

When the push buttons PB1 and PB2 are pressed simultaneously, the addresses I1 and I2 turn from state 0 to state 1and as a result, power flows through the coil providing output at O1. Output at O1 operates the solenoid coil and the cylinder moves forward for the required operation.

If any one of the push buttons PB1 or PB2 is pressed, then the corresponding bit address turns to 0. Since I1 and I2 are in series, if any of them turns to 0, there is no output at coil O1 and so the solenoid is de-energised and the cylinder returns.

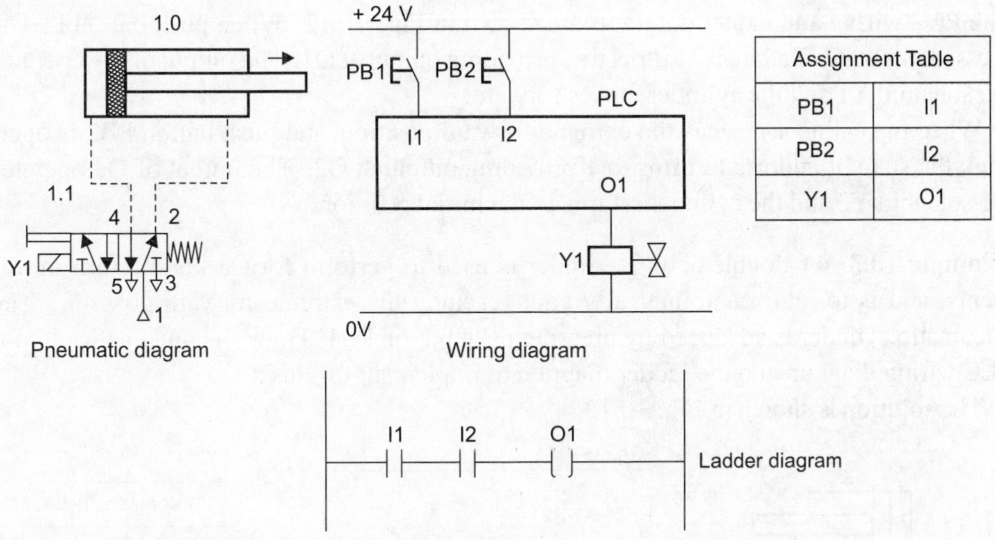

Figure 10.11 *Pneumatic diagram, wiring diagram and ladder diagram*

Example 10.2 A double acting cylinder is used to perform forward and return motion. A pneumatic cylinder is advanced by pressing push button PB1 and is returned by pressing push button PB2. Draw the pneumatic circuit, PLC wiring diagram and ladder diagram to implement this task.

The solution is shown in Fig. 10.12.

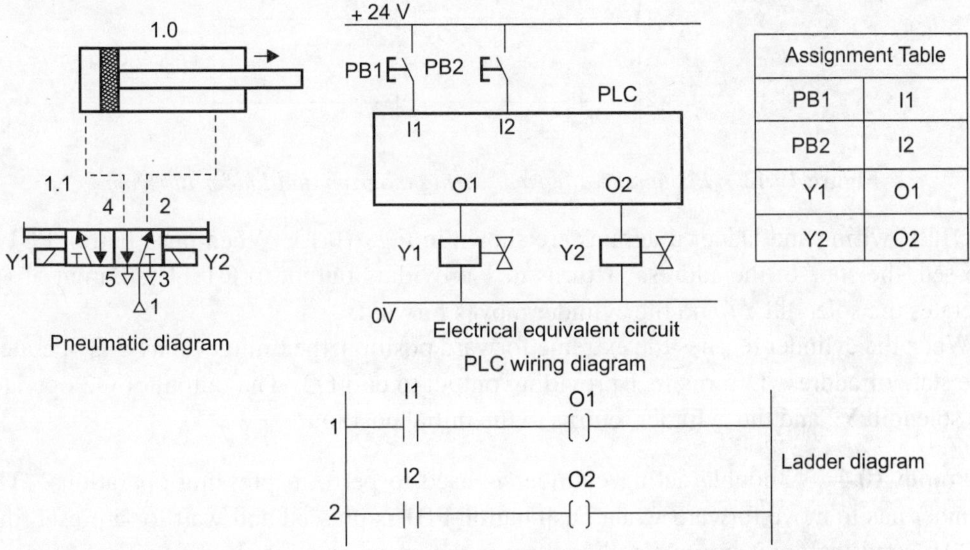

Figure 10.12 *Pneumatic diagram, wiring diagram and ladder diagram PLC*

The PLC wiring and ladder diagrams are shown in Fig. 10.12. When push button PB1 is pressed, the state of address I1 turns to 1 providing output at O1. The output of O1 operates the solenoid Y1 and the cylinder moves forward.

When the cylinder reaches the extreme forward position, and push button PB2 is operated, the state of address I2 turns to 1 providing output at O2. The output of O2 operates the solenoid Y2 and the cylinder returns to the initial position.

Example 10.3 A double acting cylinder is used to perform forward and return movements, and is to return automatically after reaching the extreme forward position. The pneumatic cylinder is advanced by pressing push button PB1. Draw the pneumatic circuit, PLC wiring diagram and the ladder diagram to implement this task.

The solution is shown in Fig. 10.13.

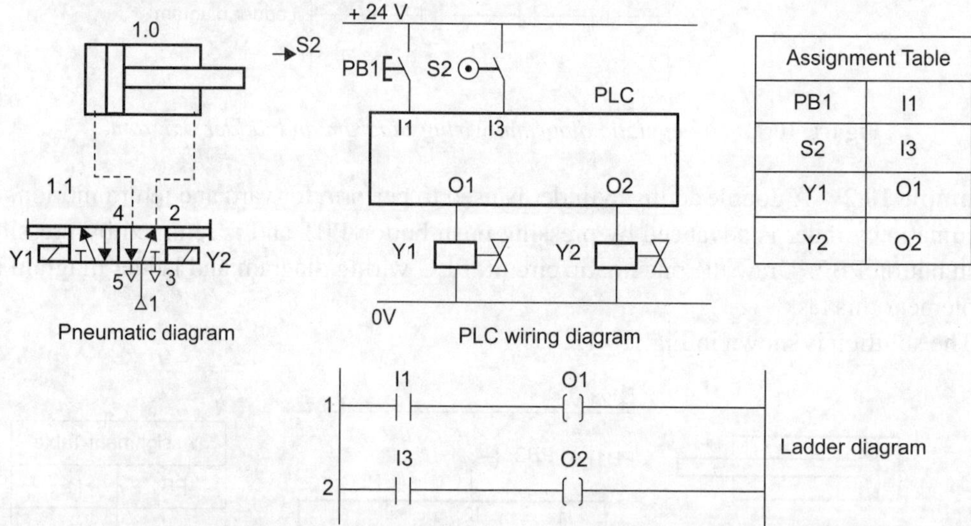

Figure 10.13 *Pneumatic diagram, wiring diagram and ladder diagram*

The PLC wiring and ladder diagrams are shown in Fig. 10.13. When push button PB1 is pressed, the state of the address I1 turns to 1 providing output to O1. The output of O1 operates the solenoid Y1 and the cylinder moves forward.

When the cylinder reaches the extreme forward position, the limit switch S2 is operated. The state of address I3 turns to 1 providing output to coil O2. The output of O2 operates the solenoid Y2 and the cylinder returns to the initial position.

Example 10.4 A double acting cylinder is used to perform pressing operations. The cylinder has to move forward when push button PB1 is pressed and wait for a preset time of 20 seconds before it automatically returns to the initial position. Limit switch S2 is used for sensing the end of the forward motion of the cylinder. Draw the pneumatic circuit, PLC wiring diagram and ladder diagrams to implement this task.

The solution is shown in Fig. 10.14.

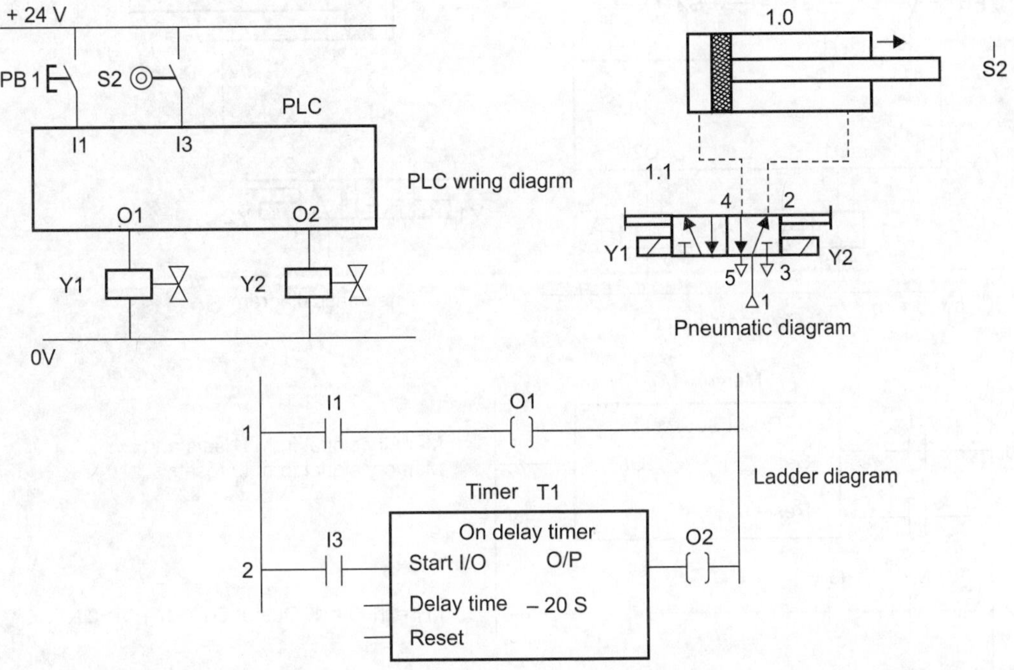

Figure 10.14 *Pneumatic diagram, wiring diagram and ladder diagram*

When PB1 is pressed, the input at address I1 is set to 1 providing output at O1 which in turn operates the solenoid coil Y1, moving the cylinder forward. When the cylinder reaches the end position, limit switch S2 is operated and as a result the state at address I3 changes to 1, starting timer T1. The signal state of timer T1 changes to 1 after 20 seconds and its output feeds O2. Coil Y2 is energised thus triggering the return motion of the cylinder.

Example 10.5 A double acting cylinder is used to perform continuous to and fro motion. The cylinder has to move forward when push button PB1 is pressed. Once the to and fro cycle starts, it should continue till the stop push button PB2 is pressed. Limit switches are used to sense the end position. Draw the pneumatic circuit, PLC wiring diagram and ladder diagram to implement this task.

The solution is shown in Fig. 10.15.

The start and stop operations can be implemented using a memory flag with address M1 that is set by PB1 and reset by PB2. The state of memory element M1 is scanned through an NO contact and is combined in series with the state of sensor S1 to implement the start and stop controls.

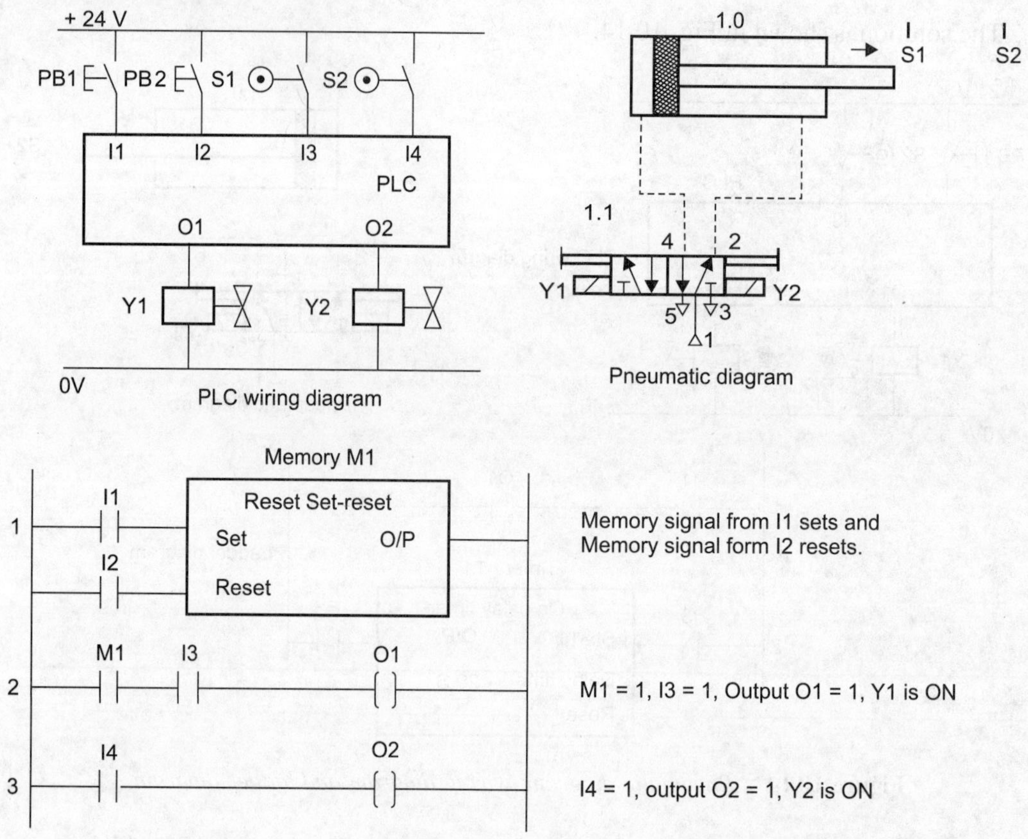

Figure 10.15 *Pneumatic diagram, wiring diagram and ladder diagram*

Example 10.6 A double acting cylinder is used to perform to and fro operations. The cylinder has to move forward when push button PB1 is pressed and continue the to and fro motion for 10 cycles. Draw the pneumatic circuit, PLC wiring diagram and ladder diagram to implement this task.

The solution is shown in Fig. 10.16.

Fully automated operation of the cylinder is achieved by using limit switches S1 and S2. Start and stop operations are implemented using a memory flag with address M1 that is set by PB1 at address I1 and reset by the NC contact of a down counter. The state of the memory flag M1 scanned by the NO contact (rung 2) is combined in series with the state of sensor S1 to implement the start and stop controls.

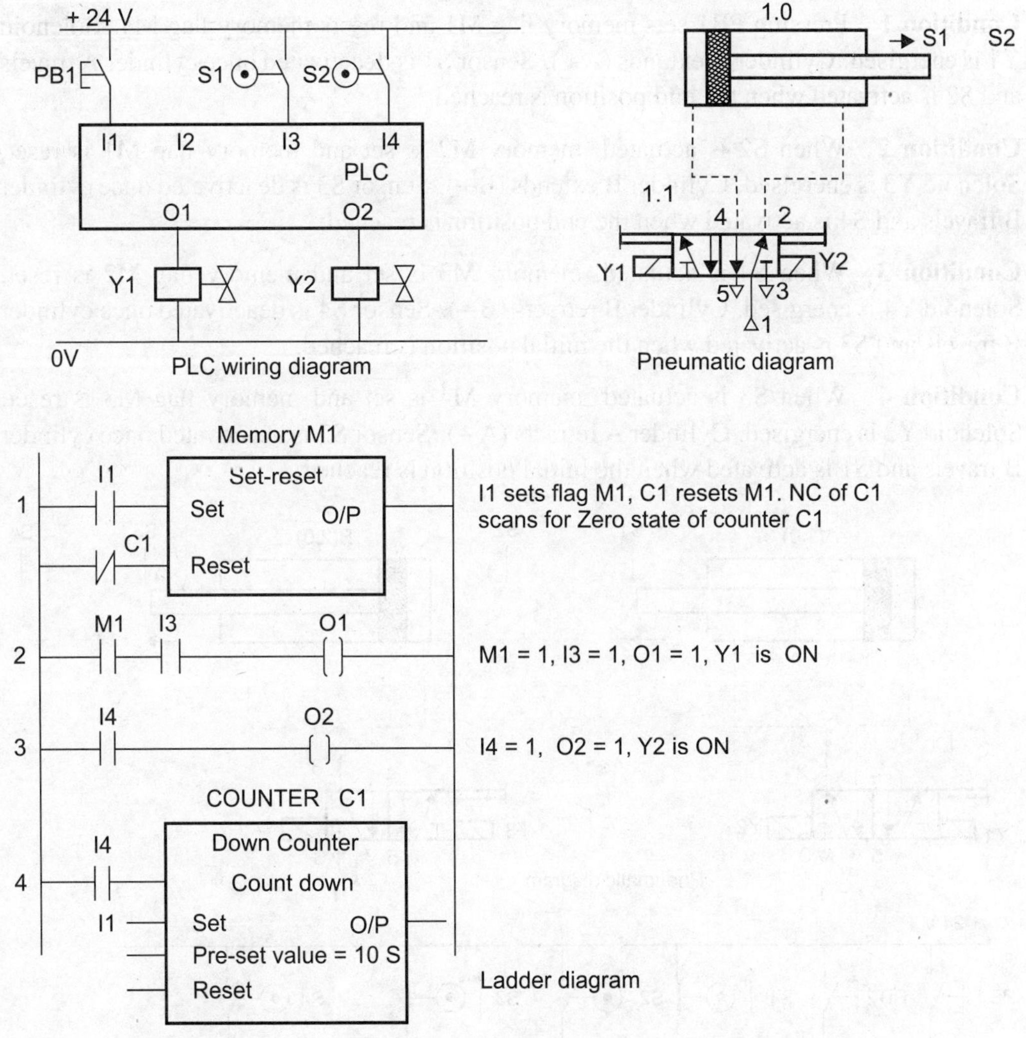

Figure 10.16 *Pneumatic diagram, wiring diagram and ladder diagram*

Example 10.7 Draw the pneumatic circuit, PLC wiring diagram and ladder diagram to implement the sequence A + B + B − A−.

The solution is shown in Fig. 10.17 and Fig. 10.18.

In this sequence circuit, PB2 is used to initiate the program. Pressing PB2 causes the last memory state M4 to set and all other memory flags M1, M2 and M3 to reset. Initially S1 and S3 are actuated and generate outputs. This is shift register method.

Condition 1 Pressing PB1 sets memory flag M1 and resets memory flag M4. Solenoid Y1 is energised. Cylinder A extends (A+). Sensor S1 is deactivated once cylinder A travels and S2 is activated when the end position is reached.

Condition 2 When S2 is actuated, memory M2 is set and memory flag M1 is reset. Solenoid Y3 is energised. Cylinder B extends (B+). Sensor S3 is deactivated once cylinder B travels and S4 is activated when the end position is reached.

Condition 3 When S4 is actuated, memory M3 is set and memory flag M2 is reset. Solenoid Y4 is energised. Cylinder B retracts (B−). Sensor S4 is deactivated once cylinder B travels and S3 is activated when the initial position is reached.

Condition 4 When S3 is actuated, memory M4 is set and memory flag M3 is reset. Solenoid Y2 is energised. Cylinder A retracts (A−). Sensor S2 is deactivated once cylinder B travels and S1 is activated when the initial position is reached.

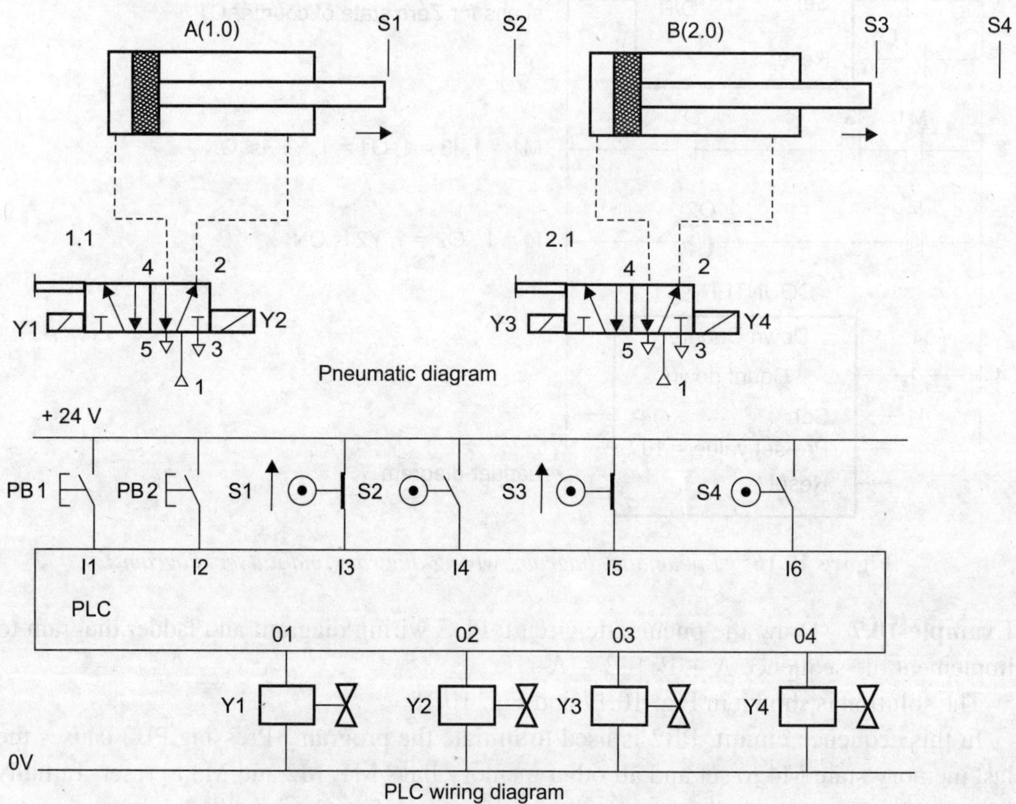

Figure 10.17 *Pneumatic diagram and the wiring diagram*

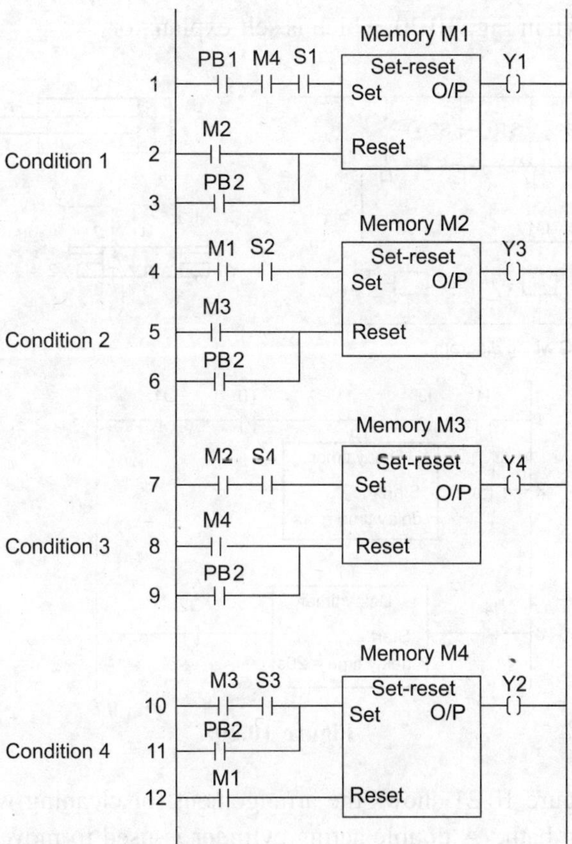

Figure 10.18 *Ladder diagram*

Example 10.8 Double-acting cylinder is used to press parts for 20 seconds (Fig. 10.19). A push button PB1 is used to control the forward stroke of the cylinder. After pressing the components for 20 seconds, the cylinder is to retract automatically. The return stroke must occur even if the start button is still actuated. A new start signal may only be effective after the initial position of the piston is reached and the push button is released. Draw a wiring diagram, pneumatic circuit and ladder diagram to implement the control task.

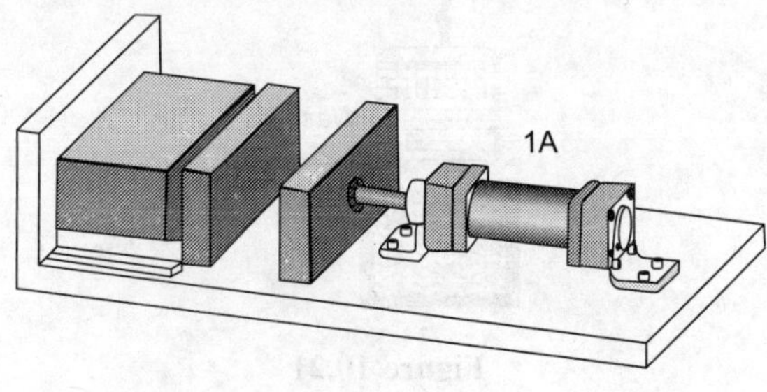

Figure 10.19

The solution is shown in Fig. 10.20, which is self explanatory.

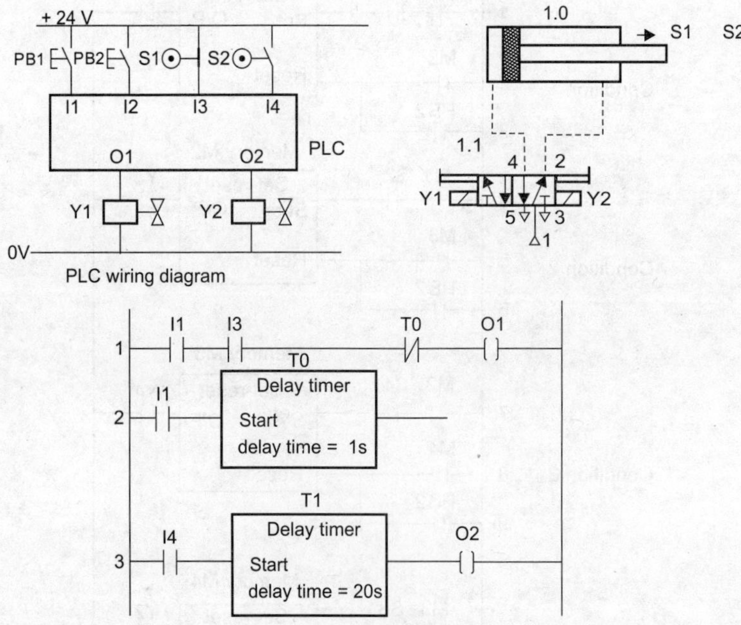

Figure 10.20

Example 10.9 Figure 10.21 shows the arrangement for cleaning washers for injection pumps in a cleaning bath. A double acting cylinder is used to move the container filled with washers up and down in the bath several times. The operator provides a start signal manually. The washing operation is turned off automatically. The washing operation is turned off automatically after a preset time. Draw a wiring diagram, pneumatic circuit and ladder diagram to implement the control task.

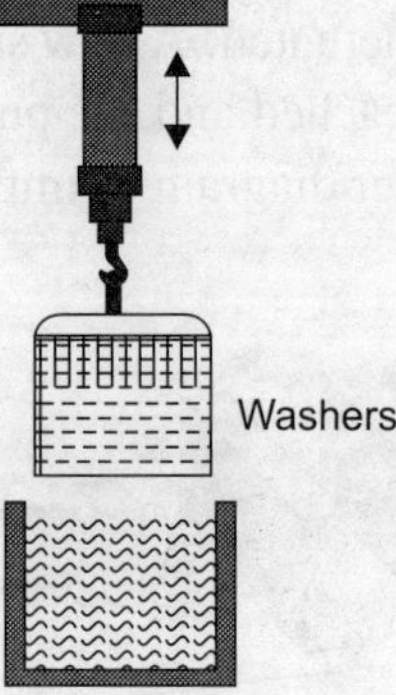

Figure 10.21

The solution is shown in Fig. 10.22, which is self explanatory.

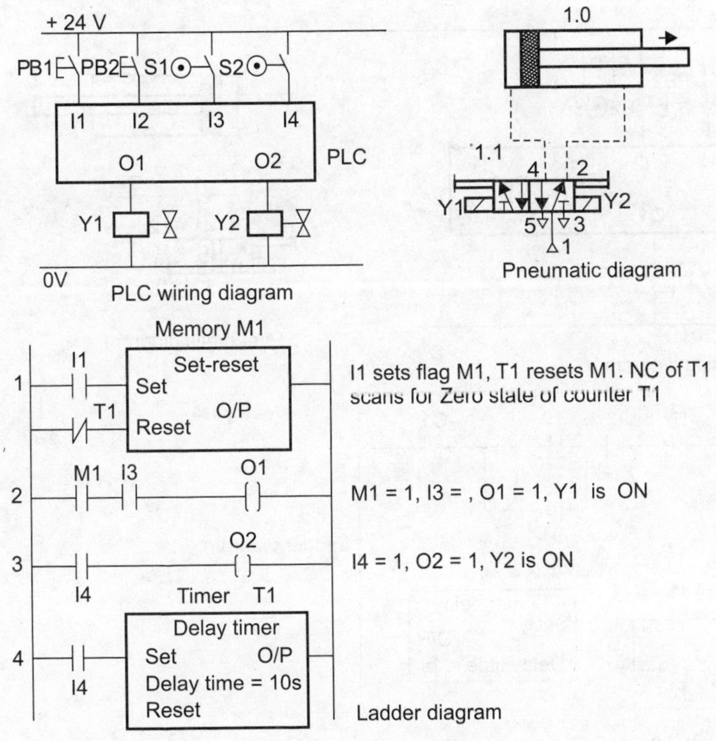

Figure 10.22 *Pneumatic diagram, wiring diagram and the ladder diagram*

Example 10.10 A double acting cylinder is used for embossing slide rules as shown in Fig. 10.23. The cylinder extends only when two push buttons are pressed simultaneously within one second to ensure the complete safety of operators. The cylinder should retract immediately if any of the push button is released. Draw a wiring diagram, pneumatic circuit and ladder diagram to implement the control task.

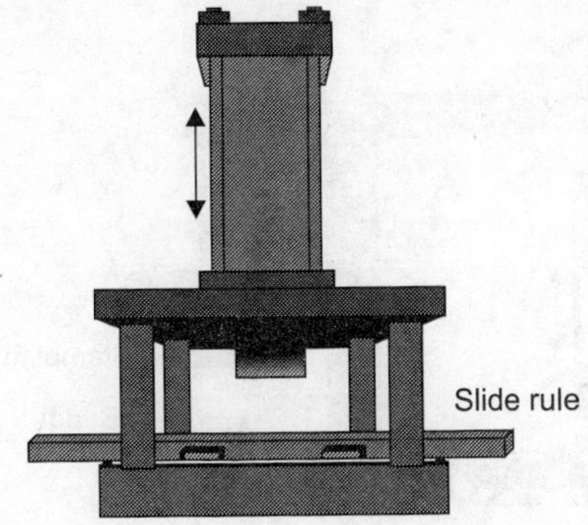

Figure 10.23

The solution is shown in Fig. 10.24, which is self explanatory.

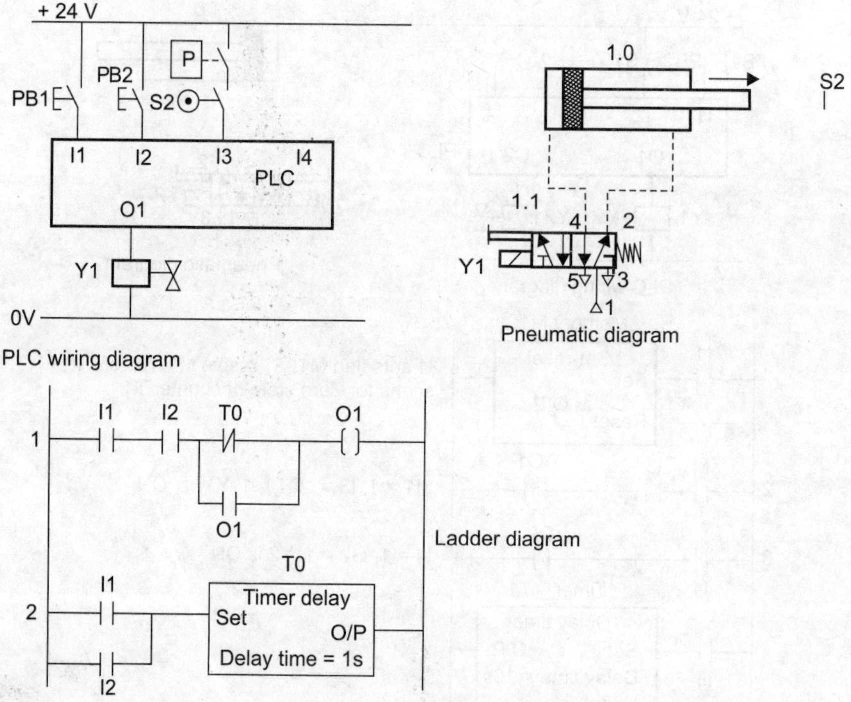

Figure 10.24 *Pneumatic diagram, wiring diagram and ladder diagram*

Example 10.11 A pneumatically controlled press with stamping die shown in Fig. 10.25 is used for producing badges from a thin metal sheet. A double acting cylinder is to extend when the push buttons PB1 and PB2 are pressed simultaneously. The cylinder is to retract automatically after reaching the forward end position and attaining a pre-set pressure to get consistent quality. The cylinder should retract immediately if the emergency push button PB3 is pressed. Draw a wiring diagram, pneumatic circuit and ladder diagram to implement the control task.

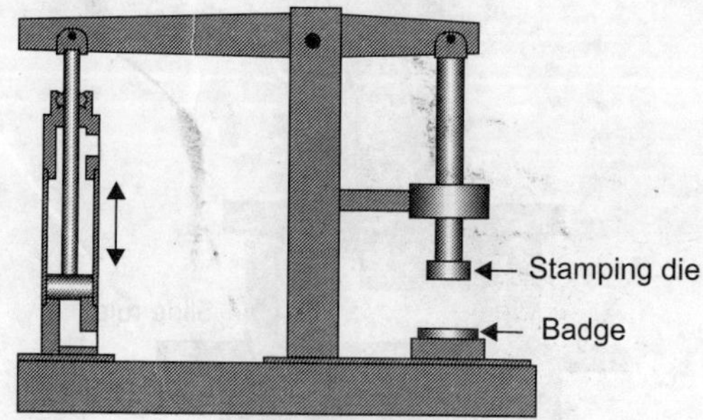

Figure 10.25

The solution is shown in Fig. 10.26, which is self explanatory.

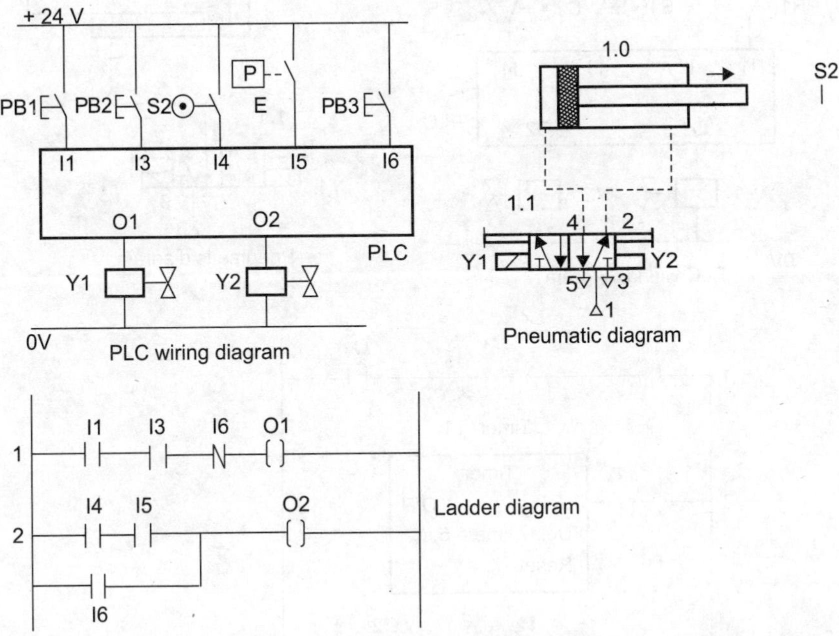

Figure 10.26 *Pneumatic diagram, wiring diagram and the ladder diagram*

Example 10.12 Double-acting cylinder is used to press together glued components (Fig. 10.27). Upon operation of a push button, the clamping cylinder slowly advances. Once the fully extended position is reached, the cylinder is to remain for a time of T = 10 seconds and then immediately retract to the initial position. A new start cycle is only possible after the cylinder has fully retracted and after a delay of 5 seconds. During this delay the finished part is manually removed and replaced with new parts for gluing. The retracting speed should be fast, but adjustable. Draw a wiring diagram, pneumatic circuit and ladder diagram to implement the control task.

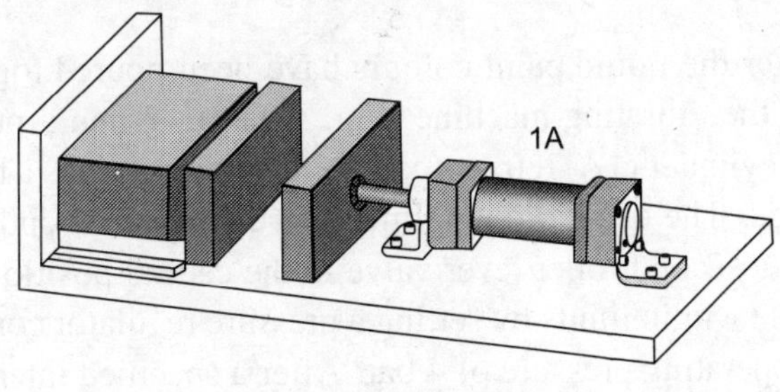

Figure 10.27

The solution is shown in Fig. 10.28, which is self explanatory.

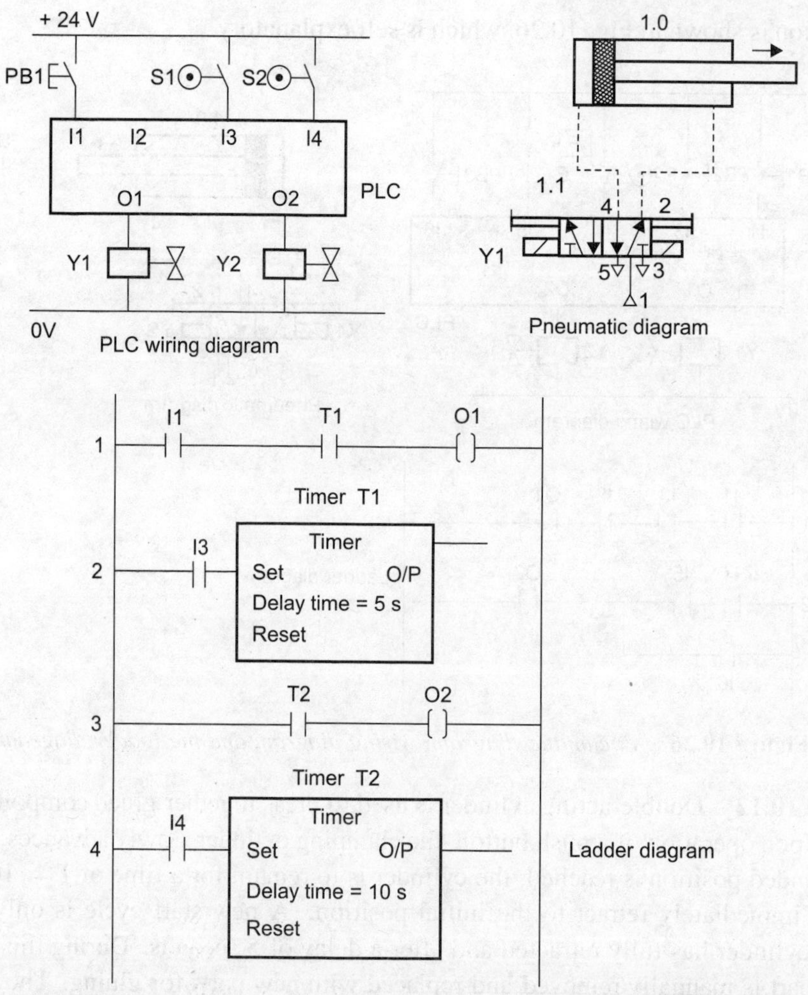

Figure 10.28 *Pneumatic diagram, wiring diagram and ladder diagram*

Example 10.13 After the liquid paint colours have been poured together into a bucket, they are mixed in by the vibrating machine (Fig. 10.29). When a push button has been pressed, the extended cylinder (1A) retracts completely and executes a to and fro movement in the rear stroke range. The oscillation is limited to the retracted end position by a roller lever valve as well as a second roller lever valve in the central position. The frequency of oscillation is adjustable within limits by setting a pressure regulator controlling the amount of air supply. Set an operating pressure of 4 bar. After a specified interval, the oscillator is switched off. The double acting cylinder extends completely and actuates the third roller lever valve. Set a vibration time of 5 seconds. Draw the pure pneumatic circuits, electro pneumatic circuit and ladder diagram.

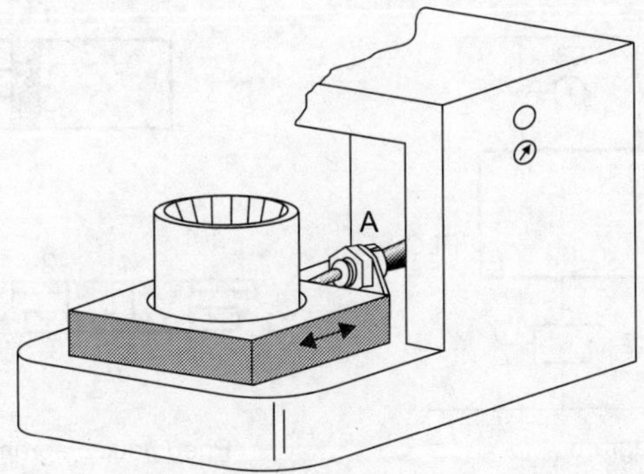

Figure 10.29

The solution is shown in Figs 10.30 and 10.31, which are self explanatory.

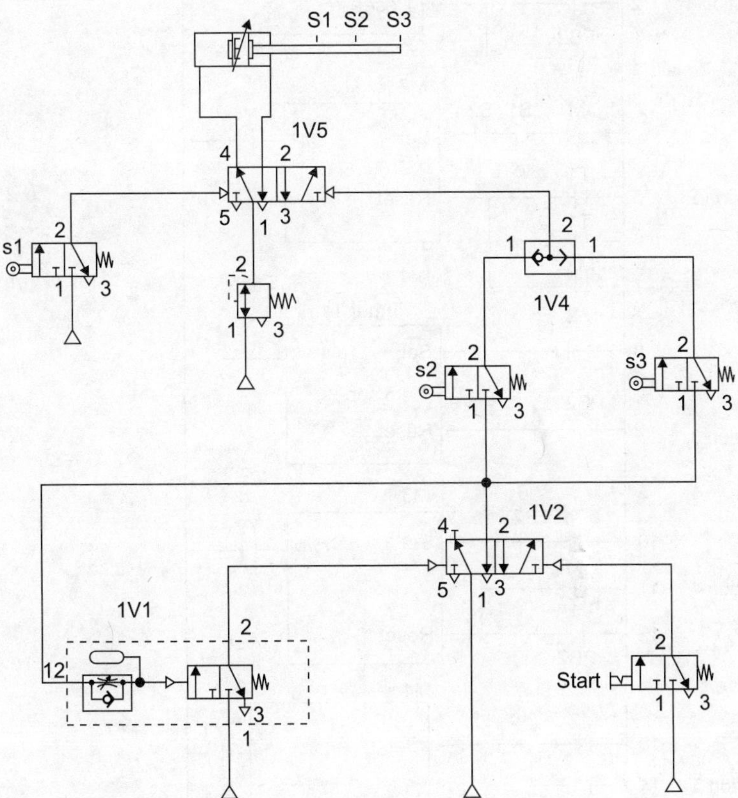

Figure 10.30 *Pneumatic circuit*

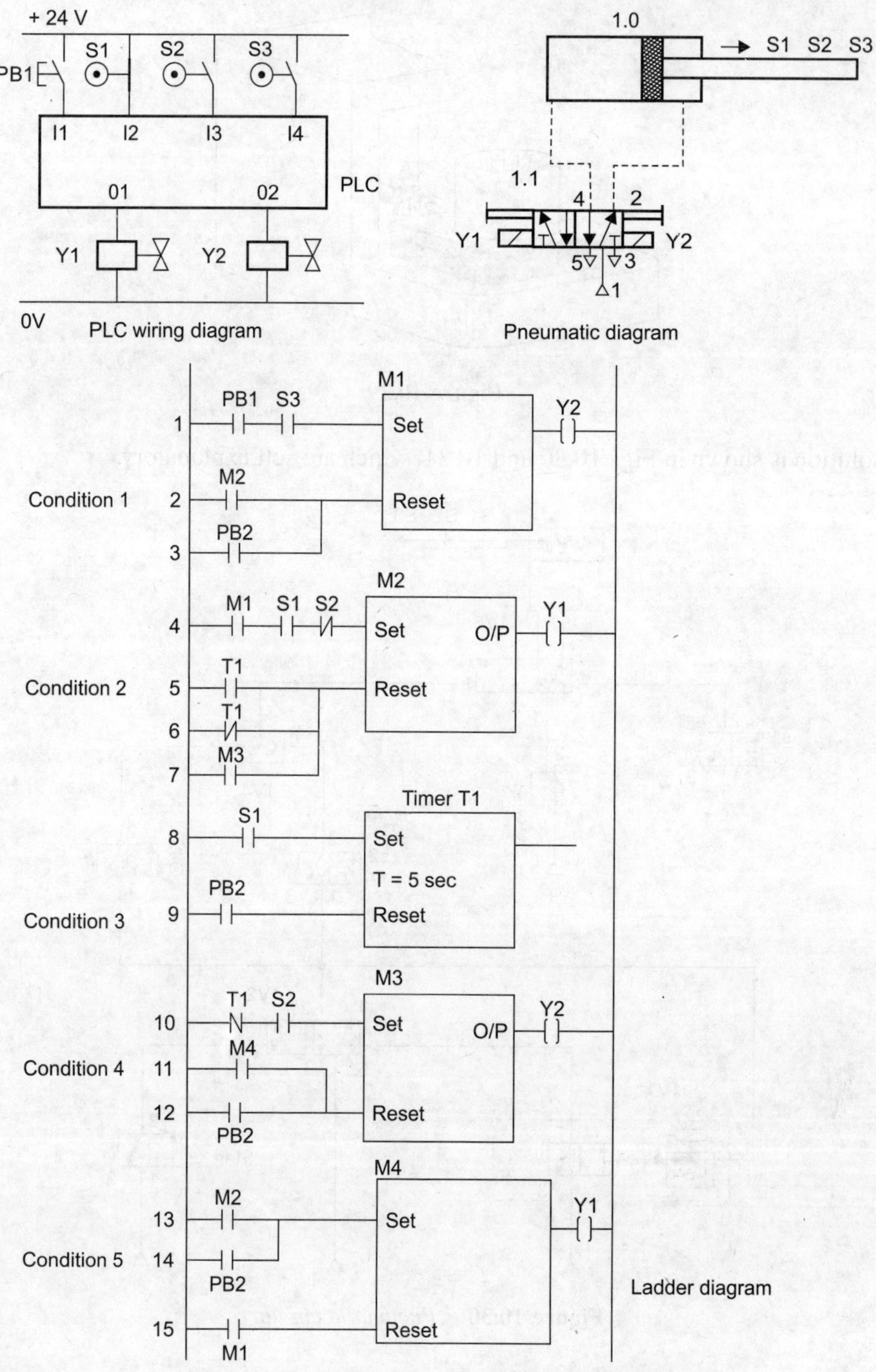

Figure 10.31 *Wiring diagram and ladder diagram*

10.10 Logic Design and PLC Circuits

Design ideas can be converted to Boolean equations either directly or using techniques discussed later. The Boolean equation form can be simplified or rearranged, and then converted into ladder logic or a circuit.

Describing in words how a controller should work is often the first step to assembling the corresponding Boolean equation. In the real world, a description of the process requires talking to the designer of the mechanical part of the system. In many cases the system does not exist yet, making this a challenging task. The next step is to determine how the controller should work. In this case it is written out in a sentence first, and then converted into a Boolean expression. The Boolean expression may then be converted to the desired form.

10.11 Areas of Application of a PLC

Every system or machine has a controller and depending on the type of technology used, the controller can be pneumatic, hydraulic, electrical or electronic, or frequently a combination thereof. Furthermore, there is a difference between hardwired and programmable logic controllers. The former is used primarily in cases where any reprogramming by the user is not a requirement and the job warrants the development of a special controller. Typically, such controllers are used in automatic washing machines, video cameras, and cars.

However if the job does not warrant the development of special controllers or if the user is to have the facility of making changes to the program, or to the setting of timers and counters, then universal controllers where the program is written to an electronic memory, is the preferred option. The PLC is such a universal controller. It can be used for different applications and via the program installed in its memory, provides the user with a simple means of changing, extending and optimising control processes.

PLCs are widely used in industry for the following reasons:

- Cost of PLC automation is lower and PLC is very versatile.
- PLC can be easily commissioned and installed.
- Programming of PLC is quite simple and ladder programming is flexible.
- Control is not hardwired and PLCs can be programmed and reprogrammed frequently.
- Monitoring of online work process is easy and troubleshooting and maintenance of PLC is convenient.
- PLCs can be classified as low cost automation devices.
- They can be used in harsh environments with high humidity and temperatures and are not affected by vibration and shock.
- They can be used to execute complex mathematical algorithms, servomotor control, stepping control, axis control, self-diagnosis, online monitoring, condition moni-

toring, system troubleshooting, communicating with other PLCs, data acquisition, networking, storage and report generation.

- PLCs are well suited for low cost automation, where frequent changes to the control requirements are expected during the operational life, e.g., in batch type production systems.

10.12 PLC Standards

In the late seventies, PLC standards focussing mainly on PLC programming were used mostly in Europe. This included non-networked PLC systems, which primarily executed logic operations on binary signals. DIN 19 239, for example, specifies a programming language which possesses the corresponding language commands for these applications. Since 1992, an international standard has come into use for programmable logic controllers and associated peripheral devices like programming and diagnostic tools, testing equipment and man-to-machine interfaces.

In 1992, IEC 1131 standards were developed as an open framework for PLC architecture. The second edition of IEC 1131 (known as IEC 61131) was published in 2003. The new IEC standard consists of five parts:

- Part 1: General information
- Part 2: Equipment requirement and tests
- Part 3: Programming languages
- Part 4: User guidelines in preparation with IEC
- Part 5: Messaging service specification

Parts 1 and 2 of this standard were adopted un-amended as European standard EN 61 131 parts 1 and 2. They are also part of a German standard. Part 3 of the new IEC deals with programming languages and defines two graphical and two textual PLC programming language standards. The standard also defines both graphical and textual sequential function chart elements to organise programs for sequential and parallel control processing. It is now possible to program PLC using following the languages:

- IL - Instruction list
- ST - Structural Text
- LD - Ladder Diagram
- FBD - Functional Block Diagram
- SFC - Sequential Function Chart

The purpose of the new standard was to define and standardise the design and functionality of PLC and the programming languages so that users had the ability to operate PLC systems from different vendors without difficulty. Several major PLC suppliers (Allen Bradley,

Klockner-Moeller, Phillips, Siemens, and Mitsubishi, to name a few) are members of an association called PLCopen, which supports IEC 1131.

Summary

1. PLC is a microcomputer with hardware and software designed to control the operations of industrial equipment and processes. A PLC can be considered as an interfacing device which receives input and energises output.
2. Many systems are wired with relays and relay controls. The wiring of control elements such as sensors and solenoids through relays is called hardwired control. Hardwired controls are cumbersome and very difficult to modify. PLCs are flexible and easy to modify.
3. A PLC structure consists of a power supply module, a central processing unit, memory, input and output modules and software.
4. The power supply for control systems is 120 /230 AC line voltage. 24V is used for internal power needs and for the input/output modules. The processor needs 5V. Input and output can be either digital or analog.
5. PLC software can be divided into the operating system and user programs. The PLC operating system is the overall software consisting of instructions and declarations to control the PLC system sources. Services of the operating system are accessed through the GUI (Graphical User Interface). User programs can be written in programming languages like ladder diagram or statement list.
6. PLC timers are software modules used for delay functions between work operations. Specific memory is assigned for timers in the system memory.
7. PLC memory elements store the intermediate results. This is done using bit memory flags in the PLC. Memory with set priority and memory with reset priority are used to implement memory functions.
8. Counters are capable of counting tasks such as incrementing or decrementing a count value when given a transitional input. There are two types of counters: up counters and down counters. An up counter increments its count value and a down counter decrements its count value.
9. PLC is a small yet powerful controller used in the automation of industrial processes and equipment. It is versatile and flexible. It can be easily adapted to applications in automotive, food industries, machine tools, petro chemicals etc.

EXERCISES

Fill in the blanks

1. The most important sections of a PLC are input section, output section and ———.
2. The structure of a PLC consists of a power supply module, CPU, I/O modules and ———.

3. User programs can be written in any standard PLC programming language like statement list and ————.
4. The AND function combines the bit addresses of inputs and produces an RLO (Results of logic operation) of ——— when all the inputs are scanned for 1.
5. The function of a timer is to provide ———— between work operations.

State whether true or false

1. Hardwired control systems are used widely when production requirements change regularly.
2. Relay controls are less expensive compared to PLC controls.
3. While processing a PLC program, the CPU scans and executes the main program cyclically.
4. The NO and NC PLC program contacts are same as the hardware NO and NC contacts.
5. The OR function combines the bit addresses of inputs and produces an RLO (Results of logic operation) of 1 when one or more of inputs are scanned for 1.

Review questions

1. What is a hardwired control? What are its disadvantages?
2. List five differences between PLC control and relay control.
3. List three input devices commonly used in PLC control.
4. Explain the working of functions of CPU, input and output module and memory in PLC.
5. Mention a few applications of PLC.
6. What are the functions of a timer in PLC?
7. Briefly explain the structure of PLC.
8. Differentiate between on-delay and off-delay timers with help of symbols.
9. What is meant by bit logic operations in relation to a PLC?
10. Describe the functions of the following PLC program elements: a) program coil, b) NC contact, c) NO contact, d) set coil, e) reset coil, f) set-reset box, g) reset-set box, h) on-delay timer, i) off-delay timer, j) up counter, k) down counter.

Problems

1. A double acting cylinder is used to open or close a control valve in a chemical industry. A double solenoid is used as the final control element. Draw a pneumatic circuit, PLC wiring diagram and ladder diagram to implement this task.

2. Sheets of foils are to be cut using cutting blades. The cutting blades are attached to a pneumatic cylinder, which is double acting. The cutting blade should advance if both pushbuttons PB1 and PB2 are pressed simultaneously. If any one of push button is released, the cutting blade should return to its home position. Draw a pneumatic circuit, PLC wiring diagram and ladder diagram to implement this task.
3. A double acting cylinder is used to extend and push a pin into a socket when a push button is pressed. The cylinder is to retract automatically after reaching the end position. The end position is sensed using a limit switch S2. Draw a pneumatic circuit, PLC wiring diagram and ladder diagram to implement this task.
4. A double acting cylinder is used to press glued components in the rubber industry. When push button PB1 is pressed the cylinder has to extend. The cylinder has to remain at the end position for 30 seconds and return automatically. The end position sensing is done using a proximity switch S2. Draw a pneumatic circuit, PLC wiring diagram and ladder diagram to implement this task.
5. Pins are to be fed to the next processing station using a double acting cylinder. The arrangement is shown in Fig. 10.19. The pins are fed continuously by the to and fro motion of the cylinder, triggered by the start push button PB1. Pressing a stop push button PB2 stops the feeding. Forward and return speeds should be adjustable independently. Draw a pneumatic circuit, PLC wiring diagram and ladder diagram to implement this task.

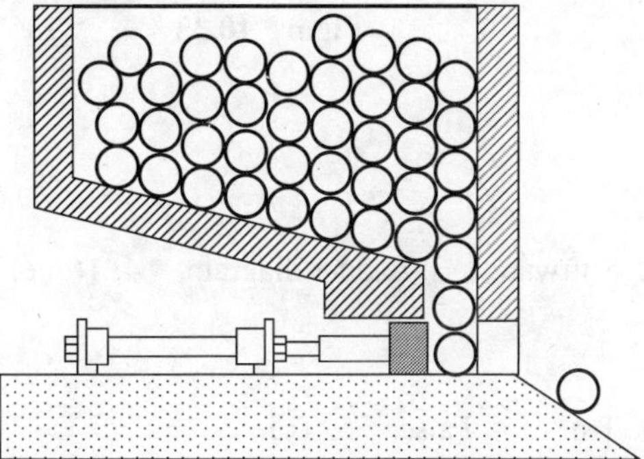

Figure 10.32

6. When a start push button is momentarily pressed, a double acting cylinder is to perform to and fro motion for 15 cycles and then to stop automatically. The cylinder should always stop in the retracted position. A 5/2 double solenoid valve is used as a final control valve. Using a down counter, draw a pneumatic circuit, PLC wiring diagram and ladder diagram to implement this task.
7. Develop ladder logic for a car door and seat belt safety system. When the car door is open and the seat belt is not buckled in place, the ignition power should not be applied. If all are safely belted, the key should start the engine.

8. Design a motor controller that has a forward and a reverse button. The motor output should only be on when one of the buttons is pushed. When both buttons are pushed the motor should not work.
9. In Fig. 10.33, the piston rod of the cylinder is to advance only if a workpiece is inserted in the workpiece retainer, a guard has been lowered and the operator presses the push button valve. Upon the release of the push button or it the guard is no longer in its lower position; the cylinder is to retract to the initial position. Draw a wiring diagram, pneumatic circuit and ladder diagram to implement the control task.

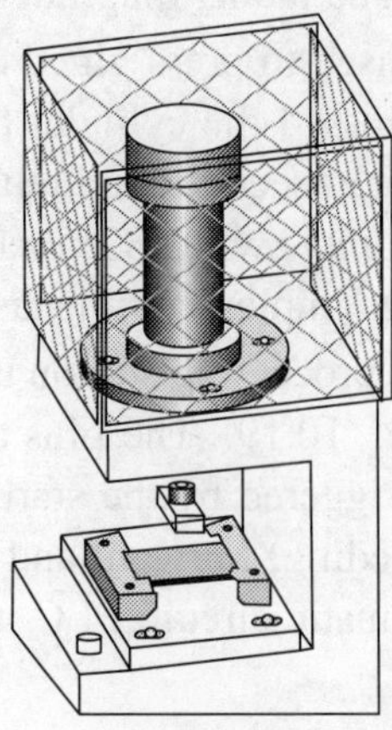

Figure 10.33

Answers

Fill in the blanks

1. Program section 2. Software 3. Ladder diagram 4. 1(one) 5. delay

True or false

1. False 2. False 3. True 4. False 5. True

11 INTRODUCTION TO FLUIDICS

Upon completion of this chapter, the reader should be able to

- Explain the meaning of fluidics
- Describe the historical development of fluidics
- List the advantages and disadvantages of fluidics
- Study the basic elements of fluidic systems
- Explain the Coanda effect
- Explain the different fluid sensors
- Design simple circuits using fluidic elements

11.1 Fluidics Defined

The word "fluidics" originates from the words "fluid" and "logic". It is the technology of controlling fluid force by means of devices, and using fluid interaction phenomena to manipulate output signals.

A major problem in hydraulic and pneumatic control systems is their susceptibility to wear and tear on components such as ball, poppet and spool valves. Another disadvantage, depending on the application, is the magnetic properties of the electrical and electronic controllers. These issues can be overcome by replacing electrical and electronics controls in certain fluid power system with fluidics. Fluidics is also known as non-moving logic controllers because they have a minimum of moving parts.

Fluidics is the technology that utilises fluid flow phenomenon in components and circuits to perform a wide variety of control functions such as sensing, logic, memory, timing, and interfacing. In other words, fluidics or fluidic logic is the use of a fluid to perform analog or digital operations similar to those performed with electronics.

Fluidics implement control functions by directing streams of gas or liquid through precision channels. The flow of media (i.e., air or liquids) in these control functions is similar to the flow of electric current through wires and contact points.

The basic distinction between conventional fluid flow engineering and fluidics is that in the former case we are normally concerned only with harnessing fluid power in an economic manner, whereas with fluidics we are thinking in terms of processing information through a fluid medium as well as transmitting power.

11.2 Historical Background

As early as 1916, Tesla invented a conduit for fluids in which flow in one direction was easier and unrestricted but flow in the opposite direction was divided into multiple channels that opposed one another and restricted the overall flow. The strong influence of electronics in this era accounts for the term "fluid diode" used to describe Tesla's valve conduit.

Later in 1933, Henri Coanda discovered the effect named after him where a jet of fluid aimed at a curved wall or surface continues to adhere to it even after the force of the control jet is removed.

Coanda noted that the entrainment of trapped air to the surface of the body over which it moved, created a partial vacuum and atmospheric pressure forced the fluid stream to cling to that surface. Even though Coanda patented a few devices based on this principle, it was not until the early 1960s that practical and reliable fluidic devices came into use. Most of these devices operate in the pressure range of 0.05 to 0.1 bar.

11.3 Advantages and Disadvantages of Fluidics

Advantages Using fluids to handle their own control offers advantages over conventional control methods. Some of the important advantages are:

1. Unlike hydraulic and pneumatic control components, fluidic devices are less susceptible to wear and tear because they have fewer moving parts.
2. Fluidic devices offer exceptional thermal and physical stability and ruggedness when compared to electronic control devices which are sensitive to temperature, shock, vibrations and radiation.
3. Fluidic devices are immune to radiation even at very high levels.
4. Fluidic devices are stable when experiencing intense vibrations and shocks.
5. Fluidic devices are simpler in construction and easy to maintain.
6. Fluidic devices are functionally reliable.
7. Fluidic devices are relatively inexpensive.
8. Fluidics can easily interface with pneumatic, electrical and other systems.
9. The space needed for mounting is minimal.
10. When fluidics use air as the working fluid, there is no problem of vibration, electrical noise, fatigue and contact contamination.

11. Since there is no arcing of switching elements, fluidic devices are safe in highly explosive and dangerous environments.
12. The actuating force required is minimal.

Disadvantages

1. The response time of fluidic devices (in the range of milliseconds) is slower compared to electronic devices, though adequate for most fluid power requirements.
2. Fluidic components normally operate at low pressure levels (0.05 to 0.1 bar). However, these low pressure signals can reliably control fluid power control systems operating up to 700 bar and provide the muscle to do useful work at rates up to several hundreds of kilowatt power.

11.4 Coanda or Wall Attachment Effect

The Coanda effect is named after Romanian engineer Henry Marie Coanda who investigated it in 1932. It is also known as the wall attachment phenomenon and can be described as follows: When a stream of fluid meets another stream, the controlling jet changes the flow of the second stream and the effect is that it adheres to the nearest surface or wall.

Figure 11.1 illustrates the Coanda effect. A free fluid jet of air or liquid passes through an orifice at a velocity high enough to produce turbulent flow. The free jet of air continues to flow in a given direction, pulling with it available fluid from its surroundings as it leaves the orifice (Fig. 11.1a).

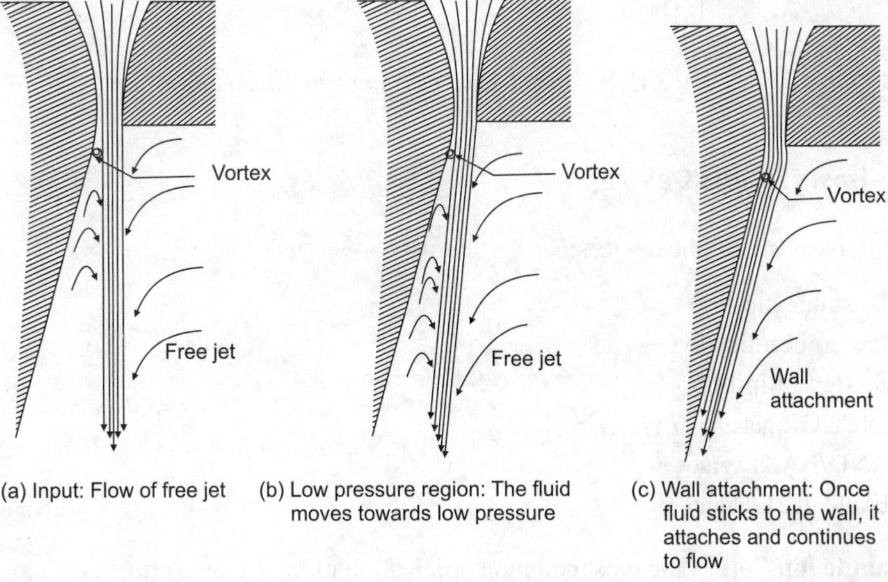

Figure 11.1 *Coanda effect*

As the jet of fluid has high kinetic energy, a low pressure region exists around it. If the volume of entraining air on one side is greater, a small vortex area of low pressure is created near the nozzle exit. This low pressure area then tends to attract the free jet, distorting it and pulling it towards the wall, because the atmospheric pressure on the other side forces the jet to cling to the surface (Fig. 11.1b).

This free jet adherence continues until a small air supply is fed to the low pressure area, thus relieving the attraction of the jet to the wall or surface. When this signal is injected, the free air then detaches itself from the wall and resumes its normal uninterrupted flow path as shown Fig. 11.1c. This attachment works as a memory function.

Figure 11.2 illustrates simple demonstration of the Coanda effect. When a finger is held near a stream of water flowing from a tap, the water attaches itself to the finger.

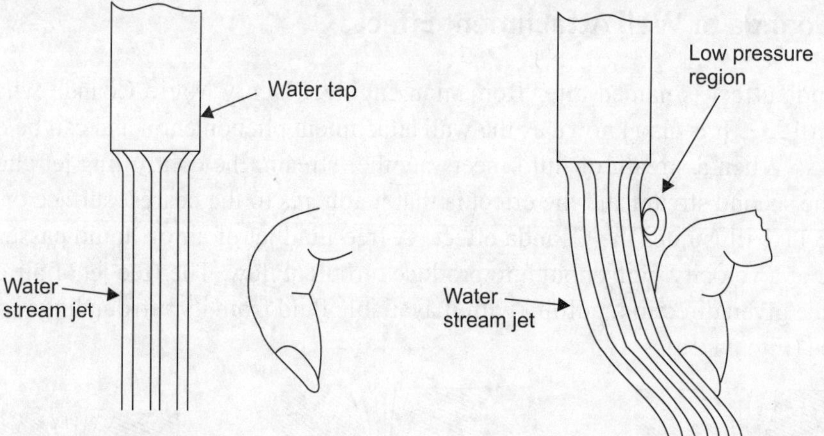

Figure 11.2 *Simple demonstration of Coanda effect*

11.5 Basic Fluidic Devices

There are basic six fluidic devices:

1. Bi-stable flip-flop
2. Preference flip-flop
3. Set-reset flip-flop
4. OR/NOR gate
5. AND/NAND gate
6. Exclusive OR gate

1. Bi-stable flip-flop The most common application of the Coanda effect is the bi-stable amplifier which is implemented by creating a Y junction in a solid piece of glass or plastic, not much larger than the size of a coin. As illustrated in Fig. 11.3, inlet air flow is from port

P_s and outlet is from O1 or O2. A small control jet applied through C1(or C2) shown makes the output alternate or flip-flop, as it is called. The output is controlled by two control jets C1 and C2.

The input energy from the input port moves either to output O1 or output O2, depending on the presence or absence of a control jet at that port.

The orifice diameter in such devices is of the order of 0.25 mm and the working pressure is 0.05 bar to 0.1 bar. The operational speed is about 1000 Hz. The device is best suited for binary logic functions (the output either exists or does not). The output from a bi-stable flip-flop unit is sufficient to control a pilot operated valve with a low pressure actuating element.

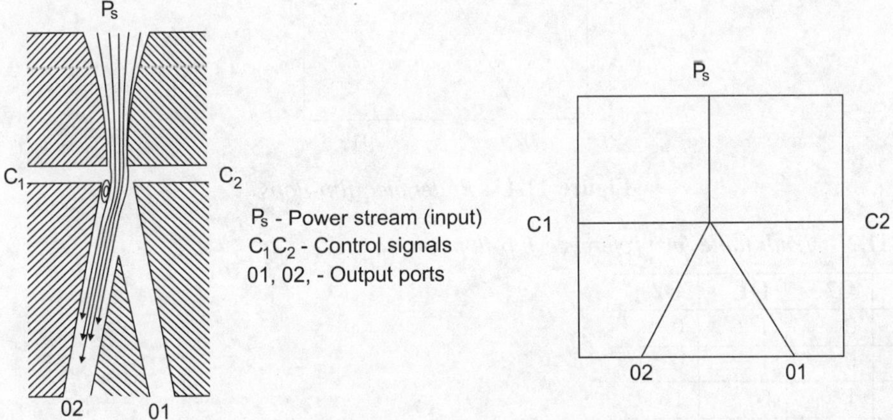

Figure 11.3 *bi-stable flip-flops*

The control signal is not only remembered by the bi-stable flip-flop device but is amplified 10:1, thus providing an output signal of sufficient strength to operate interface devices. The truth table is shown in Table 11.1. The symbol 1 denotes the presence of a signal while 0 represents the absence of a signal.

Table 11.1 *Truth table for bi-stable flip-flop*

C1	C2	O1	O2
1	0	1	0
0	0	1	0
0	1	0	1
0	0	0	1

The flip-flop has two stable states when all control signals are OFF. When both control signals C1 and C2 are turned ON simultaneously, the supply pressure splits and there are no useful ON/OFF output signals under these conditions.

2. Preference flip-flop A flip-flop with a start-up preference is used where a specific output is required when the power supply is first turned ON and all the controls are OFF.

The construction and operation is similar to a basic flip-flop. The plus (+) sign indicates that the output O1 is preferred over O2. When the control signals are OFF, the output is ON at O1(preference) and OFF at O2. The truth table is shown in Table 11.2.

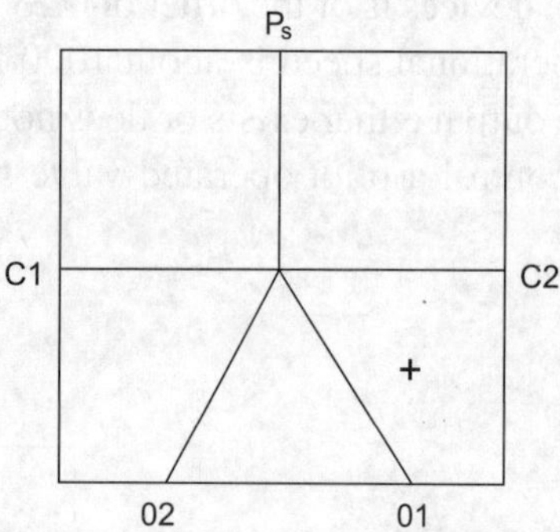

Figure 11.4 *Preference flip-flops*

Table 11.2 *Truth table for preference flip-flop*

C1	C2	O1	O2
1	0	1	0
0	0	1	0
0	1	0	1
0	0	0	1

3. Set-reset trigger (SRT) flip-flop In addition to all the capabilities of a basic flip-flop, an SRT flip-flop can also switch by applying a signal to the trigger port. The schematic diagram is shown in Fig. 11.5.

The symbol and the truth table are shown in Table 11.3. The letters S and R stand for Set and Reset respectively and T stands for Trigger. Whenever the trigger signal is applied, the output is switched.

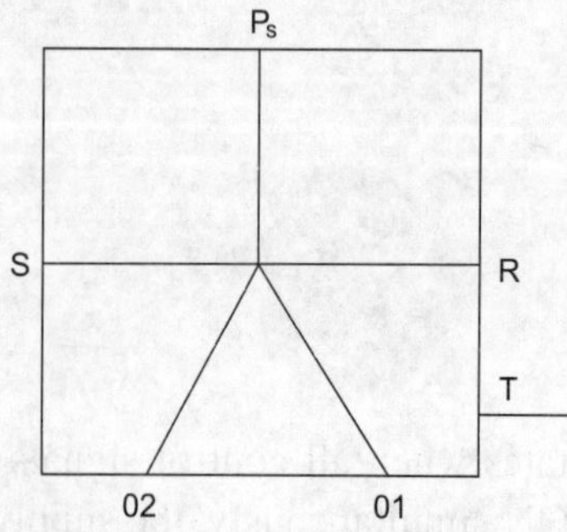

Figure 11.5 *Set-reset flip-flops*

Table 11.3 *Truth table for SRT flip-flop*

T	S	R	O1	O2
0	1	0	1	0
0	0	0	1	0
0	0	1	0	1
0	0	0	0	1
1	0	0	1	0
0	0	0	1	0
1	0	0	0	1
0	0	0	0	1

4. OR/NOR gate Further development of the basic flip-flops led to monostable devices with more than one control input. A monostable device is required to perform a monostable function analogous to the spring return function, i.e., when the control signal is removed, the device switches back to the favoured output. The basic flip-flop, the preference flip-flop and the SRT flip-flop primarily perform memory functions.

Two monostable fluidic devices are used in fluid power control:

1. OR/NOR gate
2. AND/NAND gate

The OR/NOR gate element is a basic logic building block. All the other logic functions can be obtained using OR/NOR blocks in various combinations. This device provides an output signal in response to two or more input conditions. The schematic diagram is shown in Fig. 11.6.

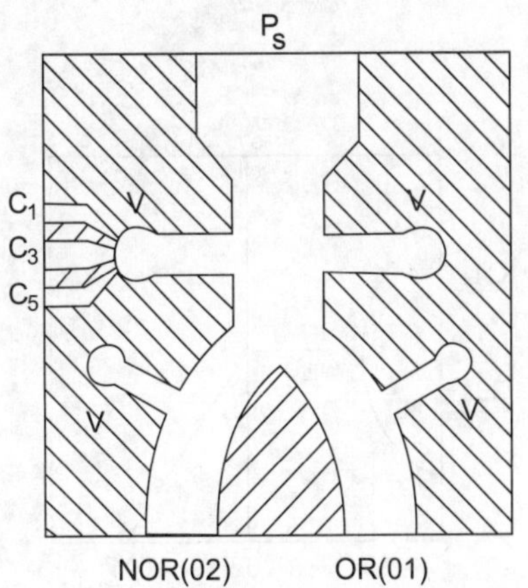

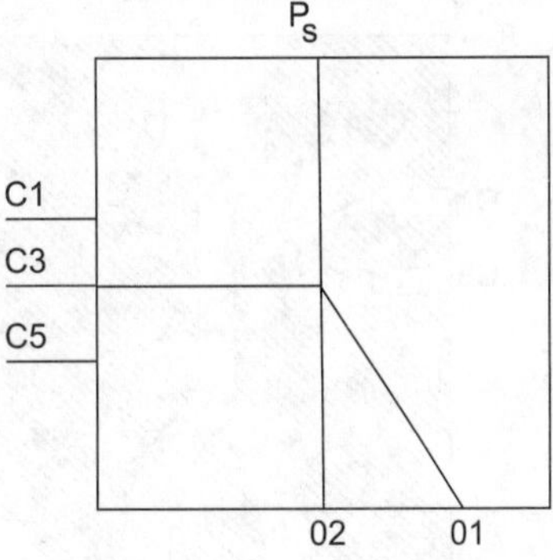

Figure 11.6 *OR/NOR gate*

OR (Port O1) – Control signal at one or any combination of the control ports

NOR (Port O2) – Control signal at none of the control ports

01 = C1 + C3, read as C1 OR C3- - - - - -OR Function

02 = $\overline{\text{C1}} \cdot \overline{\text{C3}}$, read as Complement of C1 (NOT C1) AND (NOT C3)- - - - -NOR function

A control signal at either C1 or C3 must be present to obtain an output at port O1, while no control signal at C1 or C3 is needed to get an output at the port O2. The truth table is given in Table 11.4.

Table 11.4 *Truth table for OR/NOR gate*

C1	C3	OR	NOR
		O1	O2
0	0	0	0
1	0	1	0
0	1	1	0
1	1	1	0

5. AND/NAND gate The fluidic NAND gate is used to determine the absence of all the control signals whereas the AND gate is used to determine the presence of all control signals. The construction and symbols are shown is Fig. 11.7. Both C1 and C3 control must be present to give an output at the O1 port (AND function). The absence of either or both control signals at C1 and C3 results in a stable output at the O2 port (NAND function). The schematic diagram is shown in Fig. 11.7 and the truth table is given in Table 11.5.

01 = C1 · C3 (read as C1 AND C3- - - - - -OR Function)
02 = $\overline{\text{C1}} + \overline{\text{C3}}$ (read as Complement of C1 (NOT C1) AND (NOT C3)- - - - - -NAND function)

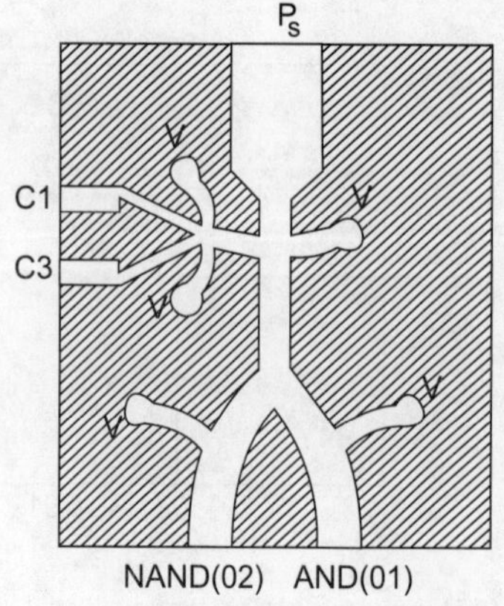

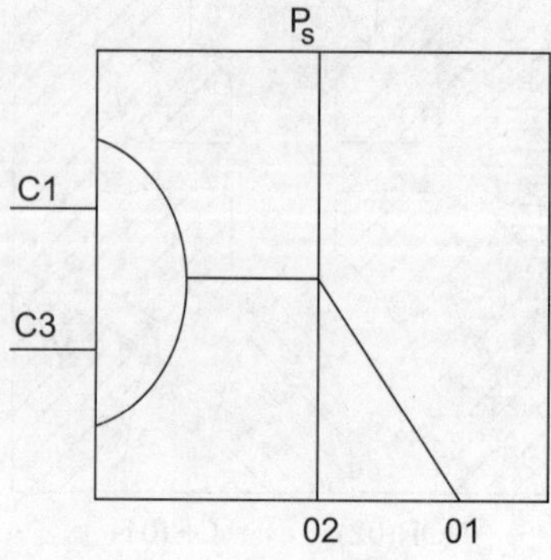

Figure 11.7 *OR/NOR gate*

Table 11.5 *Truth table for OR/NOR gate*

C1	C3	AND	NAND
		O1	O2
0	0	0	1
1	0	0	1
0	1	0	1
1	1	1	0

6. Exclusive OR gate An exclusive OR gate is a device that provides an output only when C1 or C3 or C5 or C7 is ON and is implemented by feeding the output of a jet cavity interaction into the control port of an OR gate. When control signal C1 or C3 is ON, the jet cavity interaction produces an output at O2 port. In other words, if C1 and C3 inputs are the same, then there is no output. C1 and C3 must have different inputs to get an output.

When both C1 and C3 control signals are ON, then their interaction causes them to vent to the atmosphere and hence there is no output at O2. The schematic diagram is shown in Fig. 11.8. The truth table is given in Table 11.6.

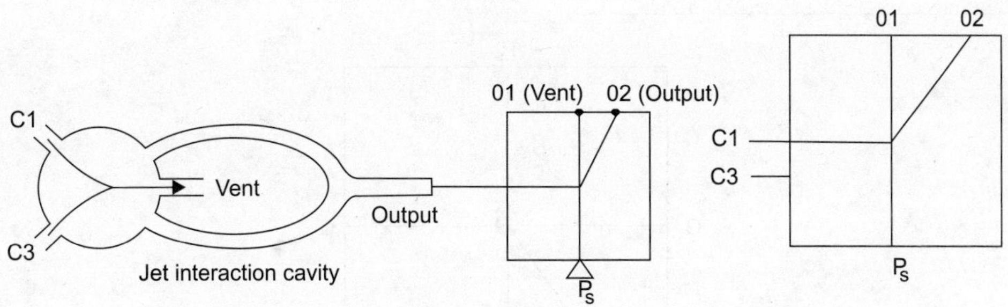

Figure 11.8 *Exclusive OR gate*

Table 11.6 *Truth table for Exclusive OR gate*

C1	C3	Vent	Exclusive OR
		O1	O2
0	0	1	0
1	0	0	1
0	1	0	1
1	1	1	0

11.6 Schmitt Trigger

The Schmitt trigger is a useful element in a fluidic circuit design and enables switching from one output port to another, depending upon which of the two control signals is stronger. For example, when control signal C1 is stronger than control signal C2, an output signal through 01 shall be available and vice versa. This can be achieved by connecting control signal C2 to a constant source (say 0.2 bar) and control signal C1 coming from an unknown pressure that is to be measured. The switch-over is then performed when the 0.2 bar signal is passed.

Schmitt trigger is often used with sensors because switching parts can be adjusted over a wide range by varying the bias pressure from 1% to 35% of the supply pressure P_s. The switching band is very narrow. The schematic diagram is shown in Fig. 11.9. The truth table is given in Table 11.7.

Table 11.7 *Truth table*

Control signal	Condition	O1	O2
1	C1 > C2	1	0
1	C1 < C2	0	1

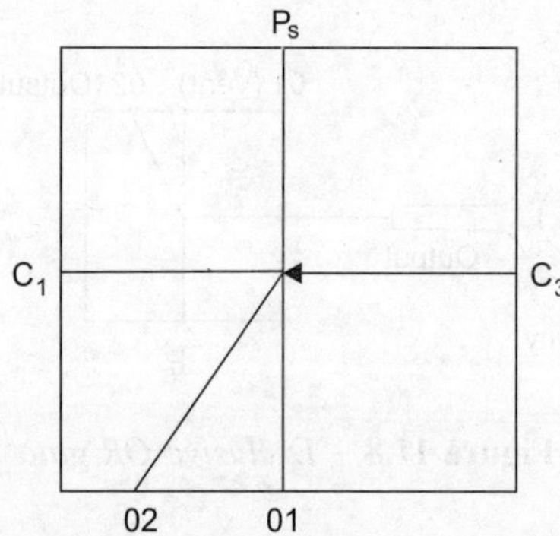

Figure 11.9

11.7 Turbulence Amplifier

The turbulence amplifier is a variation of the OR/NOR gate as shown in Fig. 11.10, and often referred to as TA. It consists of a small bore pipe conducting fluid at a low pressure, so as to achieve a laminar flow with a Reynolds number below 1500. The laminar flow discharged from the inlet pipe shoots across a gap of about 20 mm to enter the output pipe. The space between these pipes is enclosed by a cylindrical chamber whose diameter is 20 to 30 times the diameter of the inlet/outlet pipe. This chamber controls signals through several nozzles and is suitably vented to the atmosphere.

When no control signal is provided, the laminar flow of the fluid traverses from the input to the output through the open space within the chamber. But when the control signal is applied through one or more nozzles, turbulence is created between the input and output pipes, disrupting the flow to the output pipe and resulting in no output.

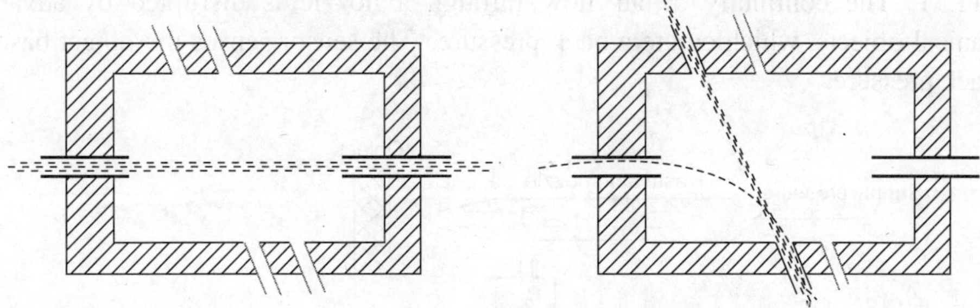

Figure 11.10 *Turbulent amplifier*

Removing the control signal re-establishes the laminar flow, resulting in output; this is the NOR logic function. The switch-off time in these devices is about 4 milliseconds while the total cycle time is about 6-7 milliseconds. By modulating the disrupting effect of the turbulence on the control signal, modulation of the output signal intensity is produced. In other words, the device is a digital ON/OFF device.

11.8 Fluid Sensors

Sensors based on fluidic principles have been extensively used in machine tools and other systems for transmitting or receiving signals. These sensors are usually pneumatic based. In machine tools the design involves detecting the presence or absence of a component, which generates a control signal in the form of small jets. The sensors then gauge the position of a component by sensing its proximity to the air jets and are known as proximity sensors.

Fluid sensors are devices that sense a change in a specific parameter, and as a result trigger a corresponding change in another parameter that can be recognised and interpreted. Various types of fluid sensors are used in machine tools and other systems to transmit and receive signals so as to control the actuator speed, position and pressure.

11.8.1 Types of fluid sensors

Fluid power systems usually include the following types of fluid sensors:

- Back pressure sensor
- Cone jet proximity sensor
- Interruptible jet sensor
- Contact sensing

Back pressure sensor The back pressure sensor works on the principle that the back pressure is created when fluid flowing though a nozzle or a restricted passage is obstructed by external forces or objects.

The construction and operation of a typical back pressure sensor is illustrated in Fig. 11.11. The continuity of air flow through a nozzle is disrupted by advancing mechanical objects, which create a back pressure. The sensor senses the object based on this back pressure.

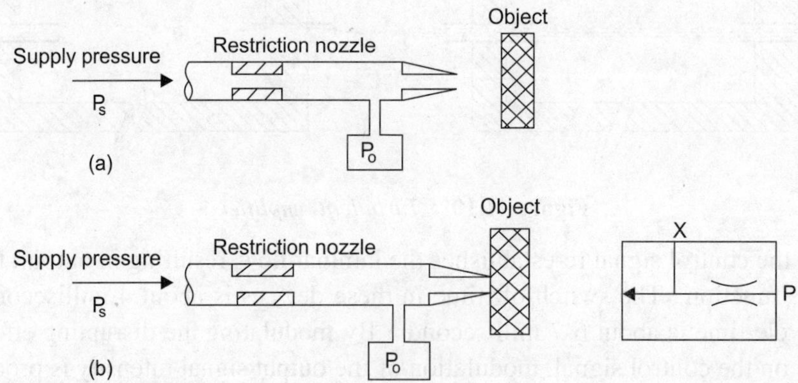

Figure 11.11 *Back pressure sensor (a) object further away (b) object is closer*

An air supply is applied to a very fine nozzle via a restriction as shown in the figure. The signal outside the nozzle feeds into a closed (non-venting) load such as an indicator. Air escapes as a fine jet from the nozzle so the pressure P_0 is lower than the supply pressure P_s because of the pressure drop across the restriction. Air loss from the jet and therefore the pressure P_0 is influenced by the gap Z between the movable object and the nozzle. The smaller the gap, the lower is the air flow and higher the pressure. The pressure difference between P_s and P_0 is an indicator of the presence or absence of the object.

Cone jet proximity sensor This sensor also works on the principle of creating a back pressure when the flow is blocked. Two jets of air flow out through an annular opening. If an object disrupts the jet, a back pressure is created which flows back into the inner nozzle and controls the other valves. The inlet pressure is limited to 0.1 to 0.2 bar. This is also called a reflex sensor.

A proximity sensor can detect objects at comparatively greater distance than a back pressure sensor. In fact, it can sense gaps 10 times larger than a back pressure sensor can, for an equivalent amount of air consumption.

This sensor operates using an annular nozzle that surrounds a sensing hole. When the sensor is supplied a pressure P_s, a conical jet forms at the tip of the nozzle which encloses a low pressure bubble within the flow cone as shown in Fig. 11.12(a). Normally the pressure at the sensing opening P_0 at the base of the cone is slightly below the atmospheric pressure.

When an object is brought near the flow cone as shown in the Fig. 11.12(b) a portion of the supply jet is reflected from the object into the low pressure bubble region. This increases the output pressure P_0. This increased output pressure permits the switching of the control element.

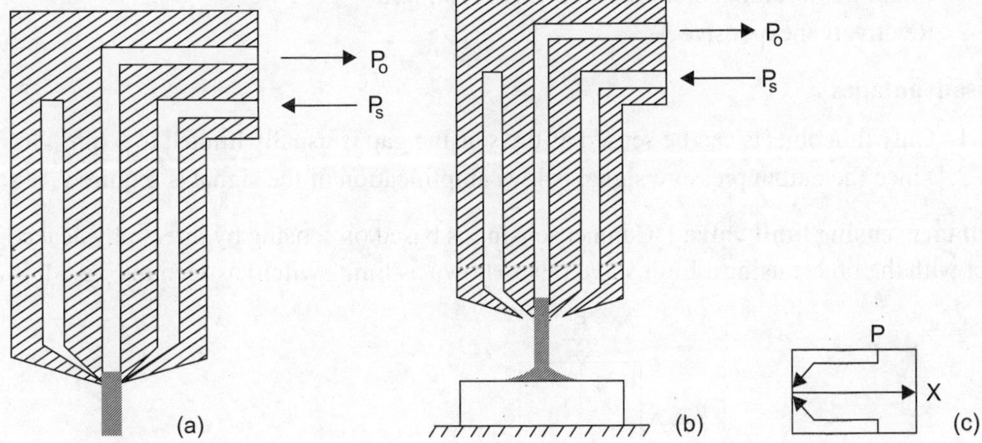

Figure 11.12 *Jet proximity sensor (a) without object, (b) with object, (c) symbol*

Disadvantages of a conical jet proximity sensor

1. Air consumption is higher than other sensors
2. Initial cost is high compared to a back pressure sensor
3. Signal amplification is required because the output pressure signal is usually weak

Interruptible jet sensor This sensor is also called a gap sensor. Low pressure is allowed to pass from the supply port to the outlet port as shown in Fig. 11.13. A mechanical component blocks the gap and therefore the air jets leaving the nozzle tip disrupts the signal at the receiver end, and therefore the downstream valve which awaits this outlet signal reverses.

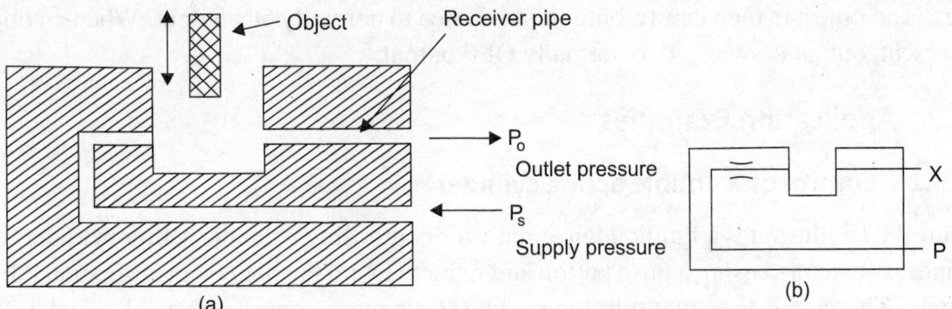

Figure 11.13 *Interruptible jet sensor (a) construction (b) symbol*

The pressure at the inlet ranges between 0.1 to 2 bar, but normal air pressure may also be used if the air is throttled before it enters the supply port.

Advantages

1. Construction and operation of the valve is simple
2. Relatively inexpensive

Disadvantages

1. Only thin objects can be sensed as the sensing gap is usually limited to 5 mm
2. Since the output pressure signal is low, amplification of the signal is required

Contact sensing limit valve Contact sensing is based on sensing by direct physical contact with the object using a limit valve (also known as limit switch) as seen in Fig. 11.14.

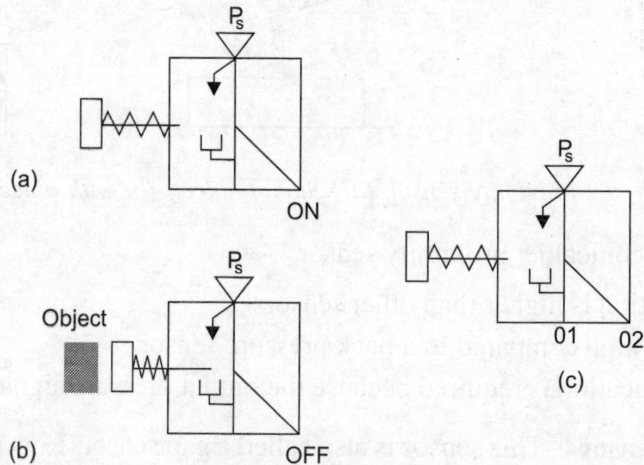

Figure 11.14 *Contact sensing limit (a) without object (b) with object (c) symbol*

The operation of a contact sensing limit switch is similar to the OR/NOR gate with the difference that mechanical input is used whereas OR/NOR gates use air as input. When there is no object in the vicinity, output is switched to normally ON output. When an object is present, output is switched to normally OFF output.

11.9 Application Examples

11.9.1 Control of a double acting cylinder

Figure 11.15 illustrates a fluidic logic circuit for controlling a double acting cylinder. The cylinder is extended using a push button and retracted with the help of an air-operated limit switch. The circuit uses two pilot operated 3/2 direction control valves (1.1 and 1.2), a preference flip-flop (1.3), a normally closed limit switch (1.4), an OR valve (1.5) and a double acting air cylinder.

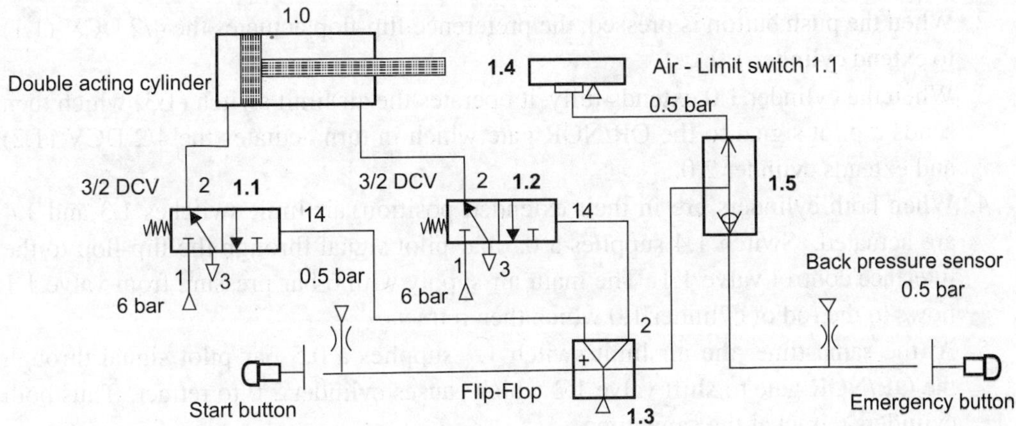

Figure 11.15 *Control of a double acting cylinder*

The start and emergency buttons represent back pressure sensors which supply 0.5 bar air to activate the preference flip-flop. The operation of the circuit is explained below.

1. When the start button is pressed by the operator, it provides an input control signal to flip-flop 1.3. Since it is a preference flip-flop, it supplies pilot air of pressure 0.5 bar to the 3/2 direction control valve (1.1).
2. When pilot line (14) of valve 1.1 is actuated, the cylinder extends.
3. When the cylinder is in the extended position, it actuates the normally closed limit switch (1.4) to supply pilot air of 0.5 bar to switch the flip-flop.
4. The output of the flip-flop (the pilot air of 0.5 bar) operates the 3/2 control valve (1.2) which then connects 6 bar pressure to the rod end of cylinder to retract it.
5. The emergency switch acts as a manual override for the limit switch and when pressed provides pilot air output from the OR valve to operate the 3/2 valve (1.2). As in step 4, the 3/2 valve (1.2) retracts the cylinder.
6. The preference flip-flop is used to ensure that the cylinder is always in the retracted position when the power is off and also when fluidic and pneumatic air supplies are turned on.

11.9.2 Control of two double acting cylinders in sequence

Figure 11.16 shows the sequence of two cylinders using fluidic control elements. The circuit uses two 4/2 direction control valves, a preference flip-flop, an OR/NOR gate, two normally closed air limit switches, and two double acting cylinders.

Operation

1. The sequence is that first cylinder 1.0 extends, then cylinder 2.0 extends, and finally both cylinders retract simultaneously.

2. When the push button is pressed, the preference flip-flop actuates the 4/2 DCV (1.1) to extend cylinder 1.0.
3. When the cylinder 1.0 extends fully, it operates the air limit switch (1.3) which then sends a pilot signal to the OR/NOR gate which in turn actuates the 4/2 DCV (1.2) and extends cylinder 2.0.
4. When both cylinders are in their extended position, air limit switches 1.3 and 1.4 are actuated. Switch 1.4 supplies a 0.5 bar pilot signal through the flip-flop to the interface control valve 1.1. The main air supply with 6 bar pressure from valve 1.1 flows to the rod of cylinder 1.0 which then retracts.
5. At the same time, the air limit switch 1.3 supplies a 0.5 bar pilot signal through the OR/NOR gate to shift valve 1.2 which causes cylinder 2.0 to retract. Thus both cylinders retract at the same time.

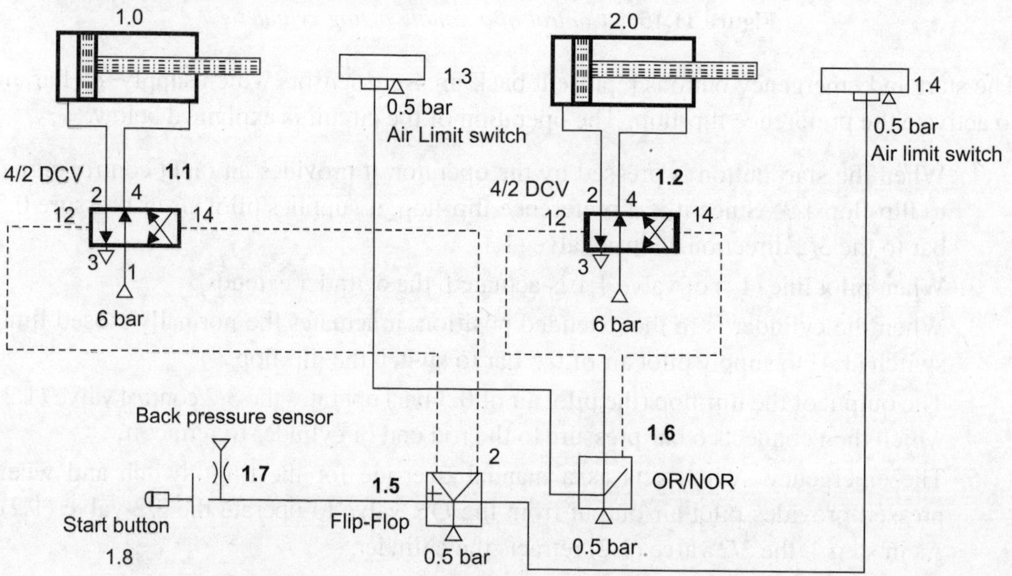

Figure 11.16 *Sequencing of double acting cylinders*

11.9.3 Continuous automatic reciprocation of a hydraulic cylinder

Figure 11.17 illustrates the fluidic logic circuit that controls the continuous reciprocation of a double-ended hydraulic cylinder. This circuit uses two OR/NOR gates, two jet sensors, a pilot operated 4/3 DCV, a preference flip-flop, a start button, a back pressure sensor and a double rod hydraulic cylinder.

Operation

1. The start button for the back pressure sensor allows 0.5 bar nozzle restricted air supply to the flip-flop. The preference flip-flop provides a pilot signal (12) to the 4/3 valve (1.1) which then shifts the cylinder towards the right.

2. When the cylinder moves to the right, it operates the interruptible jet sensor (1.3).
3. The interruptible jet sensor (1.3) causes the NOR gate to switch to output 1 to switch the flip-flop which provides a pilot signal (14) to shift the 4/3 valve (1.1), causing the cylinder to move towards the left.
4. When the cylinder rod retracts, the NOR gate (1.4) switches back to O2. But the flip-flop, being a memory valve is not switched.
5. When the cylinder rod on left end of the cylinder interrupts the interruptible jet of sensor (1.2), the NOR gate (1.6) switches to output O1. This again shifts the flip-flop output to provide a pilot signal 12 to cause the cylinder to move to the right again.
6. When the air supply is cut off, the 4/3 DCV shifts to its spring-centred position and is locked there because there is no output from the flip-flop to the pilot interface valve.
7. The cylinder reciprocates continuously in this manner until the pneumatic supply is switched off.

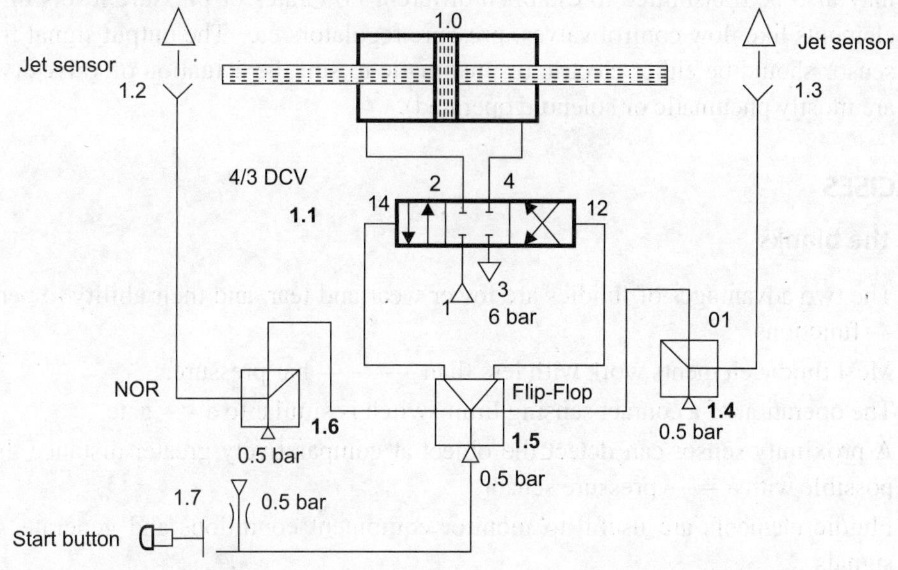

Figure 11.17 *Automatic reciprocation of a double ended cylinder*

Summary

1. A major problem with pneumatic and hydraulic systems is the susceptibility of their mechanical parts and components to wear and tear. It is therefore a challenging task for the fluid power designer to offer a totally reliable control system using a minimum of electronics.
2. An important feature of fluidics elements is that they are easily adaptable to logic functions in engineering applications. The real practical development started in year 1916 after Henri Coanda discovered the Coanda effect that "when a jet of fluid is

pushed into an inclined wall, it will stay there even after the pushing control jet is removed".

3. The most important fluidics elements include OR gates, AND gates, turbulence amplifiers, bi-stable flip-flops, NOR and exclusive OR gates.
4. Various types of pneumatic sensors are used in machine tools and other systems to transmit or receive signals. A major application of these sensors in machine tools is to sense the presence or absence of mechanical components, thereby generating a control signal in the form of small jets whose obstruction creates the signal. As most of these elements are used to sense the position of mechanical items by their proximity to the air jets, they are often called proximity sensors.
5. Fluidic elements are useful to monitor equipment conditions and generate signals. These signals are manipulated to shift the valve element of a directional control valve, which may advance or reverse the direction of a cylinder. Similarly signals may also be transmitted to establish different flow rates or pressure levels in other elements like flow control valves, pressure regulators etc. The output signal from a sensor should be either electric or pneumatic to suit the actuation of valves (which are mostly pneumatic or solenoid operated).

EXERCISES

Fill in the blanks

1. The two advantages of fluidics are lower wear and tear, and their ability to perform — functions.
2. Most fluidic elements work with less than ——— bar pressure.
3. The operation of a contact sensing limit switch is similar to a — gate.
4. A proximity sensor can detect the object at comparatively greater distance than is possible with a —— pressure sensor.
5. Fluidic elements are useful to monitor equipment conditions and generate —— signals.

State whether true or false

1. The major advantage of fluidic elements is that they are susceptible to lower wear and tear.
2. Fluidic elements work at much higher pressures compared to power pneumatics.
3. The wall attachment phenomenon is basically a memory function.
4. Most pneumatic sensors are air proximity sensors.
5. Fluidic elements are very useful in monitoring equipment conditions and generating output signals.

Review questions

1. Define fluidics.
2. List the various advantages of fluidic systems.
3. List the various disadvantages of fluidic systems.
4. Briefly explain the Coanda effect.
5. Draw the symbols for a bi-stable flip-flop, OR gate, AND gate and NOR gate.
6. Explain the working of a bi-stable flip-flop with the truth table.
7. List the advantages of fluidic pneumatic sensors.
8. Explain the principle of a back pressure sensor.
9. When are gap sensors used?
10. Compare pneumatic logic elements with fluidic logic elements.

Answers

Fill in the blanks

1. Logic 2. 1 bar gauge 3. OR/NOR 4. back 5. output

True or false

1. True 2. False 3. True 4. True 5. True

12 HYDROPNEUMATICS

Upon completion of this chapter, the reader should be able to

- Explain the use of oil in hydropneumatic circuits
- List the various types of hydropneumatic circuits
- Explain the working of an air-oil reservoir circuit
- Explain the working of hydraulic check units
- Explain the working of air-oil intensifiers
- Analyse the air-oil booster circuit
- Compare hydraulic, pneumatic and hydropneumatic systems

12.1 Introduction

While compressibility of air is the main drawback associated with pneumatic systems, it manifests mostly during traverse motion and at slow feed rates, making it difficult to maintain a constant feed rate over the course of the stroke. When an absolute steady force against a fluctuating load is required or when accuracy of the machine tool feed is critical, air compressibility is a disadvantage; the cylinder tends to slow down when load is increased and accelerate or produce jerky motion when load is suddenly decreased.

In the case of a double acting cylinder, excessive throttling of the air supply for slow speeds can result in the stick slip effect for pneumatic feed units. This is characterised by a jerky motion of the piston in the cylinder, caused by the cycle of the pressure building up to move the piston followed by the piston motion being opposed by friction caused by the flexible seals. The spasmodic progress of the piston in this manner can occur for less than a millimetre or up to several centimetres.

Therefore in pure pneumatic systems, air cylinders have to be preloaded to regulate the speed externally for precision control of actuators under fluctuating loads. This can be achieved with the help of hydraulic power transmission in the form of combined air-hydraulic or hydropneumatic systems. Hydropneumatic systems combines the advantages of both systems, namely the smooth operation under fluctuating loads of hydraulic systems and the ease and speed of air-based pneumatic systems.

12.2 Hydropneumatic Circuits

Hydropneumatic circuits can be divided into three broad categories:

1. Air-oil reservoirs
2. Hydraulic check units
3. Air hydraulic intensifiers

12.2.1 Air-oil reservoir

In air-oil reservoirs, air pressure is converted into hydraulic pressure. The circuit consists of a cylinder and piston which separates the spaces containing air and oil. When pressure is applied to the piston, it moves to displace air at a force equal to pressure multiplied by the area. Oil is incompressible and therefore facilitates regulated and smooth movement.

In air-oil systems, the output of the cylinder depends on the pressure of compressed air and is therefore much lower than that achieved by a conventional hydraulic cylinder. To achieve higher forces, the cylinder size has to be increased. A floating piston is used in air-oil systems and it restricts the mixing of air and oil. The piston also prevents the formation of froth due to turbulence and agitation of oil coming into the reservoir, which can negatively impact performance at high speeds.

Working of an air-oil reservoir

Forward stroke: When air is supplied to an air-oil reservoir, the oil flows out of the reservoir to the piston side of the cylinder, moving the piston forward. A regulator between the hydraulic cylinder and the air-oil reservoir controls the flow of oil and therefore the speed of the actuator. A schematic diagram is shown in Fig. 12.1.

Return stroke: When the 3/2 valve is released, air from the air-oil reservoir is vented to the atmosphere. The return motion of the cylinder is effected by the return spring. The return speed of the cylinder can be controlled by a regulator or a bypass check valve. Using an on-off valve on the hydraulic side of the circuit allows the piston to be stopped in any position of its travel path.

A schematic diagram of an air-oil reservoir system with double acting cylinders is shown in Fig. 12.2. Separate air-oil reservoirs feed into each side of the piston. When the DC valve is in its left mode, compressed air is fed to reservoir 1, as a result of which air pressure forces oil out into the hydraulic cylinder. A regulator controls the oil flow between the hydraulic cylinder and the air-oil reservoir. When the piston moves, the return oil enters air-oil reservoir 2. When the DC valve is its right mode, the air pushes the oil from reservoir 2, as a result of which oil flows to the rod end of the cylinder. The cylinder retracts and the return oil goes to air-oil reservoir1.

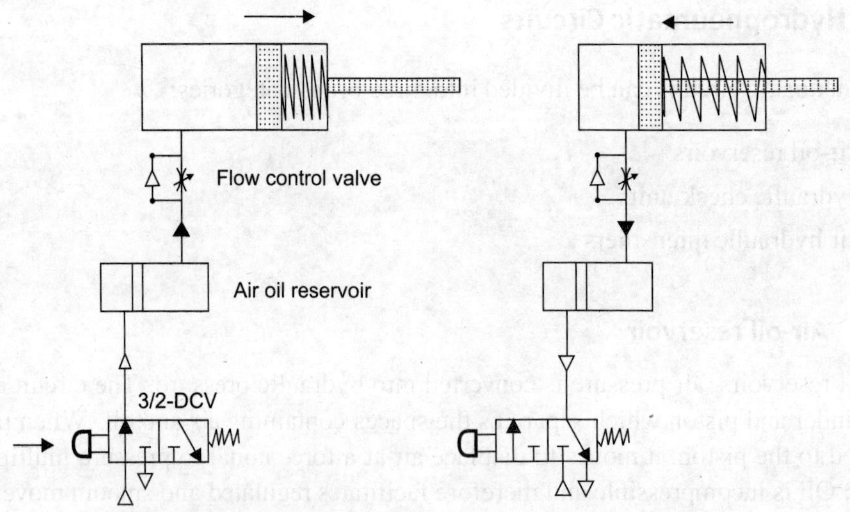

Figure 12.1 *Air-oil reservoir forward (single acting)*

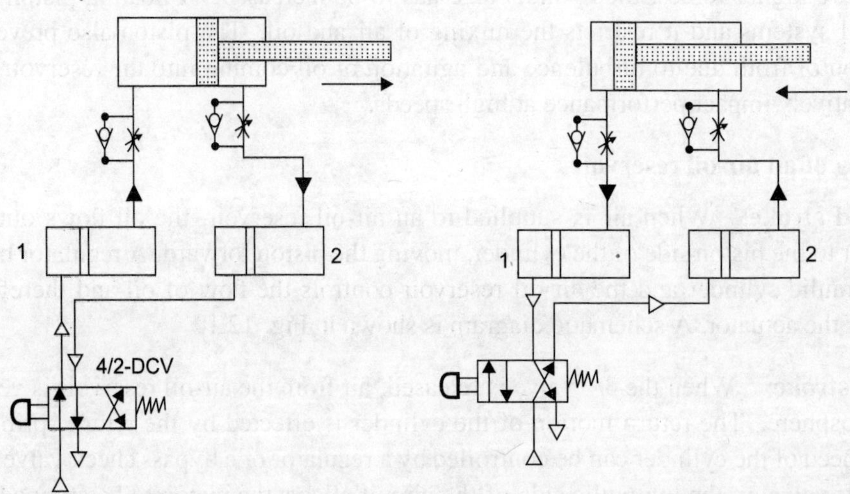

Figure 12.2 *Air-oil reservoir forward (Double acting)*

Having two air-oil reservoirs in the circuit does make the system bulky, but it is still cheaper and provides reasonable speed control and rigidity.

Application of air-oil reservoir Figure 12.3 illustrates a typical air-oil circuit for a counter-balancing application. The circuit uses an air-oil surge tank, a manually operated

3/3 DCV, an FRL unit, a flow control valve, a pressure relief valve and a cylinder. The surge tank has oil at the bottom and the air on the top.

Forward stroke When the 3/3 DCV is shifted to its upper envelope flow path configuration, the compressed air flows through the FRL unit and the valve and into the surge tank, pressurizing it. This pushes the oil out of the bottom of the surge tank and to the blind end of the cylinder through the flow control valve, extending the cylinder to lift the load. The flow control valve is used to regulate the extension speed of the cylinder.

Return stroke When the 3/3 DCV is shifted to its lower envelope flow path configuration, air leaves the surge tank and exhausts into the atmosphere through the DCV. As a result, the oil at the blind end of the cylinder returns steadily through the flow control valve, causing the cylinder to return.

The load can be stopped at any intermediate position by shifting the 3/3 DCV to its spring-centred position. This system eliminates the use of an expensive hydraulic pump and oil reservoir and the oil provides the advantage of higher output force and precise cylinder control.

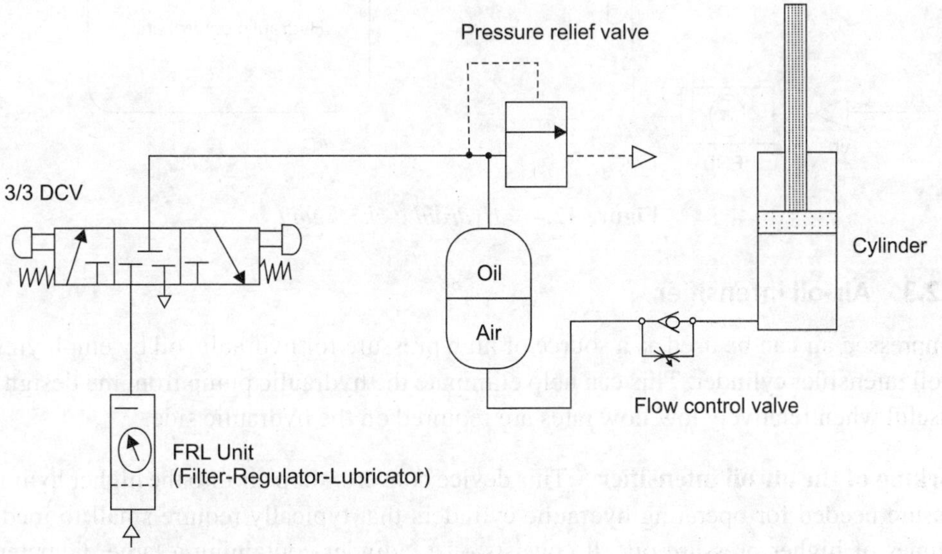

Figure 12.3 *Air-oil circuit for a counter-balancing application*

12.2.2 Hydraulic check units

Hydraulic check units have a simple design consisting of hydraulic and pneumatic systems. The unit includes two mechanically coupled cylinders mounted side by side so that one provides a check to the other. The cylinders can be arranged in tandem or in parallel.

The hydraulic component acts as a damping unit. The hydraulic cylinder has two check valves to regulate speed in both directions. It is important to note that stroke lengths of the hydraulic and the pneumatic cylinders must be identical, though they can have different bore sizes.

Starting out, the pneumatic cylinder moves fast due to the dampening effect of the hydraulic cylinder but the rest of the cylinder stroke has a controlled movement. The schematic arrangement of a hydraulic check unit is shown in Fig. 12.4.

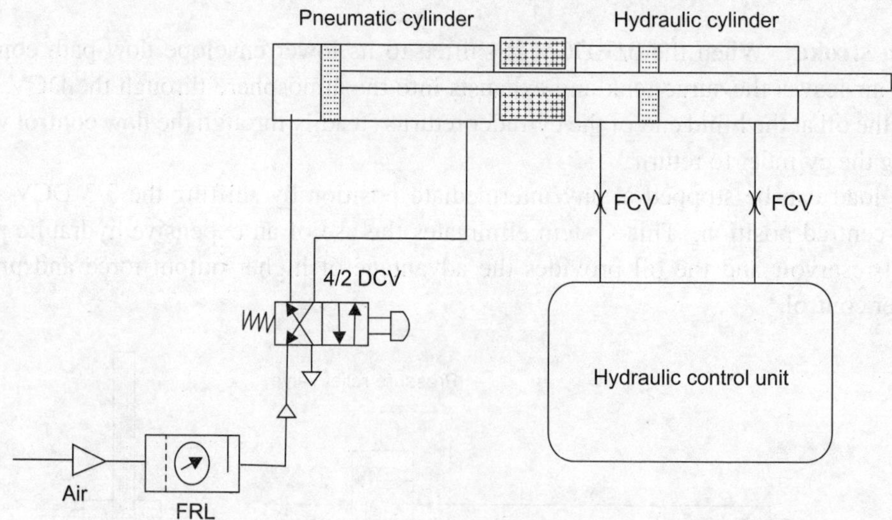

Figure 12.4 *Hydraulic check unit*

12.2.3 Air-oil intensifier

Compressed air can be used as a source of high pressure for hydraulic oil by employing an air-oil intensifier cylinder. This can help eliminate the hydraulic pump from the design and is useful when relatively low flow rates are required on the hydraulic side.

Working of the air oil intensifier This device converts shop air into the higher hydraulic pressure needed for operating hydraulic cylinders that typically require small to medium volumes of higher pressure oil. It consists of a cylinder containing a large diameter air piston driving a small diameter hydraulic piston (that is actually a long rod connected to the piston). Any shop equipped with an airline can obtain smooth, efficient hydraulic power from an air-to-hydraulic pressure booster hooked into the air line. The alternative would be a complete hydraulic system including expensive pumps and high-pressure valves. Other benefits include lower space requirements and low operating and maintenance costs.

Forward Actuation of valve 1 to the left mode directs air to the approach tank. This forces the oil through the bottom of the intensifier to the blank end of the cylinder, extending it.

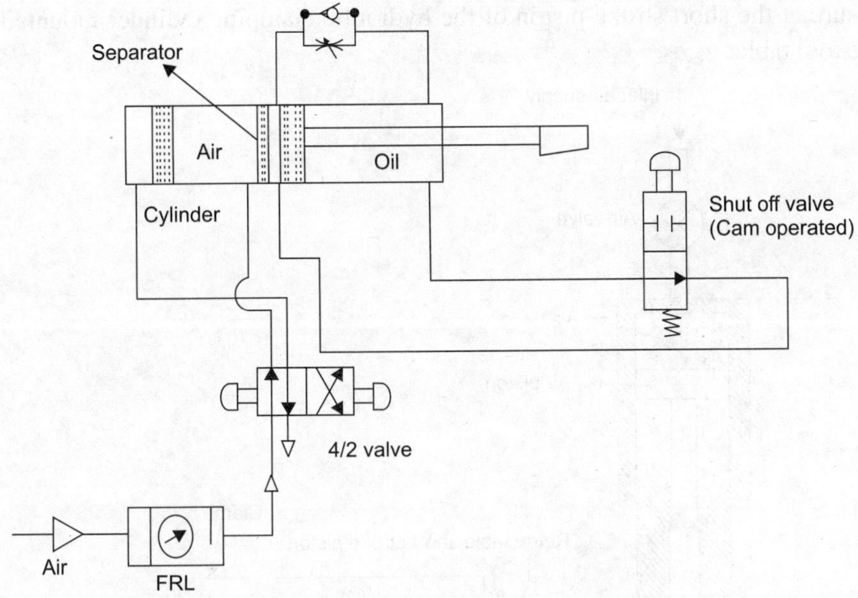

Figure 12.5 *Working of an air-oil intensifier*

Punching When the cylinder experiences a load, valve 2 is actuated to the left mode, sending air to the top end of the intensifier. The high pressure oil cannot return to the approach tank because the port is blocked by the downward motion of the intensifier piston. The cylinder therefore has the high pressure oil at the blank end to carry out the high pressure punching operation.

Retraction When valve 2 is released to the right mode, shop air is blocked and the top end of the intensifier is vented to the atmosphere. When valve 1 is released to the right mode, the air in the approach tank is vented and the shop air is directed to the return tank. This delivers oil to the rod end of the cylinder causing it to retract. The return oil from the blank end of the cylinder enters the bottom of the intensifier and flows back to the approach tank.

Application of the air-oil intensifier booster Figure 12.6 shows an application where an air-to-hydraulic pressure booster supplies high pressure oil to a hydraulic cylinder whose short stroke piston is used to clamp a workpiece to a machine tool table. Since shop air pressure normally operates at 100 psi, a pneumatically operated clamp would require an excessively large cylinder to hold the workpiece rigidly in place while it is being machined.

The air-to-hydraulic pressure booster operates as follows. Assuming that the air piston has an area of 10 cm^2 and is subjected to a 10 bar air pressure, this produces a 1000 N force on the booster's hydraulic piston. Therefore, a booster hydraulic piston with an area 1 cm^2 experiences hydraulic oil pressure of 100 bar. As per Pascal's law, this produces 100 bar oil pressure at the short stroke piston of the hydraulic clamping cylinder mounted on the machine tool table.

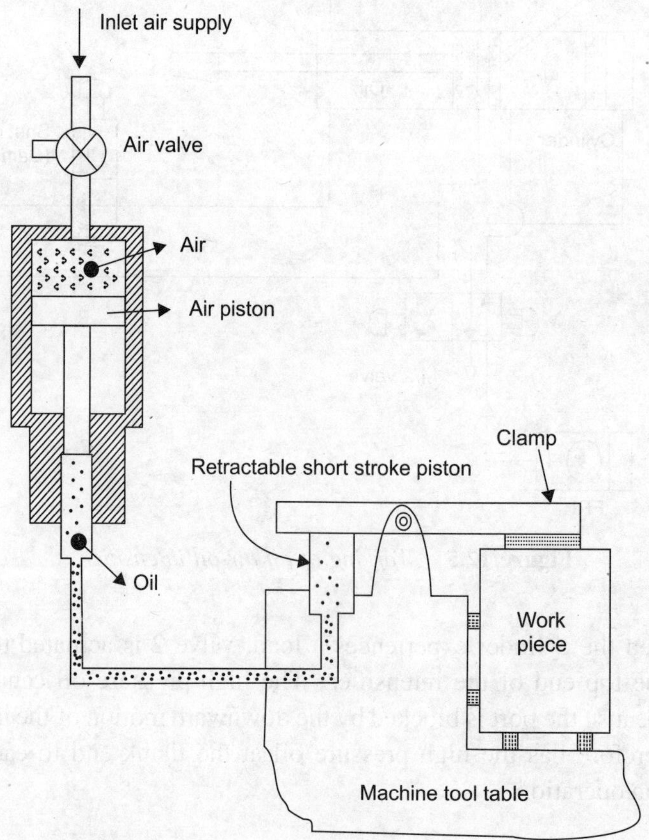

Figure 12.6 *Air-to-hydraulic pressure booster*

The pressure ratio of an air-to-hydraulic pressure booster can be found using the following equation:

$$\text{Pressure ratio} = \frac{\text{Output oil pressure}}{\text{Input oil pressure}}$$

$$= \frac{\text{Output oil pressure}}{\text{Input oil pressure}}$$

Using this equation with the earlier parameters:

$$\text{Pressure ratio} = \frac{1000 \text{ kPa}}{1000 \text{ kPa}} = \frac{10 \text{ cm}^2}{1 \text{ cm}^2}$$

For a clamping cylinder with a piston area of 0.5 cm^2, the clamping force is (1000 N/cm^2 × 0.5 cm^2) or 500 N. To provide the same clamping force of 500 N without a booster requires a clamping cylinder with a piston area of 5 cm^2, assuming 10 bar air pressure. Air-to-hydraulic pressure boosters are available in a wide range of pressure ratios and can provide hydraulic pressures up to 1000 bar using approximately 7 bar of shop air.

Example 12.1 For the pressure booster in Fig. 12.7, the following data is given:

Inlet oil pressure $(p_1) = 1$ MPa

Air piston area $(A_1) = 0.02$ m^2

Oil piston area $(A_2) = 0.001$ m^2

Load carrying capacity $= 300000$ N

Find the load required on load piston area A_3.

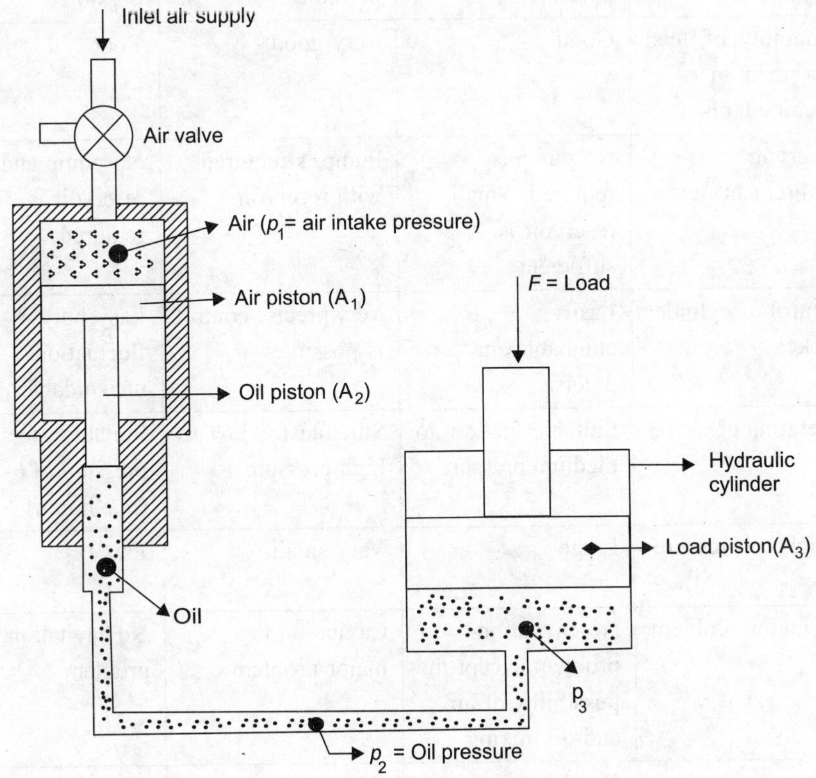

Figure 12.7

Solution

$$p_2 = \frac{p_1 A_1}{A_2} = \frac{1 \text{ MPa} \times 0.02 \text{ m}^2}{0.001 \text{ m}^2} = 20 \text{ MPa}$$

$$p_3 = p_2 = 20 \text{ MPa}$$

Also, $A_3 = \frac{F}{p_3} = \frac{300000\ \text{N}}{20 \times 10^6 \frac{\text{N}}{\text{m}^2}} = 0.015\ \text{m}^2$

12.3 Comparison of Hydropneumatic, Hydraulic and Pneumatic Circuits

Table 12.1 *Comparison between hydraulic, pneumatic and hydropneumatic systems*

No.	Parameters	Hydropneumatic	Hydraulic	Pneumatic
1	Resistance to fluctuating load	Good control	Very good control	Poor control
2	Speed of the actuator	High speed is possible	Moderate speed is possible	Very high speed is possible
3	Suitability of feed movement in machine tools.	Good	Very good	Poor
4	Reservoir requirement	No pump is required. Small reservoir is sufficient	Pump is required with reservoir.	No pump and reservoir is required
5	Control of cylinder stroke	Easily attainable and precise	Very precise control is possible	Easy but fluctuations unavoidable
6	Operating pressure	Suitable for low to medium pressure	Suitable for low to high pressure	Suitable for low pressure of 6–8 bar (maximum)
7	Weight-to-pressure ratio	Large	Very small	Large
8	Cavitation problem	No cavitation problem except the possibility of air and oil mixing	Cavitation is a major problem	No cavitation problem
9	System rigidity	Good	Very good	Poor
10	Overall cost	Low to moderate	Moderate to high	Low
11	Operating cost	Low to moderate	Moderate to high	Low
12	Maintenance	Simple	Difficult	Difficult

...contd

Table 12.1 *...contd*

13	Cushioning	Cylinder cushioning is required	Cylinder cushioning may or may not be required	Cylinder cushioning is not required
14	Air supply	Needed	Not needed	Needed
15	Oil change	Occasional change in oil may be necessary	Oil is changed as per schedule	Not needed
16	Transmitting medium	Air is the operating fluid and oil is the check medium	Uses only oil	Uses only air

Summary

1. The main drawback associated with pneumatic systems is the compressibility of air. This characteristic is not suited for fine movement control of pneumatic actuators. Hydropneumatic circuits are used where pneumatics alone are not suited, such as in the feed units of machine tools.
2. There are three basic hydropneumatic circuits: air-oil reservoirs, hydraulic check units and air hydraulic intensifiers.
3. In an air-oil reservoir, air pressure is converted into hydraulic pressure. It is made up of a cylinder and piston which separates the spaces containing air and oil. The oil used should be compatible. Leakage must be controlled because it causes loss of power. Thick oil with higher viscosity is generally not used in air-oil reservoirs.
4. Hydraulic check units are simple in design. Oil and air are kept in two separate cylinders which are coupled mechanically. The cylinder arrangement may be in parallel or in tandem.
5. Compressed air can be used as a source of high pressure in hydraulics by using an intensifier cylinder. This can be particularly useful where hydraulic pumps are not available and relatively low speeds are required on the hydraulic side. The working principle involves feeding compressed air to the blind end of the intensifier cylinder which is the air side. Forward movement of the piston rod now produces a plunger action at the smaller end of the intensifier, forcing oil under pressure from the end of the intensifier.

EXERCISES

Fill in the blanks

1. Pure pneumatic circuits cannot be used for precise movements due to ________ of air.
2. Hydraulic check systems will have separate pneumatic and ________ system.
3. Oil compatibility and oil ________ are the main concerns in the design of an air-oil system.
4. In air-oil systems, output is provided by ________ cylinder.
5. Booster circuits are used where pumps are not available and ________ is required.

State whether true or false

1. Pure pneumatic systems are used in the feed motion of machine tools.
2. The system rigidity of hydraulic systems is good compared to hydropneumatic systems.
3. Control of cylinder stroke is easy with hydropneumatic systems compared to pure pneumatic systems.
4. High speeds are possible with hydraulics when compared to hydropneumatics.
5. Pure pneumatics is recommended where the load is always fluctuating.

Review questions

1. Explain why we cannot use pure pneumatic system for machine tool feed units.
2. List the various types of hydropneumatic circuits.
3. Explain briefly the working principle of air-oil reservoir systems.
4. Draw the symbol for an air-oil reservoir and intensifier.
5. What are the important factors to be considered in air-oil reservoir systems?
6. Explain the working of a hydraulic check unit.
7. List the advantages of an intensifier circuit.
8. Explain the principle of an air-oil counter-balance circuit.
9. When are intensifier circuits used?
10. Compare pneumatic systems with hydropneumatic system.

Answers

Fill in the blanks

1. compressibility 2. hydraulic 3. leakage 4. hydraulic 5. low flow rate

True or false

1. False 2. True 3. True 4. False 5. False

13 MAINTENANCE AND TROUBLESHOOTING OF PNEUMATIC SYSTEMS

Upon completion of this chapter, the reader should be able to

- Explain the meaning of preventive maintenance and corrective maintenance
- List the various points to be considered for general maintenance
- Explain the various maintenance procedures for air intake systems, FRL units, cylinders, valves, pipes and other pneumatic components
- Explain general troubleshooting guidelines for pneumatic systems
- Explain possible malfunctions and remedies for air intake systems, FRL units, cylinders, valves, pipes and other pneumatic components
- Explain the testing procedure for pneumatic cylinders and valves

13.1 Maintenance of Pneumatic Systems

Compared to hydraulic systems, pneumatic systems are less problematic but they still need to be protected against unpredictable failures. A good working knowledge of the operation and function of pneumatic elements is important for an effective maintenance plan which includes a wide range of activities to keep the system and its components in good condition. If properly maintained, pneumatic systems can work safely and efficiently over a long service life.

Two types of maintenance activities should be planned for:

1. **Preventive maintenance (PM)** This is carried out on a regular basis at scheduled intervals. This type of maintenance is planned ahead and carried out before the failure of any components. The schedule of maintenance depends on operating cycles, complexity, time, load, etc. Installation manuals for the systems and their components will help to decide the frequency of maintenance. Spare parts are very important for both preventive and corrective maintenance. Table 13.1 shows the typical PM schedule for pneumatic systems.

Table 13.1 *Typical preventive maintenance schedule for pneumatic systems*

Equipment	Check items	Daily	Weekly	Monthly	Half yearly	Yearly	Two yearly	Operating cycles 1 million	Total operating distance	Note
Compressor	Compressor lubricant	△								Fill the oil
	Intake filter			△						Flush out
Tank	Drain Exhaust	△								
Main filter	Drain exhaust	△								
	Stains on an element				△					Change element
Air dryer	Refrigerant pressure gauge	△								
	Stains on an element	△								
	Stains on a condenser	△								Clean condenser
Filter	Drain exhaust	△								
	Stains on an element				△					Flush out
Regulator	Adjusting pressure	△								Check if there is damage on pressure gauge shell
Lubricator	Quantity of oil and level	△								Fill the oil
Solenoid valves	Leakage from exhaust port						△	△		Coat lithium saponaceous radical
Speed controller	Adjusted speed of a cylinder			△				△		
Cylinder	Air leakage from piston rod			△				△		Coat lithium saponaceous radical
	Slackness of mounting			△				△	△	

2. **Corrective maintenance** This is carried out after the failure of components and consists of tracing the fault(s), repairing, troubleshooting and re-commissioning of the systems or components.

13.2 General Maintenance Procedure

The general procedure to maintain pneumatic systems is as follows:

- Check and ensure that lines and components are free from dirt and debris for optimum performance of a pneumatic control system.
- Check for loose V-belts, worn out parts, loose bolts and nuts.
- Carry out regular checks of the FRL (Filter, Regulator and Lubricator) unit. The filter should be periodically drained and cleaned, the regulator set to the minimum required pressure and the lubricator filled with only the appropriate grade of lubricating oil.
- With the system fully charged and the engines turned off, inspect the lines and components for leaks by listening for hissing sounds (air leaving the system).
- Inspect mechanical connections for wear and tear and ensure belts and parts are properly fitted and fastened.
- Leaks and damaged components should be fixed as soon as they are discovered to minimise additional problems in the system.
- A pneumatic control system can provide years of effective service when maintenance procedures are followed on a regular basis

13.3 General Maintenance Guide

a) Maintenance of air systems and air receivers

In industries, the air compressor is positioned at a distance away from the main manufacturing /assembly shop due to air borne noise or for safety reasons. The compressed air is stored in an air receiver, from which the air is drawn into the consumer point by pipeline. The pipeline fittings and pipe joints cause pressure drop in pipeline and therefore their installation must be done carefully. Manufacturing schedules and installation procedures are to be followed for equipments like filters, cylinders, compressors, etc.
Draining of condensate from air receivers and air coolers should be automatic. Even when automatic draining is used, regular inspections are important to make sure that the unit is working effectively. Safety devices on air receivers (pressure relief valves, pressure gauges, etc.) must be maintained properly.

b) Maintenance of compressors

The following guidelines should be followed for maintenance of compressors

- The compressor intake should be taken from the outside air and should not be located in enclosed rooms.
- The open end of the intake pipe of the compressor must be well-hooded and screened to prevent rain and dirt from entering
- The intake pipe should be short as possible. The direct connections to unit is better than jointed connections

- The compressor discharge line should either be the full size of the compressor outlet or larger and it should run directly to an aftercooler. If there is no aftercooler, the line should directly run to the receiver tank.
- The piping should be sloped away from the compressor with sufficient pitch to prevent the condensate of oil drainage back into the compressor.

c) Air treatment equipment

FRL units, dryers, aftercoolers and separators should be kept clean and should be maintained as per the manufacturer's recommendation and maintenance schedule.

The important parameters for air treatment systems that should be checked are shown in Fig. 13.1.

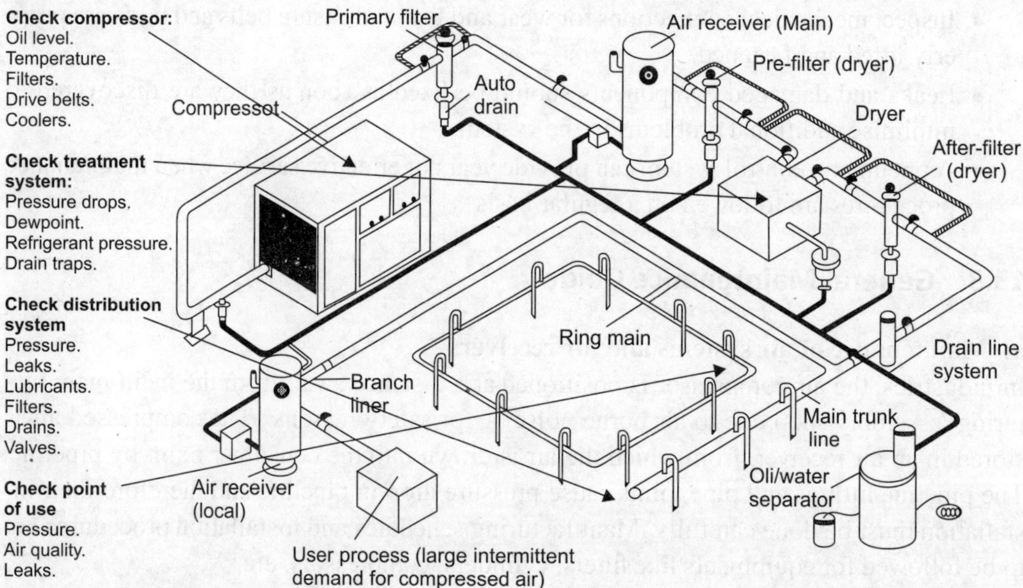

Figure 13.1 *Important parameters to be checked in air treatment systems*

Automatic drain traps Automatic drain traps are to be installed at appropriate locations. They must be inspected periodically for blockage.

d) Installation of FRL units

The filter is installed upstream from the other conditioning components. This protects the regulator from harmful contaminations and avoids fouling of the lubricator reservoir. It is a common practice to install filters in each branch. Regulators are carefully installed in the branches with the pressure settings required by the machine.

e) Leaks

Leak detection and repair is an important part of air maintenance systems. Leaks cause a pressure drop in the system, which make air tools less efficient. Compressed air is expensive and therefore leaks can contribute to overall cost increase.

The summary of all the maintenance activities related to air mains and systems is given in Table 13.2

Table 13.2 *Summary of maintenance activities of air mains and its components.*

Filter cartridges	Inspect as per the manufacturer's recommendation and replace when there is clogging or reduction in efficiency of the system.
Manual drain traps	Check the operations periodically as per maintenance schedule.
Compressor lubricant level	Inspect as per manufacturer's specification. Use proper grade of lubricating oil.
Air line filters	Replace lubricant elements as per manufacturer's specification
Water cooling systems	Check the water quality, level, temperature, flow and pressure as per maintenance schedule
System leaks	Check lines fittings, valves, disconnects and FRL units for leaks
System cleanliness	Check all air systems for dirt deposits and clean them as per specifications

f) Maintenance of air cylinders

Proper maintenance of cylinders includes the use of the right tools and should be carried out by experts in the area of pneumatic systems. Spare parts should be kept in stock and when necessary, seals should be replaced. Replacement of seals should be done for the whole cylinder rather than partial replacement, to minimise repeated repairs and interruptions to the manufacturing operations. The following precautions should be taken.

1. Protect the threads and avoiding damaging them.
2. Protect surfaces with high finishes against damage from dents, scratches and dirt.
3. Do not mix old and new parts or mislay them.
4. Take care not to scratch or cut rubber parts which will lead to leaks.
5. When seals are assembled, they should be protected from contact with sharp edges, burrs, etc.
6. Solvents that might damage the seals should not be used to clean systems parts.
7. When disassembling, a metal hammer or tools with sharp edges should not be used (use a plastic hammer).
8. Before being (re)assembled, all parts should have been cleaned and lubricated wherever necessary.

Corrective Maintenance of Air Cylinders

As soon as leaks or malfunction problems are detected in a pneumatic cylinder, it should be removed from the installation for corrective maintenance. The cylinder should be carefully disassembled, repaired and all parts should be thoroughly cleaned and checked.

1. The spare parts (seals) should all be replaced. The reuse of old seals can lead to new maintenance issues in the short term.
2. Old seals should always be checked for uneven wear, cuts, etc. for indications of bigger problems in the piston rod or tube.
3. The piston rod, bearing, tube and piston should be carefully examined for scratches, crushed parts or uneven wear signals.
4. Uneven wear in the bearing, piston or tube may be an indication that the cylinder is working out of alignment.
5. All damaged or worn out parts should be replaced or repaired so that they meet the original specifications.
6. Repaired parts should also meet the original specifications; any roughness can cause premature wearing of seals.
7. Assembly of seals should be carefully handled, and sharp tools should not be used. Any discrepancies in the assembly should be analysed for cause.
8. Seals should be lubricated with a small amount of grease when assembling.
9. The assembly of the set should be done carefully done to avoid misplacing of parts.
10. After assembly is complete, the cylinders should be subjected to bench test to certify that the problems were solved.

Preventive Maintenance of Air Cylinders

It is difficult to establish a specific interval for preventive maintenance of pneumatic equipment; however maintenance must be periodically carried out. The intervals can vary depending on the working condition of the equipment and ambient conditions (presence of dust, heat, corrosive agents, etc.).

Generally, preventive maintenance should be carried out at intervals ranging from 3 to 12 months. If the equipment is in good condition with no immediate repairs needed, the interval can be longer. On the other hand, if the equipment is older or has known problems, the interval should be shorter.

The important aspects of a preventive maintenance plan are:

- Cylinders should be carefully disassembled (often the elements are mechanically fine and need only cleaning).
- Seals should be regularly replaced with new ones, even if they are not obviously damaged or worn out.
- Metal surfaces should be carefully examined especially where piston rods and tubes move, and these parts should also be checked for scratches and burrs.

- The disassembled elements should be inspected to ensure that they are well lubricated.
- When equipment shows defects prematurely, the probable cause should be investigated and identified. Extrusion, inadequate lubrication, presence of impurities, excessive wear or grooves on piston rods, tubes or bearings are examples of root causes.
- Special care should be taken with regard to spare parts. When equipment in active use needs repair, the entire set of parts or components should be replaced. If the parts can be refurbished at the work station they can be retained as spares.
- When any cylinder parts show signs of oxidation or wear, they should be immediately replaced.An adequate stock of spare parts should always be maintained.

The testing procedure for pneumatic valves is given in Table 13.3.

Table 13.3 *Testing procedure for pneumatic cylinders*

No.	Steps to be followed
1	Connect the cylinder to a 5/2 direction control valve which is connected to the main air supply.
2	Operate the valve 10 times and observe whether the forward and return movements are smooth.
3	Leave the cylinder in the extended position. Test rear end cover and cylinder barrel connections for leaks and cushioning screws adjustments. Make sure that the tie rods are tight enough. Correct the misalignment of tie rods, if any.
4	Check the operating valve to see that air from the exhaust port can move without obstruction. If the stream is not smooth or continuous, check the rear piston seal for damage, fitting or flaws on the surface of the barrel of the cylinder.
4	Operate the valve to fully retract the piston, and check the front end cover and connections for leaks.
5	Check again for leaks and cushioning to make sure that cylinder is in good operating condition.

g) Maintenance of pneumatic valves

Pneumatic valves are normally designed for 10 million cycles of normal operation. If a valve requires servicing before 10 million cycles, a check must be carried out carefully to analyse the working cycles, the harshness of environment, etc. The following are the important points maintaining pneumatic valves:

1. **Clean air supply** Foreign material lodged in valves is a major cause of improper operation. A 5 micro rated air filter should be located close to the valve. The filter bowl should be drained regularly and if its location makes draining difficult, the filter should be equipped with an automatic drain
2. **Leaks** The main cause of valve failure is a leaking O-ring which results from abrasive particles or dust. Using clean air is important. To check whether a valve is

leaking, remove the valve from the faulty circuit and connect it to another properly functioning actuator to see how it performs. A light film of oil or soap may also be used to test for leaks.

3. **Lubricator supply rate** A lubricator should emit a fine oil mist into the air line in direct proportion to the rate of air flow. Excessive air flow can cause puddling in the valve leading to malfunction. For most applications an oil flow rate of one drop per minute is adequate though the manufacturer's instruction manual should be the final reference.
4. **Compatible lubricants** The right lubricant for the system and its operation should be used. The most suitable oils are those specifically formulated for air line service. These are generally petroleum based oils with oxidation inhibitors, an aniline point between 82 and 104 and a viscosity of ISO 32 or less is used.
5. **Cleaning the valve** If the air supplied to the valve has not been well filtered, the interior of the valve may accumulate dirt and varnish, which can affect its performance. A good commercial solvent or kerosene can be used to clean valves; chlorinated solvents or abrasive materials should not be used. Chlorine damages seals and abrasives can do permanent damage to metal parts.
6. **Electrical contacts** In the electrical circuits associated with valve solenoids, all switches or relay contacts should be kept in good condition to avoid solenoid malfunctions.
7. **Replacing worn out parts** In most cases it is not necessary to remove the valve from its installation before 10 million cycles are completed. But if the need arises, electrical power to the valve and the air supply should be turned off, and the air in the system should be exhausted before beginning any disassembly operation.

Testing procedure for pneumatic valve is given in Table 13.4.

Table 13.4 *Testing procedure for pneumatic valves*

No.	Steps to be followed
1	The valve body must be degreased using kerosene or detergent.
2	Check the valve actuating mechanism. Carefully remove the base so that threads are not damaged. Examine the valve bore, bush, valve seat and springs for pitting, score marks, scratches, and fatigue marks.
3	Check the condition of O-rings and gaskets. Remove the O-ring and examine its grooves for scratches and dirt particles. Replace the O-rings and grease the part.
4	Check whether the pilot holes of valves are clogged and need cleaning
5	If the spring or spring locating pin is corroded it should be replaced. Spring stiffness is very important and the appropriate spring as recommended by the supplier should be used.

13.4 General Troubleshooting Guide

There are 12 golden rules used in industries while trouble shooting the pneumatic system. They are:

1. Safety of the operator and machine is important
2. Use four Ws and one H, that is, what, when, where, why and how to find out the initial cause of fault.
3. Inspect the location and brainstorm with the user.
4. Understanding the working of the system is a must.
5. Observe the operation of the unit or system.
6. Check all the inputs such as air, electric current and water supply to the system.
7. Separate the subsystems from the main system or unit.
8. Make a list of the possible causes based on the user's experience.
9. List the most probable cause.
10. Check whether the listed cause is the actual cause.
11. Repair or replace the components or units.
12. Document your findings and circulate among the users of the units and equipments. Also inform the manufacturer of the unit or the component so that they can improve their design or procedure.

Troubleshooting a pneumatic system that will not start or has stopped mid-cycle should begin with the following valve checks:

1. Is electric power available?
2. Is pneumatic power available?
3. Are manual overrides engaged?
4. Are the manual controls properly positioned?
5. Is the circuit in sequence?
6. Are any lockouts engaged?
7. Is there any mechanical interference?
8. Are the solenoids actuating?

13.5 General Troubleshooting Procedure

Troubleshooting is an art but a systematic approach saves money as well as time. The functional diagram and circuit diagrams are used as troubleshooting aids. Accurate troubleshooting lies in determining the section of the control circuit that contains the fault or defective component and then selecting the component to be checked. This can be accomplished only by proper analysis of the circuit and with the aid of test equipment.

The first step in troubleshooting pneumatic systems is to understand the operation of the equipment and the associated circuits. Talking to the operator is important and saves time. Collect all the information related to the fault or failure to help find the root cause of the fault.

The second step in troubleshooting pneumatic circuits is to reduce the number of possibilities, and identifying the section of the circuit that contains the defective components. A careful check for the malfunctioning control element is important and should be done component by component rather than checking all the components at once. This second phase has to be repeated until the fault is traced and repaired.

The last step in troubleshooting is recording the fault and fault analysis to identify any recurring patterns of faults. It also serves as an important document to train newcomers. Corrective and preventive maintenance procedures should be updated with a proper record of the rectified faults.

The general troubleshooting procedure is shown in Fig. 13.2.

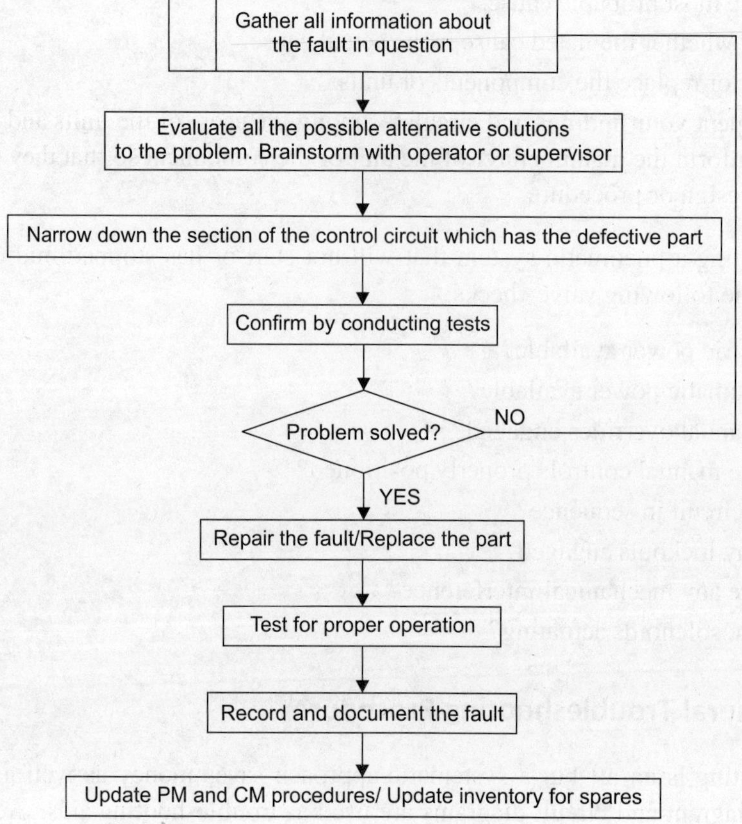

Figure 13.2 *General troubleshooting procedure*

Figure 13.3 shows the troubleshooting guide for a pneumatic cylinder with a problem of low thrust. Figure 13.4 shows the troubleshooting guide for a pneumatic cylinder with a problem of low speed.

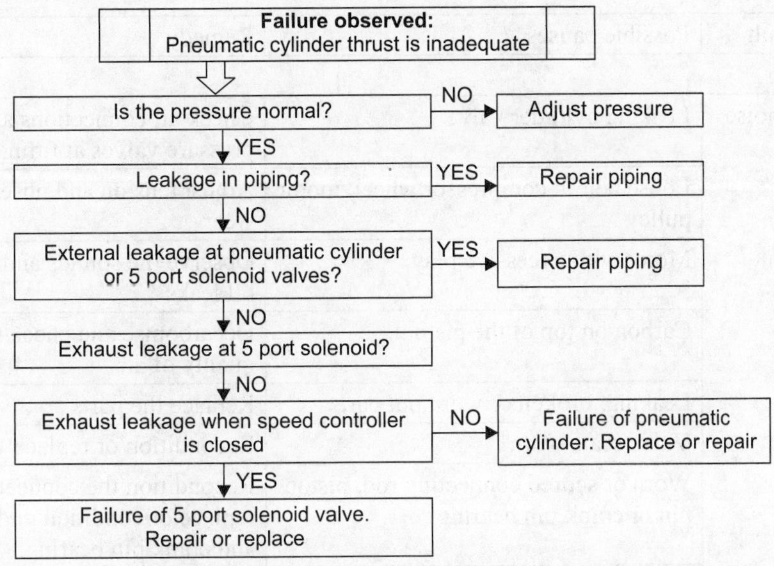

Figure 13.3 *Troubleshooting guide for pneumatic cylinder with low thrust*

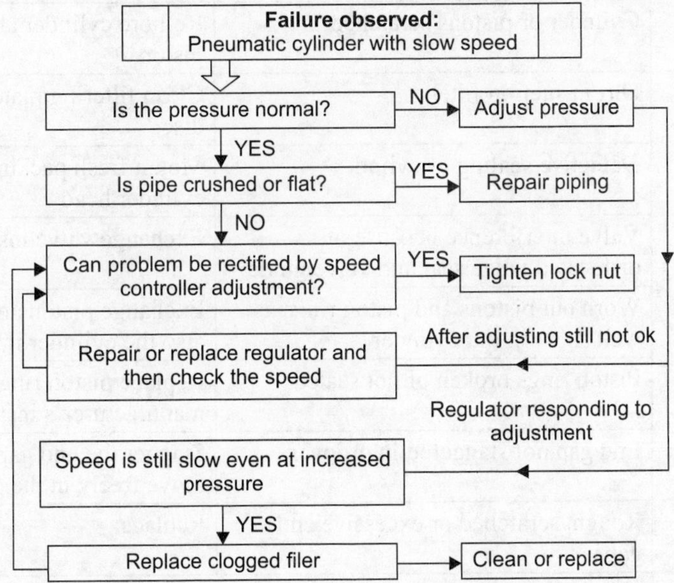

Figure 13.4 *Troubleshooting guide for pneumatic cylinder with low speed*

The general malfunctioning of various pneumatic systems with the possible causes and remedies are listed in Tables 13.5 to 13.13.

Table 13.5 *Fault diagnosis of compressor*

Trouble/Fault	Possible causes	Remedy
Compressor		
Excessive noise	Leaking cylinder valve	Check all connections and pressure valves at fittings
	Loose belt in compressor wheel, motor pulley	Adjust tension and observe for wear
	Motor with excessive play	Observe the sounds and look for oil leaks
	Carbon on top of the piston	Decarbonise and check the quality of air
	Leaking, broken or worn out parts	Replace the parts
	Valve seats worn	Recondition or replace
	Worn or scored connecting rod, piston pin or crank pin bearing	Recondition the connecting rod, replace or condition gudgeon pin and crank pin bearing.
	Defective ball bearing on crank shaft	Replace bearing
	Loose motor fan	Tighten the motor fan
	Cylinder or pistons scratched	Re-bore cylinder and replace piston.
Inadequate performance	Dirt in suction filter	Clean filtering plate and filter disc.
	Defective sealing of cylinder	Mount fresh packing of the cylinder head
	Valve interference because of dislocated valve seat and valve guide	Exchange valve insert plate
	Worn out pistons and piston rings as well as worn out cylinder	Exchange piston and rings and also the cylinder if needed
	Piston rings broken or not sealed	Replace piston rings as per manufacturer's instruction
	End gap not staggered in grooves	Stagger the end gaps, so rings can move freely in the grooves
	Rough, scratched or excessive end gaps	Replace
	Cylinders or pistons scratched or worn out	Replace

Table 13.6 *Fault diagnosis of filters*

Trouble/Fault		Possible causes	Remedy
Filters			
Flow rate at air outlet is too low		Element is clogged	Replace the element or flush out
			Use larger filter
Cannot see through the bowl		Drain rate is too high or binding foreign substances	Flush out the bowl
Drain or contaminants collect above the standard level	Flexible drain	The drain itself or clogging by foreign substances	Push the flexible drain to discharge the drainage
	Mechanical drain		Flush out the mechanical drain
Excessive pressure drop		Dirty filter element	Replace the filter element
		Filter is undersized	Replace, consult manufacturer
Contaminants carried through the filter		Elements omitted during installation	Replace the missing element
		Elements not tightened	Tighten the element to prevent bypass
		Broken elements	Replace broken parts
		Element too coarse	Replace with new element
		Broken end cap from excessive pressure	Replace and install standard particle filter ahead of coalescing filter.
Moisture in downstream air		Filter bowl contains too much water	Drain bowl. Install automatic drain
		Location is not correct	Filter may be too close to after-cooler or too high up in plant ceiling
		Body size is too large, causing low velocity and inefficient operation	Consult manufacturer
		Dew point of air is too high	Install a dryer
Air leaks from the drain outlet		Drain outlet is clogged with contaminants	Flush out the drain outlet
Plastic bowl cracked and breaking		Incompatible chemical in contact with bowl	Substitute with metal bowl. Check for compatibility.
		Excessive temperature or pressure.	Substitute with metal bowl. Consult manufacturer.
Broken bowl		Organic solvent environment	Replace with a metallic or nylon bowl

Table 13.7 *Fault diagnosis of regulator*

Trouble/Fault	Possible causes	Remedy
Regulator		
Regulator cannot reach set point	Regulator cannot reach set point	Pressure gauges are not accurate. Check calibration
	Insufficient upstream pressure	Measure and compare the inlet and out let pressure
	Incorrect control spring	Check the model number and replace
	Leakage in downstream circuit	Check fittings, valves, cylinders and regulators
	Incorrect adjusting techniques	Check the manual
Set point pressure goes too high	External loads imposing a higher pressure	Use pressure relief valves or modify the circuits, vent the regulator
	Leakage from inlet side, worn out poppet seal	Check for leakage and replace
	Non-venting regulator can aggravate pressure rise from other causes	Replace with venting regulator
Air escaping from vent hole too often	External loads imposing high pressure	Use pressure relief valves, modify circuits.
	Leakage from inlet side, worn out poppet seal	Check for leakage and replace
Pressure too low when air is flowing	Incorrect adjustment	Consult manual and manufacturer
	Setting altered	Check for vibrations, reposition or adjust and use locking devices
	Leakage in the downstream	Check connections, component seals
	Flow requirement is too high for regulator rating or plumbing	Install a larger regulator or upgrade plumbing
Chatter and vibration	Load instability from friction turning over center or reaching a natural frequency	Requires system analysis
	Regulator response to load changes is too fast or too slow	Change control springs, poppet return springs or the set point

contd...

Table 13.7 *...contd*

	Regulator design has an inherent instability in this application	Changes in spring or set point may help or it may be necessary to use another regulator
Delay or lack of reverse flow	Regulator is not designed for reverse flow	Check manufacturer's manual
	Inlet pressure exhausts too slowly	Check exhaust path for adequate capacity, check path for restricting flow control valves.
	Downstream pressure is above the set point, regulator is exhausting through vent path rather than via reverse flow	Alter system conditions to keep downstream pressure at or below set point before the reverse flow starts
Pressure regulation failure	Valve assembly is binding contaminants.	Flush out
	Sliding friction between valve assembly and bottom plug is too large.	
Air leaks out of small hole in the cover	Disk seal is broken or binding contaminants	Flush out the disk seal or replace it
	Broken diaphragm	Replace the diaphragm

Table 13.8 *Fault diagnosis of lubricator*

Trouble/Fault	Possible causes	Remedy
Lubricator		
Oil not delivered from the lubricator	Empty reservoir bowl	Refill or use automatic refill
	Clogged passageways	Disassemble and clean
	Closed needle valve	Check adjustment
	Torn restrictor	Replace the broken restrictor
	Internal parts incorrectly assembled	Disassemble and reassemble carefully
	Clogged pick-up tube inlet	Clean away debris
	Non-vertical positioning	Reposition or relocate

contd...

Table 13.8 *...contd*

Oil delivery is slow or delayed	Initial start-up delays are natural	Allow a few minutes of sustained air flows to fill internal passage way in lubricator with oil
	Internal leakage in oil passageway	Disassemble and inspect for missing, damaged or improperly assembled parts
	Clogged pick-up tube inlet	Clean away debris
Too much oil delivery	Improper needle valve adjustment	Adjust and refer manual
	Reverse flow through lubricator	Use a quick exhaust valve at the cylinder port or position the lubricator just upstream of the quick exhaust
Poor component performance even with oil delivery	Tortuous flow path for lubricator	Relocate the lubricator
	Incorrect oil type	Follow recommendations by the component manufacturer
	Possible water condensation in the oil reservoir	Maybe system hasn't been operated for long time. Manually drain the bottom of the reservoir
Reservoir bowl cracked	Oil incompatible with plastic material, excessive temperature or sun light	Use metal bowl
No lubrication	Oil level is lower than the specified	Fill oil up to recommended level
	Air consumption is low	Accelerate transfer of cylinder actuator
	Clogging of the adjusting screw of the flow guide screw and ball	Flush out
	Broken flow guide	Replace the flow guide
Broken bowl	Organic solvent	Replace bowl with metallic or nylon bowl.

Table 13.9 *Fault diagnosis of air cylinder*

Trouble/Fault	Possible causes	Remedy
Air cylinder		
With valve connected, air escapes out of vent hole	Double cup packing is leaking	Replace
	Double cup packing is loose	Disassemble and tighten
Air escapes to atmosphere at flange bushings	Cup packing is leaking	Replace
	Cup packing assembled incorrectly	Remove and reassemble correctly
	Clogged pick-up tube inlet	Clean away debris
Air escapes at piston	Groove ring is defective	Remove and replace with new
End position cushion does not respond	Lip seal on the cushioning plunger is leaking or has been fitted incorrectly	Remove and assemble correctly or install a new seal
Single acting cylinder not returning to end position	Compression spring is damaged or improperly assembled	Replace spring with one with the proper stiffness.
Misalignment of piston/piston rod or cylinder body	Excessive jerks	Align piston and piston rod or cylinder body
	Incorrect operation	Correct the operation
Cylinder fails to move the load when the valve is actuated	Binding in machine linkage	Check linkage to ensure that there is no excessive friction load
	Pressure is too low	Increase pressure or check for leaks
	Cylinder undersized	Check with designer or manufacturer and replace
	Piston rod broken	Dissemble and replace
Erratic cylinder action	Valve sticking or binding	Check for dirt or deposits
		Check for worn out parts
	Cylinder sticking or binding	Check if seals have been over-tightened
		Check for misalignment and incorrect assembly

contd...

Table 13.9 *...contd*

Cylinder body seal leak	Loose tie rods	Tighten the tie rods according to torque recommended by manufacturer
	Excessive pressure	Reduce the pressure
	Pinched or extruded seal	Check the compatibility of seal material and assembly
	Seal deterioration	Check the compatibility and alignment and assembly.
Rod gland seal leaks	Torn or worn seal	Examine the piston rod for dents and nicks. Replace the piston rod if the surface is rough
Excessive or rapid piston seal wear	Seal installed incorrectly	Check the installation instruction. Reassemble

Table 13.10 *Fault diagnosis of air motor*

Trouble/Fault	Possible causes	Remedy
Air motors		
Motor won't run	Air pressure too low	Adjust the pressure
	Insufficient lubrication	Properly lubricate the parts
Lack of power	Internal air leaks and low air pressure	Check for leaks
	Worn vanes	Chipped, cracked, loose or rough-edged vanes should be replaced.
	Worn bearing plates	If galling is present, use sand paper to remove burns
		If galling/wear is excessive, then replace the parts
Excessive vane wear	Dirty air	Clean the strainers. Strainers should be located at inlet ports to prevent the particles of dirt and rust from entering the motor.

Table 13.11 *Fault diagnosis of pneumatic valves*

Trouble/Fault	Possible causes	Remedy
Pneumatic valves		
Valve blows to exhaust	Inlet poppet not seated properly	Check for damaged poppet and replace
		Clean dirt in poppet seats
		Replace damaged poppet seat
		Disassemble and tighten
	Nicked, torn or swelling seals	Replace
	Cylinder leaks	Check for leakage, packing, assembly, orientation and damage
	Damaged spools	Replace
	Air supply pressure too low and causes the valve to actuate partially	Inspect the system for undersized supply lines, sharp bends, fittings and clogged filter element or a defective pressure regulator and adjust
	Water or oil contamination	Ensure that air is dry and that the air filter is drained regularly.
Poppet chatters	Low pilot or signal pressure	Check the pilot pressure
Spool valve action is sluggish	Swollen seals	Replace
	Varnish deposit on spools	Remove varnish using water soluble detergents
	Air supply pressure too low	Adjust and check for leak
	Low pilot pressure	Check the pilot pressure
	Poor or no lubrication	Lubricate
	Faulty silencer	Remove the silencer and check the performance
		Clean the silencer and check the performance
		Verify the size of the silencer
	Water or oil contamination	Ensure that the air supply is dry
Air flow is normal only in the actuated position	Broken return spring	Replace and check the stiffness
Solenoid buzzes	High transient voltage	Isolate solenoid circuits from main power circuits

contd...

Table 13.11 *...contd*

Sequence valve provides erratic timing	Faulty piston seal	Replace it
		Check for assembly
	Excessive lubrication	Check the system lubrication
	Fluctuating air pressure	Install a pressure regulator at the point where pressure is low
	Accumulated water	Reroute the pilot supply lines to eliminate low points
Flow control valve does not respond	Excessive lubrication	Check the system lubrication
	Incorrect installation	Install properly
	Dirt in the valve	Clean the valve
Valve leaks	Dirt	Clean
	Scored seals	Replace defective seal
	Broken seals	Replace the seals
	Groove ring damaged	Replace defective part
	Worn out valve disc	Replace defective part
	Weak or broken spring	Replace spring
	Deterioration of packing	Replace packing
	Excessive wear	Lubricate
Valve operating mechanism fails	Line pressure too low	Check for pressure drop and leaks
	Binding action of valve mechanism	Replace defective part
	Control plunger broken	Replace defective part
	Groove ring defective	Replace defective part
	Spool sticking due to swelling of seals	Replace defective part
	Burned out coil	Replace
	Broken lead to the coil	Replace or repair
	Broken parts such as pins and lever arms	Replace the part
	Corrosive damage of surface	Replace the part
	Gumming and varnish deposit	Lubricant may not have proper additives. Check with supplier

contd...

Table 13.11 *...contd*

Valve plunger does not return	Check for broken spring	Replace the spring, check the stiffness of the spring
Spool does not reverse	Lumpy bearing surface, valve does not rest uniformly	Flatten bearing surface
	Control pressure too low	Set pressure correctly
	Spool jammed	Check sealing rings
	Second control signal cannot switch due to signal conflict	Check tubing and design
Sluggishness of valves	Dirt collected in the valve especially at the groove ring	Clean the vent hole
	Blocked vent hole	Set right alignment
Valves fail to pass the rated amount of flow	Actuator mechanism not working	Align properly and replace if required.
	Loosened cam	Tighten cam
	Bent trip cam	Straighten trip cam
	Tripping mechanism worn out	Replace the defective part
The pneumatic output of solenoid valve is OFF when the relevant input device is actuated	Electrical supply is off/low voltage	Check the voltage
	Mechanical fault preventing the armature pulling in	Replace the part
	Burnt-out coil	Replace
	Failure of suppressor	Replace the part
The pneumatic output is ON when the solenoid coil is de-energised	Manual override is in the ON state	Check and correct the manual override
Loud noise from solenoid	Solenoid plunger is not seating properly	Clean and set right after reassembly
Arc at switching contacts	Failure of arc suppressors fitted within plug	Check the suppressor
Blown fuse or damaged switching circuit	Short circuit failure	Check the circuit connections

contd...

Table 13.11 *...contd*

Failure of solenoid coil	Coil loosely fixed to the solenoid stem	Fix coils firmly on solenoid stem
	Coil vibrates	Fix the coil
	Overheating of coils	Fix the coil
	Mismatched coils and stems	Use matched coils and stem

Table 13.12 *Fault diagnosis of pipelines and hoses*

Trouble/Fault	Possible causes	Remedy
Pipelines and hoses		
Pressure drop due to leakage of compressed air	Air pressure too low	Adjust the pressure
	Insufficient lubrication	Properly lubricate the parts
Pressure rise at specific points	Pipes and hoses blocked	Clean and remove dirt or foreign particles.
	Pipes and hoses bent	Correct and follow instructions of manufacturers
	Pipes and hoses not installed properly	Check the installation. Follow the installation instruction of hose manufacturer
High noise level	Silencer stops working	Check for size, clean the silencer.

Table 13.13 *Fault diagnosis of general pneumatic items*

Trouble/Fault	Possible causes	Remedy
General		
No movement	No pressure or inadequate pressure	Install or fix proper pressure sources
	No signal to the direction control valve	Correct the control circuit
	Mounting alignment failure	Correct the mounting
	Damaged piston packing	Replace the packing
Movement is not smooth	Speed is below the minimum speed	Alleviate the load variation limit
		Consider the use of new cylinder
	Mounting alignment failure	Correct the mount posture
		Change the mount type

contd...

Table 13.13 *...contd*

Damaged or deformed	Impact noise at the end of the stroke.	Improve cushion
		Decelerate the speed
		Lessen the load
Machine is working but performance is not up to the mark	Upstream pressure and flow	Fit larger pipes
		Install local receiver
		Install larger compressor
	Downstream pressure and flow	Check for twisted hose, tubes and pipe.
	Lack of lubrication	Lubricate the machine
		Fit air line lubricator
One actuator is performing poorly	Flow restriction	Adjust the flow
	Twisted tube	Repair or replace
	Misalignment	Correct
	Lack of lubrication	Lubricate adequately
	Obstruction	Remove
	Piston rod bent	Replace or repair part
	Barrel dented	Replace the part
Stoppage of machine	Electric or pneumatic supply failure	Check the main and other connections
	Limit sensor moved out of position	Adjust and tighten sensor
	Mechanical jam preventing an actuator from operating the limit switch	Release the pressure from both sides of the actuator before clearing the jam
Misalignment of components or faulty machine adjustment	Adjusting mechanism out of alignment	Readjust the mechanism
	Limit sensors moved out of positing causing short strokes	Readjust
	Failed ON sensors causing a skipped step in sequence	Check arc suppressor circuit
	Insufficient power to a stamping or processing actuator	Increase pressure to actuator or use larger actuator
Leakage	Loose joints	Tighten loose joints, fittings and glands
	Faulty or damaged fittings or ruptured pipes and hoses	Replace or repair the defective part

Summary

1. Perfect maintenance of pneumatic system starts with good design. In most pneumatic systems, problems are related to air and pipeline layout. Therefore, periodic inspection of air and pipelines is important.
2. It is imperative for the maintenance engineer to have a sound knowledge of the machine operations and the functions of the pneumatic components before maintenance work is undertaken. Pneumatic components fail due to misuse, malfunctioning and ageing. All safety procedures should be followed by everyone to prevent machine or component failure.
3. Maintenance can be classified into the breakdown or corrective maintenance (CM) and preventive maintenance (PM). Preventive maintenance is carried out on a regular schedule regardless of problems. Corrective maintenance is carried out when components fail. Proper PM and CM saves energy, time and money.
4. Leakage is an important concern for all compressed air systems. Leaks are caused by damaged or corroded pipe work, faulty seals, faulty pipe joints, poor quality fittings, tubes, incorrect installation of air line couplers, dynamic wear, cuts in tubing and open drain taps. Air leakage is the main reason for pressure drops in the system.
5. Frictional losses in pneumatic systems must be taken into account by the designer. They occur in all components between the compressor and the actuators. In addition to the cost of wasted energy in the form of heat, the frictional losses also add to maintenance and downtime costs.
6. Compressed air must be free from moisture and impurities such as dirt, dust and pipe scales. Various devices like filters, lubricators, dryers, aftercoolers etc. are installed to provide clean air. Lack of maintenance of these devices leads to system failures.
7. Proper maintenance of compressors, air receivers, air lines, FRL units, pneumatic cylinders and valves are important to increase the system efficiency of pneumatic systems.
8. Troubleshooting is an art and comes by experience. An analytical mind and a sound knowledge of the system is an essential asset of maintenance engineers. Accurate circuit analysis and understanding the working of the machines and the system helps to pinpoint the fault and save down time.
9. Fault recording and fault analysis helps to identify any recurring pattern of faults or any design/application problems.

EXERCISES

Fill in the blanks

1. Preventive maintenance is carried out on the basis of regular maintenance ————————.
2. In breakdown maintenance, troubleshooting is done after the ———— of components.
3. If the pressure drop in a filter is high, the ———— needs to be replaced.
4. If the solenoid hums loudly, then a possible cause is that the solenoid plunger not seating properly due to —- collection.
5. In servicing the air cylinder, it is good practice to replace O-rings and ————

State whether true or false

1. In pneumatic systems, air leaks are the main reason for pressure drops.
2. General maintenance procedures consist of removal of dust and dirt, checking for loose nuts and bolts, preventing air leaks, reducing frictional losses and ensuring the safety of staff.
3. Pneumatic systems are difficult to maintain when compared to hydraulic systems.
4. In general, solenoid valves are designed to provide maintenance-free service for one million cycles.
5. It is always recommended to troubleshoot all components together in a pneumatic system.

Review questions

1. Differentiate between preventive and corrective maintenance.
2. What are the advantages of preventive maintenance?
3. List the general maintenance procedures used in pneumatic systems.
4. Explain the general causes of pneumatic system failures.
5. Write a brief note on the maintenance of the compressors, air receivers, air mains, FRL units, etc.
6. Explain a method to detect air leakage.
7. What are the maintenance activities that are carried out for pneumatic valves?
8. What are the maintenance activities that are carried out for pneumatic cylinders?
9. What is the most important rule in troubleshooting of pneumatic systems?
10. Explain the possible malfunctions of valves, compressors, cylinders, and air motor and their remedies.

Answers

Fill in the blanks

1. schedules 2. failure 3. filter element 4. dust 5. packing

True or false

1. True 2. True 3. False 4. True 5. False

Index